T0181873

Lecture Notes in Artificial Intelligence   13725

Subseries of Lecture Notes in Computer Science

More information about this subseries at https://link.springer.com/bookseries/1244

Weitong Chen · Lina Yao · Taotao Cai ·
Shirui Pan · Tao Shen · Xue Li (Eds.)

Advanced Data Mining and Applications

18th International Conference, ADMA 2022
Brisbane, QLD, Australia, November 28–30, 2022
Proceedings, Part I

 Springer

Editors
Weitong Chen
The University of Adelaide
Adelaide, SA, Australia

Lina Yao
The University of New South Wales
Sydney, NSW, Australia

Taotao Cai ⓘ
Macquarie University
Sydney, NSW, Australia

Shirui Pan
Griffith University
Brisbane, QLD, Australia

Tao Shen
Microsoft
Beijing, China

Xue Li
The University of Queensland
Brisbane, QLD, Australia

ISSN 0302-9743 ISSN 1611-3349 (electronic)
Lecture Notes in Artificial Intelligence
ISBN 978-3-031-22063-0 ISBN 978-3-031-22064-7 (eBook)
https://doi.org/10.1007/978-3-031-22064-7

LNCS Sublibrary: SL7 – Artificial Intelligence

This Springer imprint is published by the registered company Springer Nature Switzerland AG
The registered company address is: Gewerbestrasse 11, 6330 Cham, Switzerland

Preface

The 18th International Conference on Advanced Data Mining and Applications (ADMA 2022) was held in Brisbane, Australia, during November 28–30, 2022. Researchers and practitioners from around the world came together at this leading international forum to share innovative ideas, original research findings, case study results, and experienced insights into advanced data mining and its applications. With the ever-growing importance of appropriate methods in these data-rich times, ADMA has become a flagship conference in this field.

ADMA 2022 received a total of 198 submissions. After a rigorous single-blind review process 192 reviewers, 76 regular papers were accepted to be published in the proceedings, 39 were selected to be delivered as oral presentations at the conference and 37 were selected as poster presentations. This corresponds to a full oral paper acceptance rate of 19.6%. The Program Committee (PC), composed of international experts in relevant fields, did a thorough and professional job of reviewing the papers submitted to ADMA 2022, and each paper was reviewed by at least three PC members. With the growing importance of data in this digital age, papers accepted in ADMA 2022 covered a wide range of research topics in the field of data mining, including machine learning, text mining, graph mining, predictive data analytics, recommender systems, query processing, analytics-based applications, and privacy and security analytics. It is worth mentioning that, firstly, ADMA 2022 organized a physical event, allowing for in-person gatherings and networking, secondly, a special inclusive workshop has been organized to enhance the experience of women non-binary and gender non-conforming in the data mining community.

We thank the PC members for completing the review process and providing valuable comments within tight schedules. The high-quality program would not have been possible without the expertise and dedication of our PC members. Moreover, we would like to take this valuable opportunity to thank all authors who submitted technical papers and contributed to the tradition of excellence at ADMA. We firmly believe that many colleagues will find the papers in this proceedings exciting and beneficial for advancing their research. We would like to thank Microsoft for providing the CMT system that is free to use for conference organization, Springer for the long-term, support and the University of Queensland and ARC Training Centre for Information Resilience (CIRES) sponsorship of the conference.

We are grateful for the guidance of the steering committee members, Osmar R. Zaiane, Jianxin Li, and Guodong Long. With their leadership and support, the conference run smoothly. We also would like to acknowledge the support of the other members of the organizing committee. All of them helped to make ADMA 2022 a success. We appreciate local arrangements from the local co-chairs, Guangdong Bai and Henry Nguyen, the time and effort of the publication co-chairs, Taotao Cai, Shirui Pan and Tao Shen, the effort in advertising the conference by the publicity co-chairs, Ji Zhang, Philippe Fournier-Viger and Grigorios Loukides, the effort on managing the Tutorial sessions by tutorial co-chairs, Tianyi Zhou and Can Wang, We would like to give very special thanks to the web chair, Shaofei Shen, Hao Yang and Ruiqing Li, for creating a beautiful website and maintaining the information. We also thank Kathleen Williamson for her contribution to managing the registration system and financial matters. Finally, we would like to thank all the other co-chairs who have contributed to the conference.

November 2022

Xue Li
Lina Yao
Weitong Chen

Organization

Steering Committee

Xue Li	University of Queensland, Australia
Osmar R. Zaiane	University of Alberta, Canada
Jianxin Li	Deakin University, Australia
Guodong Long	University of Technology Sydney, Australia

Program Committee Co-chairs

Lina Yao	University of New South Wales, Australia
Weitong Chen	University of Adelaide, Australia

Local Chairs

Guangdong Bai	University of Queensland, Australia
Henry Nguyen	Griffith University, Australia

Publicity Co-chairs

Ji Zhang	University of Southern Queensland, Australia
Philippe Fournier-Viger	Shenzhen University, China
Grigorios Loukides	King's College London, UK

Publication Chairs

Shirui Pan	Griffith University, Australia
Tao Shen	Microsoft, China
Taotao Cai	Macquarie University, Australia

Tutorial Co-chairs

Tianyi Zhou	University of Maryland, USA
Can Wang	Griffith University, Australia

Web Co-chairs

Shaofei Shen	University of Queensland, Australia
Hao Yang	University of Queensland, Australia
Ruiqing Li	University of Queensland, Australia

Industry Track Co-chairs

Lu Liu	Google, USA
Jiajun Liu	CSIRO, Australia
Sen Wang	University of Queensland, Australia

Program Committee

Abdulwahab Aljubairy	Macquarie University, Australia
Adita Kulkarni	SUNY Brockport, USA
Ahoud Alhazmi	Macquarie University, Australia
Akshay Peshave	GE Research, USA
Alan Liew	Griffith University, Australia
Alex Delis	National and Kapodistrian University of Athens, Greece
Ali Abbasi Tadi	University of Windsor, Canada
Anbumunee Ponniah	BITS Pilani, India
Atreju Tauschinsky	SAP SE, Germany
Bin Guo	Northwestern Polytechnical University, China
Bin Xia	Nanjing University of Posts and Telecommunications, China
Bin Zhao	Nanjing Normal University, China
Bo Ning	Dalian Maritime University, China
Bo Tang	Southern University of Science and Technology, China
Carson Leung	University of Manitoba, Canada
Chang-Dong Wang	Sun Yat-sen University, China
Chaoran Huang	University of New South Wales, Australia
Chen Wang	Chongqing University, China
Claudia Antunes	Universidade de Lisboa, Portugal
Clemence Magnien	Centre national de la recherche scientifique, France
David Broneske	German Centre for Higher Education Research and Science Studies, Germany
Dechang Pi	Nanjing University of Aeronautics and Astronautics, China
Dima Alhadidi	University of New Brunswick, Canada
Dong Li	Liaoning University, China
Dong Huang	South China Agricultural University, China
Donghai Guan	Nanjing University of Aeronautics and Astronautics, China
Eiji Uchino	Yamaguchi University, Japan
Ellouze Mourad	Université de Sfax, Tunisia

Elsa Negre	LAMSADE, Paris-Dauphine University, France
Farid Nouioua	University of Souk Ahras, Algeria
Fatma Najar	Concordia University, Canada
Genoveva Vargas-Solar	CNRS, France
Guanfeng Liu	Macquarie University, Australia
Guangdong Bai	University of Queensland, Australia
Guangquan Lu	Guangxi Normal University, China
Guangyan Huang	Deakin University, Australia
Guillaume Guerard	ESILV, France
Guodong Long	University of Technology Sydney, Australia
Haïfa Nakouri	ISG Tunis, Tunisia
Hailong Liu	Northwestern Polytechnical University, China
Hantao Zhao	Southeast University, China
Haoran Yang	University of Technology Sydney, Australia
Harry Kai-Ho Chan	Roskilde University, Denmark
Hongzhi Wang	Harbin Institute of Technology, China
Hongzhi Yin	University of Queensland, Australia
Hui Yin	Deakin University, Australia
Indika Priyantha Kumara Dewage	Tilburg University, The Netherlands
Jerry Chun-Wei Lin	Western Norway University of Applied Sciences, Norway
Jiali Mao	East China Normal University, China
Jian Yin	Sun Yat-sen University, China
Jiang Zhong	Chongqing University, China
Jianqiu Xu	Nanjing University of Aeronautics and Astronautics, China
Jing Du	University of New South Wales, Australia
Jizhou Luo	Harbin Institute of Technology, China
Jules-Raymond Tapamo	University of KwaZulu-Natal, South Africa
Junchang Xin	Northeastern University, China
Junhu Wang	Griffith University, Australia
Junjie Yao	East China Normal University, China
Ke Deng	RMIT University, Australia
Khanh Van Nguyen	University of Technology Sydney, Australia
Lei Duan	Sichuan University, China
Lei Li	The Hong Kong University of Science and Technology (Guangzhou), China
Li Li	Southwest University, China
Liang Hong	Wuhan University, China
Lin Yue	University of Queensland, Australia
Lizhen Cui	Shandong University, China
Lu Chen	Swinburne University of Technology, Australia

Lu Chen	Zhejiang University, China
Lu Jiang	Northeast Normal University, China
Lukui Shi	Hebei University of Technology, China
Lutz Schubert	Universität Ulm, Germany
Madalina Raschip	Alexandru Ioan Cuza University of Iasi, Romania
Maneet Singh	Indian Institute of Technology Ropar, India
Manqing Dong	University of New South Wales, Australia
Mariusz Bajger	Flinders University, Australia
Markus Endres	University of Augsburg, Germany
Mehmet Ali Kaygusuz	Middle East Technical University, Turkey
Meng Wang	Southeast University, China
Miao Xu	University of Queensland, Australia
Mirco Nanni	ISTI-CNR Pisa, Italy
Moomal Farhad	United Arab Emirates University, United Arab Emirates
Mourad Nouioua	Harbin Institute of Technology (Shenzhen), China
Mukesh Mohania	Indian Institute of Technology - Bombay, India
Nenggan Zheng	Zhejiang University, China
Nicolas Travers	Léonard de Vinci Pôle Universitaire, Research Center, France
Nizar Bouguila	Concordia University, Canada
Noha Alduaiji	Majmaah University, Saudi Arabia
Omar Al-Janabi	Universiti Sains Malaysia, Malaysia
Paul Grant	Charles Sturt University, Australia
Peiquan Jin	University of Science and Technology of China, China
Peisen Yuan	Nanjing Agricultural University, China
Peng Peng	Hunan University, China
Philippe Fournier-Viger	Shenzhen University, China
Pragya Prakash	Indraprastha Institute of Information Technology, India
Priyamvada Bhardwaj	Otto von Guericke University Magdeburg, Germany
Prof. Feng Yaokai	Kyushu University, Japan
Qing Xie	Wuhan University of Technology, China
Quan Z. Sheng	Macquarie University, Australia
Qun Chen	Northwestern Polytechnical University, China
Quoc Viet Hung Nguyen	Griffith University, Australia
Rania Boukhriss	MIRACL-FSEG, Sfax University, Tunisia
Rogério Luís Costa	Polytechnic of Leiria, Portugal
Rong-Hua Li	Beijing Institute of Technology, China
Sadeq Darrab	Otto von Guericke University Magdeburg, Germany

Sai Abhishek Sara	Indian Institute of Technology, Bombay, India
Saiful Islam	Griffith University, Australia
Salim Sazzed	Old Dominion University, USA
Sanjit Kumar Saha	Brandenburg University of Technology Cottbus – Senftenberg, Germany
Sayan Unankard	Maejo University, Thailand
Sen Wang	Griffith University, Australia
Senzhang Wang	Central South University, China
Shan Xue	University of Wollongong, Australia
Sheng Wang	Wuhan University, China
Shi Feng	Northeastern University, China
Shiyu Yang	Guangzhou University, China
Shutong Chen	Xidian University, China
Sonia Djebali	ESILV, France
Suman Banerjee	IIT Jammu, India
Sutharshan Rajasegarar	Deakin University, Australia
Tao Shen	University of Technology Sydney, Australia
Tarique Anwar	University of York, UK
Thanh Tam Nguyen	Griffith University, Australia
Tianchi Sha	Beijing Institute of Technology, China
Tianrui Li	Southwest Jiaotong University, China
Tiexin Wang	Nanjing University of Aeronautics and Astronautics, China
Tim Oates	University of Maryland Baltimore County, USA
Tung Kieu	Aalborg University, Danish
Uno Fang	Deakin University, Australia
Wei Chen	University of Auckland, New Zealand
Wei Hu	Nanjing University, China
Wei Emma Zhang	University of Adelaide, Australia
Weijun Wang	University of Goettingen, Germany
Weiwei Yuan	Nanjing University of Aeronautics and Astronautics, China
Wen Zhang	Wuhan University, China, China
Xiang Lian	Kent State University, USA
Xiangfu Meng	Liaoning Technical University, China
Xiangguo Sun	The Chinese University of Hong Kong, China
Xiangmin Zhou	RMIT University, Australia
Xiangyu Song	Deakin University, Australia
Xianzhi Wang	University of Technology Sydney, Australia
Xiao Pan	Shijiazhuang Tiedao University, China
Xiaocong Chen	University of New South Wales, Australia
Xiaohui (Daniel) Tao	University of Southern Queensland, Australia

Xiaowang Zhang	Tianjin University, China
Xie Xiaojun	Nanjing Agricultural University, China
Xin Cao	University of New South Wales, Australia
Xingquan Zhu	Florida Atlantic University, USA
Xiujuan Xu	Dalian University of Technology, China
Xueping Peng	University of Technology Sydney, Australia
Xuyun Zhang	Macquarie University, Australia
Yajun Yang	Tianjin University, China
Yanda Wang	Nanjing University of Aeronautics and Astronautics, China
Yanfeng Zhang	Northeastern University, China
Yang Li	University of Technology Sydney, Australia
Yang-Sae Moon	Kangwon National University, South Korea
Yanhui Gu	Nanjing Normal University, China
Yanjun Zhang	Deakin University, Australia
Yao Liu	University of New South Wales, Australia
Yasuhiko Morimoto	Hiroshima University, Japan
Ye Yuan	Beijing Institute of Technology, China
Ye Zhu	Deakin University, Australia
Yicong Li	University of Technology Sydney, Australia
Yixuan Qiu	University of Queensland, Australia
Yong Zhang	Tsinghua University, China
Yong Tang	South China Normal University, China
Yongpan Sheng	Chongqing University, China
Yongqing Zhang	Chengdu University of Information Technology, Australia
Youwen Zhu	Nanjing University of Aeronautics and Astronautics, Australia
Youxi Wu	Hebei University of Technology, China
Yu Liu	Huazhong University of Science and Technology, China
Yuanbo Xu	Jilin University, China
Yucheng Zhou	University of Technology Sydney, Australia
Yue Tan	University of Technology Sydney, Australia
Yuhai Zhao	Northeastern University, China
Yunjun Gao	Zhejiang University, China
Yurong Cheng	Beijing Institute of Technology, China
Yuwei Peng	Wuhan University, China
Yuxiang Zhang	Civil Aviation University of China, China
Zesheng Ye	University of New South Wales, Sydney, Australia
Zheng Zhang	Harbin Institute of Technology, Shenzhen
Zhi Cai	Beijing University of Technology, China

Zhihui Wang	Fudan University, China
Zhiqiang Zhang	Zhejiang University of Finance and Economics, China
Zhixin Li	Guangxi Normal University, China
Zhixu Li	Soochow University, China
Zhuowei Wang	University of Technology Sydney, Australia
Zijiang Yang	York University, Canada
Zongmin Ma	Nanjing University of Aeronautics and Astronautics, China

Contents – Part I

On-Device Application

Other Application

Pattern Mining

Graph Mining

Contents – Part II

Classification, Clustering and Recommendation

Multi-objective, Optimization, Augmentation, and Database

Others

Finance and Healthcare

Application of Supplemental Sampling and Interpretable AI in Credit Scoring for Canadian Fintechs: Methods and Case Studies

Yi Shen[(✉)] [iD]

Data and Analytics, Decision Insights, Equifax, Toronto, Canada
shawnie.shen@equifax.com

Abstract. Over the last decade, the fintech industry has witnessed fast growth, where online or digital lending has appeared as an alternative and flexible form of financing in addition to popular forms of financing, such as term loans, and expedited the entire lending process. High-tech online platforms have provided better user experiences and attracted more consumers and SMBs (small and medium-sized businesses). Partnerships and collaborations have also been forged between financial startups and incumbents or traditional lending institutions to further differentiate products and services to niche markets and accelerate the way of innovations. With the evolving landscape of the fintech sector, risk management tools such as credit scoring have become essential to assess credit risks such as default or delinquency based on debtor credit history or status. Many fintech companies are relatively young and sometimes serve only a small portfolio with a relatively scarce delinquency history. How can they predict default risk when making financing decisions on new applications? In this paper, we document a framework of leveraging supplemental samples of consumer or business credit information from the credit bureau that can be augmented with fintech applications for credit scoring based on theoretical and empirical studies of credit application data from a Canadian online auto leasing corporation. We also provide and compare credit scoring modeling solutions utilizing interpretable AI and machine learning methods such as logistic regression, decision tree, neural network, and XBGoost.

Keywords: Credit scoring · Sample selection bias · Machine learning · Interpretable AI · Validation

1 Introduction

Between 2010 and 2020 about 20 fintech hubs have been formulated around the globe with strong growth trends and activities or investments [1]. Canada has become home to several of the leading hubs benefiting from good quality of national regulation and mature business environment, attracting a strong talent pool and embracing technology innovation [2]. Consumers and SMBs are now taking advantage of innovation capabilities of fintechs with more flexible financing terms and programs, accelerated customer experiences from digital banking and fast and transparent funding opportunities with less fees and costs. Traditional credit risk assessment tools such as credit scoring conditional on applicants' previous credit history and current credit status are still commonly used by lending fintechs to assess borrowers' creditworthiness and delinquency risk or to make prediction of their probability of default after application. Various generic credit scores such as credit bureau scores or FICO score [3] have existed for many years to evaluate consumer or commercial credit risk and almost become universal rules of standard for underwriting. They can be easily acquired by fintech financers from credit bureaus at the point of applications and usually have good and robust performance in helping risk judgment. Sometimes customized credit scores have also been developed by lenders based on their own existing credit portfolios that make better prediction or risk classification.

One observation with this approach is that for fintechs with relatively short history of establishment, small portfolios of customers from niche markets and prudent credit decision process or market strategies, the scarcity of delinquency history or low portfolio default rate has sometimes caused generic credit scores difficult to build and unreliable overtime even they have been built. This could also be contributed by the lack of funded applications in the case of commercial credit seekers such as SMBs. Section 2 proposes a framework of supplementary sampling of consumers or businesses with similar credit quality and their proxy credit trades from credit bureau reported by various partner lenders, which is based on application credit bureau scores and default rate distributions of original fintech applications. It can be used to build credit scoring models tailored to specific fintech companies to overcome aforementioned model data issues.

In the era of big data, interpretable AI and machine learning have gained more popularity over the years and become almost ubiquitous in credit scoring literature. While traditional approach logistic regression has still been widely applied in industry practices due to its simplicity, robustness and parsimonious form and easy interpretability for regulatory purposes, many sophisticated generic classification algorithms have also been implemented in credit risk modeling and proven to provide better risk prediction and validation results in cases such as e-commerce or retail banking [4,5]. Section 3 reviews several credit risk modeling techniques including logistic regression, decision tree, constrained neural network and XGBoost for consumer or commercial credit scoring. Section 4 compares their performances based on real data analysis of application and proxy samples from a Canadian fintech company and credit bureau.

2   Supplementary Sampling

It has been well known that credit scoring model exploiting only booked applications might introduce selection bias due to reasons such as cherry picking in credit decisioning process [6]. Many reject inference methods and strategies have been evaluated under various assumptions of missing mechanism, which include statistical inference based on regression models or reclassification, variations of Heckman's 2-step sample bias correction procedure and nonparametric methods, and recently proposed machine learning methods such as SMOTE and graph-based semi-supervised learning algorithm [7]. This paper does not focus on these reject inference methods, instead we take an approach of utilizing supplementary samples and using proxy trade performances for rejects or non approvals as documented by Barakova et al. (2013) [8]. A proxy trade is a trade from similar credit product but funded by other lenders. Using massive credit bureau database, rejected or non-funded applications from a specific lender can be identified if they were able to get funds from other competitors and their performances on similar trades can also be extracted for risk modeling.

Building a credit scoring model on small application data can be unstable or unrealistic even when there could exist abundant credit attributes at the point of application due to the problem of overfitting or curse of dimensionality [9]. Independent consumers or SMBs with similar credit quality measured by credit bureau scores and their proxy consumer or commercial credit trades from the similar credit products funded by other lenders can be extracted and augmented to the original fintech credit applications so that enough bad observations and sufficient model sample size can be achieved. We prove that under certain assumptions of distributions of application credit scores, credit attributes and default rate, the conditional probability estimation of default can be unbiased using the augmented samples from both fintech applications and proxy population.

2.1 Notations

We introduce mathematical notations for credit scoring modeling as following: Let $X = \{X_1, ..., X_p\}$ be the available credit attributes from credit bureau at point of application, these are usually composed of hundreds or even thousands of predictive attributes related to applicant's credit characteristics such as payment or delinquency history in various credit products or trades, e.g., credit card, line of credit, installment loans, mortgages or telco trades; derogatory public records such as charge-off, collection or bankruptcy. They can also be related to applicant's credit appetite such as trade balance, credit limit or utilization and credit history such as number of trades opened and their ages or new credit.

Let $S = \{S_1, ..., S_s\}$ be credit scores at point of application, they could be multiple bureau risk scores, or bankruptcy scores and FICO scores.

Let $Y = \{0, 1\}$ be the indicator of default or non-default of a credit trade or loan in a performance window, e.g. 12 or 24 months after application.

Let $Z = \{F, P\}$ be the indicator of whether an application is from a specific fintech and funded by the fintech itself or by a competitor after reject inference, or an independent proxy trade from similar credit product or portfolio booked by other lenders reported to credit bureau, P represents that the trade is a proxy, F represents a trade related to a specific fintech.

Let $P(Y = 1|X, S, Z = F)$ or $P(Y = 1|X, S, F)$ be the conditional probability of default of fintech applications given application credit bureau scores and credit attributes that we want to estimate using credit scoring models.

2.2   Theories

Theorem 1. *Assuming that the distribution of applicants credit attributes related to a credit trade given default status and the same application credit bureau scores is independent of whether the trade is from a specific fintech or an independent proxy trade from credit bureau report pool, i.e., $P(X|Y, S, F) = P(X|Y, S, P) = P(X|Y, S)$, if a random sample from the proxy population has the same joint distribution of default and application credit scores as that of the applications from a specific fintech, i.e., $P(Y, S|P) = P(Y, S|F), Y = 1$ or 0, then we have the conditional probability of default as following:*

$$P(Y|X, S, F) = P(Y|X, S, P) = P(Y|X, S) \tag{1}$$

Corollary 1. *Under the same assumption as in Theorem 1, if a random sample from the proxy population has the same conditional credit score distribution $P(S|Y, P)$ and marginal default distribution $P(Y|P)$ as the fintech applications, i.e., $P(S|Y, P) = P(S|Y, F)$ and $P(Y|P) = P(Y|F)$, then we have the conditional probability of default as following:*

$$P(Y|X, S, F) = P(Y|X, S, P) = P(Y|X, S) \tag{2}$$

Corollary 2. *Under the same assumption as in Theorem 1, if a random sample from the proxy population has the same conditional default probability distribution $P(Y|S, P)$ and the marginal application credit scores distribution $P(S|P)$ as the fintech applications, i.e.,$P(Y|S, P) = P(Y|S, F)$ and $P(S|P) = P(S|F)$, then we have the conditional probability of default as following:*

$$P(Y|X, S, F) = P(Y|X, S, P) = P(Y|X, S) \tag{3}$$

Proof. See proof in Appendix.

2.3   Sampling Strategies

Based on Theorem 1 and corollaries, we propose three strategies of proxy sampling for credit scoring modeling of Fintechs' applications:

Let N be the total number of funded fintech applications and reject inferences.

Let $I(Y_i)$ be the indicator of applications being default or not default.

Let $I(S_{ik})$ be the indicator of the ith application falling into a score band k. If there exists multiple application scores from the credit bureau, the score band could be extended to a grid.

Let $I(Y_i, S_{ik})$ be the indicator of the ith application falling into an application credit score band k and being default or not default.

Strategy 1. Stratified sampling based on joint distribution of default and application credit scores $P(Y, S|F)$:

1. Find the empirical joint distribution of default and application credit scores $\hat{P}(Y, S_k|F) = \frac{\sum_{i=1}^{N} I(Y_i, S_{ik})}{N}$, where $Y_i = 1$ or 0.

2. Break the proxy population into strata based on combination of application scores bands and default status, draw a stratified sample from proxy population using proportional rate at each stratum according to $\hat{P}(Y, S_k|F)$.

Strategy 2. Stagewise sampling based on marginal default distribution $P(Y|F)$ and posterior conditional distribution of application credit scores given default status $P(S|Y, F)$:

1. Find the empirical marginal default rate $\hat{P}(Y|F) = \frac{\sum_{i=1}^{N} I(Y_i)}{N}$, $Y_i = 1$ or 0 and the empirical posterior conditional distribution of application credit scores given default status $\hat{P}(S_k|Y, F) = \frac{\sum_{i=1}^{N} I(Y_i, S_{ik})}{\sum_{i=1}^{N} I(Y_i)}$, where $Y_i = 1$ or 0.
2. Do a two stage sampling from proxy population by breaking the sample into default or not default according to empirical marginal default distribution $\hat{P}(Y|F)$ from step 1 at the 1st stage, then draw a stratified sample using proportional rate at each stratum according to $\hat{P}(S_k|Y, F)$.

Strategy 3. Stagewise sampling based on marginal application credit score distribution $P(S|F)$ and posterior conditional distribution of default given application credit scores $P(Y = 1|S, F)$:

1. Find the empirical marginal distribution of application credit score $\hat{P}(S_k|F) = \frac{\sum_{i=1}^{N} I(S_{ik})}{N}$, where S_{ik} falls in score band k and the empirical posterior conditional distribution of default given application credit scores $\hat{P}(Y|S_{ik}, F) = \frac{\sum_{i=1}^{N} I(Y_i, S_{ik})}{\sum_{i=1}^{N} I(S_{ik})}$, where $Y_i = 1$ or 0.
2. Do a two stage sampling from proxy population by breaking the sample into score bands or grids according to empirical marginal score distribution $\hat{P}(S_k|F)$ from step 1 at the 1st stage, then draw a stratified sample using proportional rate at each stratum according to $\hat{P}(Y|S_k, F)$.

It is worth to notice that using the same estimation methods of empirical distributions of default and application credit scores, the three sampling strategies are essentially equivalent.

3   Techniques of Credit Scoring

A wide variety of credit scoring techniques have been used to build credit scoring models. Hand and Henley (1997) [10] offer an excellent review of the statistical techniques used in building credit scoring models. Abdou and Pointon (2011) [11] have extensively reviewed both of the traditional and advanced techniques of credit scoring. In this paper we mainly investigate and compare four of the credit scoring techniques:

Logistic regression is a type of statistical regression analysis often used to predict the outcome of a binary target variable, e.g., default or not, conditional on a set of independent predictive variables such as the credit attributes and credit scores, it is assumed the probability of default is linked to the predictive credit attributes through a logit function [12].

Decision tree is a type of classification methods based on recursive partitioning of predictive attributes or features depending on the information gain or purity measures such as entropy or GINI after splits [13]. There have been variations of decision trees, such as CART or CHAID based on different types of target variables or splitting criteria. They are a convenient tool that can automatically handle large amounts of predictive attributes and feature selection.

XGBoost provides a regularizing gradient boosting framework for various computing languages through an open-source library [14]. The method contains rounds of iteration that create a weighted summation of learners or ensemble through gradient descent search that minimizes the empirical loss function [15]. It's a process of merging all weak classifiers together to have a model with better performance [4]. Sometimes it can also suffer from imbalanced data [16,17]. XGBoost is a slightly different version of boosting in that the optimization is not directly based on gradient descent but based on approximation. Like decision trees, XGBoost provides a convenient feature selection process through iteration and often achieves higher accuracy than a single decision tree. However, it also sacrifices the intrinsic interpretability of decision tree diagrams through aggregation of multiple tree learners.

Constraint Neural Network or NDT is a refined version of neural networks with additional constraints such as monotonic relationship between predictive attributes and target variable. A set of credit attributes are fed into multiple layers of neuron nodes through activation functions and output the final prediction through the last hidden layer. By adding monotonic constraints, it offers more interpretability for regulatory purposes and creates logical reason codes for credit decisioning while still maintaining the machine learning structure and the accuracy from artificial intelligence [18]. The invention is credited to Turner M., Jordan, L. and Joshua, A. (2021) [19], it requires stringent feature selection and normalization of data before training.

4 Empirical Studies

4.1 Data Source and Sample Facts

Auto lease application data submitted between Nov 2015 and Dec 2019 from a recently established Canadian online auto leasing corp has been analyzed. Bad or default of an auto lease is defined as 90dpd+ or worse in 24 months post application. There are 47,818 consumer applications, out of which 31,754 consumers have been funded by the fintech lender, 152 of them are defaults (0.48% bad rate); 9,093 consumer applications can also be qualified as commercial SMBs, out of which 6,035 commercial applications have been funded by the fintech lender, 33 of them are defaults (0.55% bad rate).

For the rest of non-funded applications, reject inferences are performed based on auto lease trades from credit bureau consumer or commercial trade pools opened within 2 months after the original submission of fintech application, only 559 of non-funded consumer applications can be found opened a consumer auto lease with 9 defaults (1.6% bad rate); 156 of non-funded commercial applications can be found opened a commercial auto lease with 3 defaults (1.9% bad rate).

Three benchmark consumer credit scores and three commercial credit scores have been selected from bureau: ERS2 or Equifax Risk Score predicts consumer tendency of delinquency; BNI3 or Bankruptcy Navigation Indicator predicts consumer tendency of bankruptcy; BCN9 or FICO8 score is the Fair Issac consumer credit score. BFRS2 or bankruptcy financial risk score predicts tendency of businesses to go bankruptcy in commercial trades; FTDS2 or financial trade delinquency score predicts tendency of businesses being delinquent in financial commercial trades; CDS2 or commercial delinquency score predicts tendency of businesses being delinquent in industry commercial trades.

For the consumer applications, stratified sampling strategy 2 from only ERS2 score band strata has been performed due to the lack of bad observations in some of the combined score grids, which resulted in an independant supplementary proxy sample of 104,922 consumer auto leases with 3,888 defaults (non-defaults were down-sampled 1 out of 10, 0.38% weighted bad rate). For the commercial applications, the entire commercial proxy population of 150K commercial auto leases (1.10% bad rate) has been used without further stratified sampling because of very limited fintech observations and the fintech's business strategy and risk tolerance or appetite.

4.2 Model Development and Comparisons

The fintech applications and proxy samples are matched at bureau and appended consumer and commercial credit scores and attributes at the point of application separately based on their individual categories. There are around 2K trended or static consumer credit attributes and 800 commercial credit attributes available for modeling. Before passing them to modeling, several procedures have been completed to ensure the modeling quality including data integration and cleansing, segmentation analysis, data filtering and transformation. The consumer applications have been splitted based on a credit attribute related to the worst ever rating of all credit trades from a CART tree, which results in two segments of ever 30dpd or worse, or intuitively clean or dirty.

In the model development stage, the full samples are divided into 70% training and 30% validation set for consumer and commercial applications individually. Four credit risk modeling techniques have been utilized including: logistic regression, decision tree, XGBoost and constraint neural network (NDT). The related feature selection methods for the four procedures are as following:

Logistic Regression utilizes stepwise selection according to attribute statistical significance from maximum likelihood estimation after prescreening of credit attributes. To cope with collinearity, sometime VIF (variance inflation factor) filtering has also been applied after model fit to reduce any model attributes that are highly correlated with other predictors.

Decision tree and XGBoost automatically select attributes at each partition or iteration based on information gain or variable importance when optimizing the loss functions such as impurity or negative likelihood etc. It is similar to stepwise selection in some sense but it would not update the previous selection and estimation when new attribute entered through iteration.

Constrained neural network (NDT) does not provide automatic feature selection, instead it requires preselection of attributes before model fitting. Top 20 to 50 attributes selected from XGBoost have been used based on attribute importance. Further reduction or addition of attributes seem not to improve the model performances.

The generic algorithms from machine learning and artificial intelligence usually requires tuning of hyperparameters that regularize the model fit, such as the number of features, tree depth, leaf size, learning rate, number of iterations, number of nodes, number of hidden layers, L1 and L2 regularization etc. Combination of these hyperparameters and grid search have been tested to find the best fits through cross validation.

For model interpretability, the credit attributes relative importance from the algorithms can be easily extracted.

4.3   Model Evaluation

We have used the following common performance measures for credit scoring model including:

KS (Kolmogorov-Smirnov) test statistic: It measures the separation ability of classification, which captures the maximum difference in the cumulative distributions of the good/bad samples. Larger KS represents better separation.

GINI or AUROC: The two metrics measure the discriminatory power of the classification models. The GINI Coefficient is the summary statistic of the Cumulative Accuracy Profile (CAP) chart. A ROC curve shows the trade-off between true positive rate (TPR) and false positive rate (FPR) across different decision thresholds. The AUROC is calculated as the area under the ROC curve. GINI is equal to 2AUROC-1. Larger GINI represents better discrimination.

Gains or Lift: A gain or lift chart graphically represents the improvement that a model provides when compared against a random guess. Gain is the ratio between the cumulative number of bad observations up to a decile to the total number of bad observations in the data. Lift is the ratio of the number of bad observations up to kth decile using the model to the expected number of bads up to kth decile based on a random or benchmark model.

Risk ordering and accuracy: As model prediction improves, the bad rate should improve in an orderly and predictable fashion. The estimated probability of bad decreases when the default risk decreases, so the observed bad rate should decrease as well. Z-statistics can compare the actual and predicted bad rate for each of the decile ranks so the direction of estimation discrepancies can be checked.

The model performance has been assessed on both training set and validation set by sample categories of the full (Proxy+Fintech+RI), fintech applications (Fintech+RI) and fintech funded, by applicant type of consumer or commercial and by combined or segments. The detailed comparison can be found in Table 1, Fig. 1 and 2. Key findings of validation include:

Table 1. Separation and discriminatory power

Consumer

Proxy+Fintech+RI

Segment	KS Combined Train	KS Combined Validation	KS Clean Train	KS Clean Validation	KS Dirty Train	KS Dirty Validation	GINI Combined Train	GINI Combined Validation	GINI Clean Train	GINI Clean Validation	GINI Dirty Train	GINI Dirty Validation
BCN9	36.33	37.18	35.45	35.53	37.63	36.3	47.75	46.22	42.24	42.01	49.62	44.82
ERS2	36.55	37.51	36.26	35.44	35.66	35.9	47.12	45.33	41.87	40.65	47.46	44.28
BNI3	35.06	27.53	22.44	21.51	26.94	26.76	35.73	35.21	27.99	27.54	36.48	35.31
Logistic Regression	44.2	41.52	41.78	40.92	43.77	40.04	58.59	55.03	55.97	53.53	57.7	51.45
Decision Tree	54.41	36.19	55.77	35.47	49.47	31.44	69.06	43.64	70.58	41.66	63.66	39.48
XGBoost	51.45	42.02	49.42	41.68	49.65	40.69	60.7	55.86	56.44	49.46	66.1	51.96
NDT	44.17	40.87	42.53	40.4	44.34	38.55	59.83	53.29	55.66	51.49	58.75	50.24

Fintech+RI

Segment	KS Combined Train	KS Combined Validation	KS Clean Train	KS Clean Validation	KS Dirty Train	KS Dirty Validation	GINI Combined Train	GINI Combined Validation	GINI Clean Train	GINI Clean Validation	GINI Dirty Train	GINI Dirty Validation
BCN9	30.90	34.33	32.79	29.75	27.38	38.92	28.45	42.34	38.14	34.64	31.32	44.47
ERS2	33.06	31.32	38.2	26.13	31.99	32.3	38.95	37.97	36.62	32.53	33.84	40.2
BNI3	31.63	34.56	38.19	31.65	25.46	31.81	39.96	38.78	40.01	39.49	34.34	35.77
Logistic Regression	38.27	43.80	45.19	42.86	32.86	52.47	48.52	53.53	48.84	43.61	42.04	57.11
Decision Tree	53.82	47.79	59.75	43.37	46.41	45.58	65.14	53	68.62	44.36	55.98	50.58
XGBoost	48.8	45.38	47.45	44.76	51.23	48.8	60.7	55.3	68.2	54.72	62.8	58.14
NDT	38.91	43.24	36.23	44.81	36.27	45.08	45.73	48.87	42.48	35.23	45.61	53.66

Fintech Funded

Segment	KS Combined Train	KS Combined Validation	KS Clean Train	KS Clean Validation	KS Dirty Train	KS Dirty Validation	GINI Combined Train	GINI Combined Validation	GINI Clean Train	GINI Clean Validation	GINI Dirty Train	GINI Dirty Validation
BCN9	30.27	33.11	32.62	26.21	27.82	37.15	36.75	40.75	37.37	32.97	30.27	33.11
ERS2	35.17	29.28	37.79	25.41	32.55	32.71	32.55	36.59	35.76	30.88	35.17	29.28
BNI3	32.34	31.79	39.45	30.56	25.76	28.83	40.18	34.67	40.82	37.3	32.34	31.79
Logistic Regression	37.67	44.43	44.77	41.66	34.27	52.15	48.52	53.31	48.31	43.44	37.67	44.43
Decision Tree	53.73	46.38	59.7	41.58	46.18	45.79	65.38	51	69	41.74	53.73	46.38
XGBoost	48.26	43.67	47.33	44.27	51.23	49.56	60.62	54.08	56.34	47.28	48.26	43.67
NDT	36.26	42.71	36.02	43.04	36.46	44.79	45.6	46.08	42.1	31.83	36.26	42.71

Commercial

Proxy+Fintech+RI

Score	KS Train	KS Validation	GINI Train	GINI Validation
BFRS2	18.08	16.78	19.49	16.16
FTDS2	17.34	16.09	19.1	17.33
CDS2	11.44	10.39	12	10.03
Logistic Regression	31.13	28.56	43.79	40.38
Decision Tree	31.36	28.25	44.26	40
XGBoost	36.5	30.76	52.36	44.12
NDT	36.41	28.52	51.16	39.93

Fintech+RI

Score	KS Train	KS Validation	GINI Train	GINI Validation
BFRS2	24.53	17.28	22.9	4.09
FTDS2	19.55	17.33	18.9	3.8
CDS2	14.61	24.4	9.89	12.56
Logistic Regression	43.86	16.69	44.24	4.18
Decision Tree	36.88	36.3	48.83	30.71
XGBoost	40.06	24.63	46.98	16.7
NDT	40.07	22.61	53.95	14.06

Fintech Funded

Score	KS Train	KS Validation	GINI Train	GINI Validation
BFRS2	25	16.38	24.18	6.67
FTDS2	19.9	15.74	19.19	4.6
CDS2	14.75	21.56	8.48	13.45
Logistic Regression	43.57	14.43	41.59	2.81
Decision Tree	36.52	35.61	46.56	30.71
XGBoost	40.36	19.27	44.64	9.32
NDT	38.71	18.63	51.27	13.38

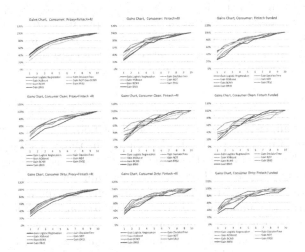

Fig. 1. Gains and lift chart:consumer scores

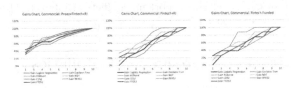

Fig. 2. Gains and lift chart:commercial scores

- Customized new scoring models almost always outperform the three benchmark bureau scores regardless of the sample categories in the validation.

- For consumer applications, XGBoost significantly outperforms the other three methods in the clean segment of Fintech applications, also has best GINI in combined segments validation. Decision tree has the best separation KS or gains at the 1st decile in the full validation data of Fintech applications. Logistic regression has the best separation KS or gains in the dirty segment of Fintech applications. NDT can have better seperation in clean segment.
- For commercial applications, there is more volatility in their performances in the fintech applications: generic algorithms always outperform logistic regression and the benchmark scores in the sample of Fintech+RI, however logistic regression and they sometimes underperform some of the benchmarks in the fintech booked applications. Decision tree seems to provide more stable separation and discriminatory power and XGBoost has the highest gains in the 3rd decile.
- For the consumer applications, the new scores seem to slightly underestimate the bad rate with most of the Z-statistics being positive but not severe (<1.96) and the risk ordering is in the right direction. For commercial applications, the new scores seem to slightly overestimate the bad rate with most of the Z-statistics being negative for fintech applications due to the higher bad rate in the proxy sample. This problem does not appear to be severe for generic algorithms, but more severe for logistic regression so further calibration may be needed even though the risk ordering is in the right direction.

5   Conclusion

Supplemental samples are effective and useful for both reject inference and predictive modeling in credit scoring. By leveraging the tailored supplemental samples, we have demonstrated that traditional credit scoring tools and innovative generic algorithms, which are from machine learning and interpretable AI, can both be utilized to assess default risk for fintech loan applications when the original volume and number of bad observations are small and not sufficient for modeling. They can provide significantly better performances than the existing benchmark scores. Generic algorithms can sometimes overfit so additional out-of-time validations need to be required for further investigation in practice.

Acknowledgements. The authors report no conflicts of interest. The authors alone are responsible for the content and writing of the paper.

Appendix. Proof of Theroem and Corollaries

Proof of Theorem 1: According to Bayes' rules, we have the conditional probability of default given applications from a specific fintech:

$$P(Y = 1|X, S, F) = P(Y = 1, S|F)P(X|Y = 1, S, F)/P(X, S|F) \quad (4)$$

In the above equation, the joint distribution of credit attributes and credit scores given applications from a specific fintech can be expanded as:

$$P(X, S|F) = P(X|Y = 1, S, F)P(Y = 1, S|F) + P(X|Y = 0, S, F)P(Y = 0, S|F) \tag{5}$$

we also have the conditional probability of default given an independent proxy trade in a proxy sample from credit bureau:

$$P(Y = 1|X, S, P) = P(Y = 1, S|P)P(X|Y = 1, S, P)/P(X, S|P) \tag{6}$$

And the joint distribution of application credit attributes and credit scores given an independent proxy trade in a proxy sample from credit bureau can be expanded as:

$$P(X, S|P) = P(X|Y = 1, S, P)P(Y = 1, S|P) + P(X|Y = 0, S, P)P(Y = 0, S|P) \tag{7}$$

Hence under the assumption $P(X|Y, S, F) = P(X|Y, S, P) = P(X|Y, S)$, if there exists a proxy sample that has the same joint distribution of probability of default and credit scores at the point of application as that of a trade related to a specific fintech, then the conditional probabilities of default from proxy or fintech are equivalent.

Proof of Corollary 1: Notice that according to Bayes' rules and Theorem 1, we have the conditional joint distributions in Eq. (4) and (5):

$$P(Y = i, S|Z) = P(S|Y = i, Z)P(Y = i|Z), i = 0 \text{ or } 1, Z = F \text{ or } P \tag{8}$$

By plugging (8) back into Eqs. (4)–(7), we have the desired result.

Proof of Corollary 2: Notice that according to Bayes' rules and Theorem 1, we have the conditional joint distributions in Eq. (4) and (5):

$$P(Y = i, S|Z) = P(Y = i|S, Z)P(S|Z), i = 0 \text{ or } 1, Z = F \text{ or } P \tag{9}$$

By plugging (9) back into Eqs. (4)–(7), we have the desired result.

References

1. Accenture: Collaborating to win in Canada's Fintech ecosystem, Accenture 2021 Canadian Fintech Report (2021). https://www.accenture.com/_acnmedia/PDF-149/Accenture-Fintech-report-2020.pdf
2. Goulard, B., Lake, K.T., Reynolds M.: Canadian Fintech Review, Torys LLP, November 2021. https://www.torys.com/our-latest-thinking/publications/2021/11/canadian-fintech-review
3. Fair-Isaac: FAQs-About-FICO-Scores-Canada-2019.pdf (2019). https://www.ficoscore.com/ficoscore/pdf/FAQs-About-FICO-Scores-Canada-2019.pdf

4. Tian, Z.Y., Xiao, J.L., Feng H.N., Wei, Y.T.: Credit risk assessment based on gradient boosting decision tree. In: 2019 International Conference on Identification, Information and Knowledge in the Internet of Things (IIKI2019)

5. Ma, Z., Hou, W., Zhang, D.: A credit risk assessment model of borrowers in P2P lending based on BP neural network. PLoS ONE **16**(8), e0255216 (2021). https://doi.org/10.1371/journal.pone.0255216

6. Hand, D.J.: Reject inference in credit operations: theory and methods. In: The Handbook of Credit Scoring, pp. 225–240. Glenlake Publishing Company (2001). /art00177

7. Kang, Y., Cui, R., Deng, J., Jia, N.: A novel credit scoring framework for auto loan using an imbalanced-learning-based reject inference. In: 2019 IEEE Conference on Computational Intelligence for Financial Engineering & Economics (CIFEr), pp. 1–8. IEEE (2019)

8. Barakova, I., Glennon, D., Palvia, A.: Sample selection bias in acquisition credit scoring models: an evaluation of the supplemental-data approach. J. Credit Risk **9**, 77–117 (2013)

9. Surrya, P.D., Radcliffea, N.J.: Why size does matter in credit scoring. In: Proceedings of Credit Scoring and Credit Control V, Edinburgh (1997) (1997)

10. Hand, D.J., Henley, W.E.: Statistical classification methods in consumer credit scoring: a review. J. R. Stat. Soc. A **160**, 523–541 (1997)

11. Abdou, H., Pointon, J.: Credit scoring, statistical techniques and evaluation criteria: a review of the literature. Intell. Syst. Account. Finance Manag. **18**(2–3), 59–88 (2011)

12. Hosmer, D.W., Lemeshow, S.: Applied Logistic Regression, 2nd edn. Wiley, New York (2000)

13. Breiman, L., Friedman, J.H., Olshen, R.A., Stone, C.J.: Classification and Regression Trees. The Wadsworth, Belmont (1984)

14. Chen, T., Guestrin, C.: XGBoost: a scalable tree boosting system. In: Krishnapuram, B., Shah, M., Smola, A.J., Aggarwal, C.C., Shen, D., Rastogi, R. (eds.) Proceedings of the 22nd ACM SIGKDD International Conference on Knowledge Discovery and Data Mining, San Francisco, CA, USA, 13–17 August 2016, pp. 785–794. ACM (2016). arXiv:1603.02754. https://doi.org/10.1145/2939672.2939785

15. Hastie, T., Tibshirani, R., Friedman, J.H.: Boosting and additive trees. In: Hastie, T., Tibshirani, R., Friedman, J.H (eds.) The Elements of Statistical Learning, 2nd edn., pp. 337–384. Springer, New York (2009). https://doi.org/10.1007/978-0-387-84858-7_10. ISBN 978-0-387-84857-0

16. Buja, A., Stuetzle, W., Shen, Y.: Loss functions for binary class probability estimation: structure and applications, Technical report, The Wharton School, University of Pennsylvania, January 2005

17. Shen, Y.: Loss functions for binary classification and class probability estimation, Ph.D. dissertation, The Wharton School, University of Pennsylvania (2005)

18. McBurnett, M., Sembolini, F., Turner, M., Jordan, L., Hamilton, H., Torres, S.R.: Comparative Analysis of Machine Learning Credit Risk Model Interpretability: Model Explanations, Reasons for Denial and Routes for Score Improvements, Credit Scoring and Credit Control XVII, University of Edinburgh, UK, August 26 2021 (2021)

19. Turner, M., Jordan, L., Joshua, A.: Machine-learning techniques for monotonic neural networks, Equifax, US patent 11010669 (2021)

A Deep Convolutional Autoencoder-Based Approach for Parkinson's Disease Diagnosis Through Speech Signals

Rania Khaskhoussy[(⊠)] and Yassine Ben Ayed

MIRACL: Multimedia Information System and Advanced Computing Laboratory, University of Sfax, National Engineering School of Sfax (ENIS), BP 1173, 3038 Sfax, Tunisia
`rania.khaskhoussy@enis.tn`, `yassine.benayed@isims.usf.tn`

Abstract. Parkinson's Disease (PD) is a neurodegenerative disease that primarily manifests through cognitive, motor and speech disorders. But it has been proven that voice changes in Parkinson's patients are among the symptoms that appear early. In this research paper, we propose a speech processing based approach for early Parkinson disease detection. Our approach evaluate the use of a deep convolutional autoencoder to extract the deep features from raw speech of PD patients and healthy subjects. Then, a classification step with MultiLayer Perceptron (MLP) which uses these deep features to build up a PD discriminant model. For an evaluation step we use a UCI dataset, our proposed approach achieve an accuracy of 95.52%, which is better than the related works accuracies using the same dataset. This prove that our system can be strongly recommended to monitor the progression of PD.

Keywords: Parkinson's disease · Autoencoder · Deep features · Raw speech · MLP

1 Introduction

Parkinson's disease was first described in 1817 under the term trembling paralysis, in a book entitled Essay on shaking palsy, by the british physician James Parkinson. PD is the most common neurodegenerative disease after alzheimer's disease, characterized by the progressive disappearance of certain neurons in the brain. The main consequence of this neuronal disappearance is the decrease in the production of dopamine (a molecule that allows neurons to communicate with each other), in a region essential to the control of movements [1]. For this, PD is considered the second leading cause of motor, cognitive and language disorders in adults after stroke [2,3].

© The Author(s), under exclusive license to Springer Nature Switzerland AG 2022
W. Chen et al. (Eds.): ADMA 2022, LNAI 13725, pp. 15–26, 2022.
https://doi.org/10.1007/978-3-031-22064-7_2

PD affects approximately 5 million people worldwide and is expected to affect 9 million by 2030 [4,5]. As its prevalence grows with age, PD touch 1.5% of the population over the age of 65. Notably, men are 1.5 times more likely than women to be affected and the typical age of PD diagnosis is 60 years old [6].The clinical diagnosis of Parkinson's disease is based on the presence of slowness of movement associated with a motor manifestation of rigidity, tremor at rest and postural instability [7]. Unfortunately, these symptoms appear after an 80% loss of dopaminergic neurons in the human brain [8]. The causes of PD remain unknown till now, but it is deduced a priori from a combination of environmental and genetic predisposing factors. As for the treatment of Parkinson's disease, there is no complete treatment, the existing drugs are intended to treat only the symptoms, they do not prevent the progression of the disease. A major research challenge is therefore, to find ways to detect the PD earlier, in order to eventually be able to slow down, or even stop, its progression from the start.

Notably, It has been proven [9,10] that voice change is one of the first symptoms that appear in the individual with suspected Parkinson's disease and even before clinical symptoms. Voice analysis can then provide indications on the correlation of speech parameters and the severity of Parkinson's disease. In this research paper, we propose a new speech analysis-based approach for detecting PD. The novelty of our approach lies in the extraction of new deep features that has not already been extracted from the used dataset. As a matter of fact, these deep features are extracted from two types of voice recordings of PD patients and healthy subjects, using deep convolutional autoencoder. Finally, to evaluate the performance of our neural model in terms of PD detection, we use three evaluation metrics which are accuracy, precision and recall.

This paper is laid out as follows: Firstly, in Sect. 2, we detail the works existing in the literature that use the human voice as a predictive marker of PD. Subsequently, in Sect. 3, we explain our proposed approach. In Sect. 4, we exhibit the experimental setup and the corresponding results. Eventually, in Sect. 5, we summarize this research paper and introduce some pertinent perspectives for future works.

2   Related Works

Notably, It has been demonstrated that speech disorders are among the first signs of Parkinson's disease [9,10]. Several studies have investigated the use of voice analysis to highlight the correlation between speech parameters and the severity of PD. In [11], the authors proposed a neural model combining the deep neural network with a stacked autoencoder to diagnose Parkinson's disease. Two databases containing voice recordings of PD patients and healthy people, were used to demonstrate the effectiveness of the proposed model. Relying upon the obtained accuracy 93.79%, the authors concluded that the deep neural network is a very effective classifier for PD diagnosis.

An optimization of cuttlefish algorithm was undertaken by [12], to select the optimal subset of features extracted from speech and voice datasets of parkinsonian patients and healthy subjects. For PD classification the authors applied

K-Nearest Neighbour (KNN) and Decision Tree (DT) classifiers on the reduced features. Their work's proved that the optimized cuttlefish algorithm surpass the traditional cuttlefish algorithm with an accuracy of 92.19%. In the same context, [13] employed the Tunable Q-factor Wavelet Transform (TQWT) to extract from voice recordings of PD patients, features have higher frequency resolution than the classical discrete wavelet transform. The authors evaluated the performance of these features using several machine leaning classifiers. The study concluded the feasibility of detecting PD using speech signals, thanks to the achieved accuracy 86% by SVM classifiers, and confirmed that TQWT performs better than others feature extraction techniques in PD classification.

In an investigation conducted by [14], an acoustic analysis was performed for PD detection. The authors analyzed 44 acoustic features, obtained from patients with PD and controls, using various machine learning algorithms. The investigation disclosed that light gradient boosting algorithm provide the best accuracy rate 88% using only seven acoustic features. In another approach, [15] introduced a prediction model to identify parkinsonian patients from healthy. The proposed model is based on the sparse autoencoder for selecting from an initial features set only the relevant features. For the classification step, six machine learning techniques were applied on the reduced features set. The study concluded that Linear Discriminant Analysis (LDA) present the best accuracy amounting to 91% compared to other classifiers.

Speech signals continue to be extensively adopted to assess PD, in a study by [16], a neural model based on convolutional neural network was proposed for discriminating PD patients from healthy subjects. The proposed model captures the spectrogram's texture features and generates high-resolution spectrograms for sample augmentation, in order to overcome the limited amount of existing patient data. The proposed method displayed a good performance in terms of sample augmentation and achieved a classification accuracy of 91.2%. In another study, [17] suggested a new approach based on two types of ensemble learning methods namely, stacking classifier and voting classifier for PD detection. The study corroborated the feasibility of these two ensemble learning in PD classification and proved that stacking classifier outperforms the voting classifier well with an accuracy of 92.2%.

Recently, other approaches have been explored for the detection of PD. [18] reported a deep CNN model based on transfer learning so as to identify PD patients from healthy people. The results revealed that the proposed model, with a fine-tuning architecture, provides good performance in terms of PD detection through the recorded accuracy rate 91.17%. In [19] a population of feature vectors was evaluated by the Genetic Algorithm (GA) to maximize the accuracy of PD diagnosis. Based on the classification by SVM, the study concluded that the best accuracy 91.18% was achieved with a reduced vector of 15 features. For more works related to PD detection, the reader may consult reviews [20, 21].

3   Proposed Approach

For detecting PD, our proposed approach relies on a direct analysis of the raw signals obtained from a dataset containing speech samples for PD patients and healthy subjects. As illustrated in Fig. 1, the raw signal is fed to a convolutional autoencoder, which convert it into a deep features vector. Then these deep features are given as input to MultiLayer Perceptron (MLP) to build the final model for PD detection.

3.1   Dataset

The dataset used in this study was obtained from [22]. As shown in Table 1, this dataset is composed of two subsets which are intended for training and testing our neural model.

The first subset is composed of 40 participants where 20 are PD patients (6 women and 14 men) and 20 are healthy people (10 women and 10 men). For each participant, 26 voice recordings including the three vowels (a, o and u), numbers and words, are recorded with an MC-1500 microphone in wav format.

With the same recording devices, two sustained vowels (a and o) are acquired three times from 28 participants (14 PD patients and 14 healthy people), To form the second subset.

Table 1. Dataset information.

Data	Number of patients	Diagnosis time (years)	Age	Recorded data
Train	20 PD	0–6	43–77	/a/o/u/ + words numbers
	20 HC	–	45–83	
Test	14 PD	0–13	39–79	/a/o/
	14 HC	–	39–79	

3.2   Deep Convolutional AutoEncoder (DCAE)

DCAE is a neural network designed to learn a representation of a dataset, with the aim of reducing the dimension of this set. It makes it possible to extract relevant features from the input whose goal is to reconstruct them in an unsupervised way. As depicted in Fig. 1, the DCAE architecture is composed of two parts: the encoder and the decoder.

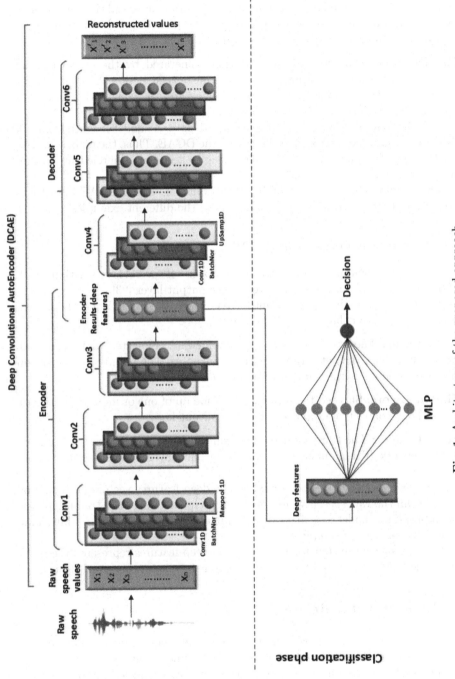

Fig. 1. Architecture of the proposed approach

The Encoder: Is made up of a repeated succession of three neuron layers namely, the convolution layer, the batch normalization layer and the max-pooling layer. These layers process the data $X[x_1, x_2, ..., x_n]$ (raw speech values) in order to build new so-called encoded representations (encoder results).

The Decoder: Process the representations generated by the encoder in an attempt to reconstruct the original data. The reconstruction $X'[x'_1, x'_2, ..., x'_n]$ of this initial data is done by three layers which are, the convolution layer, the batch normalization layer and the upsampling layer.

The differences between the reconstructed data X' and the initial data X allow to measure the error L(X, X') made by the DCAE. Thus, the main objective of DCAE is to minimize L(X, X') while trying to keep only the relevant features (deep features).

The training consists in modifying the parameters of the DCAE in order to reduce the reconstruction error L measured on the different examples.

3.3   MultiLayer Perceptron (MLP)

MLP is a formal neural network classifier composed of several layers within which information flows from the input layer to the output layer [23].

MLPs are generally organized into three layers: the input layer, the hidden layer and the output layer.

Input Layer: The first layer is completely forward connected and receives all input values from the network for learning purposes. The number of neurons in this layer is equal to the number of input data.

Hidden Layers: Succeed the input layer, made up of one or more intermediate layers. They connect the input layer to the output layer.

Output Layer: The third layer or the result layer. It gives the result obtained by the network. The number of neurons in the this layer is equal to the number of classes;

For PD detection, MLP is trained on the deep features vectors extracted by DCAE from the raw speech values of the participants. After the training phase, the appropriate model is defined to determine the decision (class: parkinsonian or healthy) of a given sample.

Figure 1 shows the structure of MLP: the deep features represents the input layer, one hidden layer and one output neuron since it is a binary classification.

4   Experimental Results

In this part, we list all experiments that we conducted on the used dataset and the obtained results. The major target of these experiments is to demonstrate, on the one hand, the effect of using the raw speech signal for the diagnosis of Parkinson's disease and on the other hand, the importance of our neural model compared to the existing work models developed for PD detection.

Several evaluation metrics can be used to assess the performance of our proposed approach. The simplest way is to evaluate the model using Confusion Matrix, Accuracy, Precision and Recall which are defined as follows:

Confusion Matrix: Is a tool for measuring the performance of classification models with 2 or more classes [24]. In the binary case (two classes), the confusion matrix is a table with 4 values representing the different combinations of actual values and predicted values as shown in the Table 2.

Table 2. Confusion Matrix

Actuel class	Predicted class	
	Positive	Negative
Positive	TP	FN
Negative	FP	TN

$$Accuracy = (\frac{TP + TN}{TP + FP + FN + TN}) \tag{1}$$

$$Precision = (\frac{TP}{TP + FP}) \tag{2}$$

$$Recall = (\frac{TP}{TP + FN}) \tag{3}$$

where:

TP = True Positive (Parkinsonian classified as Parkinsonian),
TN = True Negative (Healthy person classified as healthy person),
FP = False Positive (Healthy person classified as Parkinsonian),
FN = False Negative (Parkinsonian classified as healthy person)

DCAE Setup: The DCAE architecture used in our work is made up of six 1D convolutional layers, six batch normalization layers, three 1D max-pooling layers and three 1D upsampling layers. The activation function are all ReLU and the kernel size is 3 in all convolutional layers. The size of pooling layer is 2, optimizer algorithm is RMSprop, batch size is 64 and the validation split is equal to 0.2. This architecture is split between encoder and decoder as follows:

Encoder: It has 3 convolution blocks, each block has a convolution layer followed by a batch normalization layer and max-pooling layer.

Decoder: It has 3 convolution blocks, each block has a convolution layer followed by a batch normalization layer and upsampling layer.

Table 3 highlights the results obtained by DCAE after 200 epochs of training. Departing from this table, we conclude that our DCAE model present a good performance for the training set by an accuracy rate of 91.64% and for the testing set by an accuracy of 96.01%.

Table 3. Reconstruction results of DCAE model

Data	Accuracy (%)	Reconstruction error
Train	91.64	0.0103
Test	96.01	0.0050

Figure 2 shows the training loss of DCAE during the 200 epochs, we can detect that the proposed DCAE present a low reconstruction error of 0.01 and 0.005 respectively for the training and testing sets.

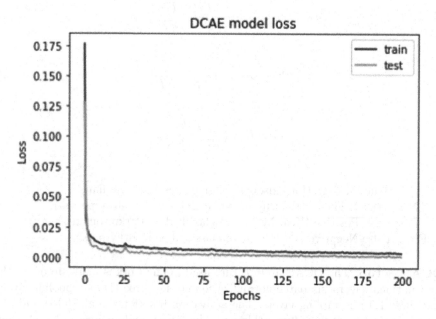

Fig. 2. Curves of training and testing loss of DCAE model

MLP Setup: The MLP architecture used in our approach relies upon 3 layers: input layer, one hidden layer and output layer composed of a single neuron. The activation function is ReLU for the input and hidden layers and sigmoid for the output layer. Further, the number of epochs chosen to train the MLP is 150 epochs.

As illustrated in Fig. 3 the MPL classifier receives as input the deep features vectors extracted by DCAE from the raw signal to build the PD discriminant model. Then this model was evaluated using accuracy, precision and recall.

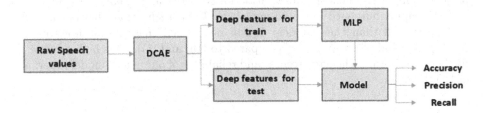

Fig. 3. DCAE features and MLP based system architecture

To calculate these evaluation metrics we first generate the confusion matrix as indicated in Table 4, where the PD patients and healthy people were labeled with 0 and 1 respectively. It is clear from these results that from 28 participants there are 13 PD patients and 12 healthy people were correctly classified and 3 others (1 PD patient and 2 healthy people) were wrongly classified.

Table 4. Confusion matrix using MLP

Classifier	Confusion matrix		
MLP		0	1	Sum
	0	13	1	14
	1	2	12	14
	Sum	15	13	28

Table 5 summarizes the classification results obtained by the MLP classifier. It is clear that the proposed model yielded good evaluation metrics values such as an accuracy amounting to 95.52%, precision of 0.93 and recall of 0.94, which means that our approach exhibited good performance in terms of Parkinson's disease detection.

Table 5. Classification results using MLP

Accuracy (%)	Precision	Recall
95.52	0.93	0.94

For further assessment, we referred to other approaches that used the same dataset and others approaches that develop autoencoder-based models to detect PD. The comparative results are shown in Table 6. The accuracy achieved by our approach is better than those yielded in related works approaches. It can be detected that compared to Xu et al [16] and Gupta et al [12], who use the same dataset, the accuracy rose from 91.2% to 95.52% for [16] and from 92.19% to 95.52% for [12]. The same observation for Xiong et al [15] and Caliskan et al [11], who apply autoencoder-based models for PD classification, that our model performed well than their models, it can be noticed that the accuracy grew between 1% and 4% respectively compared to [15] and [11]. This high accuracy further corroborates the effectiveness and reliability of our neural model in terms of PD detection.

Table 6. Comparison with related works

Methods		Classifier	Accuracy (%)
UCI dataset	Xu et al [16]	S-DCGAN-ResNet50	91.2
	Gupta et al [12]	KNN and DT	92.19
AE-based models	Xiong et al [15]	LDA	91
	Caliskan et al [11]	DNN	93.79
UCI dataset	**Proposed**	**MLP**	**95.52**

5   Conclusion

In this research paper, we set forward an approach based on the analysis of speech signals for Parkinson's disease detection. Specifically, we proposed a neural model based on recent advances in deep learning, such as deep convolutional autoencoder, for PD detection.

In the first step, we used deep convolutional autoencoder to extract deep features from raw speech signals of PD patients and healthy subjects, and in the second step, we invested the MultiLayer Perceptron to build the appropriate model using these deep features.

We demonstrated empirically that the proposed model achieved excellent results, reaching 95.52%. This confirms that our approach can detect PD using speech signals and without having a medical test.

In future work, we plan to experiment our model with other large datasets to generalize our approach. Another outstanding research direction is to use other features and classifiers to further improve PD classification results.

References

1. Pagonabarraga, J.: Apathy in Parkinson's disease: clinical features, neural substrates, diagnosis, and treatment. Lancet Neurol. **14**(5), 518–531 (2015)
2. Christopher, G.G.: The history of Parkinson's disease: early clinical descriptions and neurological therapies. Cold Spring Harb. Perspect. Med. **1**(1), a008862 (2011)
3. Khaskhoussy, R.: An I-vector-based approach for discriminating between patients with Parkinson's disease and healthy people. In: Fourteenth International Conference on Machine Vision, pp. 69–77 (2022)
4. Dorsey, E.R.: Projected number of people with Parkinson disease in the most populous nations, 2005 through 2030. Neurology **68**(5), 384–386 (2007)
5. Khaskhoussy, R.: Speech processing for early Parkinson's disease diagnosis: machine learning and deep learning-based approach. Soc. Netw. Anal. Min. **12**(1), 1–12 (2022)
6. De Lau, L.M.: Epidemiology of Parkinson's disease. Lancet Neurol. **5**(6), 525–535 (2006)
7. Jankovic, J.: Parkinson's disease: clinical features and diagnosis. J. Neurol. Neurosurg. psychiatry **79**(4), 368–376 (2008)
8. Fearnley, J.M.: Ageing and Parkinson's disease: substantia Nigra regional selectivity. Brain **114**(5), 2283–2301 (1991)
9. Harel, B.: Variability in fundamental frequency during speech in prodromal and incipient Parkinson's disease: a longitudinal case study. Brain Cognit. **56**(1), 24–29 (2004)
10. Postuma, R.B.: How does parkinsonism start? Prodromal parkinsonism motor changes in idiopathic REM sleep behaviour disorder. Brain **135**(6), 1860–1870 (2012)
11. Caliskan, A.: Diagnosis of the Parkinson disease by using deep neural network classifier. IU J. Electr. Electron. Eng. **17**(2), 3311–3318 (2017)
12. Gupta, D.: Optimized cuttlefish algorithm for diagnosis of Parkinson's disease. Cognit. Syst. Res. **52**, 36–48 (2018)
13. Sakar, C.: A comparative analysis of speech signal processing algorithms for Parkinson's disease classification and the use of the tunable Q-factor wavelet transform. Appl. Soft Comput. **74**, 255–263 (2019)
14. Karabayir, I.: Gradient boosting for Parkinson's disease diagnosis from voice recordings. BMC Med. Informat. Decis. Mak. **20**(1), 1–7 (2020)
15. Xiong, Y.: Deep feature extraction from the vocal vectors using sparse autoencoders for Parkinson's classification. IEEE Access **8**, 27821–27830 (2020)
16. Xu, Z.: Parkinson's disease detection based on spectrogram-deep convolutional generative adversarial network sample augmentation. IEEE Access **8**, 206888–206900 (2020)
17. Younis, T.: A comparative study of Parkinson disease diagnosis in machine learning. In: The 4th International Conference on Advances in Artificial Intelligence, pp. 23–28 (2020)
18. Karaman, O.: Robust automated Parkinson disease detection based on voice signals with transfer learning. Expert Syst. Appl. **178**, 115013 (2021)
19. Soumaya, Z.: The detection of Parkinson disease using the genetic algorithm and SVM classifier. Appl. Acoust. **171**, 107528 (2021)
20. Loh, H.: Application of deep learning models for automated identification of Parkinson's disease: a review (2011–2021). Sensors **21**(21), 7034 (2021)

21. Saravanan, S.: A systematic review of artificial intelligence (AI) based approaches for the diagnosis of Parkinson's disease. Arch. Comput. Methods Eng., 1–15 (2022)
22. Khaskhoussy, R.: Detecting Parkinson's disease according to gender using speech signals. In: International Conference on Knowledge Science, Engineering and Management, pp. 414–425 (2021)
23. Khaskhoussy, R.: Automatic detection of Parkinson's disease from speech using acoustic, prosodic and phonetic features. In: International Conference on Intelligent Systems Design and Applications, pp. 80–89 (2019)
24. Nancy, P.: Discovery of gender classification rules for social network data using data mining algorithms. In: Proceedings of the IEEE International Conference on Computational Intelligence and Computing Research, pp. 808–812 (2011)

Mining the Potential Relationships Between Cancer Cases and Industrial Pollution Based on High-Influence Ordered-Pair Patterns

Juanjuan Shu[1], Lizhen Wang[1,2(✉)], Peizhong Yang[1], and Vanha Tran[3]

[1] Yunnan University, Kunming 650091, China
lzhwang@ynu.edu.cn
[2] Dianchi College of Yunnan University, Kunming 650213, China
[3] FPT University, Hanoi 155514, Vietnam
hatv14@fe.edu.vn

Abstract. Co-location pattern mining aims to discover the relationships between spatial features. Traditional co-location patterns are based on clique relationships and only consider the prevalence of patterns. However, pollution sources and cancer cases do not satisfy the clique relationship, and users focus on the influence of pollution sources on cancer cases. Therefore, we propose high-influence ordered-pair patterns to study their relationships. First, we measure the influence of pollution sources on cancer cases. Then, to efficiently mine high-influence ordered-pair patterns, we propose a basic algorithm with two pruning strategies and an optimizing algorithm based on participating instances. Extensive experiments on real and synthetic datasets show that our mining results are more reasonable than existing algorithms and can provide guidance for cancer prevention. Moreover, our algorithm is also highly efficient and scalable.

Keywords: Spatial data mining · High-influence ordered-pair pattern · Participating instances · Cancer prevention

1 Introduction

Currently, effective cancer treatment has not been identified. The World Health Organization recommends that prevention is better than treatment. To prevent cancer, we must figure out the carcinogenic factors. Although the factors are not fully understood, studies [4,7] have shown that exposure to industrial pollution is one of them. Therefore, this paper attempts to analyze the influence of pollution sources on cancer cases to provide guidance for cancer prevention.

Spatial co-location pattern mining is a vital branch of spatial data mining. Many co-location pattern mining algorithms are proposed, such as join-less approach [10], tree-based approach [8], clique-based approaches [1,2], and so on. Unfortunately, existing co-location pattern mining algorithms have some problems when studying the relationships between cancer cases and pollution sources.

W. Chen et al. (Eds.): ADMA 2022, LNAI 13725, pp. 27–40, 2022.
https://doi.org/10.1007/978-3-031-22064-7_3

First, co-location patterns are based on a clique division model that requires pairwise features in a pattern in proximity, but in real life, pollution sources affecting a cancer case may be far away from each other. Therefore, Co-location patterns are unsuitable for expressing the relationship between pollution sources and cancer cases. Second, the traditional measure typically uses the participation index, which focuses on the frequency of patterns and cannot reflect the influence of pollution sources on cancer cases. Lei et al. [5] propose a function to measure the influence. However, the function is designed intuitively and a bit unconvincing. Besides, the fact that different pollution sources have different influences on cancer cases has been ignored. Third, the neighbor relationship between instances is generally determined by a given distance threshold. However, it is difficult to give a suitable distance threshold in a specific application context. Fourth, the join operation and storage of table instances are the most time-consuming and space-consuming in traditional algorithms. To tackle this problem, Yang et al. [9] propose a co-location pattern mining method based on participating instances, which shows both space and time improvements.

Based on the above discussion, this paper proposes high-influence ordered-pair patterns to study the influence of pollution sources on cancer cases. Its main contributions are summarized as follows:

(1) A measure of the influence of pollution sources on cancer cases is given while considering the distance attenuation effects, the influence superposition effects, and the types of pollution sources.
(2) We design an adaptive k-nearest neighbor algorithm based on local density to obtain neighbor relationships between pollution sources and cancer cases.
(3) To efficiently mine high-influence ordered-pair patterns, we propose a basic algorithm with two pruning strategies and an optimizing algorithm based on participating instances.
(4) Comprehensive experiments on real and synthetic datasets demonstrate the effectiveness, efficiency and scalability of the proposed algorithm.

The rest of this paper is organized as follows: Sect. 2 gives the relevant definition of high-influence ordered-pair patterns; Sects. 3 and 4 present the basic algorithm and the optimizing algorithm for mining the high-influence ordered-pair patterns; Sect. 5 shows the experimental evaluation and Sect. 6 concludes the paper.

2   High-Influence Ordered-Pair Pattern

To classify spatial features, we define the features influencing other features as **influence features,** and the features influenced by influence features as **reference features.** Given a spatial feature set F, the set of influence features is denoted as $IFS = \{p_i | p_i \in F\}$, and a spatial object $p_i.s$ of an influence feature p_i is called an **influence instance**; the set of reference features is denoted as $RFS = \{c_j | c_j \in F, c_j \notin IFS\}$, and a spatial object $c_j.t$ of a reference feature c_j is called a **reference instance**. In this paper, pollution sources affect cancer cases, so pollution source features are influence features, while a specific pollution source is an influence instance; the features of cancer cases are reference features, and a specific cancer case is a reference instance.

To reveal the relationship between influence features and reference features, we propose the concept of high-influence ordered-pair patterns. First, we give the definition of ordered-pair influence patterns.

Definition 1 *(Ordered-Pair Influence Pattern). Given a spatial feature set $F = IFS \cup RFS$ where IFS is the influence feature set and RFS is the reference feature set. An ordered-pair consisting of a subset of IFS and a subset of RFS is an ordered-pair influence pattern, denoted as $pc = \langle IFS_{pc}, RFS_{pc} \rangle$. If the number of all features in pc is k, pc is called a k-size ordered-pair influence pattern.*

To judge whether an ordered-pair influence pattern is a high-influence ordered-pair pattern, we will give the relevant definitions of its measure.

In life, the closer a reference instance is to an influence instance, the more likely the reference instance is influenced. To find the reference instances most influenced by an influence instance, we might as well find the closest reference instances to the influence instance. Referring to [3], we design an adaptive k-nearest neighbor algorithm based on local density to obtain the closest reference instances to a certain influence instance. The method is as follows: for an influence instance $p.s$, first set the value of k_{max} by the user, which decides the maximum number of the neighbors of $p.s$; then, find the k_{max} nearest reference instances of $p.s$ and store them in $neighArr(p.s)$ from near to far; third, for each reference instance $c.t$ in $neighArr(p.s)$, calculate $F(k) = k/(dist(p.s, c.t))^2$ where k is the index of $c.t$ in $neighArr(p.s)$ and $dist(p.s, c.t)$ is the distance between $p.s$ and $c.t$; finally, get the value of k when $F(k)$ is the maximum. At the time the local density of reference instances around $p.s$ is the maximum, so we select the k nearest instances as the most ones influenced by $p.s$, and they are also the neighbors of $p.s$.

Figure 1(a) is an example spatial dataset, including the reference feature set $RFS = \{A, B, C\}$, the influence feature set $IFS = \{a, b\}$, and the corresponding instances, such as a reference instance B.2. If a reference instance and an influence instance have a neighbor relationship, they are connected by a solid line in the figure. For example, a.1 has a neighbor relationship with B.1.

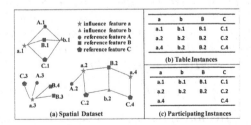

Fig. 1. (a) A spatial dataset (b) Table instances (c) Participating instances

The following work is to measure the influence of an influence instance on a reference instance. Obviously, this influence attenuates with the increase in distance. Therefore, we intend to propose a function to model the dependence between the influence of an influence instance on a reference instance and their distance. Besides, this influence depends on the types of influence features. So We need to find different functions for different influence features. To find such a function of a certain influence feature, our method is outlined as follows: first get a set of scatter points describing the dependence, then select possible curve equations by curve fitting, and finally calculate the undetermined coefficients in the curve equation to get the final expression.

To describe the dependence, the abscissa of the scatter points should be a set of distances, and the ordinate of the scatter points is a set of the corresponding quantifications of the influences at distances in the abscissa. A reference instance is a neighbor of an influence instance, then the reference instance has an influence on the influence instance. Therefore, for a certain influence feature p_i, the influence of an influence instance $p_i.j$ on a reference instance at the distance d_i can be quantified as the density of the neighbors of the instances of p_i at the distance d_i. Based on the limit idea, the density at the distance d_i can be approached by the average density in a small range $[d_i - minInterval, d_i + minInterval)(minInterval$ is a very small number). The average density of the neighbors of the instances of p_i in a small range $[d_i - minInterval, d_i + minInterval)$ is equal to the number of the neighbors of the instances of p_i whose distances are $[d_i - minInterval, d_i + minInterval)$ divided by the area; and the result is regarded as a quantification of the influence at a distance d_i. The quantifications of influences at the distances $d_1, d_2, ..., d_n$ are obtained by the above method and normalized to $[0, 1]$.

Next, we select possible curve equations to fit the set of scatter points. MATLAB's function "nlinfit" calculates the undetermined coefficients in the curve equation and returns the mean squared error (MSE). The smaller the value of MSE, the better the fitting of the curve function. An influence function to measure the influence of an instance of the influence feature p_i on a reference instance at the distance $dist$ is denoted as $Fun(p_i, dist)$. Table 1 shows the expressions and corresponding MSE of the influence functions of a real data set.

Based on the influence functions, we can calculate the influence of an influence instance on a reference instance and give the formal definition below.

Definition 2 (Single Influence). *Given a reference instance $c_j.t$ having a neighbor relationship with an influence instance $p_i.s$, and the influence function $Fun(p_i, d)$. $dist(p_i.s, c_j.t)$ is the distance between between $p_i.s$ and $c_j.t$, and $R(p_i.s, c_j.t)$ represents that they have a neighbor relationship. The influence of $p_i.s$ on $c_j.t$ is denoted as $SI(p_i.s \rightarrow c_j.t)$ and calculated as follows:*

$$SI(p_i.s \rightarrow c_j.t) = Fun(p_i, dist(p_i.s, c_j.t))(R(p_i.s, c_j.t)) \tag{1}$$

Definition 3 (Row Instance). *Given an ordered-pair influence pattern $pc = \langle IFS_{pc}, RFS_{pc} \rangle$, a set of influence instances I_p consisting of one instance of each feature in IFS_{pc}, and a set of reference instances I_c consisting of one instance*

Table 1. Influence functions and mean square errors for different influence features

Influence features	Influence functions	Mean square error (MSE)
a	$Fun(a, dist) = e^{-29.4045 * dist}$	0.0178
b	$Fun(b, dist) = e^{-29.3099 * dist}$	0.0078
c	$Fun(c, dist) = e^{-32.868 * dist}$	0.0131
d	$Fun(d, dist) = e^{-33.9258 * dist}$	0.0131
e	$Fun(e, dist) = e^{-42.8105 * dist}$	0.004
f	$Fun(f, dist) = e^{-39.609 * dist}$	0.0026
g	$Fun(g, dist) = e^{-37.2546 * dist}$	0.0067
h	$Fun(h, dist) = e^{-33.8281 * dist}$	0.0098
i	$Fun(i, dist) = e^{-62.7482 * dist}$	0.0023

of each feature in IFS_{pc}; if any instance of I_p and any instance of I_c have a neighbor relationship, then $\langle I_p, I_c \rangle$ is a row instance of pc, denoted as $RI(pc)$.

Definition 4 (Table Instance). *The collection of all row instances of an ordered-pair influence pattern pc is called the table instance of pc, denoted as* $TI(pc)$.

Definition 5 (Superimposed Influence). *Given a reference instance $c_j.t$, the set of influence instances $I_p = \{i_1, i_2, ..., i_m\}$ having neighbor relationships with $c_j.t$, and the influence feature set $IF_p = \{p_1, p_2, ..., p_n\}$ of I_p. The superimposed influence of $c_j.t$ is represented as $SII(c_j.t)$ and calculated as follows:*

$$SII(c_j.t) = 1 - \prod_{1 \leq i \leq n, p_i \in IF_p} (1 - max_{p_i.s \in I_p}(SI(p_i.s \rightarrow c_j.t))) \qquad (2)$$

Particularly, when I_p is all influence instances having neighbor relationships with $c_j.t$, at the time the superimposed influence of $c_j.t$ is the maximum and called **max superimposed influence** of $c_j.t$, denoted as $MSII(c_j.t)$.

Definition 6 (Influence Sum of a Feature). *Given a reference feature c_j and all its instances $\{c_j.t | 1 \leq t \leq |c_j|\}$, the influence sum of the feature c_j is defined as the sum of **max superimposed influence** of reference instances of c_j, which is denoted as $FIS(c_j)$. The formula is as follows:*

$$FIS(p_i) = \sum_{j=1}^{|c_j|} MSII(c_j.t) \qquad (3)$$

Definition 7 (Influence Ratio). *Given an ordered-pair influence pattern $pc = \langle IFS_{pc}, RFS_{pc} \rangle$, the table instance $TI(pc)$ of pc, and a reference feature c_j in RFS_{pc}. The influence ratio of c_j in pc is defined as the sum of the superimposed influence of the non-repeating instances of c_j in $TI(pc)$ divided by the* influence sum of the feature c_j, denoted as $FIR(c_j, pc)$. The formula is as follows:

$$FIR(c_j, pc) = \sum_{c_j.t \in \pi_{c_j}(TI(pc))} SII(c_j.t)/FIS(c_j) \tag{4}$$

where $\pi_{c_j}(TI(pc))$ is the non-repeating instances of c_j in $TI(pc)$.

Definition 8 (Influence Index). *Given an ordered-pair influence pattern* $pc = \langle IFS_{pc}, RFS_{pc} \rangle$, *the influence index of pc is the minimum of the influence ratio of the reference feature in* RFS_{pc}, *denoted as* $PII(pc)$.

Definition 9 (High-Influence Ordered-Pair Pattern (HIOPP)). *If the influence index of an ordered-pair influence pattern pc is no smaller than the influence threshold* $minPII$, *pc is called as a high-influence ordered-pair pattern.*

3   Basic Algorithm for Mining HIOPPs

3.1   Property Analysis of HIOPP

The influence index does not meet the downward closure property[1], which can be used to prune; so we analyze some properties of the influence index to enhance mining efficiency.

Lemma 1 (Conditional Monotonicity). *If ordered-pair influence patterns have the same influence features, the influence index is anti-monotone as the size of patterns increases.*

The proofs of Lemma 1 and the following lemmas are at the document (see Footnote 1). Based on Lemma 1, if $PII(pc') < minPII$, pc' and its supersets with the same influence features as those of pc' are pruned.

Definition 10 (Limit Influence Index). *Given an ordered-pair influence pattern* $pc = \langle IFS_{pc}, RFS_{pc} \rangle$, *and the table instance* $TI(pc)$ *of pc. The limit influence ratio of* $c_j \in RFS_{pc}$ *is defined as the sum of max superimposed influence of the non-repeating instances of* c_j *in* $TI(pc)$ *divided by the influence sum of* c_j. *And the limit influence index of pc is the minimum of the limit influence ratio of the reference features in* RFS_{pc}, *denoted as* $LII(pc)$. *The formula is as follows:*

$$LII(pc) = min_{c_j \in RFS_{pc}} \left(\sum_{c_j.t \in \pi_{c_j}(TI(pc))} MSII(c_j.t)/FIS(c_j) \right) \tag{5}$$

Lemma 2. *The limit influence index of a pattern is an upper bound of the influence index of the pattern.*

---

[1] https://github.com/juanjuanShu/paper/blob/main/supplement.pdf.

Lemma 3. *The limit influence index is anti-monotone as the size of patterns increases.*

If $LII(pc) < minPII$, we conclude $PII(pc) \leq LII(pc) < minPII$ by Lemma 2, so pc can be pruned, and supersets of pc can also be pruned based on Lemma 3.

3.2   Description of Basic Algorithm

Since the limit influence index satisfies the anti-monotone property, we use an apriori-like algorithm to mine patterns. However, the join method of ordered-pair influence patterns is slightly different. Given two k-size ordered-pair influence patterns $pc_1 = <IFS_1, RFS_1>$, $pc_2 = <IFS_2, RFS_2>$, where IFS_1 and IFS_2 are m-size, RFS_1 and RFS_2 are n-size, and $m+n = k$; the method of generating a $(k + 1)$-size candidate pattern pc by joining pc_1 and pc_2 is as follows:

1) **(back join)** if $IFS_1 = IFS_2$ and the first $(n - 1)$ influence features of RFS_1 and RFS_2 are the same; then $pc = <IFS_1, RFS_1 + RFS_2>$ where "+" represents join operation.
2) **(front join)** if $RFS_1 = RFS_2$ and the first $(m - 1)$ influence features of IFS_1 and IFS_2 are the same; then $pc = <IFS_1 + IFS_2, RFS_1>$.

Preprocessing: steps 1–3 lay the foundations for subsequent calculations.

Generating 2-size HIOPP (step 4): step 4 first generates 2-size candidate patterns and their table instances based on RN; then, for each candidate pattern pc, calculate its limit influence index $LII(pc)$ and its influence index $PII(pc)$, if $LII(pc) \geq minPII$, add pc to LP_k, if $PII(pc) \geq minPII$, add pc to P_k.

Generating $(k + 1)$-size HIOPP (steps 5–10): based on Lemmas 2 and 3, the patterns that do not meet $LII(pc) \geq minPII$ are pruned; but based on Lemma 1, only the patterns meeting $PII(pc) \geq minPII$ can generate candidate patterns by back join. So step 6 generates $(k + 1)$-size high-influence ordered-pair patterns P_{k+1}^1 and the patterns LP_{k+1}^1 by front join the patterns of LP_k; step 7 generates $(k + 1)$-size high-influence ordered-pair patterns P_{k+1}^2 and the patterns LP_{k+1}^2 by back join the patterns of P_k. The union of P_{k+1}^1 and P_{k+1}^2 is $(k + 1)$-size high-influence ordered-pair patterns P_{k+1}. If LP_{k+1} is not empty $(P_{k+1} \subseteq LP_{k+1})$, continue to mine higher size patterns.

4   Optimizing Algorithm for Mining HIOPPs

In the basic algorithm, the join operation and storage of table instances are the most time-consuming and space-consuming. Participating instances reduce significantly memory in storing dense data and avoid the join operation. Therefore, we intend to replace table instances with participating instances. Next, we will explore if the influence index can be calculated by participating instances and how to obtain efficiently participating instances.

Algorithm 1. Basic Algorithm

Input: (1) influence features and corresponding instances $IFDS$;(2) reference features and corresponding instances $RFDS$;(3) the parameter k_{max} of the adaptive k-nearest neighbor algorithm based on local density; (4) influence index threshold $minPII$.
Output: a set of high-influence ordered-pair patterns
Variables: (1) RN: the set of all neighbor pairs; (2)k: the size of ordered-pair influence patterns; (3) LP_k: set of k-size ordered-pair influence patterns whose limit influence index is greater than $minPII$; (4) P_k: set of k-size high-influence ordered-pair patterns;
Method:

1: RN= adaptive_k_nearest_neighbor($IFDS,RFDS,k_{max}$)
2: generate influence functions
3: calculate max superimposed influence of all influence instances and influence sum of all influence features
4: set $k = 2$, P_k, LP_k=gen_2_HIOPP($RN,minPII$)
5: **while** $LP_k \neq \emptyset$ **do**
6:     P_{k+1}^1, LP_{k+1}^1=gen_HIOPP_by_front_join($LP_k,minPII$)
7:     P_{k+1}^2, LP_{k+1}^2=gen_HIOPP_by_back_join($P_k,minPII$)
8:     $P_{k+1} = P_{k+1}^1 \cup P_{k+1}^2$, $LP_{k+1} = LP_{k+1}^1 \cup LP_{k+1}^2$,$k = k + 1$
9: **return** $\{P_2, ..., P_{k+1}\}$

4.1 Feasibility of Participation Instances

The calculation of the influence index does not necessarily require the table instance. For example, in Fig. 1(a), the influence index of $pc =$ $<\{a, b\}, \{B, C\}>$ is calculated as $PII(pc) = min(FIR(B, pc), FIR(C, pc))$; to calculate $FIR(B, pc) = \sum_{B.t\in\{B.1,B.2\}} SII(B.t)/FIS(B)$, we only need to know the instances of B participating in the pattern pc and the neighbors of the instances of B in pc; the same is true for $FIR(C, pc)$. It can be seen that to calculate the influence index of a pattern, we only need to know the instances of the features participating in the pattern and the neighbors of the instances.

Definition 11 (Participating Instance). *Given an ordered-pair influence pattern* $pc = \langle IFS_{pc}, RFS_{pc}\rangle$, *if a row instance containing the instance* $f_i.j$ *of pc can be found,* $f_i.j$ *is a participating instance of the feature* f_i *in pc. The set of participating instances of* f_i *in pc is denoted as* $PIS(f_i, pc)$.

Definition 12 (Group Neighbors). *The instances of the feature* f_p *having neighbor relationships with an instance* $f_i.j$ *are the group neighbors of* $f_i.j$ *grouped by* f_p, *denoted as* $groupN(f_i.j, f_p) = \{f_p.q|R(f_i.j, f_p.q)\}$, *where* $R(f_i.j, f_p.q)$ *represents that* $f_i.j$ *has a neighbor relationship with* $f_p.q$.

$groupN(f_i.j, f_p)$ represents all instances of f_p having neighbor relationships with $f_i.j$; $PIS(f_p, pc)$ is the instances of f_p in pc; so $PIS(f_p, pc) \cap groupN(f_i.j, f_p)$ is the all instances of f_p having neighbor relationships with $f_i.j$ in pc. Therefore, we can calculate the influence index of a pattern by participating instances of the pattern and group neighbors.

4.2   Obtaining Participating Instances

The direct way of getting the participating instances of pc is to find those satisfying the definition of participating instances in all instances of all features in pc. Such a search space is huge, so it is necessary to filter the search space.

(1) The search space is filtered as candidate participating instance

Definition 13. *Given ordered-pair influence patterns pc, pc_1, pc_2 while pc is generated by joining pc_1 and pc_2. The intersection of the participating instance set $PIS(f_i, pc_1)$ and $PIS(f_i, pc_2)$ is the candidate participating instance set of f_i in pc, denoted as $CPIS(f_i, pc) = PIS(f_i, pc_1) \cap PIS(f_i, pc_2)$.*

Lemma 4. *The participating instances of f_i in an ordered-pair influence pattern pc must be included in $CPIS(f_i, pc)$, i.e., $PIS(f_i, pc) \subseteq CPIS(f_i, pc)$.*

Based on Lemma 4, the complete $PIS(f_i, pc)$ can be obtained by searching in $CPIS(f_i, pc)$ instead of all instances of f_i, which improve the search efficiency.

(2)The search space is filtered by the definition of a row instance

The method is as follows: 1) for each influence instance $f_i.j$ in candidate participating instances must have neighbor relationships with at least one reference instance in $CPIS(f_p, pc)$ for all $f_p \in RFS_{pc}$, or $f_i.j$ can be removed from $CPIS(f_i, pc)$; 2) for each reference instance $f_p.q$ in candidate participating instances, $f_p.q$ must have neighbor relationships with at least one influence instance in $CPIS(f_i, pc)$ for all $f_i \in IFS_{pc}$, or $f_p.q$ can be removed from $CPIS(f_p, pc)$.

(3) Some of the patterns whose influence indexes are smaller than the influence index threshold are pruned in advance

Lemma 5. *For an ordered-pair influence pattern pc and the feature f_i in pc, $CFIR(f_i, pc) = \sum\limits_{f_i.j \in CFIR(f_i, pc)} SII(f_i.j)/FIS(f_i)$ is an upper bound of $FIR(f_i, pc)$.*

Based on Lemma 5, if $CFIR(f_i, pc)$ of any feature in pc is smaller than $minPII$, pc can be pruned, because $PII(pc) \leq FIR(f_i, pc) \leq CFIR(f_i, pc) < minPII$.

(4) Verify each candidate participating instance

If a row instance RI of pc contains $f_i.j$, $f_i.j$ is a participating instance in pc, and other instances of RI are also confirmed as the participating instances in pc. The instances having the potential to form row instances of pc with $f_i.j$ are called search space of $f_i.j$ on f_p in pc, denoted as $Oss(f_i.j, f_p, pc)$. $Oss(f_i.j, f_p, pc)$ is initialized as the candidate participating instance set $CPIS(f_p, pc)$ of f_p in pc. The verification is divided into two phases. The first phase is verifying all reference instances in candidate participating instances, and the second phase is verifying influence instance that are still unverified in Phase I.

Phase I: For each reference instance $f_i.j$ in candidate participating instances, considering that the influence instances must be the neighbors of the reference instances in a row instance, its search space $Oss(f_i.j, f_p, pc)(f_p \in IFS_{pc})$ is updated to $CPIS(f_p, pc) \cap groupN(f_i.j, f_p)$; then we use a backtracking algorithm to search a row instance containing $f_i.j$.

Phase II: For each influence instance $f_i.j$ in unverified candidate participating instances, its search space $Oss(f_i.j, f_p, pc)(f_p \in RFS_{pc})$ is updated to $groupN(f_i.j, f_p) \cap PIS(f_p, pc)$. Then we also use a backtracking algorithm to search a row instance containing $f_i.j$.

Algorithm 2 shows how to obtain the participating instances of a k-size pattern pc, where $groupNS$ represents group neighbors and PI_{k-1} is participating instances of $(k-1)$-size patterns. The details of "searchRI" are at the document (see Footnote 1).

Algorithm 2. gen_parti_instance_by_search($pc =< I, R >, PI_{k-1}, groupNS$)

1: **for** $f_i \in (I \cup R)$ **do:** get candidate participating instance set $CPIS(f_i, pc)$
2: **for** $f_i \in (I \cup R)$ **do:** calculate $CFIR(f_i, pc)$, **if** $CFIR(f_i, pc) < minPII$, return $\emptyset$
3: **for** $f_i \in (I \cup R)$ **do:** initialize $PIS(f_i, pc) = \emptyset$
4: **for** $f_i \in (I \cup R)$ **do**
5:     **for** $f_i.j \in CPIS(f_i, pc)$ **do**
6:         **if** $f_i.j \notin PIS(f_i, pc)$ **then**
7:             RI =searchRI($f_i.j$, pc, $groupNS$, candidate participating instance set of features in pc)
8:             **if** $RI \neq \emptyset$ **then**
9:                 **for** $f_i.j \in RI$ **do:** $PIS(f_i, pc)$.insert($f_i.j$)
10:                **end if**
11:           **end if**
12:      **end for**
13: **end for**
14: **return** $\{PIS(f_i, pc) | f_i \in (I \cup R)\}$.

---

5   Experiments

We evaluate the effectiveness and performance of our algorithms on real and synthetic data sets. For the real data set, the cancer cases are from the treatment records of a tertiary hospital in Yunnan Province, and the pollution sources come from the public platform for self-monitoring information of the emission units in Yunnan Province[2]. The related details of the real data set are shown in Table 2. For the synthetic data sets, we generate data simulating data distribution with influence instances and reference instances. Experiments are conducted on a Linux system with Intel(R) Xeon(R) CPU @ 2.60 GHz and 128 GB memory.

---

[2] https://wryjc.cnemc.cn/gkpt/mainZxjc/530000.

Table 2. The details of the real data set

Type of data	Number of features	Number of instances
Data of cancer cases	29	23290
Data of pollution source	9	5273

5.1 Effectiveness of Mining Results

We evaluate the effectiveness of the optimizing algorithm (HIOPP_OPT) by comparing the mining results of it and Lei's algorithm [5] (Lei) on the real data set. The parameter k_{max} of the adaptive k-nearest neighbor algorithm based on local density is set to 9(the maximum distance between instances is approximately equal to 0.2), while the distance threshold in Lei is set to 0.2; the measure of influence index adopts their own definitions and the influence threshold is uniformly 0.3. The mining results are sorted in descending order of influence index, and the top-10 patterns are shown in Table 3.

Table 3. Top-10 patterns mined by HIOPP_OPT and Lei on the real data set

Patterns by HIOPP_OPT	PII	Patterns by Lei's algorithm	PII
<{chemical plant, waste management plant}, {prostate cancer}>	0.668	{chemical plant, waste management plant, prostate cancer}	0.567
<{chemical plant, waste management plant}, {breast cancer}>	0.631	{waste management plant, prostate cancer}	0.511
<{chemical plant, waste management plant}, {Bladder Cancer}>	0.616	{waste management plant, anal cancer}	0.500
<{chemical plant, metal processing plant}, {sarcoma}>	0.611	{chemical plant, waste management plant, Bladder Cancer}	0.489
<{waste management plant}, {prostate cancer}>	0.610	{waste management plant, food processing plant, prostate cancer}	0.466
<{chemical plant}, {prostate cancer}>	0.609	{chemical plant, waste management plant, skin cancer}	0.462
<{chemical plant, waste management plant}, {stomach cancer}>	0.603	{waste management plant, Bladder Cancer}	0.460
<{waste management plant, non-metal manufacturing plant}, {prostate cancer}>	0.600	{chemical plant, waste management plant, anal cancer}	0.456
<{chemical plant, waste management plant}, {thyroid cancer}>	0.595	{waste management plant, thyroid cancer}	0.450
<{metal processing plant}, {sarcoma}>	0.590	{chemical plant, food processing plant, Bladder Cancer}	0.438

First, we can see that Lei uses co-location patterns to represent the mining results, while HIOPP_OPT uses high-influence ordered-pair patterns. As already discussed, co-location patterns are unsuitable for expressing the relationships between cancer cases and pollution sources. Besides, high-influence ordered-pair patterns clearly distinguish influence features from reference features, which are easier to understand. **Second**, the influence indexes of the top-10 patterns of HIOPP_OPT are higher than those of Lei. In the influence ratio defined by Lei, the numerator is the superimposed influence of the instance, and the denominator is the number of instances. This unreasonable measurement leads to lower value of the influence ratio. **Third**, the pollution sources involved in the top-10 patterns of two algorithms are different. In addition to the same pollution sources, HIOPP_OPT also involves metal processing plants and non-metal processing plants, while Lei involves food processing plants. This is because we consider the types of pollution sources when measuring the influence of pollution sources on cancer cases. In reality, metal processing plants have a higher cancer risk than food processing plants, so our mining results are more reasonable. **Fourth**, our mining results are supported by related studies. For example, the studies in [11] and [6] support <{chemical plant, metal processing plant}, {sarcoma}> and <{chemical plant}, {prostate cancer}> respectively.

5.2   Performance Evaluation

In this part, we evaluate the performance of HIOPP_OPTI by comparing it with the basic algorithm (HIOPP) and Lei. For comparative fairness, we modify Lei's measure of influence index and method of obtaining neighbor relationship of instances consistent with those of HIOPP_OPTI.

Time and Space Consumption Comparison. Experiments are conducted on the real data set with different influence index thresholds and k, where k is the parameter of the adaptive k-nearest neighbor algorithm based on local density. Figure 2(a) shows the performance comparisons with different influence index thresholds where k is fixed at 6. As the influence index threshold increases, the execution time of Lei increases rapidly while that of HIOPP_OPTI increases steadily, as does the memory consumption. When the influence threshold reaches 0.40, Lei and HIOPP crash due to insufficient memory. Figure 2(b) shows the performance comparisons with different k where the influence index threshold is fixed at 0.45. Obviously, Lei and HIOPP are very sensitive to changes of k, and Lei and HIOPP crash due to insufficient memory when $k \geq 7$. Through the

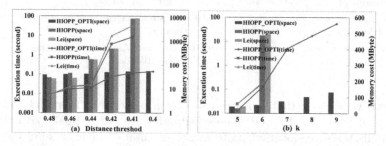

Fig. 2. Performance comparison on the real data set: (a) by distance threshold (b) by k

above experiments, we can find that HIOPP improves in execution time, but HIOPP_OPTI greatly improves in execution time and memory consumption.

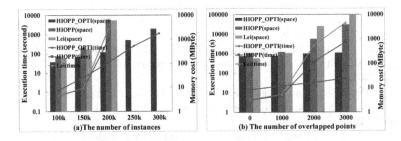

Fig. 3. Scalability test on the synthetic data sets

Scalability. We examine the scalability of HIOPP_OPTI on the synthetic data sets. In Fig. 3(a), with the number of instances ranging from 100,000 to 300,000, the time and space of HIOPP_OPTI increase relatively gradually, while Lei and HIOPP algorithms even cannot end up the programs successfully when the number of instances exceeds 250,000. Figure 3(b) shows the performance comparison with different numbers of overlapped points. The number of overlapped points controls the density of reference instances around the influence instance. The larger the number of overlapped points, the denser the data and the greater the probability that an instance appears in multiple row instances. HIOPP_OPTI runs longer when the number of overlapped points is less than 1000, but HIOPP _OPTI performs better as the number of overlapped points increases. This is because participating instances are more advantageous for storing dense data; moreover, when instances overlap in multiple row instances, all participating instances can be verified by searching for as few row instances as possible.

6   Conclusion

In view of the deficiencies of existing co-location pattern algorithms in studying the relationship between pollution sources and cancer cases, this paper presents high-influence ordered-pair patterns and a more effective measure of the influence of pollution sources on cancer cases. Moreover, we develop basic and optimizing algorithms to mine high-influence ordered-pair patterns. Extensive experiments demonstrate that our mining results are more reasonable, and our algorithm significantly improves time and space. However, our study still ignores the complexity of the real world, and in future work, we should further consider the effects of wind direction and water flow on pollutant diffusion.

Acknowledgements. This work is supported by the National Natural Science Foundation of China (61966036, 62062066), and Yunnan University Postgraduate Technological Innovation Project(2021Y175).

References

1. Yao, X., Peng, L., Yang, L., Chi, T.: A fast space-saving algorithm for maximal co-location pattern mining. Expert Syst. Appl. **63**(C), 310–323 (2016)
2. Bao, X., Wang, L.: A clique-based approach for co-location pattern mining. Inf. Sci. **490**, 244–264 (2019)
3. Bing, Z., Zuqiang, M., Liangliang, S., Hongli, L.I., Computer, S.O.: Adaptive k neighbor algorithm based on local density and purity. J. Guangxi Acad. Sci. **33**(1), 19–24 (2017)
4. García-Pérez, J., et al.: Residential proximity to industrial pollution sources and colorectal cancer risk: a multicase-control study (MCC-Spain). Environ. Int. **144**, 106055 (2020)
5. Lei, L., Wang, L., Zeng, Y., Zeng, L.: Discovering high influence co-location patterns from spatial data sets. In: ICBK, pp. 137–144. IEEE (2019)
6. Ramis, R., Diggle, P., Cambra, K., López-Abente, G.: Prostate cancer and industrial pollution. Environ. Int. **37**(3), 577–585 (2011)
7. Terrell, K.A., St Julien, G.: Air pollution is linked to higher cancer rates among black or impoverished communities in Louisiana. Environ. Res. Lett. **17**(1), 014033 (2022)
8. Wang, L., Zhou, L., Lu, J., Yip, J.: An order-clique-based approach for mining maximal co-locations. Inf. Sci. **179**(19), 3370–3382 (2009)
9. Yang, P., Wang, L., Wang, X., Zhou, L.: SCPM-CR: a novel method for spatial co-location pattern mining with coupling relation consideration. IEEE Trans. Knowl. Data Eng. (2021). https://doi.org/10.1109/TKDE.2021.3060119
10. Yoo, J.S., Shekhar, S.: A joinless approach for mining spatial colocation patterns. IEEE Trans. Knowl. Data Eng. **18**(10), 1323–1337 (2006)
11. Zambon, P., et al.: Sarcoma risk and dioxin emissions from incinerators and industrial plants: a population-based case-control study (Italy). Environ. Health **6**(1), 19 (2007)

Finding Hidden Relationships Between Medical Concepts by Leveraging Metamap and Text Mining Techniques

Weikang Yang[1], S. M. Mazharul Hoque Chowdhury[2], and Wei Jin[3(✉)]

[1] Department of Computer Science, North Dakota State University, Fargo, USA
yangvigoo@gmail.com

[2] Department of Computer Science and Engineering, Daffodil International University, Dhaka, Bangladesh
mazharul2213@diu.edu.bd

[3] Department of Computer Science and Engineering, University of North Texas, Denton, USA
wei.jin@unt.edu

Abstract. Text is one of the most common ways to store data in this computerized world. At a glance, it may seem that those data are not interconnected. But in reality, data can have hidden connections. Therefore, in this research, a new model has been presented that can find hidden relationships between two medical concepts by using MetaMap and appropriate text-mining techniques. Specifically, the model creates a new comprehensive index structure and can find cross-document hidden links connecting topics of interest that most existing approaches have ignored. Experiments show the effectiveness of the proposed model in discovering new connections between topics.

Keywords: Biomedical text mining · Hidden relationship discovery · Cross-document knowledge discovery

1 Introduction

If we consider the modern world, it is not possible to imagine it without the use of text. Therefore, text data plays a very important role in every sector of our lives, and most of the time a piece of text type data may contain some connection with other texts. In some cases, those connections are clearly visible and, in some cases, relations remain hidden. On the other hand, when people are interacting with text while reading books or newspapers or communicating with others via text messages, they try to make some connections between different things and try to get hidden information based on the text [1]. However, it is difficult for people to find all the connections based on a little information only. In some cases, a sentence indicating "A implies B" in one article and another sentence "B implies C" in another article, and people generally could not link them together to establish the relationship that "A may imply C", especially when they are facing a large volume of data. This idea was explained by Swanson in 1999 and he proposed a model named 'ABC model' to solve it [2].

W. Chen et al. (Eds.): ADMA 2022, LNAI 13725, pp. 41–52, 2022.
https://doi.org/10.1007/978-3-031-22064-7_4

The primary focus of this research is motivated by Swason's ABC model, and we extend it to a cross-document discovery setting and preform a finer granularity of relationship discovery. Additionally, to cope with a large-scale document collection, we introduce a new and efficient index structure that can support multi-level knowledge discovery and we believe it will also facilitate many other downstream natural language processing applications. We further develop a new approach that will be able to predict hidden relations between two input topics effectively. This model was tested against other existing models and showed its effectiveness. Moreover, instead of a single-thread implementation, we also explored a multi-thread implementation setting in order to reduce analysis time and increase system efficiency at the same time. Response time and accuracy can both be further improved by adding threads if computer hardware resource is concerned. In this research Medline database was used as the data source because of its large volume of articles on Bio-medicine and MetaMap was used to analyze the data. Output produced from this model will be combined with Swanson's model to find hidden connections.

The reminder of this paper is structured as follows. Section 2 discusses related work. In Sect. 3, we present an overview of our methods. In Sect. 4, we present details of our proposed techniques. Sections 5 and 6 present experiments and evaluation results. And finally, Sect. 7 brings conclusion and gives directions for future work.

2   Literature Review

In this modern world, data is everything and most of them are text-type data. There are a lot of tools available for text data analysis and researchers around the world have developed tools like MicroMeSH, SAPHIRE, and MetaMap to do this job. MetaMap is a public tool developed by the National Library of Medicine (NLM), which can map text to biomedical concepts using its enormous thesaurus [3]. This has been a popular tool applied to many Information Retrieval and text-mining applications [4]. When Metamap receives a sentence, it breaks a sentence into phrases and those phrases are used to find mapping concepts to calculate a mapping score for each concept. There are already many applications of MetaMap in the field of biomedical studies. For example, Wendy Marcelo et al. developed a system to detect patients with respiratory illness by processing patients' clinical reports with MetaMap [5]. Zuccon Holloway et al. tried to automatically identify disorders mentioned in health records such as discharge summaries by using MetaMap [6]. In order to test the performance of MetaMap, Pratt and Yetisgen conduct research, which aims to compare MetaMap's capability with that of people who are familiar with the biomedical field [7]. The result shows that MetaMap is capable to identify the biomedical concepts with a 93.3% recall compared with those tagged by biomedical experts.

Knowledge discovery in biomedical texts originates from Swanson's ABC model, based on which Jin and Srihari proposed a new type of query, namely, concept chain queries, attempting to detect hidden links between concepts [2]. Researchers Wei Jin and Rohini Srihari tried to generate concept chains connecting topics of interest from counterterrorism documents. The approach used a variant of the Term Frequency (TF) - Inverse Document Frequency (IDF) weighting scheme for evaluating each potential connecting term, achieving an approximate 82.5% recall [8]. This work is different from Jin's study by using a completely different context – the biomedical domain instead of the counterterrorism corpus. Another related work was proposed by Gopalakrishnan, Kishlay et al., which introduced a new approach that created a graph-based knowledge base and then used a bi-directional search for finding paths in the graph to answer concept chain queries in the biomedical field [9].

Researcher Vishrawas Gopalakrishnan et al. worked on Hypothesis generation and there he used medical data in order to find hidden relations [16]. In their research, they also focused on the hidden relation discovery based on given keywords to develop a hypothesis on a topic. Researcher Xiaohua Hu et al. presented a model that is capable to find hidden links from Medline data and generate novel hypothesis between these concepts [17]. Padmini Srinivasan and Bisharah Libbus worked on research where they used Medline citations to find connections between dietary substances and diseases [18]. In their research, they were trying to discover the therapeutic potential of curcumin/turmeric which is commonly used in Asia. Another research was done by Kishlay Jha and Wei Jin on the extraction of novel knowledge from biomedical literature and domain knowledge [19]. In their research, they presented a meticulous analysis of how manifold statistical information measures and semantic knowledge affect the knowledge discovery procedure.

Moreover, the use of hidden relation extraction covers a large area that includes relations between flood and hydrological variables, immunological markers in multiple sclerosis, and so on. In the year 2007, M.J. Sanscartier and E. Neufeld worked on a model that will be able to find hidden relationships between different variables in a dataset in order to correct models by finding hidden contextual variables [11]. Prakash & Surendran (2013) worked on the detection of social media hidden activity based on the connection between different types of social media profiles and the network traffic of a user [12]. Milton Pividori et al. worked on the discovery of hidden relations between different qualitative and quantitative data without standardization based on their influence on the cluster using a biological dataset [13]. In 2021, Mohamad Basel Al Sawaf et al. worked on a recurrent neural network model that will be able to analyze relationships between different hydrological variables related to flood in order to predict flooding [14]. Amir Hossein Rasekh et al. presented a model that can create descriptions of programming codes that have a relation to associated text documents [15]. Those descriptions can help a user to understand the program logically with little effort.

3   Methodology

This research applied the principle of pipe and filter software architectural design [10]. As like in this architecture the original data flow through different modules or filters and in each filter data gets modified and passed through the next one. Finally, the system produces the desired output at the end of the flow.

At first two topics will be provided to the system along with a number of preprocessed documents in order to discover hidden relations. The system will look for relations using the algorithm the can express both topics. Finally, all the findings will be presented as output. As an example, consider the task stated in [20], the first task that pioneered LBD in the biological domain - determining the relationship between fish oil and Raynaud's disease and how they are related.

The system contains three main phases – The MetaMap processing phase (will process and extract documents), the preparation phase (will generate S2C – Source to Concept and fetch the title and abstract), and the output phase (will find relations using Swanson's ABC model). Findings from the output module will be presented using a Graphical User Interface (GUI). This will show the generated concept chains and relevant evidence. Figure 1 presents the data flow of the system.

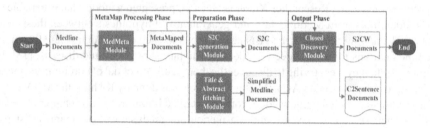

Fig. 1. Data flow diagram

4 Data Collection and Analysis

The Medline database contains a very large number of biomedical data-related citations and those were released by NLM on their official website. In this research, altogether nearly 22380000 citations were used to find hidden relations between two different topics. Here, each citation is an individual document. The Medline database includes a number of attributes such as ID, Publishing date, title, and abstract. For the Hidden Relationship Discovery title and abstract was used to find common links among topics.

4.1 Data Extraction from the Source

Data were extracted from the following link - ftp://ftp.ncbi.nlm.nih.gov/pubmed/bas eline and XML tags as < PMID >, < PubDate >, < ArticleTitle >, and < AbstractText > were used to identify useful information.

4.2 MetaMap Module – Processing Phase

The functional structure of the MedMeta module is divided into four major steps. They are:

a. Extracts useful information from the data source – Medline documents.
b. Inserts MetaMap API in the code and sends titles and abstracts to the MetaMap Server.

c. Reads from MetaMap output and builds indexes of Medline documents based on concept occurrence relationship.
d. Write index files in XML format.

The MedMeta Module consists of seven modules – MedMeta, XMLParser, MedlineCite, MetaMapProc, PostProc, OutputXML, and Thread. The sequence is presented in Fig. 2.

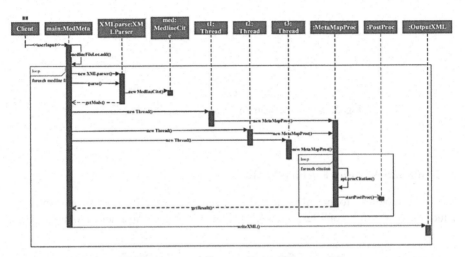

Fig. 2. Sequence diagram for MedMeta module

MedMeta's primary object is to regulate the creation of other modules and XML-Parser can process oversized XML files. In this research total data used was 93.7 GB in total, where a single file was around 200 MB. After the parsing, XMLParser stored data into MedlineCite (a placeholder module to store data) objects. MetaMapProc will communicate with server and build index to initiate post-processing in PostProc. The PostProc is divided into 3 levels of data structure. The first level is Semantic type (holds one or many Concepts objects), the second one is Concept type (holds one or many Occurrence objects) and the third one is Occurrence type (holds the entity information). OutputXML uses an XML file that will hold the data. The thread module will apply parallel processing to improve performance.

4.3   MetaMap Module – Preparation Phase

In the preparation phase, a simplified S2C document will be produced using semantic and concept type data structure to reduce access time complexity. During the data collection phase, OutputXML generated a big amount of data and those were converted to 2 level S2C to reduce size. According to Fig. 3, S2C generation module consists of four modules – S2CGenerator (forwards the document through XMLParser), XMLParser (storing data in the server), SemanticType (uniquely leveling each hash to hold the S2C relationship), and WriteXMLFile (writes relationships in the XML file). Here, S2CGenerator is responsible for forwarding the document through XMLParser. There are 133 semantic types in total and each of them holds a hash set of concepts.

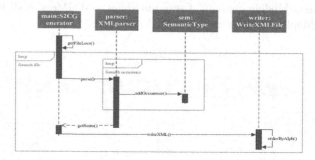

Fig. 3. Sequence diagram for S2C generation module

4.4 Title and Abstract Fetching Module

Aim of this module is to fetch the title and abstract from the original document in order to reduce the access time complexity. Figure 4 shows the title and abstract fetching module.

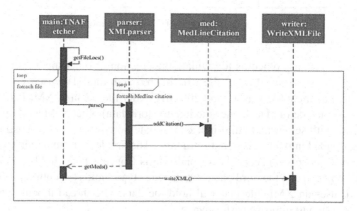

Fig. 4. Sequence diagram for title & abstract fetching module

The Title and Abstract fetching module consist of four modules. Three of them are TNAFetcher, XMLParser, and WriteXMLFile and their design are the same as their counterpart in the S2C generation module. Besides, the fourth one is MedLineCitation and it has two data fields – Title and Abstract, where data fetched from XML documents gets stored. Figure 5 shows a simplified Medline Document.

```
<MedlineCitation PMID="2447742">
  <Title>Latex agglutination test for alpha-fetoprotein in the diagnosis of premature
  rupture of the amniotic membranes (PROM).</Title>
  <Abstract>A rapid latex agglutination test for alpha-fetoprotein (AFP) was compared with
  a pH-indicator and patient history in the detection of premature rupture of the amniotic
  membranes. Of 120 patients examined, 34 had an established rupture of the membranes, and
  56 had suspected rupture. Thirty patients had no evidence of membrane rupture. The
  vaginal content was examined with a pH-indicator. Samples of vaginal content were also
  obtained to perform the latex agglutination test, and 103 of these samples were analysed
  by a radio-immunoassay (RIA) technique for AFP. Our results indicate that the latex
  agglutination test is of doubtful value due to the many inconclusive test results. The
  sensitivity of the latex test is less than 15%, and the specificity is 80%. Patient
  history combined with pH-indicator test is far more informative than the latex
  agglutination test.</Abstract>
</MedlineCitation>
```

Fig. 5. A simplified medline document.

4.5 Closed Discovery Module

This module uses user inputs, MetaMaped documents, S2C documents, and simplified Medline documents as inputs and after processing generates S2CW files and displays outputs through a Graphical User Interface. Figure 8 shows the Sequence diagram of the Closed Discovery module (Fig. 6).

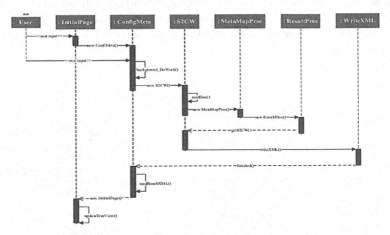

Fig. 6. Sequence diagram of closed discovery module.

The ClosedDiscovery Module consists of eight modules – InitialPage, ConfigMeta, S2CW, MetaMapProc, ResultProc, WriteXML, SentPage, and MergedSentPage. Sent-Page and MergedSentPage are not included in Fig. 8, but they are used to display additional graphical user interfaces. Here, the InitialPage of GUI takes user inputs "Topic 1" and "Topic 2" and through ConfigMeta user can select target data. ConfigMeta is also responsible for building XML documents to store sentences related to Topic 1 and Topic 2. S2CW passes the XML document to MetaMapProc, which works as a gateway to the server that return findings. ResultProc uses those findings to build concept chains by calculating Term Frequency (TF) and Inverse Document Frequency (IDF). The IDF measures the importance of the concept and is measured for concept c by –

$$IDF(c) = log_e(Total\ number\ of\ sentences\ /\ Number\ of\ sentences\ with\ concept\ c\ in\ it)$$
(1)

The assumption is that the rarer a concept appears in the sentences, the higher the IDF value is, meaning the more important the concept is to the context. Then, the ResultProc multiplies TF and IDF to get concept weights (i.e. $Weight_c = TF_c * IDF_c$). Each MetaMap may contain several concepts and ResultProc analyze them to build C2W (Concept to Weight) and C2Sents (Concept to Sentence) relationship. Together they create S2CW (Semantic to Concept Weight) relationship, based on which the weight of each concept is further normalized by:

$$Normalized\ Weight(c) = \frac{the\ weight\ of\ the\ concept\ c}{maximum\ weight\ in\ this\ Semantic\ Type} \tag{2}$$

After the normalization data will be arranged in the descending order and by merging S2CW for topic 1 (S2CWA) and S2CW for topic 2 (S2CWC) intermediate level linking (common things between Topic 1 and 2) S2CWB will be generated. At this point, WriteXML will build XML files for S2CWA, S2CWB, S2CWC, C2SentsA, and C2SentsC using DocumentBuilderFactory Java API. SentPage is responsible for viewing all sentences related to a certain concept and MergedSentPage will list out all sentences related to concept chains A-B and B-C. Figure 7 represents the GUI which takes the input from the user and visualize the output and Fig. 8 represents linking sentences between topic 1 and 2.

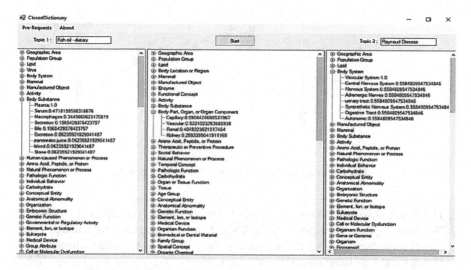

Fig. 7. GUI for visualizing relationship between topics 1 and 2

5   System Evaluation

During the development of MedMeta module a multi-thread program (3 threads) was used to communicate with the MetaMap server instead of single thread to evaluate system performance. An experiment was conducted to find the impact of threads against

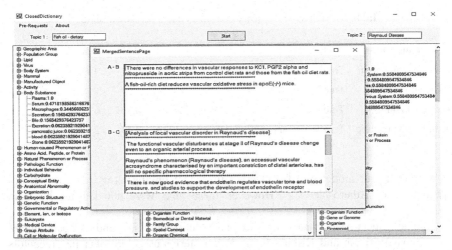

Fig. 8. GUI for sentences that are creating the relationship

Table 1. Performance comparison

Number of threads	Processing time	Average memory usage	Average CPU usage	Performance compare to single thread	Performance (%)
1	145 s	98 MB	25%	1	
2	87 s	110 MB	40%	1.67	85%
3	62 s	128 MB	73%	2.34	78%

performance with a system that has eight-core i7-4790k CPU, 16GB RAM and Windows Pro 10 x64 OS. The number of threads varies from one to three (Table 1).

According to the table it is clearly visible that with the increase of the number of threads performance increases greatly, where for 1 thread it took 145 s and for 3 threads it took 62 s. Moreover, 3 threads achieve 2.34 times improved performance over a single thread. It shows, in case of hardware limitations three-threads is a preferred option.

6   Result Evaluation

In order to justify the accuracy of the model, we used the widely adopted evaluation queries in the related literature, e.g., comparison was made between this research and Gopalakrishnan and Kishlay's study [9] using the gold standard query pair "Fish-oil" and "Raynaud's Disease". Gopalakrishnan and Kishlay had found connecting words such as platelet aggregation, vascular reactivity, blood viscosity and Prostaglandin. Our model not only finds all these impronat connecting terms, but also assigns each a high rank in its associated semantic types. Table 2 presents our partial result of evaluating the above query.

Table 2. Evaluating the query pair: fish oil and raynaud disease

Connecting concepts	Find?	Weight	Semantic type rank
Platelet aggregation	True	0.87	2
Vascular reactivity	True	1.27	1
Prostaglandin	True	0.309	6
Arthritis rheumatoid	True, it appears as "Rheumatoid Arthritis"	0.43	2

Moreover, our model also found other important linking terms such as "Hemo-dynamic" and "Atherosclerosis". However, these links were not detected by them [9].

Another query was made for the pair "Schizophrenia" and "Phospholipase A2". Gopalakrishnan and Kishlay found linking terms as chlorpromazine, receptors dopamine, prolactin, arachidonic acid, phenothiazines, and norepinephrine [9]. Table 3 shows how our framework behaves in this case. This model was able to find relevant data as them and again this system found interesting linking terms, such as "PGE2", which [9] could not detect.

Table 3. Evaluating the query pair: Schizophrenia and Phospholipase A2

Connecting concepts	Find?	Weight	Rank in the semantic type
Chlorpromazine	True	0.22	11
Receptors dopamine	True	0.33	5
Prolactin	True	1.17	2
Arachidonic acid	True	1.18	1
Norepinephrine	True	0.3	11

For the standard query pair Migraine and Magnesium, Gopalakrishnan and Kish-lay had found connecting words such as propranolol, adenosine triphosphate, calcium, ergotamine, serotonin, norepinephrine, adenine nucleotides, and epinephrine [9]. Table 4 shows how our framework behaves.

Table 4 indicates that the system has also found all those eight connecting concepts found in Gopalakrishnan and Kishlay's model. Our system also found some other linking terms that were not found in the previous study, such as "Insulin".

Table 4. Evaluating the query pair: Migraine and Magnesium

Connecting concept	Find?	Weight	Rank in the semantic type
Propranolol	True	0.41	8
Adenosine triphosphate	True	0.52	10
Calcium	True	1.50	1
Ergotamine	True, the system found "Ergot Alkaloids"	0.34	5
Serotonin	True	1.49	1
Norepinephrine	True	0.79	6
Adenine nucleotides	True	0.32	14
Epinephrine	True	0.49	13

7 Conclusion and Future Work

The goal of this research is to find the hidden connections between concepts of interest. This will assist researchers in the early discovery of hypotheses that may lead to important findings by initiating a deep understanding of the hidden information in the biomedical sector. Experiments show that the developed model can find meaningful logical connections between topics of interest and can visualize discovered hypotheses in a user-friendly way. In future work, we will expand the chains to multiple levels, which we call concept graph queries, and MetaMaped files and other additional domain-specific resources will be combined for easier access and more comprehensive knowledge discovery.

References

1. Belkin, N.J.: Interaction with texts: Information retrieval as information seeking behavior. In: Information Retrieval. p. 55–66 (1993). 10.1.1.50.6725
2. Swanson, D.R.: Complementary structures in disjoint science literatures. In: Proceedings of the 14th Annual International ACM SIGIR Conference on Research and Development in Information Retrieval, ACM Press, Chicago, IL, pp. 280–289 (1991). https://doi.org/10.1145/122860.122889
3. Aronson, A.R.: Effective mapping of biomedical text to the UMLS Metathesaurus: the MetaMap program. In: Proceedings of AMIA Annual Symposium, pp. 17–21 (2001). https://pubmed.ncbi.nlm.nih.gov/11825149/
4. Kay Deeney. MetaMap - A Tool for Recognizing UMLS Concepts in Text. U.S. National Library of Medicine (2017). https://metamap.nlm.nih.gov/
5. Chapman, W.W., Fiszman, M., , Dowling, J.N., Chapman, B.E., Rindflesch, T.C.: Identifying respiratory findings in emergency department reports for biosurveillance using MetaMap. Studies in Health Technology and Informatics, **107**(Pt 1), pp. 487–91 (2004). https://pubmed.ncbi.nlm.nih.gov/15360860/
6. Zuccon, G., Holloway, A., Koopman , B., Nguyen, A.: Identify disorders in health records using conditional random fields and metamap. In: Proceedings of the CLEF 2013 Workshop on Cross-Language Evaluation of Methods, Applications, and Resources for eHealth Document Analysis, pp. 1–8 (2013). https://eprints.qut.edu.au/62875/

7. Pratt, W., Yetisgen-Yildiz, M.: A study of biomedical concept identification: MetaMap vs. people. In: AMIA Annual Symposium Proceedings, pp. 529–33 (2003). https://pubmed.ncbi.nlm.nih.gov/14728229/

8. Jin, W., Srihari, R.K.: Knowledge discovery across documents through concept chain queries. In: Proceedings of the Sixth IEEE International Conference on Data Mining – Workshops (ICDMW'06), pp. 448–452 (2006). https://doi.org/10.1109/ICDMW.2006.105

9. Gopalakrishnan, V., Jha, K., Jin, W., Zhang, A.: A survey on literature based discovery approaches in biomedical domain. In: Journal of Biomedical Informatics, **93**, 103141 (2019). doi: https://doi.org/10.1016/j.jbi.2019.103141

10. Philipps, J., Rumpe, B.: Refinement of pipe-and-filter architectures. In: Wing, J.M., Woodcock, J., Davies, J. (eds.) FM 1999. LNCS, vol. 1708, pp. 96–115. Springer, Heidelberg (1999). https://doi.org/10.1007/3-540-48119-2_8

11. Sanscartier, M.J., Neufeld, E.: Identifying hidden variables from contextspecific independencies. In: Proceedings of the Twentieth International Florida Artificial Intelligence Research Society Conference, pp. 472–477 (2007). 10.1.1.329.7687, Florida, USA

12. Prakash, D., Surendran, S.: Detection and analysis of hidden activities in social networks. International Journal of Computer Applications (0975–8887), **77**(16), 34–38 (2013). https://doi.org/10.5120/13570-1404

13. Pividori, M., Cernadas, A., de Haro, L.A., Carrari, F., Stegmayer, G., Milone, D.H.: Clustermatch: discovering hidden relations in highly diverse kinds of qualitative and quantitative data without standardization. Bioinformatics **35**(11), 1931–1939 (2019). https://doi.org/10.1093/bioinformatics/bty899

14. Sawaf, M.B.A., Kawanisi, K., Jlilati, M.N., Xiao, C., Bahreinimotlagh, M.: Extent of detection of hidden relationships among different hydrological variables during floods using data-driven models. Environ. Monit. Assess. **193**(11), 1–14 (2021). https://doi.org/10.1007/s10661-021-09499-9

15. Rasekh, A.H., Arshia, A.H., Fakhrahmad, S.M., Sadreddini, M.H.: Mining and discovery of hidden relationships between software source codes and related textual documents. Digital Scholarship in the Humanities. **33**(3), 651–669 (2018). https://doi.org/10.1093/llc/fqx052

16. Gopalakrishnan, V., Jha, K., Zhang, A., Jin, W.: Generating hypothesis: Using global and local features in graph to discover new knowledge from medical literature. In: Proceedings of the 8th International Conference on Bioinformatics and Computational Biology, Las Vegas, Nevada, USA. pp. 23–30 (2016). 978–1–943436–03–3

17. Hu, X., Zhang, X., Yoo, I., Zhang, Y.: A semantic approach for mining hidden links from complementary and non-interactive biomedical literature. In: Proceedings of the Sixth SIAM International Conference on Data Mining, Bethesda, MD, USA, pp. 200–209 (2006). https://doi.org/10.1137/1.9781611972764.18

18. Srinivasan, P., Libbus, B.: Mining MEDLINE for implicit links between dietary substances and diseases. In Bioinformatics. **20**, i290–i296 (2004). https://doi.org/10.1093/bioinformatics/bth914

19. Jha, K., Jin, W.: Mining novel knowledge from biomedical literature using statistical measures and domain knowledge. In: Proceedings of the 7th ACM International Conference on Bioinformatics, Computational Biology, and Health Informatics (BCB '16). Association for Computing Machinery, New York, NY, USA, pp. 317–326 (2016). https://doi.org/10.1145/2975167.2975200

20. Swanson, D.R.: Fish oil, raynaud's syndrome, and undiscovered public knowledge. Perspect. Biol. Med. **30**(1), 7–18 (1986). https://doi.org/10.1353/pbm.1986.0087

Causality Discovery Based on Combined Causes and Multiple Causes in Drug-Drug Interaction

Sitthichoke Subpaiboonkit$^{(\boxtimes)}$ ⓘ, Xue Li ⓘ, Xin Zhao ⓘ, and Guido Zuccon ⓘ

The University of Queensland, Brisbane, Australia
{s.subpaiboonkit,x.li,x.zhao,g.zuccon}@uq.edu.au

Abstract. We consider the problem of automatically detecting drug-drug interactions, i.e., the occurrence of an adverse reaction caused by the co-administration of two or more drugs, from reports of suspected cases. This is an important problem because of the health implications that correctly identifying drug-drug interactions has, and because the automatic detection of such cases would enable analysis and monitoring at scale. Current automated methods are based on associations and correlation relationships between datapoints, and thus fail to identify casual relationships. In this context, we propose a novel approach that specifically identifies the combined causes and multiple causes of drug-drug interactions, along with the actual direction of the casual relation. The method is empirically validated on a real-world adverse effect dataset and contrasted against current methods for automatic drug-drug interaction.

Keywords: Causality discovery · Bayesian network · Drug-drug interaction

1 Introduction

Adverse Drug Reaction (ADR) is an unwanted and harmful effect caused by consuming a drug. A drug-drug interaction (DDI) is an ADR caused by taking a drug together with one or more other drugs (i.e., drug co-administration). For example, aspirin consumed together with warfarin may cause excessive bleeding [10]; we denote this as aspirin + warfarin → *Bleeding*. That is, a DDI is a drug-drug interaction that causes an adverse effect, i.e. an increase of toxicity.

ADRs are a major cause of morbidity and mortality. More than hundreds of thousands of people die and 770,000 people are injured because of ADRs every year in the United States alone [16]. The approximate annual cost related to ADRs is $136 billion [5]. It is estimated that about 30% of ADRs incidents are DDIs [5,25]. It is important that DDIs are detected early when new drugs are released. DDIs can be identified via clinical trials (randomized control trials); however, it is infeasible to identify all possible DDIs using clinical trials as this would require testing for all possible drug combinations. These trials are indeed costly (both financial and temporally) [15].

W. Chen et al. (Eds.): ADMA 2022, LNAI 13725, pp. 53–66, 2022.
https://doi.org/10.1007/978-3-031-22064-7_5

The Spontaneous Reporting System process has been established with the aim to support the discovery of ADRs (including DDIs) that have not previously been detected by clinical trials. In this process, information related to drug usage and related ADRs reports are collected from healthcare authorities, healthcare providers, drug manufacturers and patients. This kind of report facilitates drug safety surveillance by means of providing reporting capabilities for ADRs [18] and DDIs [31] analysis. One such system is the FDA Adverse Event Report System (FAREs) [3]. However, a considerable amount of new reports are submitted every day, and they are often too many for experts to manually analyse [29] to determine if the reported event provides valid evidence of a casual relation between drug intake and adverse event. It has been suggested that computational methods and in particular data mining algorithms would be capable of analyse this data and discover new DDIs [29]. The majority of the existing computational methods for pharmacovigilance specific to DDIs discovery are based on statistical association, correlation or classification methods; these include association rule [11,24], logistic regression [28], bi-clustering [12] and support vector machines [8]. Although these models identify association and correlation between data points, they fail to unveil causal relations between data points. For example, these methods are capable of modelling a correlation between the co-administration of drug X and Y and the adverse reaction Z – but this does not necessarily mean that it is the drug co-administration that causes the ADR. Indeed, in the task of DDIs identification, it is important to identify causal relations between drugs co-administrations and adverse reactions: these comprise of causal events and their direct consequences [9].

Causal Bayesian Network (CBN) is a computational method that has been used to represent and discover causal relationships among data based on conditional dependencies between variables in a direct acyclic graph [23,30]. However, this method is computationally inefficient and unfeasible for high-dimensional, large-scale observational data [6]. The Causal Association Rule Discovery (CARD) [5] (an extension of the CBN approach with constraint-based causality adjusted to reduce scope of considered data and time complexity) has been applied to the problem of discovering the causes of a DDI using the concept of V-structure—the relationship describing two causes that are marginally independent become dependent when their common effect is given [30]. V-structures can be classified into two groups: multiple causes and the combined cause (see Sect. 2 for further details). Knowing which group a DDI relates too is important as it may facilitate or scope further pharmacokinetic and pharmacodynamic investigations in order to understand the reasons or mechanisms involved in the DDI: a process that is necessary for drug safety surveillance [22]. However, CARD's results cannot discriminate between the two groups without further input from experts; this is because the CBN does not hold a direct representation of the concept of combined cause. While there are other causality-based methods to discover DDI [27,32], they require domain knowledge of an existing drug causing an adverse effect to discover other drugs causing the adverse effect. This requirement is a core difference between previous methods and our proposed method that does not require preliminary domain knowledge or expert intervention. In addition, previous methods do not consider the difference between multiple causes and combined causes. We would not include those methods in our study.

In this paper, a novel constraint-based method is proposed to:

1. discover the combined causes in DDI that produce a target adverse effect and can discriminate between combined cause and multiple causes. Our method is based on the concept of V-structure. We further develop a heuristic algorithm to discover combined causes in DDIs and consider an instantiation of the algorithm in the scenario of two drugs being the cause of a target adverse effect: this reduces the time complexity of the problem;
2. identify the multiple causes of a DDI by using the V-structure concept to first identify all DDIs and then from these deduct the combined-cause DDIs.

The proposed method is applied to real-world data and the output is evaluated using a database of known DDIs and compared with the state-of-the-art method.

In this study, the term "DDI" represents drug-drug interaction causing a target adverse effect.

2 Background

2.1 Combined Causes and Multiple Causes in DDI

The aforementioned combined causes and one or more causes of DDI can be simply described in an example in Fig. 1 and 2, respectively. From Fig. 1 (multiple causes) a, b, either drug A or drug B individually causes an adverse effect C, and their interaction of drug A and B causes an adverse effect C. From Fig. 1 c, drug A and drug B individually cause an adverse effect C, and their interaction causes an adverse effect C. From Fig. 2 (combined causes), neither drug A nor drug B individually causes adverse effect C, but their interaction causes an adverse effect C.

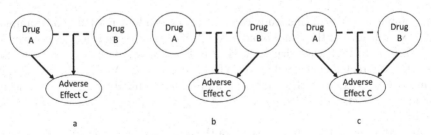

Fig. 1. The multiple causes, where solid arrows represent causal relationship and dashed lines denote the interaction of two variables

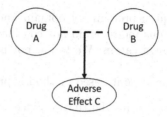

Fig. 2. The combined causes, where solid arrows represent causal relationship and dashed lines denote the interaction of two variables

The Bayesian constraint-based causality (CBC) methods that discover local causal discovery for a target variable instead of discovering a complete Bayesian graph in CBN are proposed to efficiently handle high-dimensional data. For example, approaches related to Markov blanket concept [1,2,4,7] and V-structure [5] or common effect [30]. Both of them mainly focus on multiple causal variables that cause a target-effect variable. However, the Markov blanket still suffers exponential time complexity when conditional variable sets are considered. It also has a causal direction ambiguity problem, while V-structure-related approaches can indicate causal direction.

To our knowledge, Causal Association Rule Discovery (CARD) [5] is the only CBC approach that is proposed to discover causal DDI without preliminary domain knowledge's input data. This method's algorithm uses association rules to do data preprocessing and filtering, and applies the Bayesian network's V-structure model or common effect to discover causation. V-structure is a causality property that represents two causal variables that are marginally independent and are dependent when knowing their common effect variable as a fixed target. This method can discover causal multiple drugs that cause a target adverse effect. However, it cannot separate the combined causes of which individual causes do not cause the target effect from multiple causes (at least one cause causes a target adverse effect) due to the graph representation limitation of CBN.

2.2   Limitations of CBN

The combined causes of this study can't be discovered by CBN [19]. For example, in CBN, if drug A is the cause of adverse effect C then there is a directed causal edge from A to C. If each drug A and drug B alone is not the cause of adverse effect C, no directed edge can be drawn from A or B to C. As a result, by CBN, it is impossible to examine the combined causes, interacting drugs A and B causing the adverse effect C of which individual A or B alone does not cause the adverse effect C [26].

According to DDI, finding the case of the combined causes is difficult because all individual drugs are independent of the adverse effect, which provides no hint for experts to discover those combined causes of the adverse effect. Even though the existing method can discover DDI that their adverse effect is caused by either one or more drugs or no individual drug, those causal drug roles cannot be identified from the outcome without experts' help. Knowing the types of DDI discovery may help us to understand the pharmacokinetic and pharmacodynamic properties of discovering DDI.

Another concept of combined causes is mentioned in casualty discovery approaches by the integration of association rules with partial association test [14] and with retrospective cohort studies [17]. Even though their practical methods can identify causation, their combined causes' concept is similar to multiple causes or not the specific combined cause concept as we propose. To our knowledge, The relevant combined causes' concept in causality is introduced in the algorithm called multi-level HITON-PC (MH-PC) [19]. This work has modified HITON-PC [1] which is a heuristic constraint-based algorithm ordinarily used to discover local causal structures with a fixed target effect. However, the outcome of MH-PC has the same drawback as the HITON-PC which is the mixture between direct parent and children of a fixed target because of causal direction ambiguity problem. Therefore, it tends to be not practical for real-world data or complex data. In addition, those mentioned methods have not been applied to solve the DDI problem before.

3   Proposed Method

This novel study aims to discover combined causal drugs and identify multiple causal drugs of a target adverse effect. For DDI identification studies, consideration of any two drugs is the majority [5]. In this study, we focus on both a combined cause and multiple causes consisting of two drugs and design the algorithm that suits our aim.

Notation: In this paper, we use upper case letters to represent random variables (such as X), multiple uppercase letters to represent the combined variables (such as $X\&Y$), bold uppercase letters denote a set of variables (such as S). This study uses only binary variables, i.e., the values of each variable are either 0 or 1.

Conditional Independence [23]: Let's X, Y and Z are random variables. X and Y are conditionally independent when Z is given if joint probability $P(.)$ function is as follows

$$P(X|Y, Z) = P(X|Z) \tag{1}$$

where $P(Y, Z) > 0$ (if $C = \emptyset$, it is called unconditional independence).

For the notation in this paper, we use $I(X, Y)$ to represent X, Y that are independent; $I(X, Y)|S$ represents the conditional independence of X and Y given S; $Dep(X, Y)|S$ represents the dependence of X and Y when bold S is given. To do the statistical test of independence, we use Chi-square [20].

Causation: Let $\rightarrow$ represents the left-side variable causing right-side variable. e.g $(X, Y) \rightarrow Z$ represents multiple causes of X and Y that cause Z, and $(X\&Y) \rightarrow Z$ denotes X and Y that are combined causes of Z.

To find only all combined variable set, there are $(2^n - 1)$ patterns where n is the total variable. The identification of their causality also costs exponential time. Therefore, it is not practical to discover causality in all combined causes. In this study, we demonstrated our approach with two variables in order to tackle large size and high-dimensional data.

The V-Structure or the Common Effect: X and Y are independent $(I(X,Y))$. X and Y together cause Z if X and Y are dependent when Z is given. Z is the common effect of multiple causes, X and $Y((X,Y) \rightarrow Z)$.

V-structure is a robust method to discriminate and discover local causal structures. The structure can confirm the causality and causal direction, which is not statistically equivalent to any other structures. V-structure includes a combined cause structure; however, it cannot be used to explicitly identify the combined cause because the traditional V-structure property does not explain the property of the combined cause. The definition of combined cause is as follows.

Definition 1 (Combined cause of one effect). *Let S be a set of multiple variables. S is a combined cause of Z if S causes Z, and any variables or any of its subsets of S are not a cause of Z.*

For example, $X, Y, W \in S$. S is a combined cause of Z if $X, Y, W, X\&Y, X\&W$ and $Y\&W$ do not cause Z, but $X\&Y\&W$ cause Z.

According to Definition 1, we propose the definition of a two-variable combined cause of one effect based on V-structure in Definition 2.

Definition 2 (Two-variable combined cause of one effect based on V-structure). *X and Y are independent $(I(X,Y))$. $X\&Y$ together are a combined cause of Z if X is independent of Z $(I(X,Z))$; Y is independent of $Z(I(Y,Z))$; and X and Y are dependent when Z is given $(Dep(X,Y)|Z)$. Z is the common effect of the combined cause, X and Y. $(X\&Y \rightarrow Z)$.*

According to Definition 2, the additional rule of independence for each cause variable and effect is to identify that the individual cause of a combined cause alone does not cause the effect. The properties of definition that are X is independent of Z or $(I(X,Z))$ and Y is independent of Z or $I(Y,Z)$ are to prove that there is no individual cause of the adverse effect.

Combined Cause
Problem Statement: Let $D = \{D_1, D_2, .., D_n\}$ represent n drugs taken indicators, where D_i is the binary indicators that denote a patient taken the drug $i^t h$. If drug D_i is taken, then $D_i = 1$; if not, then $D_i = 0$. Let A denotes the binary indicators of occurrence of adverse effect in a patient. If adverse effect A occurs, then $A = 1$.

The input consists of binary variables of drugs $D = \{D_1, D_2, ...D_n\}$ as predictor drugs and binary variable of adverse effect A.

The output is $O \in D$, a set of drugs that are combined cause of adverse effect

Pruning Steps: If A is the target adverse effect, and D_i and D_j are the combined cause, then D_i and D_j satisfy 1. $I(D_i, A)$ and $I(D_j, A)$ 2. $I(D_i, D_j)$ 3. $I(D_i, D_j)|A = 1$ does not hold or $Dep(D_i, D_j)|A = 1$.

If any D_n do not satisfy those rules, they will be pruned.

Algorithm: Combined Cause Discovery in DDI

1: Input: Predictor drugs $D = \{D_1, D_2, ..., D_n\}$ and adverse effect A
2: Threshold of dependency value th = the critical value of Chi-square when significance level $\alpha = x$
3: Output: Combined causes' set O
4: $O = \emptyset$; //initialise output variable as an empty set
5: $T = \emptyset$; //temporary variable as a copy of combined pairs i.e. $D_i \& D_j$ in D
6: **for each** $D_n \in D$ **do**
7:     **if** $Dep(D_i, A)$ with Threshold = th **then**
8:         $D = D - \{D_i\}$;
9: **for each** pair from combination matching of any two variables $\in D$ i.e.$D_i \& D_j$ **do**
10:     **if** $I(D_i, D_j)$ with Threshold = th **then**
11:         $T = T \cup \{D_i \& D_j\}$;
12: **for each** $\{D_i \& D_j\}$ in T **do**
13:     **if** $Dep((D_i, D_j) \mid A = 1)$ with Threshold = th **then**
14:         $O = O \cup \{D_i \& D_j\}$;

Our Proposed Algorithm: Combined Cause Discovery in DDI According to our algorithm, Combine Cause Discovery, at the first step, any drugs in D will be removed from D if they are dependent on the adverse effect A (line 6–8) because they are against the combined cause definition. This definition proves that there is no individual cause of the adverse effect. This removal of unrelated drugs will greatly reduce the time spent on the next step. Line 9–11 explains that all combinations of any 2 variables in D will be matched in order to check their independence to each other. If they are independent, they will be the candidate of the combined cause (the dependent cases will be pruned). This check is based on the definition of the V-structure and combined cause. The last step is to iteratively check whether those candidate combined causes are dependent when A = 1 is given or not. If they are, then they are the combined causes of the adverse effect A. We do not consider the case of A = 0 because it is less meaningful. It is less reasonable if the causality discovery comes from the absence of an adverse effect that is in the public interest.

The algorithm time complexity can be divided into 3 parts, the first part is from line 6–8, which is n times where n is the number of drugs. After the first part, the number of drugs should be reduced to m where $m \leq n$ from pruning drugs that relate to the individual cause of the adverse effect. The second part (line 9–11) is the matching combination drugs that costs $\frac{(m(m-1))}{2}$. After the second part, the number of the matched pair will be reduced if some of them are found to be dependent on each other, which breaks the property of combined causes. However, the worst case is pruning cannot be done to any drugs, then the computation time to do the third part (line 12–14) is equal to less than $\frac{(m(m-1))}{2}$. In total computation time of the worst case will be $\frac{n+(n(n-1))}{2} + \frac{(n(n-1))}{2} = O(n^2)$.

Algorithm: Multiple-Cause Discovery in DDI

1: Input: Predictor drugs $D = \{D_1, D_2, ..., D_n\}$,adverse effect A and combined cause's set O from algorithm Combined Cause Discovery
2: Threshold of dependency value th = the critical value of Chi-square when significance level $\alpha = x$
3: Output: Multiple-Cause's set M
4: $M = \emptyset$; //initialise output variable as an empty set
5: $T = \emptyset$; //temporary variable as a copy of drug pairs i.e. $D_i \& D_j$ in D
6: **for each** pair from combination matching of any two variables $\in D$ i.e.$D_i \& D_j$ **do**
7:      **if** $I(D_i, D_j)$ with Threshold = th **then**
8:          $T = T \cup \{D_i \& D_j\}$;
9: **for each** $\{D_i \& D_j\}$ in T **do**
10:      **if** $Dep((D_i, D_j) \mid A = 1)$ with Threshold = th **then**
11:          $M = M \cup \{D_i \& D_j\}$;
12: $M = M$ - O

Multiple Causes

To find multiple causes, we will find DDIs using the original V-structure and then deduct them with the combined cause set identified by our Combined Cause discovery algorithm.

Problem Statement: Let $D = \{D_1, D_2, .., D_n\}$ represent n drugs taken indicators, where D_i is the binary indicators that denote a patient taken the drug i^th. if drug D_i is taken, then $D_i = 1$; if not, then $D_i = 0$. Let A denotes the binary indicators of occurrence of an adverse effect in a patient. If an adverse effect A occurs, then $A = 1$.

The input consists of binary variables of drugs $D = \{D_1, D_2, ...D_n\}$ as predictor drugs, binary variable of adverse effect A and Combined cause's set, which is the outcome from Combined Cause Discovery algorithm.

The output is $M \in D$, a set of drugs that are multiple causes of an adverse effect.

Pruning Steps: If A is the target adverse effect, and D_i and D_j have causal relationship, then D_i and D_j satisfy

1. $I(D_i, D_j)$. 2. $I(D_i, D_j)|A = 1$ does not hold or $Dep(D_i, D_j)||A = 1$.

If any D_n do not satisfy those rules, they will be pruned.

Our proposed algorithm: Multiple Cause Discovery in DDI The line 6–11 follows the properties of V-structure concept. In the first process, any possible drug combination pairs of D will be selected if they are independent of each other (line 6–8). The drug pairs that are dependent are not considered which will reduce the time computation of the next process. The second process (line 9–11) is iterative selecting all drug pairs that are dependent when adverse effect $A = 1$ is given. From the second process, the outcome is DDIs with causal relationships based on V-structure. The last step (line 12) shows multiple causes' set M from the deduction of all DDIs' set with the combined cause's set, O. The computation time is similar to that of our Combined Cause Discovery algorithm or $O(n^2)$.

4   Empirical Evaluation

We used the Food and Drug Administration (FDA) Adverse Event Report System (FAERS) from the USA which is reported quarterly as data to experiment with our combined cause discovery algorithm. We selected FAERS in XML format that consisted of approximately 300,000 patients' data. In the whole dataset, we used two kinds of data, drug usage and adverse effects. There were approximately 2,000 adverse effects and 4,000 drugs. We considered all types of adverse effects. However, FAERS has some drawbacks, for example, missing data, the small sample size in some reports, duplicates of reports, high dimensional data and sparse data [5, 21].

To solve these drawbacks, we followed some related studies [5, 11]. Duplicate reports were deleted if they had similar data to at least eight drugs or adverse effects and demographic information of patients. Reports that did not have any adverse effects or drug information were not considered. Drugs and adverse effects that appeared in at least 5 reports were considered. According to the drug name, we did entity standardisation.

For the threshold of statistical independence of a test, we use 95% of the confidence interval. The threshold of frequency support of DDIs used in this experiment is 50 [13] to balance between the number of discovered DDI and variation of content, such as drugs and adverse effects.

Evaluation: There is no ground-truth data for DDI to evaluate the outcome from the predictions. In other works related to data mining and statistical tests for DDI and pharmacovigilance studies [5, 11], randomly selected 100 discovered rules are manually reviewed by experts and pharmacists to evaluate the accuracy. However, the knowledge of experts is limited and cannot evaluate more than half of the discovered drug combinations that are not known to interact with. Alternatively, we select two reliable pharmaceutical databases that some pharmacists use to study the information to evaluate the experiment in this study with selected FAERS data. Two reliable pharmaceutical drug-drug interaction databases (drug databases) are MedicinesComplete (MedComp)[1] and Drugs.com[2], were selected to perform.

We randomly select 100 DDI predictions from FAERS data that are discovered by using our proposed method and CARD.

The prediction outcome will be classified into 3 categories to evaluate the proportion of each category in the total prediction outcome. 1. "Correct" DDI prediction is the predicted DDI that is found in at least one of the drug databases. 2. "Incorrect" prediction is the predicted DDI that is confirmed from both of the drug databases to have no interaction or not cause any adverse effects or cause different side effects. 3. "Unknown" DDI is predicted DDI that does not match any DDI in the drug databases; however, this does not mean that they are incorrect. They have not yet confirmed or discovered the causality by clinical or biological methods. To evaluate the effectiveness of our method and CARD, we measure the precision of DDI discovery. In this study, precision is $\frac{TP}{(TP+FP)}$, where TP is the number of "Correct", and FP is the number of "Incorrect". In this research with real-world data, recall cannot be used because "Unknown" could not yet identify to be false negatives (one of the requirements to calculate recall) or true negatives.

Furthermore, we also compared the proportion results of the discovered combined cause and multiple causes between our method and CARD.

5   Results and Discussions

From the Table 1, evaluated by checking those interactions with the MedComp and Drug.com databases, our method found existing drug-drug interaction 24% while CARD found 21%. Incorrect predictions are 25% and 29% for our method and CARD, respectively. The unknown outcomes are 51% for our method and 50% for CARD. Our result precision is 0.49, while that of CARD is 0.42.

Some of our results are shown in Table 2. For example of our results, RANITIDINE + NAPROXEN are predicted to cause drug ineffectiveness, which is similar to what Drug.com shows, while MedComp reveals that there are few studies that show no significant signs of interaction between RANITIDINE + NAPROXEN when consumed together. An example in the unknown group from Table 2 is FLUOXETINE + CLARITIN that is predicted to cause pain, which is still very difficult to identify.

The proportion of combined cause predicted by our method is around 25% and that of multiple causes is around 75%, while CARD could not discriminate between them (Table 3).

Our results from combined cause predictions tend to be that each drug is not the cause of that adverse effect or the effect is very trivial; however, the

---

[1] MedicinesComplete published in Pharmaceutical Press and the Royal Pharmaceutical Society:https://www.medicinescomplete.com/mc/alerts/current/drug-interactions.htm.

[2] Data sources from Micromedex, Multum and Wolters Kluwer database:https://www.drugs.com/drug_interactions.php.

adverse effect happens when those drugs are consumed together. For a simple example from Table 1, taking SYMBICORT alone does not reduce its own effectiveness and consuming HUMALOG alone is not the cause of its ineffectiveness, while when they are consumed together HUMALOG effectiveness is reduced. This shows that our proposed definition for the combined cause can work and represent the characteristics of the combined cause in real-world data.

Table 1. Outcome comparison based on three prediction evaluations, e.g. correct, incorrect and unknown comparison.

	Prediction outcome (%)			Precision
	Correct	Incorect	Unknown	
Our method	24	25	51	0.49
CARD	21	29	50	0.42

Table 2. The examples of drug-drug interactions and their adverse effect predicted by our combined cause discovery model and found in selected reliable pharmaceutical databases.

Drug-drug interaction predicted by our model	Adverse Effect	Drug Databases that confirm the predicted DDI
SYMBICORT + HUMALOG	Drug ineffective	MedComp, Drug.com
RANITIDINE + NAPROXEN	Drug ineffective	Drug.com
PROZAC + MIRALAX	Heart palpitations	Drug.com
METOPROLOL + ALBUTERAL	Asthma	MedComp, Drugs.com
ASPIRIN + VOLTAREN	Gastrointestnal haemorrhage	MedComp, Drugs.com
FLUOXETINE + ZYRTEC	Drowsiness	Drugs.com
FLUOXETINE + CLARITIN	Pain	–

Table 3. The comparison of DDI cause type prediction

	Percentage of DDI cause type prediction	
	Our Method	CARD
Combined causes	25	0
Multiple causes	75	0

6   Conclusion

Combined causes consist of two or more causes that cause the effect where no individual causes the effect, while multiple causes have one or more causes affect the effect. They are useful to discover some information that is hard to judge by experts or professionals, e.g. combined drugs causing an adverse effect in the DDI problem. Causal combined drugs' discovery in DDI is difficult for experts because it lacks hints to discover other combined drugs and those drugs tend to not depend on each other and the adverse effect. In addition, discovering them automatically without expert help will improve and accelerate further DDI discovery studies based on pharmacokinetics and pharmacodynamics. To find combined causes and multiple causes, would cost exponential computation time. To our knowledge, there have been very few works to discover combined causes and multiple causes that are based on causality discovery. In addition, the causal Bayesian Network, the main framework approach for representing causality, cannot discover the combined causes with the original definition.

In this study, we proposed a new definition that is embedded in the V-structure, a specific local Bayesian network structure, to discover combined causes and multiple causes in DDI. However, DDI with many drugs needs exponential time to discover its causality because of its high dimension of data. Therefore we consider two-combined drug cases to handle data with large size and high dimensions with our algorithm that has $O(n^2)$ time complexity. The outcomes from our proposed method showed that it can successfully discover combined causes and multiple causes in real-world data of DDI that the state-of-the-art could not perform.

In the future, we will extend the proposed algorithm to handle combined drugs that consist of more than two drugs given an adverse effect. The concept will be (1) Each candidate's drug usage has to be statistically independent of the target adverse effect. (2) All pairs of the combination of candidate drugs' usage from the candidate combined cause have to be independent of each other. (3) All drug usage of candidate combined causes has to be dependent on when the target adverse effect is given. The computation in (1) is n, (2) is $\frac{n!}{(n-r)! * r!} * \frac{r*(r-1)}{2}$, and (3) is r where n is the number of all drugs and r is the number of combined drugs' size.

References

1. Aliferis, C.F., Statnikov, A., Tsamardinos, I., Mani, S., Koutsoukos, X.D.: Local causal and Markov blanket induction for causal discovery and feature selection for classification part i: algorithms and empirical evaluation. J. Mach. Learn. Res. **11**(Jan), 171–234 (2010)
2. Aliferis, C.F., Statnikov, A., Tsamardinos, I., Mani, S., Koutsoukos, X.D.: Local causal and Markov blanket induction for causal discovery and feature selection for classification part ii: analysis and extensions. J. Mach. Learn. Res. **11**(Jan), 235–284 (2010)

3. Bate, A., Evans, S.: Quantitative signal detection using spontaneous ADR reporting. Pharmacoepidemiol. Drug Saf. **18**(6), 427–436 (2009)
4. Bühlmann, P., Kalisch, M., Maathuis, M.H.: Variable selection in high-dimensional linear models: partially faithful distributions and the pc-simple algorithm. Biometrika **97**(2), 261–278 (2010)
5. Cai, R.: Identification of adverse drug-drug interactions through causal association rule discovery from spontaneous adverse event reports. Artif. Intell. Med. **76**, 7–15 (2017)
6. Chickering, D.M., Heckerman, D., Meek, C.: Large-sample learning of Bayesian networks is NP-hard. J. Mach. Learn. Res. **5**, 1287–1330 (2004)
7. Cooper, G.F.: A simple constraint-based algorithm for efficiently mining observational databases for causal relationships. Data Min. Knowl. Disc. **1**(2), 203–224 (1997). https://doi.org/10.1023/A:1009787925236
8. Elath, H., Dixit, R.R., Schumaker, R.P., Veronin, M.A.: Predicting deadly drug combinations through a machine learning approach (2018)
9. Freedman, D.: From association to causation: some remarks on the history of statistics. J. Soc. Fr. Stat. **140**(5), 5–32 (1999)
10. Hansen, M.L., et al.: Risk of bleeding with single, dual, or triple therapy with warfarin, aspirin, and clopidogrel in patients with atrial fibrillation. Arch. Intern. Med. **170**(16), 1433–1441 (2010)
11. Harpaz, R., Chase, H.S., Friedman, C.: Mining multi-item drug adverse effect associations in spontaneous reporting systems. In: BMC Bioinformatics, vol. 11, pp. 1–8. BioMed Central (2010)
12. Harpaz, R., Perez, H., Chase, H.S., Rabadan, R., Hripcsak, G., Friedman, C.: Biclustering of adverse drug events in the FDA's spontaneous reporting system. Clin. Pharmacol. Ther. **89**(2), 243–250 (2011)
13. Ibrahim, H., Saad, A., Abdo, A., Eldin, A.S.: Mining association patterns of drug-interactions using post marketing FDA's spontaneous reporting data. J. Biomed. Inform. **60**, 294–308 (2016)
14. Jin, Z., Li, J., Liu, L., Le, T.D., Sun, B., Wang, R.: Discovery of causal rules using partial association. In: Data Mining (ICDM), 2012 IEEE 12th International Conference on, pp. 309–318. IEEE (2012)
15. Kovesdy, C.P., Kalantar-Zadeh, K.: Observational studies versus randomized controlled trials: avenues to causal inference in nephrology. Adv. Chronic Kidney Dis. **19**(1), 11–18 (2012)
16. Lazarou, J., Pomeranz, B.H., Corey, P.N.: Incidence of adverse drug reactions in hospitalized patients: a meta-analysis of prospective studies. JAMA **279**(15), 1200–1205 (1998)
17. Li, J., Le, T.D., Liu, L., Liu, J., Jin, Z., Sun, B.: Mining causal association rules. In: Data Mining Workshops (ICDMW), 2013 IEEE 13th International Conference on, pp. 114–123. IEEE (2013)
18. Liu, M., et al.: Determining molecular predictors of adverse drug reactions with causality analysis based on structure learning. J. Am. Med. Inform. Assoc. **21**(2), 245–251 (2014)
19. Ma, S., Li, J., Liu, L., Le, T.D.: Mining combined causes in large data sets. Knowl.-Based Syst. **92**, 104–111 (2016)
20. McHugh, M.L.: The chi-square test of independence. Biochemia medica: Biochemia medica **23**(2), 143–149 (2013)
21. Norén, G.N., Orre, R., Bate, A., Edwards, I.R.: Duplicate detection in adverse drug reaction surveillance. Data Min. Knowl. Disc. **14**(3), 305–328 (2007). https://doi.org/10.1007/s10618-006-0052-8

22. Palleria, C.: Pharmacokinetic drug-drug interaction and their implication in clinical management. J. Res. Med. Sci. **18**(7), 601 (2013)
23. Pearl, J.: Causality. Cambridge University Press, Cambridge (2009)
24. Qin, X., Kakar, T., Wunnava, S., Rundensteiner, E.A., Cao, L.: Maras: signaling multi-drug adverse reactions. In: Proceedings of the 23rd ACM SIGKDD International Conference on Knowledge Discovery and Data Mining, pp. 1615–1623. ACM (2017)
25. Quinn, D., Day, R.: Drug interactions of clinical importance. Drug Saf. **12**(6), 393–452 (1995)
26. Spirtes, P., et al.: Causation, Prediction and Search. MIT press, Cambridge (2000)
27. Subpaiboonkit, S., Li, X., Zhao, X., Scells, H., Zuccon, G.: Causality discovery with domain knowledge for drug-drug interactions discovery. In: Li, J., Wang, S., Qin, S., Li, X., Wang, S. (eds.) ADMA 2019. LNCS (LNAI), vol. 11888, pp. 632–647. Springer, Cham (2019). https://doi.org/10.1007/978-3-030-35231-8_46
28. Van Puijenbroek, E.P., Egberts, A.C., Meyboom, R.H., Leufkens, H.G.: Signalling possible drug-drug interactions in a spontaneous reporting system: delay of withdrawal bleeding during concomitant use of oral contraceptives and itraconazole. Br. J. Clin. Pharmacol. **47**(6), 689–693 (1999)
29. Ventola, C.L.: Big data and pharmacovigilance: data mining for adverse drug events and interactions. Pharm. Ther. **43**(6), 340 (2018)
30. Waldmann, M.R., Martignon, L.: A Bayesian network model of causal learning. In: Proceedings of the twentieth annual conference of the Cognitive Science Society, pp. 1102–1107 (1998)
31. Xiang, Y., et al.: Efficiently mining adverse event reporting system for multiple drug interactions. AMIA Summits Transl. Sci. Proc. **2014**, 120 (2014)
32. Zhan, C., Roughead, E., Liu, L., Pratt, N., Li, J.: Detecting high-quality signals of adverse drug-drug interactions from spontaneous reporting data. J. Biomed. Inform. **112**, 103603 (2020)

An Integrated Medical Recommendation Mechanism Combining Promote Product Singular Value Decomposition and Knowledge Graph

Yibo Sun[1], Chenlei Liu[2], Xue Tong[2], and Bing Hu[3(✉)]

[1] University of Queensland, St Lucia, QLD 4072, Australia
[2] Post Big Data Technology and Application Engineering Research Center of Jiangsu Province, Post Industry Technology Research and Development Center of the State Posts Bureau (Internet of Things Technology), Nanjing University of Posts and Telecommunications, Nanjing, China
[3] Broadband Wireless Communication Technology Engineering Research Center of the Ministry of Education, Nanjing University of Posts and Telecommunications, Nanjing, China
hubing@njupt.edu.cn

Abstract. Due to the increasingly serious problem of information overload, it is difficult for medical staff to accurately screen out data conducive to patient consultation when dealing with massive amounts of diagnostic information. In this paper, we propose a hybrid recommendation mechanism integrating the improved Singular Value Decomposition and Knowledge Graph for medical information. It can effectively analyze the characteristic relationship between patients and medical item information and give appropriate medical recommendation scheme information based on patients' interests and preferences. We design the Promote Product Singular Value Decomposition (PPSVD) algorithm and integrate knowledge graph technology to construct the medical system's recommendation model. Finally, the experimental results show that the integrated medical recommendation model proposed has higher accuracy and usability than any other algorithms.

Keywords: Medical recommendation system · Integrated algorithm · PPSVD · Knowledge graph

1 Introduction

Nowadays, the medical industry has become more intelligent, which has greatly promoted the development of medical information technology but also led to the geometric growth of medical information. The dramatic increase in the diversity of data has a significant impact on the efficiency of patient care. When faced with a large number of patient treatment options, it is hard for doctors to quickly and efficiently select the most appropriate healthcare solutions for patients. Therefore, effectively recommending the appropriate medical treatment, physician, and rehabilitation plan to patients among the huge amount of medical and health information data has become a issue in the medical field. Medical recommendation systems can establish binary correspondence between patients and medical information based on the information needs of patient visits combined with the main features of medical information data, and recommend various medical information data to applicable patients using specific information filtering techniques. However, the current recommendation systems [1, 2, 8] still have some problems. For instance, the sparse behavioral data matrix between users and items in the coordinated filtering recommendation algorithm based on user behavior, the cold start problem when recommending new users or new items. When the number of users and items grows dramatically, the execution time efficiency of the recommendation algorithm decreases, and the accuracy of the recommendation decreases dramatically [3]. When new users and new items appear in the system When new users and items appear in the system, the sudden generation of detailed information can affect the stability of the recommendation system [4].

In this paper, we serves the following main contributions: (1) We design a novel integrated medical recommendation mechanism by establishing a binary correspondence between patients and medical items based on the complex and diverse patient and medical item information data in the medical field. (2) We redesign the PPSVD algorithm to perform data dimensionality reduction on the scoring matrix and adopt the knowledge mapping to determine the most suitable medical item fields for patients based on the scoring matrix's reduced dimensionality.

2   Proposed Methodology

This section will elaborate on the proposed integrated medical recommendation mechanism.

2.1   Promote Product Singular Value Decomposition Algorithm

Obtain Prediction Scores. This mechanism includes multiple patient and medical item types. These two data types will be represented using two multidimensional vectors, respectively. Define them as $P = [P_1, P_2, P_3, ..., P_p]$, $M = [M_1, M_2, M_3, ..., M_m]$, where p denotes the total number of patients, and m denotes the total number of medical items. The relationship model between patients and medical items is established, where P_i denotes the $i - th$ patient, MI_j denotes the $j - th$ medical item, and $D_{i,j}$ denotes the rating of patient i on medical item j.

Build Preference Matrix. The PPSVD relies on a matrix of patient preference scores for medical items. Based on this matrix, the model creates a predictive score for the patient, and a preference matrix R is defined as follows. Herein, $R_{(p_i, M_j)}$ represents the preference score of patient i for medical item j in R.

$$R = \begin{bmatrix} R_{p_1,M_1} & R_{p_1,M_2} & R_{p_1,M_3} & \cdots & R_{p_1,M_m} \\ R_{p_2,M_1} & R_{p_2,M_2} & R_{p_2,M_3} & \cdots & R_{p_2,M_m} \\ \vdots & \vdots & \vdots & \vdots & \vdots \\ \vdots & \vdots & \vdots & \vdots & \vdots \\ R_{p_p,M_1} & R_{p_p,M_2} & R_{p_p,M_3} & \cdots & R_{p_p,M_m} \end{bmatrix} \tag{1}$$

Remove Edge Data. We design an improved method to remove the redundant edge data in the preference matrix and reduce the distribution range of noise data. The preference matrix R obtained by preprocessing has the characteristics of high dimensionality, many edge data, and extensive noise data. By taking the total number of patients as the main dimension of the preference matrix R, the data on the diagonal of the preference matrix R is called singular in the traditional SVD method. A one-dimensional vector is created for these singular values, defined as $Q = [Q_1, Q_2, Q_3, ..Q_p]$. The removal based on the singular value vector Q can effectively ensure the accuracy of the removed edge data. The improved idea of this thesis is to set a noise data boundary value, called E, based on the singular value vector, and E is defined as.

$$E = \sqrt{\frac{\sum_{i=1}^{p} \left(Q_i - \overline{Q}\right)^2}{m}} \tag{2}$$

In the above equation, $\overline{Q}$ is the mean value of the vector data set Q. In order to remove the edge data in the preference matrix and reduce the distribution of noisy data, it is necessary to determine whether the data is one of these two categories. The determination is based on E, which denotes the preference matrix critical value. The specific determination rules are as follows by comparing the values in the matrix R with the magnitude of E.

- If $R_{i,j} \geq E$ in R, retain the data in the row and column where $R_{i,j}$ is located.
- If $R_{i,j} < E$ in R, delete the data in the rows and columns where $R_{i,j}$ is located.

Predict Preference Scores. According to the critical value of the preference matrix, it can be achieved to remove the edge data in the matrix and reduce the distribution range of the edge data, to reduce the time complexity of the recommendation model and improve the accuracy of the recommendation results. The new preference matrix R' is defined, and the parameter TOP-K is defined to indicate the dimension of the matrix R'. Herein, the parameter TOP-K is much smaller than the dimension p of the initial matrix R. Based on R', a prediction of patients' preference scores for unrated medical items is established.

First, the matrix R' is decomposed into two elements: the patient factor matrix and the medical item factor matrix. Defined as $U_{(p*k)}, I_{(u*k)}$, respectively, where the value of k is the dimension of the matrix R', indicating the total number of attributes for which patients have potential preferences for items. Based on the degree of patients' preferences for these k item attributes and the present rate of these potential attribute factors in a certain type of medical item, the patient's preference score for that type of medical item can be determined, and the preference score is defined by the following method.

$$SS_{u,i'} = p_u * I_i + \mu * (\delta_u + \delta_i) \qquad (3)$$

Optimize Error Values. In Eq. 4, the set deviation error parameters $\delta_u, \delta_{ui}, \mu$ are used to reduce the error of the prediction scores. Considering that the custom parameter values still have some limitations, the Least Squares is introduced to continue the optimization. Where, $(SS_{u,i'} - SS_{u,i})^2$ denotes the model prediction error of the patient preference score, which is trained by square minimization of four parameters, the factor vectors p_u and I_i, and the error parameters δ_u and δ_i. The method is defined as follows.

$$\text{error} = \min\left[\sum_{u,i \in R'} (SS_{u,i'} - SS_{u,i})^2 + \mu * \left(p_u{}^2 + I_i{}^2 + (\delta_u + \delta_i)^2\right)\right] \qquad (4)$$

After the second round of optimization training on the prediction error, the prediction scores of the target patients were set based on the historical preference scores of patients similar to the target patients. The final prediction method was obtained as follows.

$$S_{u,i} = \frac{\sum_{i=1}^N \left(SS_{u,i}\overline{SS_u}\right) \cdot \left((SS_{v,i} - \overline{SS_v})\right)}{\sqrt{\sum_{i=1}^N \left(SS_{u,i} - \overline{SS_u}\right)^2} \cdot \sqrt{\sum_{i=1}^N SS_{v,i} - \overline{SS_v}}^2} + \text{error} \qquad (5)$$

In the above prediction process, the PCC theorem is used to measure the similarity between two patients. Similar patients v of patient u are derived according to equation $\sum_{i=1}^N \left(SS_{u,i} - \overline{SS_u}\right) \cdot \left((SS_{v,i} - \overline{SS_v})\right) / \sqrt{\sum_{i=1}^N \left(SS_{u,i} - \overline{SS_u}\right)^2} \cdot \sqrt{\sum_{i=1}^N SS_{v,i} - \overline{SS_v}}^2$. Herein, $SS_{u,i}$ and $SS_{v,i}$ both denote the patient's rating of item i, $\overline{SS_u}$ and $\overline{SS_v}$ are the average ratings of patients u and v for medical item i, and N denotes the total number of medical items in the preference matrix. Thus, the predictive scoring model can be obtained for patients' ratings on medical items.

Based on the above, we can derive a process for the predicted scores of patients on items based on PPSVD as Fig. 1.

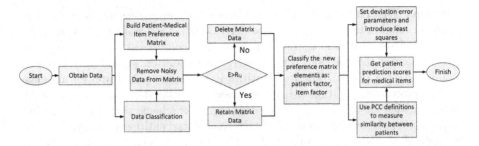

Fig. 1. Prediction score process based on PPSVD

Based on the PPSVD, patients' predicted preference score model for medical items had been constructed. We introduce the knowledge graph and integrates "medical semantic" information into the recommendation model, adding medical items, semantic association information between patients, item similarity, and other attributes in the knowledge graph to the recommendation model. It adds the semantic association information and item similarity between medical items and patients in the knowledge graph to the recommendation model. In addition, it highly utilizes the implicit information of medical items and patients in the recommendation system to give similar semantic recommendation solutions to achieve fast and efficient medical recommendations.

2.2 Knowledge Graph to Recommendation

We apply knowledge graph to the medical recommendation to achieve the fusion of patient and medical item entity semantics with the recommendation model and choose to jointly train the knowledge graph technology in the recommendation model based on improved multi-task learning.

Design Entities and Relationships Structure. The knowledge graph is constructed based on the data in the traditional recommendation knowledge base. The established triad is composed of: head entities, relationships among entities, and tail entities, without considering the issue of entity attributes in the knowledge graph. The knowledge graph TransE representation can be well applied to deal with one-to-one relationships, but in dealing with complex relationships still has shortcomings. In this paper, the TransR knowledge representation method can be used to deal with complex relationships in knowledge graphs. According to this method, the triad of attributes of entities is vectorized to obtain a vector representation of the triad: $T_u = (h, r, a)|h \in History_u$. Herein, $History_u$ denotes the vector of historical medical item entity sets associated with patients, T_u denotes the patient u's history information contained in the knowledge graph triad in the medical item entity set, h denotes the entity vector of the knowledge graph base, a denotes the entity attributes in the knowledge graph base, and r denotes the relationship vector between entities and attributes in the knowledge graph base.

Construct Discriminative Model. Combining the entities and attributes in the vector triad in the knowledge KI to derive the hierarchical model of patients' interest preference attributes, another important component of the triad is the entity, which is the core subject of the recommendation scheme in the whole recommendation system. The item entities in the knowledge graph KI are classified as positive correlation entities and negative related entities. The positive related entities are subjects of the recommendation scheme that highly matches the patient's needs and interests. The negative related entities are subjects of the recommendation scheme less relevant to the patient's recommendation scheme.

For several vector triples in the knowledge graph KI, if we want to determine the positive and negative relevance of the entities in (h, r, t), the idea of model setting in this paper is to replace the head entity h in a certain vector triple with an arbitrary random entity h_{random} in the knowledge graph KI, and replace the tail entity t with an arbitrary random tail entity t_{random}. Moreover, the newly generated triple $(h_{random}, r, t), (h, r, t_{random})$ do not exist in the knowledge graph KI. Suppose both random head entity h and tail entity t in the knowledge graph KI can satisfy the requirements. In that case, the entities are considered to be positively correlated entities. Otherwise, they are negatively correlated entities. Based on the random head entity and tail entity selected above, the method is established to distinguish the positively and negatively correlated entities in the knowledge graph KI. TransR knowledge representation method is an extension of TransE representation method, which considers that Each entity in the graph KI has multiple attributes. Before establishing the transition relationship r between entities, the entities should be mapped to the corresponding relationship space. This model defines W_h, W_t vectors and sets the edge function f. The function f is obtained as the squared value of the entity vector modulus, which is expressed as

$$f(h, t) = \left\| h - t - \left(W_h^T \cdot h \cdot W_h - W_t^T \cdot t \cdot W_t \right) \right\|^2 \qquad (6)$$

Based on the above edge function f, the definition gives the final method model for determining the positive and negative related entities in the knowledge graph KI, which is represented as follows.

$$\text{class}_{h,t} = \begin{cases} \max\left(0, f(h,t) - f\left(h_{\text{random}}, t\right)\right) - f(h,t) + f\left(h_{\text{random}}, t\right) \\ f(h,t) - f\left(h, t_{\text{random}}\right) - \left(\max\left(0, f(h,t) - f\left(h, t_{\text{random}}\right)\right)\right) \end{cases} \qquad (7)$$

Herein, the $class_{h,t}$ model is executed with the preconditions that $(h, r, t) \in T_u$, $(h_{random}, r, t) \notin T_u$ and $(h, r, t_{random}) \notin T_u$. For the head entities h and h_{random} in the knowledge graph KI, calculate the two If $f' > f''$ and $f_1 > f_2$, the entities h and t are considered as positively correlated entities. Otherwise, they are judged as negatively correlated entities. The model for determining positive and negative related entities in the knowledge graph KI is thus obtained.

Construct Interest Level Model. The idea of applying knowledge graph is to abstract the interaction process between entities such as hospital patients and medical items as relationships in knowledge graphs. Although the semantic and structural information of patients and medical items in the knowledge graph is considered in establishing KI positive and negative related entity models, the above recommendation model does not do a detailed analysis of the association between patients and KI in the medical system. Therefore, this subsection will analyze the relationship between patients and KI in the medical system in detail based on the recommended methods established in the above two subsections and establish a method for determining the patient preference level based on this model. The specific design is as follows:

$$W_{a,T_a} \mid = \frac{\exp\left(h_a{}^T \cdot r_a\right)}{\sum_{(h,r,a)\in T_u} \exp(h \cdot r)} \tag{8}$$

where the calculation of the weight of entity attribute a relies on the head entity vector h. The head entity associated with attribute a is denoted as h_a, and the relationship vector r associated with attribute a is denoted as r_a. For any vector triplet, the dot product values of the head and tail entity vector of all triplets $(h, r, a) \in T_u$ containing the a attribute are calculated and exponentiated. Than two values are divided to obtain the weight values of the a attribute in all medical item entity attributes of patient u. After obtaining the weights of entity attributes, the interest preference level of patient u for attribute a is determined based on this as Eq. 9.

$$\text{Interest }_{u \to a} = \sum_{(h,r,a) \in T_u} a * \frac{\exp\left(h_a{}^T \cdot r_a\right)}{\sum_{(h,r,a)\in T_u} \exp(h \cdot r)} \tag{9}$$

$Interest_{(u} \to a)$ denotes the interest preference value of patient u for entity attribute a. In this paper, we set three rank parameters to take values: $\gamma_1, \gamma_2, \gamma_3$, where: $\gamma_1 = \text{Interest }_{u \to a} / \exp\left(h_a{}^T \cdot r_a\right)$, $\gamma_2 = \text{Interest }_{u \to a} / \sum_{(h,r,a)\in T_u} \exp(h \cdot r)$ and $\gamma_3 = \text{Interest }_{u_u \to a} / W_{a,T_a}$.

If $Interest_{(u} \to a)$ is less than γ_1, the interest preference level of patient u for attribute a is considered as "not interested". If $Interest_{(u} \to a)$ is between γ_1 and γ_2, the interest preference level of patient u for attribute a is considered as "interested". If $Interest_{(u} \to a)$ is greater than γ_3, the interest preference level of patient u for attribute a is considered to be "very interested". Thus the interest preference level of patient u for attribute a is obtained based on the knowledge graph KI model.

Integrated Algorithm for Medical Recommendations. This subsection integrates the above methods to give the final recommendation model. The overall flow chart of the integrated recommendation algorithm is designed as Fig. 2.

The pseudo-code design of the integrated recommendation algorithm is as Algorithm 1.

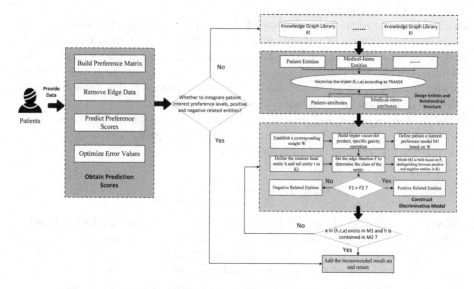

Fig. 2. The process of integrating recommendation algorithm.

3   Experiment

In order to verify the recommendation performance of the integrated algorithmic recommendation model proposed in this paper, experiments are designed in this section for validation.

3.1   Environment and Dataset

We utilize an Ubuntu 16.04 server with an Intel Core i7-9750H 2.60 GHz CPU, Nvidia GTX1080 GPU, and 32 GB of RAM to conduct experiments. We select the dataset from website CMS.gov (Centers for Medicare Services), which displays massive information data in the medical field, including hospital, patient care, medical item expenditure, medical item quality and medical item rating of hospital system. Centers for Medicare Services dataset contains 1002 medical patients, 4300 medical disease information, and 3223 medical item rating information.

3.2   Evaluation Metrics

The evaluation metrics of the selected recommendation system are precision and mean-reverse ranking (MRR) indicators. Precision is mainly used to describe the proportion of the score records of patient users and medical items in the final medical recommendation scheme given in the medical recommendation system, which is shown as Eq. 10.

Algorithm 1. Integrated recommendation algorithm incorporating PPSVD and knowledge graph

---

Require: Patient u, medical item i, vector triplet (I_h, I_r, I_t) in Knowledge Graph KI
Ensure: Recommendation Set $SET = [SET_0, SET_1, SET_2, \cdots, SET_K]$
  1: $D = 0$;
  2: **for** Random(u,i) in Range($S_{u,i}$.length) **do**
  3:     **if** $Interest_{u \to i} < Interest_{u \to a} / \exp\left(h_a{}^T \cdot r_a\right)$ **then**
  4:         break;
  5:     **else if** $Interest_{u \to i} > Interest_{u \to a} / \exp\left(h_a{}^T \cdot r_a\right)$ && $Interest_{u \to i} <$ $Interest_{u \to a} / \sum_{(h,r,a) \in T_u} \exp\left(h_a{}^T \cdot r_a\right)$ **then**
  6:         break;
  7:     **else**
  8:         **if** $f(h,t) > f(h_{random}, t)$ && $f(h,t) > f(h, t_{random})$ **then**
  9:             **if** SET.length $== K$ **then**
10:                break;
11:             **else**
12:                SET.append((u,i));
13:             **end if**
14:         **else**
15:             break;
16:         **end if**
17:     **end if**
18: **end for**
19: **return** Recommendation Set: SET

$$\text{Precision} = \frac{\sum_{u \in U} |R_u \cap T_u|}{\sum_{u \in U} |R_u|} \tag{10}$$

where R_u represents the final set of medical recommendation scheme list presented to patient u based on the model of integrated recommendation algorithm. T_u represents the medical information in the training set of patient u.

The meaning of MRR is to rank the recommendation solutions finally given by the integrated recommendation algorithm model established in this paper in the order of taking the inverse ranking and using it as the accuracy, and then averaging all the data afterwards, and the formula is introduced as Eq. 11.

$$\text{MRR} = \frac{1}{m} \sum_{i=1}^{m} \left(\sum_{V_j \in \text{test}(u_i)} \frac{1}{\text{rank}(u_i, V_j)} \right) \tag{11}$$

where m represents the number of medical patients in the data set. V_j represents the medical recommendation items with higher compliance in the final list of medical recommendations generated in the integrated recommendation model. $test(u_i)$ represents the set of medical items related to patient u in the dataset, and $rank(u_i, V_j)$ represents the position of medical information V in the final recommendation solution for patient user u_j, i.e., the position of the first medical item in the final recommendation list in the medical item class in dataset.

3.3   Results and Analysis

In the integrated medical recommendation model established in this paper, the set error parameter μ is an important factor affecting the prediction score. By setting different μ parameter values, this experiment analyzes the index value comparison between the comparison mentioned above algorithms and the integrated recommendation algorithm established in this paper. Based on the set values of different μ parameters, experiments are carried out for the above-mentioned experimental comparison algorithms in turn, and the two set experimental indicators are compared. Among them, the experimental objects consist of six methods: CKE [9], CMF [5], RIPPLENET [6], DKN [7], CDFM [2] and ours.

According to the different values of μ, we get the Precision and MRR values under the corresponding μ values for each method. The comparisons of each method are as Fig. 3 and 4.

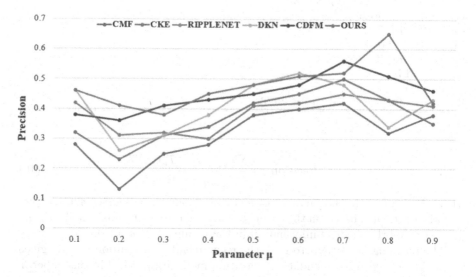

Fig. 3. Comparison results of precision

Through the above experimental comparison, we can find from the experimental results that under different values of the error parameter. The six types of recommendation methods show different values of Precision accuracy. The maximum value of CMF is 0.42. The maximum value of CKE is 0.5, the maximum value of the RIPPLENET is 0.45, the maximum value of DKN is 0.52, the maximum value of CDFM is 0.56, and the maximum value of our method is 0.65. Data comparison shows that the overall recommendation performance of our method is higher than that of the previous five methods. In the experimental results based on the MRR index, the maximum value of our method is 0.62. The comparison of the experimental results shows that our recommendation method

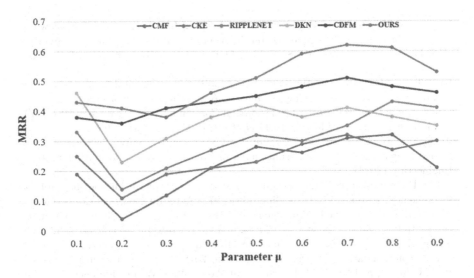

Fig. 4. Comparison results of MRR

has higher recommendation accuracy, which verifies the accuracy and usability of the recommendation model established in this paper.

4 Conclusion

In this paper, we propose a medical recommendation mechanism based on an integrated algorithm. We firstly represent an improved PPSVD algorithm and a knowledge graph method. Then, we design an integrated medical recommendation algorithm based on both to realize the recommendation of medical programs. Finally, we confirm the performance of the integrated medical recommendation method proposed through experiments with other recommendation algorithms.

Acknowledgements. This work is supported by the National Natural Science Foundation of China (No. 61972208) and Jiangsu Postgraduate Research and Innovation Plan (No. KYCX20_0761).

References

1. Basu, D., Kashid, S., Pawar, S., Datta, D.: An integrated detection and treatment recommendation framework for breast cancer using convolutional neural networks and TOPSIS. In: 2020 IEEE 17th India Council International Conference (INDICON), pp. 1–7 (2020). https://doi.org/10.1109/INDICON49873.2020.9342266
2. Chen, J., He, P., Pan, F.: A preferred music recommendation method for tinnitus personalized treatment based on signal processing and random forest. In: 2021 6th International Conference on Intelligent Computing and Signal Processing (ICSP), pp. 470–473 (2021). https://doi.org/10.1109/ICSP51882.2021.9408902

3. Inayatulloh, I.: Proposed it governance for hospital based on TOGAF framework. In: 2021 International Conference on Information Management and Technology (ICIMTech), vol. 1, pp. 476–480 (2021). https://doi.org/10.1109/ICIMTech53080.2021.9534992

4. Isril, R.A., Junaedi, D., Herdiani, A.: Mobile-based hospital recommendation according to patient needs using saw method (case study: Banda aceh). In: 2020 8th International Conference on Information and Communication Technology (ICoICT), pp. 1–5 (2020). https://doi.org/10.1109/ICoICT49345.2020.9166365

5. Li, F., Xu, G., Cao, L., Fan, X., Niu, Z.: CMF: coupled matrix factorization for recommender systems. In: International Conference on Web Information Systems Engineering (2014)

6. Wang, H., et al.: RippleNet: propagating user preferences on the knowledge graph for recommender systems. In: Proceedings of the 27th ACM International Conference on Information and Knowledge Management, CIKM 2018, pp. 417–426. Association for Computing Machinery, New York (2018). https://doi.org/10.1145/3269206.3271739

7. Wang, H., Zhang, F., Xie, X., Guo, M.: DKN: deep knowledge-aware network for news recommendation. In: Proceedings of the 2018 World Wide Web Conference, WWW 2018, pp. 1835–1844. International World Wide Web Conferences Steering Committee, Republic and Canton of Geneva, CHE (2018). https://doi.org/10.1145/3178876.3186175

8. Yu, X., Zhan, D., Liu, L., Lv, H., Xu, L., Du, J.: A privacy-preserving cross-domain healthcare wearables recommendation algorithm based on domain-dependent and domain-independent feature fusion. IEEE J. Biomed. Health Inform. 26(5), 1928–1936 (2022). https://doi.org/10.1109/JBHI.2021.3069629

9. Zhang, F., Yuan, N.J., Lian, D., Xie, X., Ma, W.Y.: Collaborative knowledge base embedding for recommender systems. In: Proceedings of the 22nd ACM SIGKDD International Conference on Knowledge Discovery and Data Mining, KDD 2016, pp. 353–362. Association for Computing Machinery, New York (2016). https://doi.org/10.1145/2939672.2939673

Web and IoT Applications

Joint Extraction of Entities and Relations in the News Domain

Zeyu Li[1(✉)] [iD], Haoyang Ma[2], Yingying Lv[1], and Hao Shen[1]

[1] State Key Laboratory of Media Convergence and Communication, Communication University of China, Beijing, China
lizeyu@cuc.edu.cn
[2] North China Institute of Computing Technology, Beijing, China

Abstract. Extracting entities and relationships between entities from news text information is the core task of building news knowledge graphs. In recent years, with the rise of knowledge graphs, the joint extraction of entity relationships has become a research hotspot in the field of natural language processing. Aiming at the problem that there are many entities in news text data and overlapping relationships between entities are common, this paper first proposes a labeling strategy around the central entity, which transforms the extraction of entities and relationships into sequence labeling problems. After that, this paper also proposes a joint extraction model, which is based on pre-trained language and combined with the improved Bi-directional Long Short-Term Memory (BiLSTM) and Conditional Random Field (CRF) model to achieve entity and relationship extraction. The experimental results on two public news datasets show that our proposed joint extraction model has different degrees of improvement in accuracy and recall compared with other popular joint extraction models. The F1 value on NYT and DuIE both achieved the highest values, reaching 71.6% and 81.4%, which proves that the method proposed in this paper is effective.

Keywords: Joint extraction · Deep learning · Relation overlap

1 Introduction

In modern society, with the rapid development of network media and of information networks. With the exponential growth of data resources, how to extract valuable information from massive unstructured or semi-structured news data has become an important issue in the field of natural language processing. Therefore, the knowledge graph, which can be rich in objective world knowledge and store information in a structured form [1], has gradually become a research hotspot in the field of news. By building a knowledge map in the news field, the rapid development of intelligent information service technology in the news industry can be promoted, thereby bringing greater economic benefits and better news dissemination effects.

The triplet (h, r, t) composed of entities and the relationships between entities is the key semantic information in the knowledge graph, where h represents the head entity and t represents the tail entity, r represents the relationship that exists between entities.

W. Chen et al. (Eds.): ADMA 2022, LNAI 13725, pp. 81–92, 2022.
https://doi.org/10.1007/978-3-031-22064-7_7

How to extract entities and relationships from a large amount of news text data are the two core tasks of constructing and updating news knowledge graphs. There are usually a large number of entities in the text information in the news domain, and the problem of overlapping relationships is common. For example, in the short text "Alibaba company founded by Jack Ma is located in Hangzhou", there are two overlapping relation triples (Jack Ma, created, Alibaba) and (Alibaba, located in, Hangzhou). Therefore, if one wants to improve the accuracy of relation extraction in the field of news, in addition to improving the accuracy of named entity recognition, one must also be able to accurately identify overlapping relationships between entities. There are currently three joint extraction methods, which are method based on shared parameters, the method based on sequence annotation, and the method based on graph structure. These joint models fall into two paradigms. The first normal form can be expressed as $(h, t) \rightarrow r$, which first identifies all entities in a sentence, and then performs relation classification based on each extracted entity pair. However, these methods need to enumerate all possible entity pairs, and relation classification may suffer from redundant entities. Bekoulis et al. [4] proposed another paradigm, denoted as $h \rightarrow (r, t)$, which first detects head entities and then predicts the corresponding relations and tail entities. Compared with the first paradigm, the second paradigm can jointly identify entities and all possible relationships between them at one time, which can better solve the problem of overlapping entity relationships [5, 6]. Therefore, in order to effectively extract entities and overlapping relationships while reducing redundant information extraction, this paper proposes a central entity-oriented labeling strategy, which transforms the entity-relationship joint extraction task into a sequence labeling task to distinguish entities at different positions. Then, based on the theoretical basis of the second joint extraction paradigm, bidirectional long short-term memory (BiLSTM) and conditional random field (CRF), this paper proposes a RoBERTa-BiLSTM*-CRF entity-relationship joint extraction model, which is based on the RoBERTa pre-training model [7] and improved BiLSTM-CRF. The F1 values on the open-source datasets NYT and DuIE published in the news field reached 0.716 and 0.814. The experimental results confirmed the effectiveness and feasibility of the method in this paper.

2 Research Status

Since the concept of entity relation extraction task was proposed, after more than 20 years of continuous research, there have been relatively rich research results [2]. Traditional entity relationship extraction generally adopts the pipeline method, which divides named entity recognition and relationship extraction into two independent subtasks, and directly extracts the relationship between entities on the basis of entity recognition has been completed. Miwa used text features to construct feature vectors, and Kambhatla et al. [8] used the lexical and semantic features of text to extract relationships through the maximum entropy model. Because the construction of feature vectors requires researchers to have a large amount of linguistic knowledge, and features need to be manually designed, a lot of work is required to preprocess the data, which may also lead to the transmission of errors, so the accuracy of the early extraction model is not too high. Deep learning can automatically learn the features in the text and can have high accuracy, so it has gradually become the mainstream method for researching row entity and relationship extraction in recent years. At present, the relationship extraction methods based on deep learning can be roughly divided into pipeline relationship extraction methods and joint relation-ship extraction methods.

The pipeline relationship extraction method based on deep learning still de-composes the relationship extraction task into two independent subtasks, such as Nayak et al. [9] Perform classification and transform the relation extraction task into a classification prediction task [10]. Although the pipeline extraction method based on deep learning is easy to implement and has strong flexibility, the error in the process of entity extraction may be transmitted to the process of relation classification, resulting in the transmission of errors. Therefore, Miwa et al. [3] used the neural network model to solve the joint entity-relation extraction task for the first time, and integrated the two tasks into the same model through the method of sharing parameters, but the two tasks are still separate processes, resulting in a lot of redundant remaining information. In order to solve this problem, Zheng et al. [11] designed a novel labeling method, which extracts entities and relationships at the same time, transforms the extraction problem into a labeling task, and uses neural networks to model, avoiding the complexity of feature engineering. Wang et al. [12] proposed a new joint learning model based on the graph structure, which can enhance the correlation between related entities using a loss function of paranoid weight. However, these three joint extraction methods all face the problem of overlapping relationships in the extraction process. In order to effectively solve this problem, in recent years, several studies have tried to apply the Attention mechanism and pre-training models to joint extraction tasks. Liu et al. [13] proposed an attention-based joint relation extraction model, which designed a supervised multi-head self-attention mechanism as a relation detection module to separately learn the associations between each relation type to identify overlapping relations and relationship types. The joint extraction models proposed in the past two years basically introduce the Attention mechanism [14] and different pre-training models [15] to solve the problem of overlapping relationships in texts.

3 Methodology

This paper proposes a RoBERTa-BiLSTM*-CRF entity-relationship joint extraction model based on pre-trained embedding vectors, which map the sentences to the vector space and input the model, and which use an improved BiLSTM net-work to capture the semantic information in the sentence. Then combined with the CRF, for each input sentence, the predicted entity and related annotation sequence is obtained as the output of the model. The overall framework of the model is shown in Fig. 1.

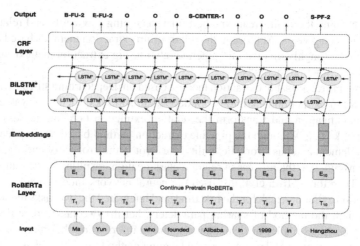

Fig. 1. Model framework

3.1 Labeling Strategy for Central Entities

The labeling strategy proposed in this paper is to divide labels into three parts: boundary information, relation classification, and entity order, as shown in Fig. 2. The boundary information part uses the 'BIESO' labeling method to represent the boundary information of a single word and character in the entity, B represents the first word in the entity head, and I represent the word in the middle of the entity, and E represents the last word of the entity, S represents an entity with only one word or character, and O represents other words that are not entities. The relation classification part is composed of pre-defined relation type labels. Furthermore, we specifically propose a new relation class "CENTER", which is the highest-level relation class to label the narrative body, the central entity, in news texts. The entity order is the serial number marked by a number to distinguish the internal head and tail entities of the triplet. The serial number of the central entity as the head entity of the triplet is fixed as '1', and the other entities are the tail entities in the triplet. The serial number is '2'.

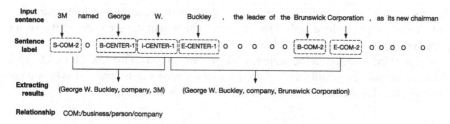

Fig. 2. An example of text labeling

First necessary to obtain entities according to entity boundaries and to determine entity relationships and all possible combinations according to the nearest neighbor principle. When the entity-relationship category is not "CENTER", find the entities with the closest distance, the same relationship category, and different entity sequence numbers in the text before and after the entity to form entity-relationship triples. When the entity-relationship category is "CENTER", the entity-relationship triplet is formed by finding entities with different entity positions in the two text directions before and after and matching them.

3.2 RoBERTa Presentation Layer

In view of the fact that the distance between entity pairs with existing relationships is relatively far in news text data [19], in order to facilitate the construction of the input of the RoBERTa representation layer, we define each input news text sentence as a word sequence d of length m. So $d = (w_1, w_2, w_3, \ldots, w_m)$, the RoBERTa model maps each word $w_i(1 \leq i \leq m)$ in d to a word vector of dimension n, and the transformed vector represents the sequence V such as Formula 1.

$$V = (v_1, v_2, v_3, \ldots, v_m), v_i \in \mathbb{R}^n (1 \leq i \leq m)$$ (1)

3.3 Improved BiLSTM* Layer

Due to its design characteristics, the LSTM is very suitable for processing text data. However, due to the tanh function used by the LSTM neural network, the gradient disappears during the training process of the network. Therefore, this paper proposes a new activation function RTLU, which is used to replace the tanh function inside the LSTM neuron. The function expression is shown in Formula 2.

$$f(x) = \begin{cases} x, x \geq 0 \\ tanhx, x < 0 \end{cases}$$ (2)

Specifically for the neuron x, since the ReLU function does not have the problem of neuron saturation when x is in the positive interval, it can easily solve the gradient disappearance problem caused by the soft saturation of the tanh activation function when x is in the positive interval. When x is in the negative range during the backpropagation process, the tanh function will not cause the problem that the weight is not updated due to the gradient being 0, which can well solve the neuron death problem caused by the ReLU function not updating the weight.

In order to improve the performance of the model, this paper replaces the tanh function of the LSTM neuron with the RTLU activation function. The internal structure of the improved LSTM neuron LSTM* is shown in Fig. 3.

Output of the hidden layer is h_{t-1}. Then the control formula of the forgetting gate at time t is:

$$f_t = \sigma(W_f \cdot [h_{t-1}, V_t] + b_f)$$ (3)

where W_f is the weight matrix of the forget gate, b_f is the bias term of the forget gate, $[h_{t-1}, V_t]$ represents the splicing of two vectors, σ represents the activation function, f_t represents how many times t-1 should be retained unit status.

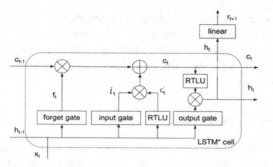

Fig. 3. Internal Structure of the improved LSTM*

Similarly, the input gate control formula at time t is:

$$i_t = \sigma(W_i \cdot [h_{t-1}, V_t] + b_i) \tag{4}$$

where W_i is the weight matrix of the input gate, and b_i is the input gate bias term. The input of the unit state at time t is jointly determined by the output at time t-1 and the input at time t. The formula is as follows:

$$c'_t = \text{RTLU}(W_c \cdot [h_{t-1}, V_t] + b_c) \tag{5}$$

where W_c is the weight matrix, b_c is the bias term. The unit state c_t at time t is determined by the unit state c_{t-1} at time t-1 and the calculation results of formulas (3), (4), (5), and the formula is as follows:

$$c_t = f_t \odot c_{t-1} + i_t \odot c'_t \tag{6}$$

where $\odot$ represents the multiplication of the corresponding position elements in the matrix. Then the control formula of the "output gate" at time t is:

$$o_t = \sigma(W_o \cdot [h_{t-1}, V_t] + b_o) \tag{7}$$

where W_o is the weight matrix, b_o is the bias term. The final output of the forward LSTM* at time t is obtained by multiplying the unit state and the judgment condition obtained by the output gate. The formula is as follows:

$$\overrightarrow{h_t} = o_t \cdot \text{RTLU}(c_t) \tag{8}$$

The hidden layer of BiLSTM* also saves the forward output vector $\overrightarrow{h}$ and the reverse output vector $\overleftarrow{h}$ . .The output of the improved BiLSTM* encoding layer at time t is:

$$H = \overrightarrow{h_t} + \overleftarrow{h_t} \tag{9}$$

Table 1. Dataset size.

Dataset	NYT	DuIE
Total number of sentences	61385	70128
Total number of relationships	68312	76254
Number of relationship types	24	27
Train set sentences	42970	49089
Test set sentences	12277	14026
Validation set sentences	6138	7013

Table 2. Experiment environment.

Hardware equipment	CPU: Intel I7–9700 3.00 GHz
	RAM: 32.0 GB
Software	Windows 10 64bit Python 3.6.0
	tensorflow 1.14.0 keras 2.4.3 gensim 3.4.3 jieba 0.42.1

4 Experiment

4.1 Experimental Data and Experimental Environment

To verify the effectiveness of the model proposed in this paper for entity relation extraction in the news domain, it is tested on public datasets NYT [20] and DuIE [21]. These two open-source datasets contain multiple relational triples, making them ideal for evaluating models for extracting overlapping relational triples. This paper filters out overly complex long sentences with more than 100 words in the two datasets and some sentences that are not closely related to the news domain, and randomly divide them into the training set, test set, and validation set according to the ratio of 7:2:1. The specific information of the two datasets is shown in Table 1. The experimental environment of this experiment is shown in Table 2.

4.2 Evaluation Standard

In the entity-relationship joint extraction experiment, the evaluation criteria used are international standards, including the precision (P), recall (R) and F1 value. The extracted relation is considered correct when both the category and head and tail entities of the relation are correct. The formula parameters are defined as follows.

$$P = \frac{TP}{TP + FP} \tag{10}$$

$$R = \frac{TP}{TP + FN} \tag{11}$$

where TP is the number of correct identifications, FP is the number of irrelevant identified objects, and FN is the number of unidentified objects that exist in the dataset.

It is usually necessary to comprehensively consider the harmonic mean of precision and recall, that is F1 value, which is defined as follows.

$$F_1 = \frac{2 * P * R}{P + R} \tag{12}$$

4.3 Experimental Parameters

In the model training process, 20% of the data is used as the validation set to adjust the parameters during training. After several experiments and fine-tuning, the experimental parameters set for the NYT and DuIE datasets are slightly different.

During the training process, the batch_size is 64, the number of LSTM units is 256, and the maximum text length is 128. The RoBERTa model and the BERT model used in pre-training are both Google's open-source pre-training models, both of which are 12-layer and hidden. The layer is 768-dimensional and adopts the 12-head mode, while the Glove model is trained on the open source news corpus of Sogou Lab. Its feature vector is 400-dimensional, and the remaining parameters are the default values in gensim. In addition, the Adam optimizer [22] with initial learning rates of 1e-3 and 1e-5 is used to learn 100 epochs on the training sets of NYT and DuIE, respectively, with dropout sizes of 0.5 and 0.2 to speed up training and prevent overfitting. Obtain the best F1 value model on the validation set.

4.4 Experimental Design

In order to prove the effectiveness of the model in this paper, we have compared it with the joint relation extraction models in recent years. The baseline models used in this paper for comparison are as follows:

GraphRel: The model proposed by Fu et al. [23] is a joint extraction model based on graph structure. The model divides the overall joint entity relationship extraction into two stages. Both stages use a bidirectional graph convolutional neural network for feature extraction and prediction.

Glove-BiLSTM-CRF: Hu et al. [6] proposed a joint learning model that can identify overlapping relationships between entities, using a parameter sharing method to achieve joint extraction tasks, by allowing the two tasks to share except the last layer of relationship classifiers The neural network parameters of all layers use the connection between tasks to optimize the effect of the two tasks.

TETI: Based on the encoder-decoder structure, Chen et al. [24] fused entity category information to construct the entity-relationship joint extraction model FETI. The prediction of the head and tail entity categories is added in the decoding stage and constrained by an auxiliary loss function so that the model can use the entity category information more effectively.

RoBERTa-BiLSTM-CRF: Li et al. [18] proposed this new model for the evaluation object extraction task, and this paper uses this model for the joint extraction of entity relations, aiming to verify the effectiveness of our model improvement.

CASREL: Wei et al. [16] proposed a joint extraction method of sequence labeling based on the BERT [17] pre-training model, and implemented a Cascade Binary Tagging Framework that is not troubled by the overlapping triple problem. The relation in triple is modeled as a function that maps a head entity to a tail entity, rather than treating it as a label on an entity pair.

4.5 Result Analysis

In order to verify whether the RoBERTa-BiLSTM*-CRF model proposed in this paper can effectively extract overlapping relationships in news texts, experiments were conducted on the open-source English NYT news dataset and Chinese DuIE dataset. The results are shown in Table 3.

It can be seen that on the two news datasets, the RoBERTa-BiLSTM*-CRF model proposed in this paper has achieved the highest F1 value for relation ex-traction, indicating that our model is superior at the joint entity relation extraction task of news texts. From the comparison of indicators, the precision rate of each model is generally higher than the recall rate. The extraction effect of sever-al joint extraction models based on character vectors is significantly better than the Glove-BiLSTM-CRF model based on word vectors and the GraphRel model based on bi-RNN and GCN. Specifically, the RoBERTa-BiLSTM-CRF model based on character vectors and the model proposed in this paper has improved F1 values on the NYT dataset by 15.7% and 18.2%, respectively, compared with the Glove-BiLSTM-CRF model based on Glove word vectors. The F1 value on the DuIE dataset has increased by 15.9% and 18.1% respectively, which also shows that adding the RoBERTa pre-trained language model can obtain dynamic word vectors according to the context information of words, and use a self-attention mechanism to obtain bidirectional semantic features, which greatly improves the effectiveness of entity relation extraction.

The proposed model in this paper improves the F1 value of the RoBERTa-BiLSTM-CRF model by 2.14% and 1.88% on the NYT and DuIE datasets, respectively, indicating that the comprehensive performance of the BiLSTM*-CRF model is due to the BiLSTM-CRF model, indicating that the improved LSTM neural Meta can effectively improve the effect of entity-relationship joint extraction task. Both CASREL and our model based on the BERT training model are joint extraction methods based on sequence annotation, but CASREL is not as good as our model, but both models are better than the RoBERTa-BiLSTM-CRF model. It shows that the RTLU activation function proposed in this paper has a positive effect on entity relation extraction.

In addition, this paper also explores the different effects of different parameters on the joint extraction performance of our model. First, set the batch_size to 32 and the number of LSTM units to 256, and test the impact of dropout changes on the F1 value of the comprehensive evaluation index of the model, as shown in Table 4. As the value of Dropout increases, the F1 value of the model will first increase to the highest value, and then as the Dropout value continues to increase, the value of F1 will decrease instead.

Afterward, the dropout values on the NYT dataset and the DuIE dataset are set to 0.5 and 0.2 respectively, and the batch_size is uniformly set to 32 to test the impact of the change in the number of LSTM units on the comprehensive evaluation index F1

Table 3. Comparison with baseline model.

Models	NYT			DuIE		
	P	R	F1	P	R	F1
GraphRel	0.632	0.597	0.614	0.705	0.683	0.694
Glove-BiLSTM-CRF	0.611	0.602	0.606	0.692	0.687	0.689
TETI	0.686	0.653	0.669	0.758	0.754	0.756
CASREL	0.715	0.695	0.705	0.806	0.798	0.802
RoBERTa-BiLSTM-CRF	0.713	0.689	0.701	0.792	0.786	0.789
RoBERTa-BiLSTM*-CRF	**0.724**	**0.709**	**0.716**	**0.823**	**0.805**	**0.814**

Table 4. Effect of dropout on RoBERTa-BiLSTM*-CRF model.

Dropout	NYT			DuIE		
	P	R	F1	P	R	F1
0.1	0.715	0.699	0.707	0.819	0.803	0.811
0.2	0.716	0.701	0.708	**0.823**	**0.805**	**0.814**
0.3	0.719	0.704	0.711	0.821	0.802	0.811
0.4	0.722	0.706	0.714	0.817	0.799	0.808
0.5	**0.724**	**0.709**	**0.716**	0.812	0.791	0.801
0.6	0.720	0.707	0.713	0.811	0.785	0.798

value of the model, as shown in Table 5. It can be seen that in the RoBERTa-BiLSTM*-CRF model, increasing the number of LSTM units in the model will improve the model performance to a certain extent, but as the number of LSTM units continues to increase, the model will have problems with overfitting and increase the cost of model training. The overfitting effect of deep neural networks can be reduced by adding dropout, but the larger the dropout, the more information is discarded, and the model performance will gradually decrease. Therefore, selecting the appropriate number of LSTM units and dropout values can effectively improve model performance and alleviate the over-fitting problem caused by too few training samples. It can be seen from Table 4 andTable 5 that on the NYT dataset and DuIE dataset, when the dropout of the joint.

entity-relation extraction task is 0.5 and 0.2, the F1 value reaches the highest. When the number of LSTM units is 256, the F1 value reaches the highest, and the entire model is optimal at this time. To sum up, the RoBERTa-BiLSTM*-CRF model proposed in this paper has the highest F1 value for the joint entity-relation extraction task on the English and Chinese news domain datasets compared with other methods. Compared with other joint extraction methods based on sequence annotation, the F1 value of the model in this paper can be improved by 1.5%3.2%, which shows the superiority of the LSTM improvement strategy in this paper.

Table 5. Effect of the number of LSTM units on model.

LSTM units	NYT			DuIE		
	P	R	F1	P	R	F1
64	0.718	0.702	0.710	0.816	0.795	0.805
128	0.721	0.708	0.714	0.822	0.803	0.812
256	**0.724**	**0.709**	**0.716**	**0.823**	**0.805**	**0.814**
512	0.720	0.704	0.712	0.819	0.800	0.809

5 Conclusion

In this paper, we propose an entity-relationship joint extraction method for news domain text information and a central entity-oriented labeling strategy, which transforms the entity-relationship joint extraction task into a sequence of labeling tasks. The experimental results show that the model proposed in this paper can effectively extract the overlapping relationship between entities and entities in the text data in the news domain with the help of the central entity-oriented annotation strategy. However, in the process of extraction, it is also a problem to be solved that the way of entity semantic distinction is too simple. Therefore, future work considers improving the accuracy of multiple relation extraction between the same entity pair by further improving the central entity-oriented annotation strategy.

References

1. Pujara, J., Miao, H., Getoor, L., Cohen, W.: Knowledge graph identification. In: Alani, H., et al. (eds.) ISWC 2013. LNCS, vol. 8218, pp. 542–557. Springer, Heidelberg (2013). https://doi.org/10.1007/978-3-642-41335-3_34
2. Li, D., Zhang, Y., Li, D., et al.: Review of entity relation extraction methods. J. Computer Res. Dev. **57**(7), 1424–1448 (2020)
3. Miwa, M., Bansal, M.: End-to-end relation extraction using LSTMs on sequences and tree structures. In: Proceedings of the 54th Annual Meeting of the Association for Computational Linguistics, pp. 1105–1116. Association for Computational Linguistics, Strasbourg (2016)
4. Bekoulis, G., Deleu, J., Demeester, T., et al.: Joint entity recognition and relation extraction as a multi-head selection problem. Expert Syst. Appl. **114**, 34–45 (2018)
5. Zhao, T., Yan, Z., Cao, Y., Li, Z.: Entity relative position representation based multi-head selection for joint entity and relation extraction. In: Sun, M., Li, S., Zhang, Y., Liu, Y., He, S., Rao, G. (eds.) CCL 2020. LNCS (LNAI), vol. 12522, pp. 184–198. Springer, Cham (2020). https://doi.org/10.1007/978-3-030-63031-7_14
6. Hu, Y., Yan, H., Chen, C.: Joint entity and relation extraction for constructing financial knowledge graph. J. Chongqing University Technol. (Natural Science) **34**(5), 139–149 (2020)
7. Liu, Y., Ott, M., Goyal, N., et al.: Roberta: A robustly optimized bert pretraining approach. arXiv preprint arXiv:1907.11692 (2019)
8. Kambhatla, N.: Combining lexical, syntactic, and semantic features with maximum entropy models for information extraction. In: Proceedings of the ACL Interactive Poster and Demonstration Sessions, pp. 178–181. Association for Computational Linguistics, Strasbourg (2004)

9. Nayak, T., Ng, H.T.: Effective attention modeling for neural relation extraction. In: Proceedings of the 23rd Conference on Computational Natural Language Learning (CoNLL), pp. 603–612. Association for Computational Linguistics, Strasbourg (2019)

10. Zhang, D., Peng, D.: ENT-BERT: entity relation classification model combining bert and entity information. J. Chinese Computer Syst. **41**(12), 2557–2562 (2020)

11. Zheng, S., Wang, F., Bao, H., et al.: Joint extraction of entities and relations based on a novel tagging scheme. In: Proceedings of the 55th Annual Meeting of the Association for Computational Linguistics, pp. 1227–1236. Association for Computational Linguistics, Strasbourg (2017)

12. Wang, S., Yue, Z., Che, W., et al.: Joint extraction of entities and relations based on a novel graph scheme. In: Twenty-Seventh International Joint Conference on Artificial Intelligence, pp. 4461–4467. AAAI Press, Menlo Park (2018)

13. Liu, J., Chen, S., Wang, B., et al.: Attention as relation: learning supervised multi-head self-attention for relation extraction. In: Proceedings of the Twenty-Ninth International Joint Conference on Artificial Intelligence, pp. 3787–3793. Springer (2021). https://doi.org/10.24963/ijcai.2020/524

14. Lai, T., Cheng, L., Wang, D., et al.: RMAN: Relational multi-head attention neural network for joint extraction of entities and relations. Applied Intelligence **52**(3), 3132–3142 (2021)

15. Qiao, B., Zou, Z., Huang, Y., et al.: A joint model for entity and relation extraction based on BERT. Neural Computing Appl. **34**(5), 3471–3481 (2022)

16. Wei, Z., Su, J., Wang, Y., et al.: A novel cascade binary tagging framework for relational triple extraction. In: Proceedings of the 58th Annual Meeting of the Association for Computational Linguistics, pp. 1476–1488. Association for Computational Linguistics, Strasbourg (2020)

17. Devlin, J., Chang, M. W., Lee, K., et al.: Bert: Pre-training of deep bidirectional transformers for language understanding. arXiv preprint arXiv:1810.04805 (2018)

18. Li, Z., Yu, T., Shen, H.: Research on Opinion Targets Extraction of Travel Reviews Based on RoBERTa Em-bedded BILSTM-CRF Model. In: 2021 International Conference on Culture-oriented Science & Technology (ICCST), pp. 114–118. IEEE, Piscataway (2021)

19. Fu, R., Li, J., Wang, J., et al.: Joint extraction of entities and relations for domain knowledge graph. J. East China Norm. Univ. Nat. Sci. **2021**(5), 24–36 (2021)

20. Riedel, S., Yao, L., McCallum, A.: Modeling relations and their mentions without labeled text. In: Balcázar, J.L., Bonchi, F., Gionis, A., Sebag, M. (eds.) ECML PKDD 2010, LNCS, vol. 6323, pp. 148–163. Springer, Heidelberg (2010)

21. Li, S., et al.: DuIE: a large-scale chinese dataset for information extraction. In: Tang, J., Kan, M.-Y., Zhao, D., Li, S., Zan, H. (eds.) NLPCC 2019. LNCS (LNAI), vol. 11839, pp. 791–800. Springer, Cham (2019). https://doi.org/10.1007/978-3-030-32236-6_72

22. Kingma, D., Ba, J.: Adam: A Method for Stochastic Optimization. Computer Science (2014)

23. Fu, T., Li, P., Ma, W.: Graphrel: Modeling text as relational graphs for joint entity and relation extraction. In: Proceedings of the 57th Annual Meeting of the Association for Computational Linguistics, pp. 1409–1418. Association for Computational Linguistics, Strasbourg (2019)

24. Chen, R., Zheng, X., Zhu, Y.: Joint entity and relation extraction via fusing entity type information. Comput. Eng. **48**(3), 46–53 (2022)

Event Detection from Web Data in Chinese Based on Bi-LSTM with Attention

Yuxin Wu[1], Zenghui Xu[2], Hongzhou Li[3(✉)], Yuquan Gan[4],
Josh Jia-Ching Ying[5], Ting Yu[2], and Ji Zhang[6(✉)]

[1] School of Computer Science and Technology,
Nanjing University of Aeronautics and Astronautics, Nanjing, China
[2] Zhejiang Lab, Hangzhou, China
{xuzenghui,yuting}@zhejianglab.com
[3] School of Life and Environmental Sciences,
Guilin University of Electronic Technology, Guilin, China
homzh@163.com
[4] School of Telecommunication and Information Engineering,
Xi'an University of Posts and Telecommunications, Xi'an, China
ganyuquan@xupt.edu.cn
[5] Department of Management Information Systems,
National Chung Hsing University, Taichung 402, Taiwan ROC
[6] School of Mathematics, Physics and Computing,
The University of Southern Queensland, Toowoomba, Australia
Ji.Zhang@usq.edu.au

Abstract. Events are important activities people are involved in real life, and the information about events may be fascinating and important for people to understand and keep abreast with the key developments of some important social and individual subjects. In the big data era, event detection methods can help people efficiently and quickly extract specific information from massive Web information. However, the existing methods usually load the entire Web page information as the input into the models, and the rich noise and irrelevant information on Web pages will seriously impact the event detection performance of these methods. Also, the existing methods mostly used static models, which fail to consider the dynamics of information on the Web. To improve the performance of event detection and classification, we propose in this paper a new method that partitions the Web pages into multiple text blocks and utilizes Bi-LSTM with the attention mechanism for fine-grained event detection from Chinese Web pages. We also propose a dynamic method that updates the data as well as the model regularly and incrementally, making our model more adaptive to the ongoing changes of the Webpage data. The experimental results show that our model outperforms existing methods in event detection in terms of detection performance, the associated computational overhead, and the ability to deal with evolving Webpage information.

Keywords: Event detection · Text blocks · Dynamic maintenance

This research is supported by the Natural Science Foundation of China (No. 62172372), Zhejiang Provincial Natural Science Foundation (No. LZ21F030001) and Zhejiang Lab (No. 2022KG0AN01).

W. Chen et al. (Eds.): ADMA 2022, LNAI 13725, pp. 93–105, 2022.
https://doi.org/10.1007/978-3-031-22064-7_8

1   Introduction

In our daily lives, people are involved in various different types of events that occur at specific times and locations, which are important information for people to understand the latest developments of some important social and individual subjects. Examples of such events in the context of academic organizations, such as universities, could be graduation ceremonies, academic visits and research seminars, etc. Events are an important carrier of information that help us recognize and understand the real world [15].

As more and more information is digitized and published online (e.g., as in Webpages), it becomes possible to collect a large amount of information regarding different types of events in order to produce a collection of events that are interesting to different cohorts of users. However, in the big data era, effectively and efficiently generating such event-based information is challenging. Firstly, there is a large number of potential Webpages, even for a single organization, that potentially contain event-based information. Secondly, too much irrelevant information makes it nontrivial to accurately obtain information about specific events, which are treated as noise and may impact event detection in an adverse manner. Last but not least, the information on Webpages are dynamic in the sense that most of them are constantly evolving. New Webpages may be created that carry event-based information that should be periodically collected and analyzed to detect events for users. The above challenges will make event detection methods time-consuming, inaccurate and lack of the ability to deal with evolving event information. Therefore, how to quickly, accurately, and dynamically detect and classify event information from massive and noisy Webpages is the key problem to be solved in this paper.

So far, researchers have developed and proposed in literature different event detection, and classification methods, which use IT and intelligence technologies to automatically detect and classify events from massive data [21]. Obviously, such automatic event detection and classification technologies can alleviate the problems caused by information overloading in the big data era and improve the efficiency of people's acquisition of event-related information. However, these methods suffer some key drawbacks as follows:

a) The methods based on pattern matching rely on specific linguistic knowledge and requires domain knowledge of experts, also the process of pattern formulation is time-consuming, laborious, and prone to errors. The method based on machine learning is limited in the extraction of deep text features. The relatively simple and shallow neural networks are prone to convergence to the local minimum, which affects event detection performance.

b) The methods based on deep learning usually take the whole text content of Webpages as the input corpus in the model training stage, which includes a large amount of irrelevant information. Therefore these methods suffer from inferior detection performance.

c) Most event detection models are static and lack the mechanism of continuous maintenance. These models stops iteratively updating the detection model after training is completed or are not able to incrementally update the model when new Webpages are generated. They can always generate new models from scratch, but it is apparently too computationally expensive and not efficient at all when dealing with large evolving information on Internet.

To address aforementioned drawbacks, this paper propose a Web-based events deep detection model(Attention with Bi-directional Long Short-Term Memory, ABiLSTM). The incremental crawler technology is used to periodically obtain the latest updates of Webpages. The BiLSTM neural network is leveraged to learn effectively the text features of Webpages which are partitioned. The Attention mechanism is also introduced in the model to assign different weights to text to allow the model to focus on the key textual information indicative of the occurrence of events. Finally, the method of dynamic incremental data maintenance is adopted, which help achieve a good efficiency in dynamic detection of events without sacrificing the detection effectiveness. Specifically, the main contributions of this paper include:

1. The textual content of Webpages is partitioned into different blocks for a finer textual granularity, which can effectively minimize noisy information and capture the key textual content, considerably contributes to the better event detection performance.
2. To effectively capture and model the textual information for event detection, the BiLSTM neural network with attention mechanism is leveraged in our model to learn effectively the text features of partitioned Webpages.
3. Our model supports the dynamic maintenance of the event detection results. By periodically retraining the model with newly incremental Webpages information, our model can iterate continuously.
4. The experimental results show that our model outperforms existing methods in event detection in terms of classification effectiveness and the ability to deal with evolving Webpage information.

2   Related Work

2.1   Pattern Matching Based Methods

Pattern matching based methods manually design the feature template of the corresponding event in advance, and then match it with the target text.

Zhihu et al. proposed an event-adaptive concept integration algorithm, which estimates the effectiveness of semantically related concepts by assigning different weights, and realizes the use of related concepts to match a target event differently and achieve better results [8]. Song Qing et al. identified key sentences in news articles according to word frequency and text clustering and performed subsequent identification and extraction of key sentences [13].

Although the structure of the event detection methods based on pattern matching is intuitive, similar to the way of human thinking, and easy for people to understand, these methods rely on the knowledge of specific domain experts, the time and labor costs are high in the process of pattern formulation, and errors are prone to occur. At the same time, the specific pattern is only for specific fields, the generalization ability of the model is generally poor.

2.2   Machine Learning Based Methods

Machine learning based methods mainly focuses on the discovery of features and the construction of classifiers, which transform event detection into classification.

Traditional Machine Learning Methods. In the past, the text classification mostly used the bag-of-words model, N-grams and its word frequency-inverse document frequency as features, and the major traditional models for text classification include naive Bayes [3], support vector machines, k-nearest neighbors, and hidden Markov models. Chieu and Ng first introduced the maximum entropy classifier into event extraction and used the classifier to detect events and their elements [2]. Zhang et al. manually designed features for Chinese news texts and used CRF to identify and extract trigger words [20]. However, the traditional machine methods, including shallow neural networks, in most cases are not very effective in modeling the textual information of Webpages for the purpose of event detection and classification particularly given the presence of ample noises in the Webpages involved.

Deep Learning Based Methods. Nowadays, with the rise of deep learning and the more efficient extraction of text features by word embedding methods, the neural network model in deep learning has been more and more applied in the field of NLP. The word vector representation model word2vec proposed by Mikolov in 2013 has achieved good text representation results [12]. Kim first used a CNN network model in 2014 to deal with the classification of textual data, and the method performed well in multiple text classification tasks [6]. Lai and niu proposed a model RCNN based on RNN concatenated CNN and achieved good results in classification [10]. Zhou et al. took advantage of the selective retention of long-sequence information by memory cells in the LSTM to extract the sequence information of the text to achieve text classification [22]. Zhang et al. compared the impact of word embeddings at different dimensions on the results of text classification [20]. Lin et al. added a self-attention mechanism to the sentence classification problem to deeply mine the internal relations between sentences and text context semantics, so as to improve the classification performance of the model [9]. The Bert model proposed by Devlin in 2018 is a two-way language model in a complete sense, which can take into account the text sequence information, context information, and grammatical context information in the whole sentence, avoiding the problem of polysemy [4]. Since then, classification based on Bert model has been applied in many fields, such as CM-BERT [18] and BERT4TC [19].

It can be seen from the above research work that, no matter whether it is based on pattern matching or machine learning event detection methods, firstly the original data are event Webpages, secondly, it is necessary to obtain relevant event features, such as trigger words [5,17], entities [7,14], dependencies [1], etc., to detect events. However, an overwhelmingly majority of Webpages for a given Website may typically be non-event Webpages, and it is necessary to detect event-type texts before classifying them as events. The non-event-type data has no clear theme, the text format is not fixed and various. Moreover, it is usually that a few key sentences, rather than the whole Web-page, indicate that this is a Webpage to represent an event. Existing work also confirmed that traditional methods of inputting the whole training corpus into the model have a poor performance, while a model with attention mechanism can be better. Therefore, we adopt in this paper the idea of text partition of Webpages to reduce the textual granularity for training our detection model in a fine-grained manners.

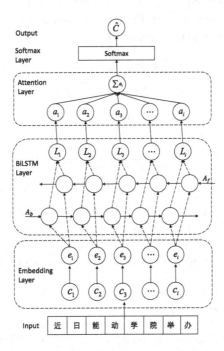

Fig. 1. Structure diagram of our model

3   ABiLSTM Model

3.1   Problem Formulation

Given a Webpage $C = (c_1, c_2, \ldots, c_i)$, classify the Webpage into a binary category in $S = (s_1, s_2)$, and then if C is an event Webpage, we further multi classify it into a category in $R = (r_1, r_2, r_3, \ldots, r_j)$. Among them, c_i is the word set of each Webpage, s_1, s_2 corresponds to event and non-event Webpages, respectively, and $r_1, r_2, \ldots, r_j$ correspond different categories of events. For example, these categories can be student life, scientific research, work updates and others in a university Website.

3.2   Static Classification Model

Figure 1 presents the overall structural architecture of our model. Before the data is loaded into model for both training and testing, the Webpage text is divided into blocks. Specifically, our model contains the following several layers:

Input and embedding layer. For the input long text $C = (c_1, c_2, \ldots, c_i)$, first map each word to the word vector of the corresponding dimension through Word2vec, and then multiply it with the weight matrix obtained by One-hot encoding, and finally get each word embedding to represent e_i.

Bi-LSTM layer. The feature vector e_i is input to the forward loop unit A_b and the reverse loop unit A_f of the LSTM, and the forward output $\overrightarrow{h_i}$ and the reverse output $\overleftarrow{h_i}$ are obtained. Finally, the output L_i of this layer is spliced by $\overrightarrow{h_i}$ and $\overleftarrow{h_i}$. The formulas involved in this step are:

$$f_t = \delta \left(W_f \cdot [h_{t-1}, x_t] + b_f \right) \tag{1}$$

$$i_t = \delta \left(W_i \cdot [h_{t-1}, x_t] + b_i \right) \tag{2}$$

$$\tilde{C}_t = \tanh \left(W_C \cdot [h_{t-1}, x_t] + b_C \right) \tag{3}$$

$$C_t = f_t * C_{t-1} + i_t * \tilde{C}_t \tag{4}$$

$$o_t = \delta \left(W_o \cdot [h_{t-1}, x_t] + b_o \right) \tag{5}$$

$$h_t = o_t * \tanh \left(C_t \right) \tag{6}$$

where i_i, f_t, and o_t are the input gate, forget gate and output gate respectively; C_{t-1}, $\tilde{C}_t$, C_t are the memory unit value at the previous moment, the candidate memory unit value at the current moment, and the memory unit state value at the current moment respectively; W and b represent the trainable parameters;

Attention layer. Then in the attention layer, we calculate the similarity between the query vector Q and the key vector K_i of each data feature. The *softmax* function is used to normalize the attention score to obtain the weight distribution. According to the weight coefficient, the Value is weighted and summed, and the calculation process of the final feature at time t is shown in formula (7):

$$\text{Attention} = \sum_{i=1}^{N} a_i \cdot V = \sum_{i=1}^{N} \frac{\exp\left(h_t\right)}{\sum_{j=1}^{N} h_j} \cdot V \tag{7}$$

where i represents the number of input text sequences, h_t is the hidden layer state sequence at time t, h_j is the hidden vector corresponding to the j feature word in the text at time t; a_i is the weight of the hidden layer state h_t at time t to h_j distributed;

Softmax layer. The output data is passed to the *softmax* layer to be transformed into a probability output, and the label value is output through the *argmax* function.

3.3   Dynamic Model Maintenance

In the era of big data, Web pages frequently update event information, and the classified Web pages may have content changes or generate a large number of new Web pages. To adapt to the dynamics of network events, ensure the effectiveness of the model and prolong the life cycle of the model, this paper adopts the strategy of dynamic maintenance for the classification model. This paper proposes to use an incremental crawler to crawl Web pages regularly, use MD5 value to quickly check whether the content of Web pages has changed from the latest time snapshot, filter duplicate Web pages, and collect new data incrementally. Compared with traditional maintenance from scratch, dynamic maintenance can effectively reduce the data collection time and model training time, making the method in this paper more lightweight.

The dynamic maintenance strategies are explored in our work. Specifically, at a certain time node t_2, we use the time period from the previous time node t_1 to t_2 to generate new text data or text data with content changes, load them into the m_1 model that has been generated on the t_1 time node, and finally output the model m_2 on the t_2 time node. The specific algorithm flow is presented in Algorithm 1.

4   Experimental Evaluation

4.1   Dataset and Experimental Setup

The dataset of this paper comes from the texts on all Web pages of Nanjing University of Aeronautics and Astronautics (http://www.nuaa.edu.cn). Comprehensively crawl [11,16]the Website through the crawler tool, take NUAA news Web page and NUAA Graduate School Web page as the starting URL, and set the list of allowed crawling domain names to ensure that the crawlers are all running under a specific domain name (nuaa.edu.cn), filtering irrelevant URLs. During the crawling process, the crawler tool will first compare the MD5 value of the current Web page with the existing MD5 value in the database. If there is no repetition, grab all text content on the Web page, including navigation bar, title, text, time, and other features, and finally save it in TXT file format with the file name MD5 value; If there are duplicates, skip the current Web page.

Dataset 1 of the two classification problems in this paper has a total of 7956 text data, including 2135 event-type text data and 5821 non-event-type text data. Dataset 1 is divided at a ratio of 7:3, of which 70% of the data is used as the training set, with a total of 5569 records; 30% of the data is used as the test set, with a total of 2387 records. Event-type text data, such as school news, is marked as 1; non-event-type text data such as personal introduction, navigation bar, rules, legal text, is marked as 0.

Dataset 2 of the multi-classification problem in this paper has a total of 8618 text data, which can be divided into 5 categories according to the category dimension, including 1761 text data for campus life, 1232 text data for academic research, and 3215 text data for communication and visit. 2,105 pieces of work dynamic text data and 305 pieces of other text data; According to the time dimension, it can be divided into 4-time nodes, and the interval between each time node is 60 days. At the first time node t_1, 6678 pieces of data are collected, and at the second time node t_2, 1002 pieces of data are added. 565 pieces of data are added at the third time node t_3, and 373 pieces of data are added at the fourth time node t_4. Category 1 events in dataset 2 are for students campus life, focusing on students'daily life; Category 2 events are for scientific research, focusing on academic exchanges between teachers in various colleges in the school; Category 3 events are for school work, focusing on relevant work in the school; Category 4 events are for research cooperation and exchange, focusing on exchanges with organizations outside the University; Category 5 events are for others that do not belong to any of the aforementioned four categories.

The experiments are performed on Windows 10 professional platform, using Python 3 as the compiler language and Pycharm community integrated development environment.

4.2 Chinese Text Preprocessing

First, manually label dataset 1 and dataset 2, so that the data corresponds to the label one by one. Next, the data is cleaned to remove invalid characters, punctuation marks, etc. Then, the long text in dataset 2 is divided into several blocks, the size of each text block is adjusted by setting the number of words in the block, and finally, Chinese word segmentation is realized by the Jieba word segmentation tool.

4.3 Sensitivity Analysis

We carry out the sensitivity study of our proposed model and several mainstream baseline methods, including SVM, TextCNN, Bi-LSTM,Bert, in terms of precision rate(P), recall rate(R) and F1-score as the evaluation metrics.

We study the effectiveness of all the methods under varying block sizes, denoted by k. In the dataset 2 experiments, the block size, k, is varied from 25 words to 100 words. Table 1 presents the F1-Score of different methods under different block sizes. The table shows that different block sizes have an impact on the classification results, with k = 75 words being the most appropriate value for our model. The block size is 25 words, the text block does not contain the complete event information to properly train the model; when the size is 100 words or above, the noises in the text block increases, which adversely affects the model learning.

Table 1. The F1-Score of the algorithm under different block sizes

Models	k = 25 words	k = 50 words	k = 75 words	k = 100 words
SVM	77.6%	81.5%	84.5%	83.2%
TextCNN	79.6%	83.1%	88.6%	85.1%
BiLSTM	80.2%	85.6%	89.2%	86.8%
Our model	**81.4%**	**86.8%**	**90.3%**	**88.2%**

Algorithm 1. Dynamic Maintenance Model Algorithm Process

Input: New long text data or changed text data generated between t_1 and t_2 time nodes $C = (c_1, c_2, \ldots, c_i)$, $c_i (i \in [1.i])$ is a single Chinese word;

Output: Event class label R;

1: Load the ABiLSTM model m_1 output at time t_1;
2: Input the word c_i into the Embedding layer and convert it into the embedded word $e_i = W^{wrd} v^i$, $W^{wrd} \in R^{d^w |V|}$ The output is the embedded word vector of the long text $emb_s = \{e_1, e_2, \ldots, e_n\}$;
3: The text embedding vector $emb_s = \{e_1, e_2, \ldots, e_n\}$ is used as the input of the m_1 model, and the output is:

$$L_i = \text{LSTM}(e_i), i \in [1, n]$$

4: Input the L_i generated by the previous layer to the Attention layer, and obtain implicit information through nonlinear transformation, and the output is:

$$a_i = \sum \alpha_i L_i = \sum \frac{\exp(h_i^T h_w)}{\sum_t \exp(h_i^T h_w)} L_i$$

5: Use a_i as the input of the softmax layer to get the probability of the corresponding category of the current input long text:

$$R = \text{argmax}\, \text{soft}\hat{\text{max}}(a_i)$$

6: Output ABiLSTM model m_2 at time t_2
7: **return** R

4.4 Effectiveness Analysis

In this section, we carried out two experiment that compares the performance of models based on dataset 1 and dataset 2.

Table 2. Models performance based on dataset 1

Models	Event-type	Non-event-type
SVM	86.9%	82.5%
TextCNN	94.3%	93.2%
LSTM	95.1%	94.2%
Our model	96.2%	97.2%

Compared with non-event-type data, the event-type data of dataset 1 has obvious keyword characteristics and certain event elements, at the same time the boundary between the two is clear, and the overlapping part is less. As can be seen from the Table 2, the baseline models have achieved good classification results, especially the F1 value of our model for event-type data is 96.2%.

As can be seen from the Table 3, the classification effect of deep learning model is better than that of machine learning model. This is because the neural network can extract text features at a deeper level and learn more key information. The RNN neural network is better than CNN because the BiLSTM unit

Table 3. Models performance based on dataset 2

Models	Mic_Precision	Mic_Recall	Mic_F1-score	s/epoch
SVM	0.763	0.778	0.77	–
TextCNN	0.821	0.855	0.838	58
LSTM	0.845	0.875	0.86	78
BiLSTM	0.871	0.902	0.886	108
Bert	**0.902**	**0.932**	**0.917**	**361**
Our model	**0.887**	**0.921**	**0.893**	**134**

can retain or discard the current content features according to their importance on the one hand, and effectively extract the connection between the long-term sequences in this article; on the other hand, from the two Read the sequence data in each direction, capture the information of the left and right contexts, connect and summarize to complete the prediction, and extract the semantic information of the text at a deeper level. The introduction of the attention mechanism has helped the model to be slightly improved. The role of keywords is highlighted by assigning weights, and the model is more targeted for learning.

Finally, the classification effect of the model in this paper is slightly behind the BERT model, because Bert is a pre-training model, and the scale of the data set used in its training is hundreds to thousands of times that of this paper. A larger data set can enable deeper information on text features. The capture of semantic information is closer to the reading comprehension of the human brain; next, a transformer with very good parallel performance and a deeper and wider scale can be used in the selection of encoder, so that words at any position can ignore the restrictions caused by direction and distance, and have the opportunity to integrate each word directly. Although Bert is an excellent pre-training model, and the classification effect of the model is better, even its fine-tuning consumes huge computational resources. The Bert model takes 361s to train an epoch, while the ABiLSTM model takes 134s, which is only 1/3 of that of Bert when the difference in classification effect is not very large. Therefore, based on the experimental results, this paper selects the ABiLSTM model to complete the multi-classification problem in this paper.

4.5   Dynamic Maintenance Comparison

In this section, we compared the performance of the two model maintenance strategies, the incremental maintenance and the maintenance from scratch.

Table 4. The F1-Score of the algorithm under the two maintenance methods

Maintenance mode	t_1	t_2	t_3	t_4
Incremental maintenance	0.823	0.855	0.87	0.879
Maintenance from scratch	0.834	0.874	0.887	0.893

Table 5. The cost of computing resources under the two maintenance methods

Maintenance mode		$0\tilde{\ }t_1$	$t_1\tilde{\ }t_2$	$t_2\tilde{\ }t_3$	$t_3\tilde{\ }t_4$
Incremental maintenance	h/data collection time	26	3	2	2
	s/epoch	78	20	20	16
Maintenance from scratch	h/data collection time	26	30	33	35.5
	s/epoch	78	105	120	138

It can be seen from Table 4 that the model classification effect of maintenance from scratch is better than that of incremental maintenance. This is because the amount of data input to the model is different between the two methods except at time t_1. For example, at the t_2 time node, 7680 pieces of data are input to the model maintenance from scratch, while only 1002 pieces of data are provided by incremental maintenance. But what contrasts sharply with the less obvious classification effect is the time spent on data collection for both approaches. Maintenance from scratch, when crawling at the beginning, while searching for URLs and saving texts, it often takes more than one day to obtain a complete data set of the current time node. As shown in the Table 5 incremental maintenance takes a lot of time to crawl for the first time. For the rest of time, the crawler will compare the MD5 value first, and crawl if there is no repetition. The number of newly generated Web pages per unit time interval is not large, so it takes less time. Considering the computational resource overhead and classification effect, this paper maintains data in a dynamic incremental manner.

5   Conclusion and Future Work

In this paper, a Bi-LSTM deep learning model with the attention mechanism is proposed for event detection. The Webpage textual content is partitioned into blocks with a fixed size, whereby the key information contributing to event detection is extracted for training our model in order to achieve better detection accuracy. Incremental model maintenance strategies are applied to continuously maintain our model using incrementally crawled Webpages information to achieve efficient model maintenance.

In the future, we will carry out more comprehensive experiments by creating additional real-life datasets from organizations other than universities. An interactive system will be developed for presenting the detected events in a more visualized and interactive manner across different platforms. We will also extend the model to deal with other languages, such as English, where such adaption, we believe, is straightforward.

References

1. Can, E.F., Manmatha, R.: Modeling concept dependencies for event detection. In: Proceedings of International Conference on Multimedia Retrieval, pp. 289–296 (2014)

2. Chieu, H.L., Ng, H.T.: A maximum entropy approach to information extraction from semi-structured and free text. Aaai/iaai **2002**, 786–791 (2002)
3. Cui, W.: A Chinese text classification system based on naive bayes algorithm. In: MATEC Web of Conferences, vol. 44, p. 01015. EDP Sciences (2016)
4. Devlin, J., Chang, M.W., Lee, K., Toutanova, K.: BERT: pre-training of deep bidirectional transformers for language understanding. arXiv preprint arXiv:1810.04805 (2018)
5. Ding, N., Li, Z., Liu, Z., Zheng, H., Lin, Z.: Event detection with trigger-aware lattice neural network. In: Proceedings of the 2019 Conference on Empirical Methods in Natural Language Processing and the 9th International Joint Conference on Natural Language Processing (EMNLP-IJCNLP), pp. 347–356 (2019)
6. Kalchbrenner, N., Grefenstette, E., Blunsom, P.: A convolutional neural network for modelling sentences. arXiv preprint arXiv:1404.2188 (2014)
7. Kumaran, G., Allan, J.: Text classification and named entities for new event detection. In: Proceedings of the 27th Annual International ACM SIGIR Conference on Research and Development in Information Retrieval, pp. 297–304 (2004)
8. Li, Z., Yao, L., Chang, X., Zhan, K., Sun, J., Zhang, H.: Zero-shot event detection via event-adaptive concept relevance mining. Pattern Recogn. **88**, 595–603 (2019)
9. Lin, Z., et al.: A structured self-attentive sentence embedding. arXiv preprint arXiv:1703.03130 (2017)
10. Liu, P., Qiu, X., Huang, X.: Recurrent neural network for text classification with multi-task learning. arXiv preprint arXiv:1605.05101 (2016)
11. Lu, H., Zhan, D., Zhou, L., He, D.: An improved focused crawler: using web page classification and link priority evaluation. Math. Probl. Eng. **2016**, 1–10 (2016)
12. Mikolov, T., Chen, K., Corrado, G., Dean, J.: Efficient estimation of word representations in vector space. arXiv preprint arXiv:1301.3781 (2013)
13. Qing, S., Ying, Z., Pengzhou, Z.: Research review on key techniques of topic-based news elements extraction. In: 2017 IEEE/ACIS 16th International Conference on Computer and Information Science (ICIS), pp. 585–590. IEEE (2017)
14. Sayyadi, H., Hurst, M., Maykov, A.: Event detection and tracking in social streams. In: Proceedings of the International AAAI Conference on Web and Social Media, vol. 3, pp. 311–314 (2009)
15. Sims, M., Park, J.H., Bamman, D.: Literary event detection. In: Proceedings of the 57th Annual Meeting of the Association for Computational Linguistics, pp. 3623–3634 (2019)
16. Singh, B., Gupta, D.K., Singh, R.M.: Improved architecture of focused crawler on the basis of content and link analysis. Int. J. Mod. Educ. Comput. Sci. **9**(11), 33 (2017)
17. Tong, M., et al.: Improving event detection via open-domain trigger knowledge. In: Proceedings of the 58th Annual Meeting of the Association for Computational Linguistics, pp. 5887–5897 (2020)
18. Yang, K., Xu, H., Gao, K.: CM-BERT: cross-modal BERT for text-audio sentiment analysis. In: Proceedings of the 28th ACM International Conference on Multimedia, pp. 521–528 (2020)
19. Yu, S., Su, J., Luo, D.: Improving BERT-based text classification with auxiliary sentence and domain knowledge. IEEE Access **7**, 176600–176612 (2019)
20. Zhang, C., Hong, S., Zhang, P.: The research on event extraction of Chinese news based on subject elements. In: 2016 IEEE/ACIS 15th International Conference on Computer and Information Science (ICIS), pp. 1–5. IEEE (2016)

21. Zhong, Z., Jin, L., Feng, Z.: Multi-font printed Chinese character recognition using multi-pooling convolutional neural network. In: 2015 13th International Conference on Document Analysis and Recognition (ICDAR), pp. 96–100. IEEE (2015)
22. Zhou, C., Sun, C., Liu, Z., Lau, F.: A C-LSTM neural network for text classification. arXiv preprint arXiv:1511.08630 (2015)

Sentiment Analysis of Tweets Using Deep Learning

Jaishree Ranganathan[✉] and Tsega Tsahai

Middle Tennessee State University, Murfreesboro, TN 37132, USA
Jaishree.Ranganathan@mtsu.edu

Abstract. The coronavirus pandemic has caused a worldwide crisis and a drastic change in day-to-day life activities. Worldwide, people use social media platforms to share and discuss their opinions about the situation. Twitter is one such platform for public conversation around the coronavirus pandemic, the spread of disease, vaccination, non-pharmaceutical interventions, and many other discussions. In this study, we use Twitter social medial data for sentiment analysis. The tweets are collected based on covid-19 related hashtags. This work presents a deep learning-based framework for sentiment analysis using DistilBERT, a distilled version of Bidirectional Encoder Representation from Transformers (BERT), Convolutional Neural Network (CNN), and Long Short Term Memory (LSTM). The results show that transformer-based pre-processing and fine-tuning yield better performance results. The DistilBERT model yields the highest accuracy of 91.46% compared to the CNN and LSTM models.

Keywords: BERT · Coronavirus · Convolutional Neural Network · Long Short Term Memory · Tweets · Sentiment analysis

1 Introduction

Covid-19 disease outbreak has caused changes in several aspects of people's lives worldwide. There are protocols in place for medical and public health systems due to the risks posed by the coronavirus disease outbreak. Historical data and computer technologies have been helpful in the decision-making process during such infectious disease outbreaks in the past [2]. The emergence of web 2.0 has led to more user-generated content and usability for end-users compared to the earlier web 1.0. Some of the popular platforms of user-generated content on the web include blogs, forums, and social networking sites like Twitter and Facebook [1]. Thus there is an increase in the number of social media platforms and the users of these forums worldwide. A tremendous amount of user-generated content is available on Twitter, Facebook, and other venues like e-commerce websites and news portals. Such information is considered an asset of data by businesses, individuals, and other entities looking for timely feedback. These resources are extantly used to understand general public opinions. Some of the most common applications of sentiment analysis include consumer product reviews, customer service, and stock markets [9]. It is evident that such social media platforms have

W. Chen et al. (Eds.): ADMA 2022, LNAI 13725, pp. 106–117, 2022.
https://doi.org/10.1007/978-3-031-22064-7_9

also become significant sources in reflecting real-world events [5]. For instance, social media has become a blazing platform with many discussions and opinions related to coronavirus worldwide.

Twitter is one of the popular microblogging platforms in everyday life for people regardless of their geographic location [21]. Twitter data has played a significant role in the past for several epidemics, outbreak predictions, and monitoring public health [29]. Content on such web platforms includes a variety of raw, unstructured data formats, including text, images, video, and audio. Advancement in computer technologies and techniques, including artificial intelligence and machine learning, provides the power to process such unstructured data and gain valuable insights [8]. Natural language processing (NLP) is a subfield of artificial intelligence constantly gaining attention in almost every domain, including public health, business, and education. Sentiment analysis is one of the trending and most studied research areas in natural language processing and machine learning. In this study, we use tweets collected using hashtags related to coronavirus. We present a deep learning-based framework for sentiment analysis of tweets. We also present a comparative study of results obtained using several deep learning models. The models include convolutional neural network (CNN), long short-term memory (LSTM), and distiled version of bidirectional encoder representation from transformer (BERT). The rest of the paper is organized as follows; Sect. 2 describes the related work, Sect. 3 - describes data collection and pre-processing, Sect. 4 - describes methodology, Sect. 5 - describes experiments and results, and Sect. 6 - conclusions.

2   Related Work

2.1   Sentiment Analysis on Tweets

Sentiment analysis of text, especially tweets, is a trending area of research with a range of machine learning models for automated text sentiment classification. However, it is important to have a labeled dataset for supervised classification. Several works in the literature use emoticons [11], dictionary or lexicons [23] as source of labeling data. Then use traditional machine learning classifiers like Naive Bayes, Maximum Entropy, SVM, Decision Tree, Random Forest, and Decision Table Majority.

Similarly, most of the work applying deep learning models like neural networks uses pre-existing annotated datasets for training and testing purposes. For instance, [16] use standard Twitter sentiment datasets like Stanford Twitter Sentiment Test (STSTd), SE2014 from SemEval2014, Stanford Twitter Sentiment Gold (STSGd), Sentiment Evaluation Dataset (SED), Sentiment Strength Twitter Dataset (SSTd). They achieved the highest accuracy of 87.62% using Glove embedding and Deep Convolutional Neural Network (CNN) on the STSTd dataset. Also in [7], the authors use Stanford Sentiment Treebank (SSTb) movie reviews and Stanford Twitter Sentiment corpus (STS) Twitter messages. They present Character to sentence CNN for sentiment prediction and achieve the highest accuracy of 86.4% on the Twitter sentiment corpus. Another study [17], uses STS Gold Dataset and Movie Review data for training the CNN model and achieve an accuracy of 75.39%.

Some of the works use SemEval datasets. For example, [12] utilize SemEval2014 Task9-SubTask B full data, the SemEval2016 full data Task4 and the SemEval2017 development data for the model train and test. They use Glove and word2vec embedding for CNN and LSTM and achieve the highest accuracy of 59% using multiple CNN and bi-LSTM networks; and [31] use benchmark sets from the SemEval 2015 dataset, which are manually annotated into positive, negative, and neutral labels. They show that deep convolutional neural networks with pre-trained vectors like Glove achieve an F1 score of 64.85%. Similarly, research indicates the use of other pre-annotated datasets for the sentiment classification of tweets. For instance, [26], use the train and dev corpora from Twitter'13 to 16 for training and Twitter'16-dev as a development set. They use Lexical, part-of-speech, and sentiment embeddings to train the CNN model with the fusion input and achieve an F-1 score of 63%. Authors in [34] present a word-character-based CNN and utilize pre-annotated Twitter corpus. They achieve the highest accuracy of 0.8119 for polish language sentiment classification.

Recently transformer-based models for sentiment classification on tweets have been explored. For instance, [22] use the BERT base model for the classification of positive and negative sentiments of Italian tweets based on the SENTIPOLC 2016 corpus. They achieve an F-score of 0.75. In [10], the authors propose a target-dependent BERT model for sentiment classification and achieve an accuracy of 77.31% on the pre-existing twitter dataset.

2.2   Sentiment Analysis on Coronavirus Related Tweets

Covid-19 caused an unprecedented impact worldwide, leading to public health emergencies, lockdowns, and other protective protocols. It is important to understand public opinions and concerns in such scenarios. Social media plays a critical role during such disease outbreaks. This section describes the existing literature that uses tweets related to coronavirus disease to perform sentiment analysis.

Several studies collect Twitter data using covid-related keywords. However, the need for annotated data for supervised classification is a limitation. Most studies use pre-existing tools or lexicons to label the collected tweets for further classification using a machine learning model. For instance, [4] use TextBlob and Afinn to label the tweets for supervised classification into positive, negative, and neutral categories. They propose a fuzzy rule-based model with 79% F1-score and compare bag-of-words and Doc2Vec using several classifiers, including Naive Bayes, Support Vector Machine (SVM), Ensemble models, multinominal, Bernoulli, and logistic regression classifiers. Similarly, [19] scrape tweets based on the keyword "coronavirus" and use VADER to label the tweets into positive, negative, and neutral. They use Long Short Term Memory (LSTM) and Artificial Neural Network (ANN). The neural network models achieve an accuracy of 84.5% and 76%, respectively. Authors in [25] crawl twitter data with keyword "COVID-19" using rapid miner tools and use Naive Bayes classification; Another study [13], extract tweets using keywords related to lockdown and annotate the data using TextBlob and Vader lexicons. They use eight different classifiers and achieve 84.4% as the best accuracy using LinearSVC classifier with unigram features; Similarly in [28], the authors use social distancing keywords to extract tweets and use SentiStrength to annotate the data. They further utilized SVM for sentiment classification with an accuracy of 71% for positive, negative, and neutral sentiments and 81% with only positive and negative tweets; Authors Villavicencio et al. [32] manually annotate tweets and employ Naive Bayes classification on English and Filipino tweets. Their model yield 81.77% accuracy.

Authors Sitaula and Shahi [30] use publicly available Nepali tweets dataset for sentiment classification. They use feature extraction with a multi-channel convolutional neural network on Nepali covid-19 related tweets, where they achieve an accuracy of 71.3%. Similarly, [15] train the deep learning model using the annotated Sentiment140 dataset. And then test the model with a new set of data related to covid tweets collected by the authors.

Research using Twitter data is most commonly used for real-time trend analysis. So, it is necessary to collect tweets based on specific keywords related to the event in consideration. Annotated data may not always be available in a similar domain. In such scenarios, the model may not be able to generalize for current new data, especially Twitter being the most active platform with millions of users. Also, existing works use word embeddings for deep learning-based classification. We collect Twitter data related to coronavirus disease using related hashtags. In this work we apply pre-processing for tweets explained in Sect. 3 and use the transformer based tokenizer (Sect. 4.1). We annotate data using RoBerTa base Twitter sentiment model [18]. Finally we present a deep learning-based framework for sentiment classification of text data.

3   Data Collection and Pre-Processing

Data Collection: The dataset for the experiments are tweets collected using the Tweepy library. We collect tweets using hashtags related to coronavirus between August 31, 2021 to November 26, 2021. The hahstags for the data collection is listed in the Table 1. Initial raw data collection was part of [24].

Table 1. Hashtags used in Data Collection.

Hashtags
#coronavirus, #covid, #COVID19, #corona,#pandemic, #coronaviruspandemic #staysafe, #washyourhands, #disinfectant, #handwashing, #mask, #ppe, #covidvaccine, #covidvaccination, #vaccine, #coronavirusvaccine, #boostershots, #boostershot, #pfizerbooster, #deltavariant, #coviddeltavariant, #SARSCoV2, #muvariant, #mu, #gammavariant, #gamma, #delta, #stayhome, #quarantine, #lockdown, #stayathome, #socialdistancing

Pre-processing: Pre-processing is an essential step in natural language processing applications. It is the process of converting the raw form of text into a form suitable for specific tasks and retaining the original information.

The most common form of text pre-processing is lowercase every single token of the input text. Here the tweet text is lowercased to avoid sparsity in the dataset. Following that, we utilize Python's Regular Expression library to locate the URLs, and user mentions in the tweets and replace them with the words url and user respectively. Hashtags are essential in tweets, as users express their opinions or feelings through hashtags. So, we tokenize and replace the hashtags by using Python's wordninja library. For example, the hashtag #stayhome will be converted to (stay home).

Twitter users also express their emotions towards a topic using emoticons (emojis). Since we build a sentiment analysis model, we consider the emotion expressed through emoticons. However, we cannot use emoticons as input to text classification models. We use Regular Expression and Python's demoji library to convert the emoticons into phrases. Finally, to reduce the noise of the tweets, stop words were removed by using Pythons Natural Language Toolkit (NLTK). We remove the stop words as they add little to no value to a text. Figure 1 shows the model pipeline in this work.

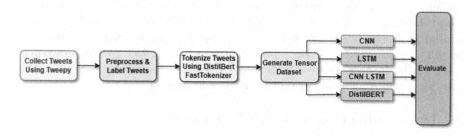

Fig. 1. Model framework.

Data Labeling: We present a supervised learning model framework in this work. We need labeled data for the classification model training and evaluation. We utilize the TextClassificationPipeline from Huggingface [33] to create a labeled dataset using the model [18]. The dataset now contains three class labels: positive, negative, and neutral.

Additional Pre-processing: Tweets are limited in the number of characters. In this process, we remove tweets with less than five words after initial pre-processing. In order to use a balanced dataset for the learning models, we perform random undersampling of the majority class (neutral and negative). Finally, we have 487,998 tweets in the dataset for use in the experiments. The sentiment distribution of this dataset is shown in Table 2. Then we shuffle the dataset to maintain the distribution of positive, negative, and neutral tweets across the train and test sets for further experiments.

Table 2. Sentiment distribution of tweets in the final dataset.

Class label	Number of instances
Positive	161,998
Negative	163,000
Neutral	163,000

4 Methodology

This paper presents a deep learning-based framework for classifying tweets into positive, neutral, and negative classes. Section 2.2 we see that most of existing studies on coronavirus related tweets are based on traditional machine learning models. Based on Sect. 2.1, where many deep learning models have proven to provide better results, in this study we experiment using Convolutional Neural Network (CNN), Long Short-Term Memory (LSTM), and Distiled version of Bidirectional Encoder Representation Transformer (DistilBERT) models. This section describes further text processing and the model architectures.

4.1 Text Tokenization and Padding

It is imperative that for the learning models to deal with text data, it needs some form of tokenization to convert the text to numerical representation. We tokenize the tweet text with FastTokenizer from the uncased DistilBERT model [27]. The tokenizer vocabulary contains 30,522 words. This tokenizer uses the concept of word piece tokenizer. The tweet text is of varying lengths. We use 100 as the maximum sequence length. The tokenized tweets are then truncated or padded for uniformity. These tokens are then mapped to their corresponding numerical IDs. Following that, the labels are one-hot encoded. The encoded label with corresponding tokenized tweets are converted into input pipelines by utilizing tensor flow tf.data.Dataset [20] API. This API creates an iterable dataset that is input to our text classification models.

4.2 Convolutional Neural Network Model (CNN)

Convolutional Neural Networks (CNN) is a type of feed-forward neural network that makes use of several hidden layers [3]. The hidden layers are typically the convolution, pooling, activation, dropout, and dense layers.

As shown in Fig. 2, the model starts with an input layer, followed by an embedding layer, three convolution layers with kernel sizes 3, 4, and 5, respectively, 32 filters, and relu as the activation function. These layers are followed

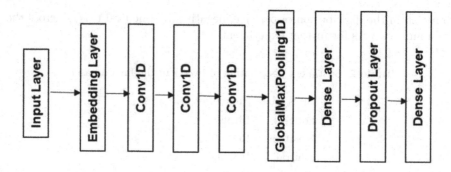

Fig. 2. CNN architecture.

by a 1D GlobalMaxPool layer, a Dense layer with 128 filters and relu activation function, a Dropout layer, and a final Dense layer with softmax activation to perform the classification into three classes. This model is compiled with the catagorical_crossentropy loss function and the Adam optimizer with a learning rate of 0.00001.

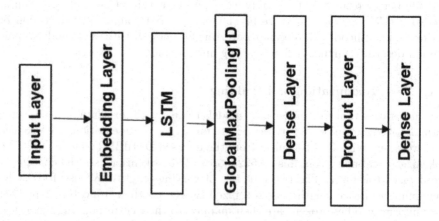

Fig. 3. LSTM architecture.

4.3   Long Short-Term Memory (LSTM)

Long Short-Term Memory (LSTM) [14] is a type of Recurrent Neural Network (RNN) that is sequential and uses long memory for activation functions in hidden layers. LSTM has the ability to handle the vanishing gradient problem present in RNN.

The structure (See Fig. 3) of the LSTM includes an input layer, an embedding layer, followed by an LSTM layer. The output of this LSTM layer is then pooled using the 1D global max-pooling layer and passed to a dense layer with 64 units and the relu activation function and a dropout layer. Finally, a dense layer with softmax activation function. Similar to CNN model, the LSTM model is also compiled with the catagorical_crossentropy loss function and the Adam optimizer with a learning rate of 0.00001.

4.4 CNN-LSTM

This model utilizes a 3-layer 1-dimensional CNN with three different kernel size (3, 4, and 5 respectively) and relu activation function with 128 filters and a single layer of the LSTM network. Followed by the pooling, dense layer with relu activation function, dropout layer and final dense layer with softmax activation function. Figure 4 shows this architecture where the input is directed to an embeding layer and then the layered CNN and LSTM. This model is compiled with the catagorical_crossentropy loss function, Adam optimizer, and with learning rate of 0.00001.

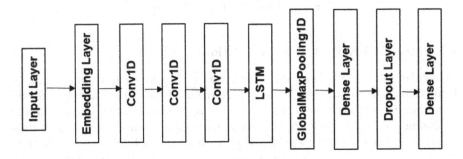

Fig. 4. CNN-LSTM architecture.

4.5 Distiled Bidirectional Encoder Representation from Transformer (DistilBERT)

Bidirectional Encoder Representation from Transformers (BERT) is a fine-tuning-based deep bidirectional language representation model. The BERT model is pre-trained on unlabeled data over different pre-training tasks [6]. In this study, we use DistilBERT, a general-purpose pre-trained version of BERT that retains 97% of the language understanding capabilities [27]. DistilBERT uses knowledge distillation during the pre-training phase to reduce the model's size. We fine-tune the model using the distilbert-base-uncased transformers model from Huggingface [33]. The model includes the input layer with input ids and attention mask, followed by the distilBERT main layer, global max pool layer, dropout and final dense layer. The model uses 0.00002 as learning rate with Adam optimizer and categorical_crossentropy loss function. (refer Fig. 5).

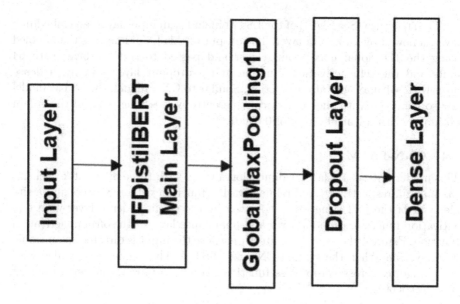

Fig. 5. DistilBERT architecture.

4.6   Stratified K-Fold Cross Validation

In regular k-fold cross-validation, the data is partitioned into approximately k equal folds. Train and validation is performed on the partitioned data for 'k' iterations. In each iteration, the k − 1 fold is used for training the model and the remaining one fold for validation. The accuracy obtained in each iteration is then averaged to get the overall model accuracy. In stratified k-fold cross validation, the data is arranged in such a way that each fold is a good representation of the whole dataset. In this paper, we use 5-fold stratified cross-validation. Further details are described in Sect. 5.

5   Experiments and Results

The proposed models are trained and tested on Google Colab. We use the dataset from Sect. 3 and apply the tokenization and padding from Sects. 4.1, for the models. To train and evaluate the models, we use stratified k-fold cross-validation with k = 5. The model performance is shown using the average validation accuracy of the final epochs from each of the five folds in the cross-validation. For example, the CNN-LSTM model was trained for ten epochs. The validation accuracy that is calculated on the 10th epoch is the one that was utilized to calculate the average validation accuracy.

As it can be seen in Table 3, the DistilBERT model achieved the highest accuracy of 91.46%. This may be due to the fact that attention masks were added to the input of the DistilBERT model along with the input ids. The attention masks allow the layers of the model to differentiate between the tokens that are the results of padding and the ones that are not. The next best-performing model was the CNN model with an accuracy of 85.17%, followed by the CNN-LSTM model with an accuracy of 84.30%. The LSTM model achieved an accuracy of 83.96%. It can be observed that CNN has better performance than LSTM as the two models with the least accuracies involve LSTM layers.

Table 3. Average accuracies and hyperparameters.

Model	Average accuracy	#Epochs
Convolutional Neural Network (CNN)	85.17%	20
Long Short Term Memory (LSTM)	83.96%	30
CNN-LSTM	84.30%	10
DistilBERT	91.46%	2

6   Conclusions

Public health emergencies provide a drastic change in peoples day to day activities. It is essential to monitor such situations to mitigate future emergencies. Social media platforms are major source of information disemination all over the world. People share their opinion and feeling in every aspect. Especially the pandemic and the non-pharmacheutical intervention heightened public emotions and distress. To capture people's emotions and to understand the spread of disease and public health protocols, it is important to process such information using computational models, especially because of the complex nature of the data. Natural language processing and sentiment analysis help us understand such public perceptions. We collected data using covid-related hashtags and presented a deep learning model framework for the sentiment classification of tweets into positive, negative, and neutral classes. We observe the best performance of 91.46% with the DistilBERT model. We use stratified 5-fold cross-validation, and the average accuracy obtained is shown in Table 3 for all the models. In this digital era, big data is obvious in every domain. Especially social media data, with millions of users all around the world. The transformer distilBERT model is computationally complex. In future we plan to integrate distributed computing using Spark for the transformer model to improve the computational time for big data sets.

References

1. Agichtein, E., Castillo, C., Donato, D., Gionis, A., Mishne, G.: Finding high-quality content in social media. In: Proceedings of the 2008 International Conference on Web Search and Data Mining, pp. 183–194 (2008)

2. Alamoodi, A., et al.: Sentiment analysis and its applications in fighting COVID-19 and infectious diseases: a systematic review. Expert Syst. Appl. **167**, 114155 (2021)

3. Albawi, S., Mohammed, T.A., Al-Zawi, S.: Understanding of a convolutional neural network. In: 2017 international Conference on Engineering and Technology (ICET), pp. 1–6. IEEE (2017)

4. Chakraborty, K., Bhatia, S., Bhattacharyya, S., Platos, J., Bag, R., Hassanien, A.E.: Sentiment analysis of COVID-19 tweets by deep learning classifiers-a study to show how popularity is affecting accuracy in social media. Appl. Soft Comput. **97**, 106754 (2020)

5. De Choudhury, M., Counts, S., Czerwinski, M.: Identifying relevant social media content: leveraging information diversity and user cognition. In: Proceedings of the 22nd ACM Conference on Hypertext and Hypermedia, pp. 161–170 (2011)

6. Devlin, J., Chang, M.W., Lee, K., Toutanova, K.: BERT: pre-training of deep bidirectional transformers for language understanding. arXiv preprint arXiv:1810.04805 (2018)

7. Dos Santos, C., Gatti, M.: Deep convolutional neural networks for sentiment analysis of short texts. In: Proceedings of COLING 2014, the 25th International Conference on Computational Linguistics: Technical Papers, pp. 69–78 (2014)

8. Drus, Z., Khalid, H.: Sentiment analysis in social media and its application: systematic literature review. Proc. Comput. Sci. **161**, 707–714 (2019)

9. Feldman, R.: Techniques and applications for sentiment analysis. Commun. ACM **56**(4), 82–89 (2013)

10. Gao, Z., Feng, A., Song, X., Wu, X.: Target-dependent sentiment classification with BERT. IEEE Access **7**, 154290–154299 (2019)

11. Go, A., Bhayani, R., Huang, L.: Twitter sentiment classification using distant supervision. CS224N Proj. Rep. Stanford **1**(12), 2009 (2009)

12. Goularas, D., Kamis, S.: Evaluation of deep learning techniques in sentiment analysis from Twitter data. In: 2019 International Conference on Deep Learning and Machine Learning in Emerging Applications (Deep-ML), pp. 12–17. IEEE (2019)

13. Gupta, P., Kumar, S., Suman, R., Kumar, V.: Sentiment analysis of lockdown in India during COVID-19: a case study on Twitter. IEEE Trans. Comput. Soc. Syst. **8**(4), 992–1002 (2020)

14. Hochreiter, S., Schmidhuber, J.: Long short-term memory. Neural Comput. **9**(8), 1735–1780 (1997)

15. Imran, A.S., Daudpota, S.M., Kastrati, Z., Batra, R.: Cross-cultural polarity and emotion detection using sentiment analysis and deep learning on COVID-19 related tweets. IEEE Access **8**, 181074–181090 (2020)

16. Jianqiang, Z., Xiaolin, G., Xuejun, Z.: Deep convolution neural networks for Twitter sentiment analysis. IEEE Access **6**, 23253–23260 (2018)

17. Liao, S., Wang, J., Yu, R., Sato, K., Cheng, Z.: CNN for situations understanding based on sentiment analysis of twitter data. Proc. Comput. Sci. **111**, 376–381 (2017)

18. Loureiro, D., Barbieri, F., Neves, L., Anke, L.E., Camacho-Collados, J.: TimeLMs: diachronic language models from Twitter. arXiv preprint arXiv:2202.03829 (2022)

19. Mansoor, M., Gurumurthy, K., Prasad, V., et al.: Global sentiment analysis of COVID-19 tweets over time. arXiv preprint arXiv:2010.14234 (2020)

20. Murray, D.G., Simsa, J., Klimovic, A., Indyk, I.: tf.data: a machine learning data processing framework. arXiv preprint arXiv:2101.12127 (2021)

21. Murthy, D.: Twitter. Polity Press Cambridge (2018)

22. Pota, M., Ventura, M., Catelli, R., Esposito, M.: An effective BERT-based pipeline for twitter sentiment analysis: a case study in Italian. Sensors **21**(1), 133 (2020)

23. Ranganathan, J., Hedge, N., Irudayaraj, A.S., Tzacheva, A.A.: Automatic detection of emotions in Twitter data: a scalable decision tree classification method. In: Proceedings of the Workshop on Opinion Mining, Summarization and Diversification, pp. 1–10 (2018)

24. Ranganathan, J., Tsahai, T.: Analysis of topic modeling with unpooled and pooled tweets and exploration of trends during COVID. Int. J. Comput. Sci. Eng. Appl. (IJCSEA) **11** (2021)

25. Ritonga, M., Al Ihsan, M.A., Anjar, A., Rambe, F.H., et al.: Sentiment analysis of COVID-19 vaccine in Indonesia using naïve bayes algorithm. In: IOP Conference Series: Materials Science and Engineering, vol. 1088, p. 012045. IOP Publishing (2021)

26. Rouvier, M., Favre, B.: SENSEI-LIF at SemEval-2016 task 4: polarity embedding fusion for robust sentiment analysis. In: SemEval@ NAACL-HLT, pp. 202–208 (2016)

27. Sanh, V., Debut, L., Chaumond, J., Wolf, T.: DistilBERT, a distilled version of BERT: smaller, faster, cheaper and lighter. ArXiv abs/1910.01108 (2019)

28. Shofiya, C., Abidi, S.: Sentiment analysis on COVID-19-related social distancing in Canada using Twitter data. Int. J. Environ. Res. Public Health **18**(11), 5993 (2021)

29. Singh, R., Singh, R., Bhatia, A.: Sentiment analysis using machine learning technique to predict outbreaks and epidemics. Int. J. Adv. Sci. Res **3**(2), 19–24 (2018)

30. Sitaula, C., Shahi, T.B.: Multi-channel CNN to classify nepali COVID-19 related tweets using hybrid features. arXiv preprint arXiv:2203.10286 (2022)

31. Stojanovski, D., Strezoski, G., Madjarov, G., Dimitrovski, I.: Twitter sentiment analysis using deep convolutional neural network. In: Onieva, E., Santos, I., Osaba, E., Quintián, H., Corchado, E. (eds.) HAIS 2015. LNCS (LNAI), vol. 9121, pp. 726–737. Springer, Cham (2015). https://doi.org/10.1007/978-3-319-19644-2_60

32. Villavicencio, C., Macrohon, J.J., Inbaraj, X.A., Jeng, J.H., Hsieh, J.G.: Twitter sentiment analysis towards COVID-19 vaccines in the Philippines using naïve bayes. Information **12**(5), 204 (2021)

33. Wolf, T., et al.: Transformers: state-of-the-art natural language processing. In: Proceedings of the 2020 Conference on Empirical Methods in Natural Language Processing: System Demonstrations, pp. 38–45. Association for Computational Linguistics (2020). https://doi.org/10.18653/v1/2020.emnlp-demos.6, https://aclanthology.org/2020.emnlp-demos.6

34. Zhang, S., Zhang, X., Chan, J.: A word-character convolutional neural network for language-agnostic twitter sentiment analysis. In: Proceedings of the 22nd Australasian Document Computing Symposium, pp. 1–4 (2017)

Cyber Attack Detection in IoT Networks with Small Samples: Implementation And Analysis

Venkata Abhishek Kanthuru[1,2,3], Sutharshan Rajasegarar[2,3(✉)],
Punit Rathore[4], Robin Ram Mohan Doss[2,3], Lei Pan[2,3], Biplob Ray[5],
Morshed Chowdhury[2,3], Chandrasekaran Srimathi[1], and M. A. Saleem Durai[1]

[1] VIT Vellore, Vellore, India
[2] Centre for Cyber Security Research and Innovation (CSRI), Waurn Ponds,
Australia
[3] Deakin University, Geelong, Australia
srajas@deakin.edu.au
[4] Indian Institute of Science, Bangalore, India
[5] Central Queensland University, Melbourne, Australia

Abstract. Securing Internet of Things networks from cyber security attacks is essential for preventing data loss and safeguarding backbone networks. The resource-constrained nature of the sensor nodes used in the IoT makes them vulnerable to various attacks. Hence, it is important to monitor network traffic information to accurately and promptly identify threats. In this paper, using a machine learning-based framework for learning and detecting such attacks in an IoT network from the network data is proposed. Further, a real IoT network consisting of Raspberry Pi sensor nodes and ZigBee communication modules is built for implementing two cyber attacks. The network traffic information for normal and attack scenarios is collected to evaluate the attack detection performance of learning-based models. We performed a comparison analysis with deep learning and traditional machine learning models. Our evaluation reveals that the proposed features and the machine learning framework can detect attacks with high accuracy from the network traffic information. In particular, the triplet network-based deep learning framework showed promising results in efficiently detecting the attacks from the traffic information with merely a small set of training samples.

Keywords: Attack detection · IoT networks · Triplet network · ZigBee

1 Introduction

Internet of Things (IoT) has become an important network for sensing, communication, and analysis purposes. The total number of IoT connections is expected to reach 25 billion by 2025 [13] in the world. The global IoT solutions and services market is expected to grow to USD 278.9 billion by 2024 [13]. However, the

W. Chen et al. (Eds.): ADMA 2022, LNAI 13725, pp. 118–130, 2022.
https://doi.org/10.1007/978-3-031-22064-7_10

resource-constraint nature of the IoT devices prevents the use of sophisticated security tools for securing the networks, leaving them vulnerable to security attacks. A malicious attacker can easily capture one or more of the IoT nodes in the network and launch attacks on other nodes in the same network or use it as a vehicle to reach the backbone cloud network to steal valuable information or impose attacks. Hence, it is important to monitor and detect the attacks that emerge in the IoT network accurately and promptly. The challenge here is how to detect these attacks from the network traffic information collected accurately.

To detect the attacks from the traffic information in an IoT network, a data-driven model is required to learn the normal behavior of the network and detect the attacks from the traffic data that deviate from the normal behavior. In this work, we propose using machine learning-based methods to learn normal behavior patterns and detect traffic attacks.

A real IoT network consisting of several resource-constrained sensor nodes is implemented in this work. The network is built using Raspberry Pi 3B+ devices connected to XBee S2C Zigbee radio frequency modules. One of the nodes is configured as an attacker, which can send trusted messages to the other nodes, including the coordinator. It makes detecting the attacker difficult as conventional security mechanisms rely on trust mechanisms to verify the node's identity and accept a message. To overcome this shortcoming, we have to primarily rely on the message patterns present in the communication of the nodes to weed out any malicious nodes. To identify the patterns, we log certain network flow information about each packet sent or received by the node. These logs are subsequently used to learn the model and discern any patterns in the communication.

To identify a suitable machine learning model for our detection engine, we built supervised and unsupervised algorithms, including deep learning methods for performance comparison. Our evaluation reveals that the proposed features and the machine learning models provide promising results in detecting the attacks from the network traffic information. In particular, a triplet network-based deep learning framework can detect these attacks with high accuracy, albeit requiring a low number of samples for learning the model accurately. This feature is vital in the case of attacks on IoT devices, as we often need to detect the attack as quickly as possible with small data available in the network. Next, we survey the literature on security attacks relating to IoT networks.

2 Related Work

The importance of IoT security has been emphasised in several studies and applications [2,6]. In [4], the use of Artificial Neural Networks (ANN) at IoT gateways is investigated to detect anomalies in an IoT network. Around 4000 data are collected centrally from the edge devices (10 Arduino Unos connected to ESP8266 WiFi modules and temperature sensors) at the gateway (Raspberry Pi 3B+) node. The features used include device ID, sensor value, and the timestamp of the data transmission. A random subset of these samples was used with an ANN and showed achievement of 99% accuracy with 1% false-negative rate.

In [18], a ZigBee-based IoT network was built using only three Raspberry 3B+ based sensor nodes. The sensors were programmed to send a packet to the coordinator every 35 s, containing a randomly generated number. After the message transmission, the sensor device enters sleep mode to reduce power consumption. The attacker node begins its attack by first scanning the network to identify the target device. Then, a malicious packet containing AT (attention) command is sent to a chosen target device on the network. By examining the packet data captured, it is verified that the attack worked and did not affect the other nodes of the network. In [14], a relatively larger network was implemented using ZigBee-based sensor nodes, and several attacks using AT commands were analyzed. However, a learning-based mechanism to detect those attacks was not discussed. The above analysis of the existing work on IoT systems reveals that a learning-based mechanism is more suitable than the existing solutions to detect and protect the IoT devices and the backbone network. Below we detial our implementation and evaluation.

3   System Architecture

Here, we present our IoT network realised using Raspberry Pi-based sensor nodes and the attacks implemented. The network traffic logs are collected for two scenarios, including normal and attack scenarios. Several machine learning-based models, including deep learning, are used to identify the attack from the network traffic.

3.1   Network Topology

In order to collect real data from IoT network with different real attack scenarios, we have created an IoT network testbed consisting of 11 sensor nodes. The network traffic data are collected both during the normal operation (i.e., without any attack) of the network, as well as when the attacks are carried out (using another sensor node) on to the IoT testbed. Figure 1a shows the hierarchical network topology of the implemented IoT testbed. Note that the attacks implemented and the network traffic-based attack detection mechanisms performed in this work are generic so that they can be readily generalized for use with other network typologies/configurations. The network has been realised using Raspberry Pi 3B+ devices as sensor nodes. Each Raspberry Pi node is connected with a Digi XBee S2C Zigbee Radio Frequency module to enable wireless communication using the ZigBee protocol [9]. The network formulation and the wireless communication by the ZigBee modules are controlled using a Python program, utilizing the API (application programmer interface) provided by Digi [8]. Figure 1 shows the sensor nodes used in our IoT network.

The ZigBee protocol specifies the network configurations, such as the node categories, network formation, node joining and leaving mechanisms, and packet transmission protocol. Each node in the network can be configured as one of the

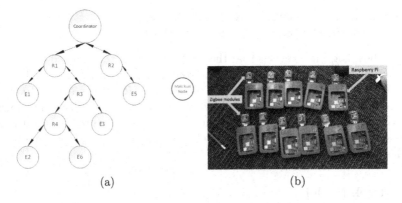

(a)   (b)

Fig. 1. (a) Network topology (b) Raspberry Pi 3B+ & ZigBee modules

three types of nodes—(i) coordinator, (ii) router, or (iii) end device. The coordinator node is usually the topmost node in the hierarchical network topology because it functions as a gateway to an external network or servers. The router node allows the data from one node to be communicated to another via itself (re-routing), enabling multi-hop communication and helping the network scale to a larger geographical area. The end nodes are positioned as the leaf nodes of the network topology. The end nodes communicate their data to the coordinator via router nodes. The implemented hierarchical network in this paper, as shown in Fig. 1a, has one coordinator (the topmost node), four routers {R1, R2, R3, R4}, five end devices {E1, E2, E3, E5, E6}, and a malicious node (shown in red). The malicious node is configured as a router and programmed to launch attacks in the network.

Data Logging Format: Each sensor device (end devices, routers, and coordinator) has been programmed to log the information of every message it receives and transmits. For instance, when sending a packet, an end device logs the XBee ID and Send-Timestamp information; when acknowledging, the end device logs the XBee ID, Send-Timestamp, and Status-Code. When receiving a packet, the router logs the Recvd-data, Timestamp, ID − > Dest.ID, and Timestamp. When receiving, the coordinator logs the Recvd-data and Timestamp.

Each node's logs can be collected by accessing the nodes individually. However, to automate the log collection process in our implementation, we programmed each node to send every log generated as a message (or appended to the routed message) to the coordinator. This setup will result in the coordinator receiving each message containing the packet progress information via the different nodes in the network along the route. In essence, the message received by the coordinator will reveal the complete information about where the packet originated from, which intermediate routes it has taken to reach the coordinator, and the date and time information. An example format of the resulting messages logged at the coordinator is shown below:

```
{E1 > R1, 18:15:10.409417, 18:15:10.524533,
 R1 > C, 18:15:10.558810, 18:15:10.779988}
```

The above log entry reveals that the packet has been transmitted from an end device E1 at time 18:15:10.409417 to the router R1, which was received at time 18:15:10.524533. Then the message was re-sent from R1 at time 18:15:10.558810 to coordinator C, and it was received by C at time 18:15:10.779988. The logging process at the central (coordinator) node is not required in practice, as it can be accessed from individual nodes separately. Here it is used for convenience and easy log collection purposes.

3.2   Attack Model

In a ZigBee network, some network parameters at a particular node can be configured from another remote node in the same network, i.e., nodes having the same network PAN-ID (personal area network ID). A mechanism called *AT commands* (attention commands) can be used to perform such configuration changes remotely. These AT commands are defined by the Zigbee protocol, which allow nodes to carry out specific configuration operations. In this work, these AT commands are exploited to launch security attacks on the network. We implemented the following two variants of redirection attacks in the network utilizing AT commands.

- Redirection of packets towards another node or an attacker node
- Redirection to an invalid node address (black hole attack)

Redirection attacks redirect the target's traffic to a new parent node. The new parent node can be selected to maximize the distance the packets need to travel, i.e., increasing the number of hops required, thereby increasing the power required to transmit the message. The other variant of the redirection is to maximize the number of connected children nodes. If many nodes are forced to connect to the same parent, the parent will be overwhelmed, and after a certain number of connections, it cannot accept anything new. The redirection attacks also function as a battery depletion attack, as the nodes need to keep forwarding the same data or send the data to a further node which requires an increase in signal power in the network.

Figure 2 shows a few attack examples. Figure 2a shows how the attacker redirects a target's communication to a new node. The malicious node sends an AT command to node E3, setting its new parent as R1. Any new messages will now be sent to R1 directly instead of via R3. Figure 2b shows what happens when an attacker sets the parent of the target node to an invalid address using AT commands. Node E3 that now tries to send a message to the new parent address will fail and will not be able to send any new messages successfully.

Data Collection: The two attacks described above have been implemented, and the network traffic data are collected, as explained earlier. These data/logs are labeled as "attack". To allow the model to learn the normal traffic behavior, we

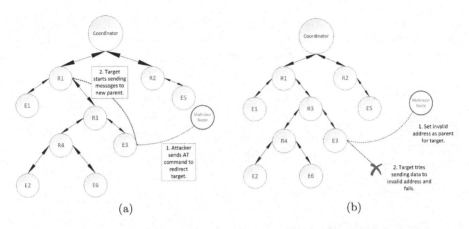

Fig. 2. Redirection attacks: (a) redirecting to a new parent node (b) redirecting to an invalid parent node address

have also collected data while allowing the network to operate under normal conditions, i.e., non-attacking conditions. These logs are labeled as "normal".

We aim to evaluate how these security attacks can be accurately detected by monitoring the network traffic logs alone. We propose to use machine learning models to learn network traffic patterns. We assess the performance of both supervised and unsupervised machine learning-based techniques to detect the attacks. Next, we introduce our attack detection system.

4   Threat Detection System

In the IoT testbed, the messages are created at each node at randomized time intervals, between 1 and 5 s, before they are sent to the coordinator. The logs are collected from the network for approximately 36 min, that is, 18 min for attacking and 18 min for non-attacking scenarios. The traffic logs are labeled as either 0/1 (Normal/Attack). The collected data are split into windows of 5 s to derive the traffic features. Various features are computed using the data in each window, including number of unique first destinations, average packet delay, average first-hop delay, average hops, average packet size, packet count, packet delays (1st quartile-Q1, 3rd quartile-Q3, interquartile range - IQR), 1st hop delays (Q1, Q3, IQR), and Shannon entropy.

5   Modelling the Traffic Data and Evaluation

Classical Machine Learning Models and Deep Learning Models: In order to model the network traffic patterns and detect the attacks, supervised and unsupervised machine learning techniques, including deep learning algorithms, are implemented. Normal and attack/anomalous traffic data are used to train the supervised algorithms. For the unsupervised algorithms, such as one-class methods, only the normal traffic data are used for training. Table 1 shows the implemented algorithm and the results in summary.

Table 1. Evaluation using several machine learning (ML) methods

ML	TN-CNN	SVM	XGBoost	RF	TN-DNN	DNN	K-M	LSTM-AE	LSTM-En-1SVM	1SVM
Accuracy (%)	**99.40**	98.26	98.17	98.09	97.26	93.60	79.75	95.00	95.00	80.50
F1-score (%)	**98.00**	98.00	98.00	98.00	97.00	93.00	79.00	92.00	92.00	80.00

5.1 Results and Discussion

Our evaluation aims to assess the various machine learning-based models' ability to detect IoT attacks. Accuracy and F1-score are used as the metrics to report the results. In summary, the attack detection performances obtained for both supervised and unsupervised schemes are given in Table 1. The F1-scores tabulated are the averages of the weighted average F1-scores. It shows that all the models have achieved high detection performances. Among the supervised classifiers, the triplet network with CNN has achieved the highest accuracy with 99.40 %. Among the unsupervised models, both LSTM-based schemes have achieved the highest accuracy of 95%. We give detailed analysis of the different algorithms used in our evaluation and the results.

5.2 Supervised Methods

Triplet Network. The triplet-based deep learning network uses similarities between input triplets to learn the network parameters. A loss function, called the triplet loss, is used to guide the learning of the embedding. Any deep learning framework, such as CNN (convolutional neural network), LSTM (Long short term memory) and DNN (dense neural network/fully connected network) can be used as the base model for forming the triplet network (Fig. 3a). The aim is to reduce the distance of similar inputs in the latent space to form well-separated clusters. The architecture is "Online" [16], when the model's loss is computed on the fly by forming triplets from the training data embeddings. The goal is to reduce the intra-label embeddings' distances while increasing the inter-label embeddings' distances. The loss L of a triplet (a, p, n) can be defined as $L = max(d(a, p) - d(a, n) + margin, 0))$, where a is the anchor object, p is the positive object (similar to a), n is the negative object (dissimilar to a) and $d(\cdot)$ is the distance function. The margin is a constant positive number. To improve network performance, the selection of negative class objects is important. Based on the selection mechanism, the selected negatives are called hard, semi-hard, or easy negatives. Hard and semi-hard negatives mostly give better performance with good discriminating capabilities than easy negatives. In this work, we experimented with both methods and reported the one that provided the best results.

For our experiments, we implemented two varients of the triplet networks, namely CNN-based and DNN-based triplet networks, and we denote them as *triplet network - CNN (TN-CNN)* and *triplet network - DNN (TN-DNN)* based methods, respectively.

Triplet Network - CNN Based Method (TN-CNN): We first convert the input data into images to train the TN-CNN, as CNN works well with the image data. Note that the collected network traffic data are time-series data. We generated recurrence plots (images) from the time series data, considering windows of 32 measurements each at a time, producing 32×32 image for each window of measurements. Each feature of the dataset is converted into a channel of recurrence plots. The image's label is set as "attack" if at least one measurement within the window is labeled as "attack". The base model, trained using triplet loss with semi-hard negatives, comprises the following configuration.

– CNN (64, (2,2), Relu): Max Pool (2,2): Dropout 30%: CNN (32, (2,2), Relu)
– Max Pool (2,2): Dropout 30%: Flatten Layer: Dense (256): L2 Normalization

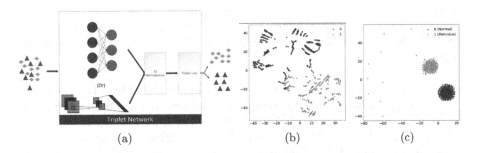

(a)	(b)	(c)

Fig. 3. (a) Triplet Network (b) t-SNE plot: TN-CNN (c) t-SNE plot: TN-DNN

We trained the base model for 100 epochs and generated a t-SNE plot of the training images' embeddings, as shown in Fig. 3c. It can be observed that the embeddings are mostly well separated and compact, indicating good classification performance. The embeddings of the training images are then fed as input to train the classifier model, consisting of a single Dense Layer with Sigmoid activation. The trained model detected the security attacks with an accuracy of 99.40% on the test image set. This is the highest accuracy obtained among the other machine learning models we compared in our evaluation.

Triplet Network - DNN Based (TN-DNN; Online Hard Triplets). Another variant of the triplet network is the triplet Network - DNN-based (TN-DNN). The base model is formed using a dense neural network (DNN) and is trained with the triplet loss using hard negatives. The base model comprises the following layers:

– Dense Layer (256, Relu): Dense Layer (128):L2 Normalization layer.

We use an Adam optimizer with a learning rate of 0.001 for optimization. We use a large batch size of 1024 to ensure a sufficient number of triplets are obtained to compute the loss. On training the model for 250 epochs, we got the embeddings as shown in the t-SNE plots in Fig. 3b. The generated embeddings are subsequently used to train a single layer DNN with binary cross-entropy loss and Sigmoid activation. The model is trained for 5000 epochs with a batch size of 128. The trained model achieved a mean accuracy of 97.26%. The mean accuracy is computed by training the model five times on different splits of the dataset. This DNN-based triplet network has shown lower accuracy than the CNN-based triplet network. The t-SNE plots show the embeddings obtained from these networks. Note that this model (TN-DNN), with parts of it trained using online hard triplet loss, performs much better than the DNN model alone, which was trained entirely using binary cross-entropy (mean accuracy of 93.60%).

SVM Support vector machine (SVM) [7] is a supervised classifier that classifies the data by constructing a hyperplane that can separate the classes at a higher dimensional space to which the data from the input space are mapped via the kernel functions. The optimal kernel and their corresponding parameters are found using 5 fold cross-validation technique. The best parameters obtained for our SVM are the best Kernel: RBF and the C parameter value of 100. According to Table 1, the SVM classifier has achieved the second-highest accuracy of 98.26%.

XGBoost XGBoost [5] is a supervised classifier implemented based on the gradient boosting principle. The optimal hyper parameters obtained from the training are: column sample by tree = 0.7, ETA=0.15, Gamma = 0.1, Maximum depth = 8, Minimum child weight = 1, and Objective function = binary:logistic. The XGBoost has achieved the third highest classification accuracy of 98.17%.

Random Forest (RF) is an ensemble of decision trees based algorithm [3] . The obtained feature importance scores are shown in Fig 4a. It shows that the feature first quartile (Q1) of packet delay is the most influential feature in classifying the attack, followed by the interquartile range (IQR) of packet delay and IQR of first-hop delay. RF achieved a detection accuracy of 98.09% on the test data.

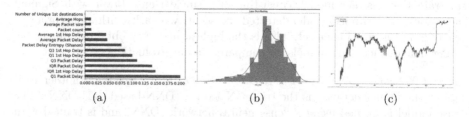

Fig. 4. (a) Feature ranking from RF (b) LSTM-AE: training loss distribution (c) LSTM-AE: MAE and threshold (red line)

K-Means Clustering (K-M) [12] finds K groups of similar points in a given data, and used as a supervised classification model to detect attacks, where K is the number of clusters. First, the training data are grouped into K clusters. Then the clusters are labeled as normal or attack (anomaly) using the label of the majority of the data points that fall inside the cluster. These labeled clusters are then used with the test data to classify them as normal or attack. The elbow method is used to choose the best K value, which is 11. This scheme achieved a detection accuracy of 79.75%, which is the lowest among all the schemes used.

Dense Neural Network (DNN). The DNN model used has the following configuration:

– Dense Layer (10, Relu): Dense Layer (8, Relu): Dense Layer (1, Sigmoid).

Adam is used as the optimizer, and the binary cross-entropy is used as the loss function. The DNN has achieved an accuracy of 93.60% for the test data.

5.3   Unsupervised Methods

LSTM-AutoEncoder (LSTM-AE). The LSTM autoencoder [17] model learns by compressing the input data (time series/sequence) signals and recreating it. During the training process, the recreated signals are compared against the original input data, and the difference between them is minimized. Any signal resulting from malicious/attack activity results in an output signal having a much higher mean absolute error (MAE) than the normal activity. It allows us to classify the input signal as "benign" (normal) or "malicious" (attack). Before the data can be presented to the model, they need to be converted to a time series having the form [input samples, time steps, features]. In our experiments, we have converted the data with 50 unit time lags for each feature, resulting in 663 input features. The classifier model comprises the following layers:

– LSTM (650, Relu): LSTM (64, Relu): Repeat Vector (64)
– LSTM (64, Relu): LSTM (650, Relu): Time Distributed Dense Layer (663)

The first two LSTM layers generate the input signal's compressed representation (at the latent space). This compressed signal is repeated and stretched back to the original input dimensions by the following LSTM and Time Distributed layers. This recreated signal is compared with the original to compute the loss MAE. All the layers are trained to try to recreate the normal activity signals so that anytime a signal associated with the malicious activity is fed, the output signal obtained will be far off from the input signal. We can classify the signals as malicious using the training data samples by setting a threshold on the loss MAE. The LSTM autoencoder uses the Adam optimizer and the mean absolute error (MAE) as the loss function. The loss distribution obtained for our data is shown in Fig. 4b. To identify the attack (malicious) traffic, we need to define a threshold for the loss. We set the threshold as the mean of loss MAE plus two times standard deviations of the loss MAE. The model achieved a detection accuracy of 95% and a weighted F1-Score of 92%. The plot of the losses associated with training and test data, with the threshold, is shown in Fig. 4c. It reveals that the selected threshold was able to separate the attack traffic from most of the normal traffic.

LSTM-Encoder One Class SVM (LSTM-En-1SVM). This model [11] is constructed by removing the decoder part of the LSTM-Autoencoder (explained in the above subsection) and connecting the model (encoder only) to a one-class support vector machine (1SVM), similar to the one presented in [10]. Before the encoder is connected to the one-class SVM, it is trained to accurately "encode" the input signal. This training is done similarly to how the LSTM Autoencoder is trained. The output of the encoder for the training data (normal traffic) is stored and used for training the one-class SVM. Once the training is complete, the model can label whether the encoded input signal corresponds to normal or malicious traffic. The LSTM-En-1SVM model has achieved a detection accuracy of 95% and a weighted F1-score of 92%. The model configuration is as follows:

– LSTM (650, Relu): LSTM (64, Relu): Repeat Vector (64): One-class SVM.

One-Class SVM (1SVM). This uses only one class data and label, particularly normal data, to learn the model. Any data that falls outside the pattern of the learned class are identified as anomalous. We used a plane-based One-Class SVM [15]. To find the optimal kernel and the parameter values, a grid search is performed with five-fold cross-validation, and the obtained values are: $C = 1$, $\nu = 0.1$, and the best kernel is RBF. 1SVM has achieved an accuracy of 80.5%.

5.4   Comparison with Relatively Larger Dataset

The above analysis used the small data collected from our IoT network with 11 sensor nodes to train the model for attack detection. The results reveals that the triplet network provides better performance among all the schemes. In order to test the performance of the triplet networks on a relatively larger dataset, we used a recent IoT dataset called IoT-23 [1] to further evaluate the triplet network. The IoT-23 dataset comprises network traffic data from 23 attack flow scenarios. Due to the large dataset size, we have sampled each of the 23 attack flow files and labeled the resulting flows as either "benign" (normal) or "attack", to enable binary classification. The total number of flows (malicious and normal) in the sampled dataset is 183,131. For the CNN-based triplet network (with online Semi-Hard Triplets), the training set used comprised 5,000 recurrence plots (2,500: "normal", 2,500: "malicious"). The testing set comprised 173,945 recurrence plots. Only the top 4 features were used to generate the recurrence plots, selected using a random forest feature selector, namely source & destination port, number of originator IP bytes, and number of originator packets sent. It reduced the computational space requirements. For the DNN-based triplet network (with online Semi-Hard Triplets), the training set comprised 5,130 samples (2,565 "normal", 2,565 "malicious"), and the testing set comprised 168,481 flows. Here, all the 11 features are included. The accuracy and F1-score obtained for the Triplet network-DNN based are 95.69% and 96%, respectively, and for the Triplet network-CNN based are 91.40% and 92%, respectively. Using a smaller training set (around 5,000 samples), the results demonstrates that the triplet network can learn from small data yet show high accuracy. The triplet network achieved above 91% accuracy (for both variants- CNN- and DNN-based Triplet networks). These results are significant since they show that the triplet network can be used for learning from minimal data and detecting attacks. When dealing with attacks in IoT networks, the objective is to detect the attack in the network as quickly as possible, and the number of samples available for new attack scenarios will be a few in practice. Hence, these results support the triplet network-based framework as a feasible system for detecting such attacks.

6 Conclusion

Securing IoT from attacks is important to safeguard backbone networks. A real IoT network consisting of Raspberry Pi sensor nodes and ZigBee communication modules is implemented. Two attacks were launched in the network by exploiting the AT commands feature of the ZigBee protocol. The network traffic data were collected for attack and normal scenarios, and a set of features was constructed to enable attack detection. The use of several classical machine learning and deep learning methods were analyzed for modeling and detecting the attacks. Evaluation reveals that the features used along with the machine learning models provide promising results in detecting the attacks from the network traffic. In particular, the Triplet network-based models perform well in detecting the attacks, even learning from small data samples.

References

1. Parmisano, A., Garcia, S., M.J.E.: Stratosphere laboratory - a labeled dataset with malicious and benign IoT network traffic. (2020).https://www.stratosphereips.org/datasets-iot23
2. Ashraf, I., et al.: A survey on cyber security threats in IoT-enabled maritime industry. IEEE Trans. Intell. Transp. Syst. 1–14 (2022)
3. Breiman, L.: Random forests. Mach. Learn. 45(1), 5–32 (2001)
4. Canedo, J., Skjellum, A.: Using machine learning to secure IoT systems. In: Proceeding of the 14th Conference on Privacy, Security and Trust, pp. 219–222. IEEE (2016)
5. Chen, T., He, T., Benesty, M., Khotilovich, V., Tang, Y.: Xgboost: extreme gradient boosting. R package version 0.4-2, pp. 1–4 (2015)
6. Chowdhury, M., Ray, B., Chowdhury, S., Rajasegarar, S.: A novel insider attack and machine learning based detection for the internet of things. ACM Trans. IoT 2(4), 1–23 (2021)

7. Cortes, C., Vapnik, V.: Support-vector networks. Mach. Lear. **20**(3), 273–297 (1995)
8. DigiXBee: Python library (2022). https://xbplib.readthedocs.io/en/latest/
9. DigiXBee: Zigbee modules (s2c) (2022). http://www.digi.com/resources/documentation/digidocs/pdfs/90001500.pdf
10. Erfani, S.M., Rajasegarar, S., Karunasekera, S., Leckie, C.: High-dimensional and large-scale anomaly detection using a linear one-class SVM with deep learning. Pattern Recogn. **58**, 121–134 (2016)
11. Ergen, T., Kozat, S.S.: Unsupervised anomaly detection with LSTM neural networks. IEEE Trans. Neural Net. Learn. Syst. **31**(8), 3127–3141 (2019)
12. MacQueen, J.: Classification and analysis of multivariate observations. In: 5th Berkeley Symposium Mathamatical Statistics Probability, pp. 281–297 (1967)
13. Markets: IoT solutions & markets (2020). https://www.marketsandmarkets.com/Market-Reports/iot-solutions-and-services-market-120466720.html
14. Piracha, W.A., Chowdhury, M., Ray, B., Rajasegarar, S., Doss, R.: Insider attacks on Zigbee based IoT networks by exploiting AT commands. In: Shankar Sriram, V.. S.., Subramaniyaswamy, V.., Sasikaladevi, N.., Zhang, Leo, Batten, Lynn, Li, Gang (eds.) ATIS 2019. CCIS, vol. 1116, pp. 77–91. Springer, Singapore (2019). https://doi.org/10.1007/978-981-15-0871-4_6
15. SchölkopfÜ, B., Williamson, R.C., SmolaÜ, A., Shawe-Taylory, J.: SV estimation of a distribution's support. Adv. Neural Inf. Process. Syst **41**, 582–588 (2000)
16. Schroff, F., Kalenichenko, D., Philbin, J.: Facenet: a unified embedding for face recognition and clustering. In: CVPR, pp. 815–823 (2015)
17. Sutskever, I., Vinyals, O., Le, Q.V.: Sequence to sequence learning with neural networks. In: NIPS, pp. 3104–3112. NIPS'14, MIT Press, USA (2014)
18. Vaccari, I., Cambiaso, E., Aiello, M.: Remotely exploiting at command attacks on zigbee networks. Secur. Commun. Netw. 1–9 (2017)

SATB: A Testbed of IoT-Based Smart Agriculture Network for Dataset Generation

Liuhuo Wan[1], Yanjun Zhang[1,2], Ruiqing Li[1], Ryan Ko[1], Louw Hoffman[1], and Guangdong Bai[1(✉)]

[1] The University of Queensland, Brisbane, Australia
g.bai@uq.edu.au
[2] Deakin University, Melbourne, Australia

Abstract. Agriculture has seen many revolutions since the booming Internet of Things (IoT) was embedded to enable the *smart agriculture* (SA) scenarios. SA integrates end devices, gateways and clouds to digitalize and automate traditional farming methods. Due to the open deployment and wide range accessibility, SA systems face a new attack surface that may lead to security and privacy concerns. It is expected that the cyber security and data science research communities will set off on constructing advanced technologies to safeguard this critical infrastructure, e.g., data-driven protection and AI-enabled defense.

In this work, we set up an SA testbed named SATB that can facilitate SA dataset generation. SATB is designed to be extensible so that it is capable of incorporating sensors (e.g., SenseCAP sensors) and protocols (e.g., LoRaWAN) that are extensively adopted in real-world SA systems. To test the usability of SATB, we use it to create a comprehensive SA network dataset for research use. With SATB, our dataset can capture data that rigorously covers the whole lifecycle of SA scenarios, from the *authentication* stage to the *runtime functioning* stage. We design five typical test cases, and SATB can generate network traces based on them. SATB also supports generating attack traces of network reconnaissance and vulnerability scanning. We show the details of our dataset collection process on SATB and conduct a preliminary statistical analysis, to enlighten potential smart use of our testbed. The collected dataset is released online to facilitate related research: https://github.com/UQ-Trust-Lab/2022-SATB.

Keywords: Smart agriculture · LoRaWan · Network · Testbed

1 Introduction

The recent booming Internet of Things (IoT) has penetrated many industries, from automotive, home automation, manufacturing to agriculture and cities. Agriculture, one of the most important activities of human civilization, is a typical area that IoT has introduced revolutions. Thanks to the digitalization technologies and innovations, we have entered an era of *smart agriculture* (SA)

© The Author(s), under exclusive license to Springer Nature Switzerland AG 2022
W. Chen et al. (Eds.): ADMA 2022, LNAI 13725, pp. 131–145, 2022.
https://doi.org/10.1007/978-3-031-22064-7_11

or *Agriculture 4.0*. Integrating connected devices with cloud and mobile applications, SA scales up production processes while consuming less natural resources. It has reshaped those traditional processes from managing cattle, monitoring climate conditions, irrigating farms to harvesting crops, greatly enhancing automation, efficiency and sustanability. A recent study reports that the IoT-enabled SA market reached 11.20 billion USD in 2021 and is expected to reach 26.65 billion in 2030 [3].

As one of the emerging paradigms that incorporate IoT into a traditional domain, SA has demonstrated its similarity in the design of the architectures and workflows to other domains. Nonetheless, it has a wider attack surface that leads to new cyber threats that have not been well studied in other IoT systems. First, in order to enable a longer range of wireless communication, SA systems usually support LoRaWAN [4] that enables miles of long-range communications. Unlike in smart home systems where Bluetooth and ZigBee are mainly used for communications and wireless signals are restricted within a narrow and trustworthy range (e.g., 10 m for Bluetooth), an attacker can easily capture or inject wireless traffic to gain credentials or hijack into established communications. Second, end devices in SA, e.g., sensors and actuators, are mostly deployed in open or rural areas. Unlike in smart city systems where boxes can be installed or camera surveillance can be deployed to protect them, the SA devices are exposed to physical attacks. The attacker may temper with or even remove the sensors. Third, the SA devices are also subject to extreme weather conditions (e.g., rain and lightning) and wild animals, which may render them unavailable.

Due to the rapid growth of SA, several studies have been conducted to analyze the security of SA systems. Various advanced data analysis techniques have been proposed to detect attacks from the huge amount of data [9,21,26,33]. Nonetheless, there is still a lack of SA domain-specific dataset that can be used by these studies for evaluation and benchmarking. Researchers have to turn to those generic IoT or industry IoT (IIoT) datasets, e.g., UNSW-NB15 [24] and TON_IoT [23].

In this work, we set off to build up an SA testbed named SATB for collecting SA domain-specific datasets. SATB follows a widely adopted architecture, which interconnects end devices, gateways and servers with the protocols used in the real-world SA scenarios. Without losing representativeness, our testbed is constructed from equipment and cloud services that have been widely deployed for commercial use. For the communication between sensors and the servers, we deploy LoRaWAN, the *de facto* protocol used in real-world SA systems.

SATB enables us to construct a preliminary dataset. As a case study, we design five test cases to drive the execution of SATB, and during the execution, we capture the communication traces. Through this, we come up with a dataset that covers the main stages of SA, from *authentication* to *runtime functioning*. This yields 671,239 records of network packets. With SATB, we are also able to conduct attacks and include the attack data into our dataset. As the first step, we have conducted a network reconnaissance with Nmap [5] and a vulnerability scanning with Open-VAS [7]. Both tools are typical attack tools of penetration testers and hackers in practice. This generates another 167,090 records of attack packets.

To explore the usability of SATB, we explore the quality of the collected dataset with some preliminary studies. With simple preprocessing, the collected traces can be formatted. We can then reveal the protocols and services that packets belong to. The LoRaWAN payloads are kept in the dataset and can be decoded with off-the-shelf parsers. Key information for intrusion detection, such as the used ports and constant strings can be revealed by our dataset.

Contributions. In summary, our work makes the following contributions.

- **An extensible and representative testbed.** We construct a testbed named SATB with commercial SA equipment and open-source software stacks. It represents the mainstream architecture and workflow of SA systems. Our testbed is extensible that other sensors or servers can be easily plugged for simulation and dataset collection.
- **Simulation of real-world attacks.** Based on our in-house testbed, we are able to take the first step of generating attack data against SA systems. We conduct network reconnaissance and vulnerability scanning on the testbed, and the attack traces can be captured by our testbed.
- **An SA network dataset.** With SATB, we build a dataset that includes 876,371 records of network packets. It covers the communications between end sensors and the servers, and contains the data of the entire lifecycle from authentication to runtime functioning.

2   Background and Related Work

2.1   Smart Agriculture

The smart agriculture (SA) technology is a concept that makes use of advanced technologies such as Internet of Things (IoT), robotics and big data, to enhance the traditional agriculture industry. These technologies can form a network of physical objects, including sensors, softwares, or other technologies. Owing to enhanced connectivity, devices become able to communicate with gateways, servers, applications as well as other devices, exchange data and perform the tasks that they are designed for.

With the empowerment of smart agriculture technologies, the farming process has been changed dramatically. In a farming process, the SA can be divided into three stages, i.e., *data gathering*, *data analysis*, and *reaction and control*. In the first stage, the sensors deployed in the farms are used to monitor the animal activities and environment such as temperature, soil moisture and humidity. They upload the data to the cloud in real time. In the second stage, the cloud side employs data-driven techniques to analyze the data, and then corresponding actions can be taken in the reaction and control stage. This can either be a real-time "trigger and action automation", or an action taken after diagnosis. Taking watering as an example, if the sensor detects that the soil moisture level is lower than expected, the water dispensers in the farm can start to work automatically and water the crops until the soil has reached the desired moisture level. It ensures that the crops will not be underwatered or overwatered.

With SA technologies, the farmers are able to monitor the environment remotely and automate the farming process with the minimal use of labour and natural resources such as water and fertilizer.

2.2 LoRaWAN

LoRaWAN [4] is a communication protocol at the data link and network layers. It works on top of LoRa radio modulation technique. A LoRaWAN architecture typically consists of four components: *end nodes, gateway, network server* and *application server*.

- **End nodes.** End nodes are typically sensors and actuators. They communicate with the gateway through the LoRa protocol.
- **Gateway.** The gateway is the bridge between end nodes and the network server. It plays a role of the access point or the relay. The gateway communicates with the end nodes through LoRa radio and gathers data from them. The data is forwarded to the network server through a network protocol compatible with the server, such as Ethernet, WiFi and 3G/4G.
- **Network server.** The network server handles the communication with the gateway and conducts data preprocessing, e.g., deduplication.
- **Application server.** The application server mainly conducts data analysis and interaction with clients. It is usually installed with computational resources to handle massive data and generates intelligent response. It also provides interfaces to the client applications, such as mobile applications and web applications. The raw data and its analysis results are displayed to users, and users' management is forwarded back to the network server and eventually end nodes for actuation. In practice, the network server and application server are mostly merged into one single server.

2.3 Related Work

Cyber Security Issues in SA and IoT. A broad range of cyber security challenges brought by the interconnectivity of SA and IoT have been noticed and disccussed [8,12,15,19,28–30]. Various threats have been revealed, including hardware/firmware-specific attacks [13,14], network-layer attacks [11,20], and cloud-specific attacks [18]. It has been demonstrated that the cyber security issues in SA have attracted attention and posed threat and damages if unsolved. Various studies have also been proposed to migitate the cyber risks at different IoT layers, from device layer [16,32], edge layer [25] to cloud layer [31].

IoT-Related Testbeds and Datasets. Sivanathan et al. [27] build up a smart environment with a variety of IoT devices, including spanning cameras, lights, plugs, motion sensors, appliances, and health-monitors. Alsaedi et al. [10] construct a TON_IoT dataset, which includes telemetry data of IoT/IIoT services, operating system logs and network traffic of IoT network. Compared to these studies, SATB mainly focuses on SA scenarios and reflects the representative workflow and working procedures. Specific analysis and data processing are also performed on the traffic. For example, LoRaWAN packets are decoded and attributes are extracted from them.

Recently, Gondchawar et al. [17] proposes a IoT-based SA framework, which consists of three end nodes and a PC to control the system. The end nodes are AVR Microcontroller Atmega 16/32, ZigBee Module, and Temperature Sensor LM35. Their testbed uses Dig Trace, SinaProg and Raspbian Operating System as the software stack. Compared to their work, the devices in our testbed are more up-to-date and can better reflect the current SA systems in practice.

3   SATB: A LoRaWAN SA Testbed

3.1   Components of SATB

We have built up a LoRaWAN Smart Agriculture Testbed named SATB to facilitate data collection. SATB consists of three major components: *end devices*, *gateway* and *server*, as shown in Fig. 1.

End devices and gateway. We deploy two SenseCAP sensors and SenseCAP gateway, as listed in Table 1. The sensors monitor the soil moisture, temperature and light intensity, and report the readings to the gateway through LoRa radio.

Server. The server our testbed uses is called ChirpStack [2], which is a LoRaWAN network server stack. It consists of the *gateway bridge*, *network server* and *application server*, which merge the modules discussed in Sect. 2.2 into one server. The gateway bridge handles the communication with the LoRaWAN gateway and converts the LoRa packets into ChirpStack Network Server compatible data format. The network server then de-duplicates the frames received by the LoRa gateways and forwards them to Application Server. The application server plays the role of the "inventory" of the LoRaWAN infrastructure. It offers a web interface for users to manage, view and organize the stored data.

3.2   Functionalities of SATB

Each end devices and the gateway have an identifier called Extended Unique Identifier (EUI), which is delivered to the end user with their packages. For a device to be recognized by ChirpStack server, its EUI has to be registered with the latter. This can be done through the web portal provided by the application server. For a device to be connected to the network server, the correct *app key* should be provided upon authentication.

The testbed is deployed in the Device Test Lab on UQ campus. The sensors are connected to the gateway wirelessly while the gateway is connected to the Internet via Ethernet. A Dell desktop installed with the ChirpStack network server is connected to the same Ethernet. Three Docker containers are deployed in the desktop, each of which hosts one module of the ChirpStack network server. They communicate with each other through Docker's single-host networking. We install a packet logger named Wireshark [1] on the desktop to capture all traffic relayed by the gateway. It also can capture the traffic among the three modules.

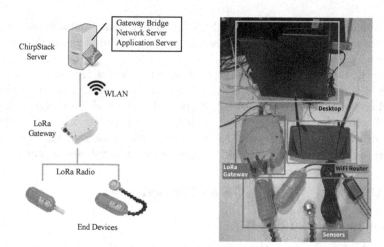

Fig. 1. The architecture and a photo of our testbed

Table 1. The IoT devices in our testbed

IoT device	Device type
SenseCAP outdoor gateway	Gateway
SenseCAP soil moisture & temp sensor	End device
SenseCAP light intensity sensor	End device

4   A Case Study: Constructing an SA Dataset with SATB

To demonstrate the usability of SATB, we use it to construct a preliminary dataset. It includes SA network traces in normal operation and attack scenarios.

4.1   Test Cases and Data Collection

To retain representativeness, we propose five application scenarios to make our dataset. For each scenario, we design a test case to drive the execution and communication of each component in our testbed. Below, we describe our test cases (TCs).

- **TC1.** When the gateway and the sensors are not registered with ChirpStack server, we connect the gateway and the sensors to the network server. In this test case, the gateway and sensors cannot be recognized by the network server.
- **TC2.** When the gateway and the sensors are not registered, we register the gateway, connect the gateway to the network server, and then connect the sensors to the network server. In this test case, the gateway can be connected but the sensors are not recognizable.

- **TC3.** When the gateway is registered and the sensors are not, we connect the gateway to the network server, register the sensor with correct EUI but with incorrect *app key*, and then connect the sensor to the network server. In this test case, the gateway can be connected but the sensors cannot.
- **TC4.** When the gateway is registered and the sensors are not, we register the sensors, connect the gateway to the network server and connect the sensors to the network server. In this test case, both the gateway and the sensors can be connected to the network.
- **TC5.** When the gateway and the sensors are both registered, we connect the gateway and the sensors to the network server. In this test case, both the gateway and the sensors can be connected to the network.

To test SATB comprehensively, we also create attack scenarios on top of the use cases. As the first step, we consider two types of attacks - reconnaissance and vulnerability scanning. For the former, we apply a widely used network scanner called Nmap [5] to scan against our testbed, with different scanning methods such as SYN Scanning and TCP Connect Scanning. For the latter, we use a popular vulnerability assessment tool OpenVAS [7] to conduct a vulnerability scanning. It has the signature of most known software vulnerabilities and checks whether the scanned system is subject to those vulnerabilities. We turn on Wireshark to log the network traffic while we execute the test cases and attacks. Filters are set to ensure that only the relevant packets are collected. The packets are saved as *.pcap* files.

4.2   Data Preprocessing

After the data collection, we further process the data. We parse each packet in the *.pcap* file to extract the features, including source ip address, destination ip address, source port, destination port, protocol information and other protocol-specific fields depending on the protocol of the packet. For example, if a packet is an MQTT packet, MQTT-specific fileds, such as the MQTT message type, remaining length, and payload field are included. After the processing, the extracted fields of the packet are exported to a *.csv* file. The main features of our dataset are listed in Table 2.

During the data process, we also apply a tool called Zeek (previously Bro-IDS) [6] on the captured packets. It takes the *.pcap* file as input and turns it into high-fidelity transaction logs. This tool is also able to find unusual and suspicious behaviours, which will be used to confirm the label of the packets captured in the attacking stage.

4.3   A Preliminary Study of the Dataset

In this section, we present a preliminary study of the collected dataset and give its basic information.

Table 2. Main features of our dataset

	Name	Description
1	src_ip	The source ip address of the packet
2	dst_ip	The destination ip address of the packet
3	src_port	The source port of the packet
4	dst_port	The destination port of the packet
5	protocol	The protocol-specific information of the packet
6	other	Other information about the packet

Packets Distribution

Overall, we have collected 876,371 packets, which take around 1.5 GBytes. They mainly cover three working stages, i.e., *authentication stage*, *runtime functioning stage*, and *attacking stage*. The majority of the packets are in runtime use (671,239 packets). The authentication stage and attacking stage have 38,042 and 167,090 packets respectively.

Authentication. There are two types of authentication in LoRaWAN: Over The Air Activation (OTAA) and Activation By Personalisation (ABP). The one used by our testbed is OTAA, which is more extensively used in practice than ABP. In OTAA activation, each time when the end devices intend to join, the session keys are different. When an end device attempts to join the network, it sends a *join-request* message which contains AppEUI, DevEUI, DevNonce and Message Integrity Check (MIC) value. The network server processes this message, and then generates two session keys and a *join-accept* message if the end device is permitted to join. The *join-accept* message is then encrypted using the *app key* and sent back to the end device as a normal downlink. If the join request is not accepted, no response will be replied.

Runtime Functioning. After the authentication, the end devices and the gateway are connected to the network. The testbed starts operating. The readings of the sensors are periodically reported to the network server, through the relay of gateway and gateway bridge. The update period can be configured through the web portal of the application server. The network server then dedupliates the received packets and formats them so that they can be usable by the application server.

Attacking. The following scanning traces are included in the collected dataset.

1. Nmap SYN Scan: It scans the ports of a target machine by sending a TCP packet with the SYN flag set.
2. Nmap TCP Connect Scan: It scans the ports of a target machine by establishing a full connection.
3. Nmap Service and Version Scan: It detects more information about the version of the services running on the target machine.
4. Nmap Operating System Scan: It detects the operating system of the target machine by sending a series of TCP, UDP packets and examining the response.

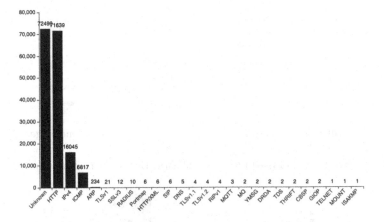

Fig. 2. Protocol types in the authentication stage

Table 3. An example of a decoded LoRaWAN payload

Feature	Description	Decoded Value
Message Type	MAC message types	Join Request
PHYPayload	LoRa Physical Payload	MHDR [1]—MACPayload [..]—MIC [4]
MHDR	MAC Header	02
MACPayload	MAC Payload of Data Messages	8DC0022CF7F110
MIC	Message Integrity Code, which ensures the integrity of the packet	24400040
MACPayload	MAC Payload of Data Messages	AppEUI [8]—DevEUI [8]—DevNonce [2]
AppEUI	64-bit unique identifier for Join Server	2410F1F72C02C08D
DevEUI	64-bit globally-unique Extended Unique Identifier (EUI-64) assigned by the manufacturer	400040
DevNonce	A random number	NA

5. OpenVAS full and deep scanning: It performs a full and deep vulnerability scanning against the target machine.

Packets of Authentication Stage

We attempt to recognize the service types of the packets collected at the authentication stage. We focus mainly on the application-layer protocols and the statistics as shown in Fig. 2. The majority of packets in the authentication stage is PGSQL, which is mainly used for PostgreSQL, an open-source relational database management system. It is integrated into the ChirpStack network server stack to store device-related events. Since all events are written into the database, the packets of this type become dominant.

Many packets have unknown types and they are shown as TCP or UDP packets. Almost all UDP packets contain LoRaWAN payloads that are not recognized by Wireshark. However, the LoRa packets can actually be extracted

(a) Authentication stage   (b) Runtime stage

Fig. 3. Word cloud of used ports in authentication stage and runtime stage

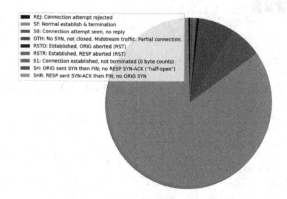

Fig. 4. The Distribution of connection state in authentication stage

from them. For example, one of the UDP packets that we have captured contains "028dc0022cf7f11024400040" as its payload, which is hex encoded. We use a LoRa packet decoder to decode the payload and the attributes can be extracted as shown in Table 3.

We also conduct an analysis on the source ports of the packets. In Fig. 3a, we use a word cloud to present the occurrence of ports in the dataset. If a port appears more frequently, its size in the graph is larger. We can observe that the ports 56062, 56064, and 51620 account for a higher proportion in the captured packets. They are mainly used by TCP.

We also analyze the connection states from the packets. As shown in Fig. 4, around 84% of the packets has a connection state of SF, which means the connection is established and terminated normally. OTH and S0 also have a higher percentage than the rest of the packets. The state OTH occurs when there is no SYN seen in the connection, it might because it is a midsteam traffic. 6% of the packets have S0 as their connection state, which means that a connection has been attempted, but no reply is seen.

Packets of Runtime Functioning Stage
Similarly, we attempt to recognize the service types of packets and the ports used in the runtime functioning stage. The results are shown in Fig. 5 and Fig. 3b.

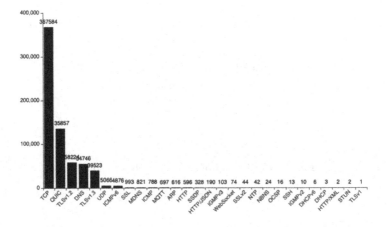

Fig. 5. Protocol types in runtime stage

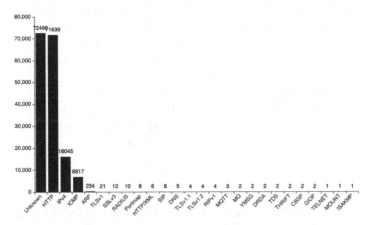

Fig. 6. Protocol types in the attacking stage

Packets of Attacking Stage

We first attempt to recognize the service types of packets in the attacking stage. The statistics is shown in Fig. 6. We can notice that the service of 43.24% of the packets are unknown (tagged as UDP and TCP packets), while 42.87% of the packets are HTTP packets. After looking into the packets, the high percentage of HTTP packets is due to the vulnerability scanning of OpenVAS. When this tool conducts scanning, it requests a number of web locations that may contain security issues such as SQL injection or buffer overflow. The word cloud of scanned web locations is shown in Fig. 7.

We also use Zeek to process the packets captured in attacking stage. There are 28,806 out of 167,090 attacking packets that are detected unusual or improper behaviors and flagged. The flags can be grouped to a list of 21 flags in total as shown in Table 4. The most frequent flags that are extracted are

Fig. 7. Word cloud of scanned web locations

Table 4. The list of attack flags

Features	
possible_split_routing	DNS_Conn_count_too_large
inappropriate_FIN	line_terminated_without_CRLF
data_before_established	unknown_HTTP_method
bad_HTTP_request	unescaped_%_in_URI
empty_http_request	partial_escape_at_end_of_URI
line_terminated_with_single_CR	data_after_reset
DNS_truncated_len_lt_hdr_len	premature_connection_reuse
DNS_label_forward_compress_offset	HTTP_version_mismatch
DNS_label_too_long	bad_HTTP_request_with_version
DNS_truncated_RR_rdlength_lt_len	double_%_in_URI
missing_HTTP_uri	

possible_split_routing, data_before_established and inappropriate_FIN. The possible_split_routing happens when the other side of the connection is not seen. The data_before_established means before a connection was established, some data is sent by a TCP endpoint. The inappropriate_FIN happens when a FIN set in the packet does not follow the RFC for TCP/IP standard.

5   Usage of the SATB Testbed

In this section, we outline a few usage of the SATB Testbed, towards the reliable operation in a smart agriculture ecosystem.

5.1   Development of Intrusion Detection Systems for SA

The rapid development and adoption of highly diverse IoT devices in smart agriculture have created increased attack surface. Therefore, building an intrusion detection system (IDS) has become an increasingly urgent issue for detecting network security breaches against SA systems. However, as the performance of an effective IDS largely relies on the availability of representative and real-life data, the lack of SA domain-specific datasets has resulted in a significant gap in the IDS development.

Our high-fidelity SATB Testbed can alleviate the gap by providing intensive sets of traffic traces for training and testing IDSs. SATB covers the complete workflow of smart agriculture system ranging from authentication to runtime usage, which represents the sum of all possible attack vectors that may be exploited. In addition, the construction of the SA dataset has demonstrated the capacity of SATB in incorporating a variety of real network intrusion scenarios against SA systems, not only reflecting the existing attacks of the time, but are also dynamic and extensible as network scenarios evolve.

5.2   Preservation of Data Privacy and Integrity for SA

The most significant feature of smart agriculture is its employment of a vast amount of sensors, devices and equipment, from which an enormous amount of data gets generated. The challenge of preserving data privacy and integrity remains open in this area due to the massiveness, complexity and heterogeneity of the generated data [19, 22].

Our SATB Testbed, which serves as the representative digital twins of the counterparts in the real-world production environment, can greatly facilitate the research in the key properties of data protection from the following perspectives: (1) the secure sharing of data with different sensitivity classifications across stakeholders in the entire supply chain, (2) the secure processing of sensitive data (e.g., agriculture purchase information) with provable privacy-preservation and verifiable prevention of information leakage, (3) the provenance based data integrity checking and verification in the smart agriculture environment.

5.3   Development of Data-Driven Applications for SA

We see the opportunities of SATB contributing to the research area in integrating artificial intelligence and machine learning technologies for the development of data-driven applications in smart agriculture. For example, the massive real-life data generated by our testbed (including time series and spatial data) can be used to train and evaluate real time monitoring systems, which are crucial to modern farms in improving their productivity and security. In addition, despite the advantages of LoRaWAN (such as low-power and long-range transmissions which are typically desirable in the smart agriculture environment), the technology has the disadvantage of low-bandwidth transmissions, which has presented an arising need to integrate edge computing to facilitate the move of data processing and analytics to end devices. Our testbed can be used to construct and test such computing framework with data on edge, extending the functionalites of smart agriculture applications using LoRaWAN for wide area coverage.

6 Conclusion

In this work, we have designed and developed an SA testbed named SATB to facilitate simulation, attack experiments and dataset generation. SATB consists of the end devices, gateways, software stacks and protocols that are widely adopted in real-world SA systems. It is designed to be extensible so that new hardware and software modules can be installed. To show SATB's usability, we used it to construct a comprehensive SA network dataset, which rigorously cover the whole lifecyle of SA scenarios. For future works, we will enrich the types of the end sensors to support other protocols. We will also embed an open gateway that enables the interception of the LoRa packets between the end sensors and the gateway.

Acknowledgment. This work is supported by The University of Queendland under the Cyber Research Seed Funding.

References

1. Analysis tool wireshark, March 2022. https://www.wireshark.org/
2. Chirpstack, March 2022. https://www.chirpstack.io
3. Internet of things in agriculture market, March 2022. https://www.emergenresearch.com/industry-report/iot-in-agriculture-market
4. Network scanner nmap, March 2022. https://lora-alliance.org/about-lorawan/
5. Network scanner nmap, March 2022. https://nmap.org/
6. Network security monitoring tool zeek, March 2022. https://zeek.org/
7. Scanner openvas, March 2022. https://www.openvas.org/
8. Abuan, D.D., Abad, A.C., Lazaro Jr, J.B., Dadios, E.P.: Security systems for remote farm. J. Autom. Control Eng. **2**(2), 115–118 (2014)
9. Al-Hawawreh, M.S., Moustafa, N., Sitnikova, E.: Identification of malicious activities in industrial internet of things based on deep learning models. J. Inf. Secur. Appl. **41**, 1–11 (2018)
10. Alsaedi, A., Moustafa, N., Tari, Z., Mahmood, A., Anwar, A.: TON_IoT telemetry dataset: a new generation dataset of IoT and IIoT for data-driven intrusion detection systems. IEEE Access **8**, 165130–165150 (2020)
11. Antonakakis, M., et al.: Understanding the Mirai botnet. In: 26th USENIX Security Symposium (USENIX Security 2017), pp. 1093–1110 (2017)
12. Barreto, L., Amaral, A.: Smart farming: cyber security challenges. In: 2018 International Conference on Intelligent Systems (IS), pp. 870–876. IEEE (2018)
13. Cojocar, L., Zaddach, J., Verdult, R., Bos, H., Francillon, A., Balzarotti, D.: PIE: parser identification in embedded systems. In: Proceedings of the 31st Annual Computer Security Applications Conference, pp. 251–260 (2015)
14. Costin, A., Zaddach, J., Francillon, A., Balzarotti, D.: A {Large-scale} analysis of the security of embedded firmwares. In: 23rd USENIX Security Symposium (USENIX Security 2014), pp. 95–110 (2014)
15. Demestichas, K., Peppes, N., Alexakis, T.: Survey on security threats in agricultural IoT and smart farming. Sensors **20**(22), 6458 (2020)
16. Feng, X., et al.: Snipuzz: Black-box fuzzing of IoT firmware via message snippet inference. In: Proceedings of the 2021 ACM SIGSAC Conference on Computer and Communications Security, pp. 337–350 (2021)

17. Gondchawar, N., Kawitkar, R., et al.: IoT based smart agriculture. Int. J. Adv. Res. Comput. Commun. Eng. **5**(6), 838–842 (2016)
18. Gruschka, N., Jensen, M.: Attack surfaces: a taxonomy for attacks on cloud services. In: 2010 IEEE 3rd International Conference on Cloud Computing, pp. 276–279. IEEE (2010)
19. Gupta, M., Abdelsalam, M., Khorsandroo, S., Mittal, S.: Security and privacy in smart farming: challenges and opportunities. IEEE Access **8**, 34564–34584 (2020)
20. Kolias, C., Kambourakis, G., Stavrou, A., Voas, J.: DDoS in the IoT: Mirai and other botnets. Computer **50**(7), 80–84 (2017)
21. Koroniotis, N., Moustafa, N., Sitnikova, E.: A new network forensic framework based on deep learning for internet of things networks: a particle deep framework. Future Gener. Comput. Syst. **110**, 91–106 (2020)
22. Mahadewa, K., et al.: Identifying privacy weaknesses from multi-party trigger-action integration platforms. In: Proceedings of the 30th ACM SIGSOFT International Symposium on Software Testing and Analysis, pp. 2–15 (2021)
23. Moustafa, N.: A new distributed architecture for evaluating AI-based security systems at the edge: network TON_IoT datasets. Sustain. Cities Soc. **72**, 102994 (2021)
24. Moustafa, N., Slay, J.: UNSW-NB15: a comprehensive data set for network intrusion detection systems (UNSW-NB15 network data set). In: 2015 Military Communications and Information Systems Conference (MilCIS), pp. 1–6 (2015). https://doi.org/10.1109/MilCIS.2015.7348942
25. Nesarani, A., Ramar, R., Pandian, S.: An efficient approach for rice prediction from authenticated block chain node using machine learning technique. Environ. Technol. Innov. **20**, 101064 (2020)
26. Popoola, S.I., Adebisi, B., Hammoudeh, M., Gui, G., Gacanin, H.: Hybrid deep learning for botnet attack detection in the internet-of-things networks. IEEE Internet Things J. **8**(6), 4944–4956 (2021). https://doi.org/10.1109/JIOT.2020.3034156
27. Sivanathan, A., et al.: Classifying IoT devices in smart environments using network traffic characteristics. IEEE Trans. Mob. Comput. **18**(8), 1745–1759 (2018)
28. Xie, F., Zhang, Y., Wei, H., Bai, G.: UQ-AAS21: a comprehensive dataset of Amazon Alexa skills. In: Li, B., et al. (eds.) ADMA 2022. LNCS, vol. 13087, pp. 159–173. Springer, Cham (2022). https://doi.org/10.1007/978-3-030-95405-5_12
29. Xie, F., et al.: Scrutinizing privacy policy compliance of virtual personal assistant apps. In: 37th IEEE/ACM International Conference on Automated Software Engineering (ASE 2022) (2022)
30. Yazdinejad, A., et al.: A review on security of smart farming and precision agriculture: security aspects, attacks, threats and countermeasures. Appl. Sci. **11**(16), 7518 (2021)
31. Zhang, H., Lu, K., Zhou, X., Yin, Q., Wang, P., Yue, T.: SIoTFuzzer: fuzzing web interface in IoT firmware via stateful message generation. Appl. Sci. **11**(7), 3120 (2021)
32. Zheng, Y., Davanian, A., Yin, H., Song, C., Zhu, H., Sun, L.: {FIRM-AFL}:{High-Throughput} greybox fuzzing of {IoT} firmware via augmented process emulation. In: 28th USENIX Security Symposium (USENIX Security 2019), pp. 1099–1114 (2019)
33. Zhou, X., Hu, Y., Liang, W., Ma, J., Jin, Q.: Variational LSTM enhanced anomaly detection for industrial big data. IEEE Trans. Industr. Inf. **17**(5), 3469–3477 (2021). https://doi.org/10.1109/TII.2020.3022432

An Overview on Reducing Social Networks' Size

Myriam Jaouadi[✉][iD] and Lotfi Ben Romdhane

MARS Research Lab LR17ES05 Higher Institute of Computer Science and Telecom
(ISITCom), University of Sousse, Sousse, Tunisia
jaouadimaryem@gmail.com, lotfi.BenRomdhane@isitc.u-sousse.tn

Abstract. Social networks are important dissemination platforms that
allow the interchange of ideas. Such networks are omnipresent in our
everyday life due to the explosive use of smartphones. Consequently,
modern social networks have reached a significant number of users, making their size huge. Thereby scaling over such large data remains a challenging task. Reducing social networks' size is a key task in social network
analysis to deal with this data complexity. Many approaches have been
developed in this direction. This paper is dedicated to proposing a new
taxonomy covering different state-of-the-art methods designed to cope
with the explosive growth of social network data. The suggested solution to the extensive generated data is to reduce the network's size. We
then categorized existing works into two main classes that reflect how
the reduced network is generated. After that, we present new directions
for reducing large-scale social network size.

Keywords: Social networks · Graph sampling · Graph coarsening

1 Introduction

Modern social networks have reached an unprecedented number of users [2] due
to their accessible handling. For example, Facebook is the first social network to
surpass 1 billion registered accounts and currently sits at 2.91 millions monthly
active users[1] that tag photos of new friends, check up on old ones, and post about
sport, politics, etc [2]. In fact, the classical methods designed for social networks
analysis become inapplicable [1]. In order to handle a such problem, several
works have been developed with the aim of reducing the network's size while
preserving its original properties. Indeed, reducing the network's size will serve
to manifold tasks such as influential nodes detection [8], communities selection
[1] and so on.

Reducing a social network's size aims at finding a representative pattern of
the network while retaining its key properties. Choosing a subset of nodes or/and

---

[1] https://www.statista.com/statistics/272014/global-social-networks-ranked-by-
number-of-users/.

© The Author(s), under exclusive license to Springer Nature Switzerland AG 2022
W. Chen et al. (Eds.): ADMA 2022, LNAI 13725, pp. 146–157, 2022.
https://doi.org/10.1007/978-3-031-22064-7_12

edges from the original network is the simplest way to form the reduced version [3]. Many works have been proposed in this direction. In this paper we provide a state-of-the-art survey on reducing social networks' size. In this detailed survey, we divide the existing models into two main categories, graph sampling and graph coarsening models. Our main concern is to investigate the existing approaches and compare them according to the preservation of the network properties. The recent applications of reducing the network 's size are also surveyed. Finally, future directions are discussed.

The remainder of this paper is organized as follows: Sect. 2 presents preliminary concepts of reducing social networks' size. Section 3 is about graph sampling methods. In Sect. 4 existing models for graph coarsening are discussed. Section 5 outlines recent applications and the final section concludes this paper and proposes some future research directions.

2   Preliminaries

2.1   Problem Definition

Since graphs are the privileged mathematical tools to model social networks, we can define the reduction of a network size as: Given an undirected social graph $G(V, E)$, with $n = |V|$ nodes and $m = |E|$ edges, the goal is to create a version G' having n' nodes such that $n' << n$. The reduced graph G' should be the most similar to the initial one G. In other words, G' conserves structural properties of G [2]. We will define graph properties in what follows.

2.2   Network Properties

In order to evaluate the efficiency of the reduction method, we should check some graph properties [5]. Indeed, preserving the original network's structure proves the success of the reduction method. Manifold properties were considered for this task, (e.g. the clustering coefficient, the degree distribution, the graph diameter,etc).

Definition 1 *(Degree Distribution [2]). One of the most relevant and simple graph properties is the degree distribution $P_{deg(k)}$ which can be defined as the fraction of nodes in the graph having the same degree k [5]. It can be described formally as:*

$$P_{deg}(k) = \frac{|\{v; deg(v) = k\}|}{n} \tag{1}$$

where $deg(v)$ is the degree of node v.

Definition 2 *(Clustering coefficient [2]). Another measure for graph properties is the clustering coefficient which quantifies the likelihood of two neighbors of a node being neighbors themselves. Clustering coefficient $CC(G)$ for a given graph*

G is defined as the ratio of the number of triangles to the number of triplets known as length two paths [6].

$$CC(G) = \frac{3 * numberTriangles}{0.5 * allTriplets} \quad (2)$$

where $allTriplets = \sum_{i=1}^{n}((|NB(u_i)| - 1) * |NB(u_i)|)$ with $NB(u_i)$ is the set of direct neighbors of u_i.

Definition 3 *(Graph diameter [3]). Diameter $D(G)$ of a graph G can be defined as the longest distance between all pair of vertices in G. More formally, it can be described as;*

$$D(G) = max_{v \in V} R(v) \quad (3)$$

where $R(v)$ is the radius of a node v, i.e. the maximum shortest path distance to all other vertices. Since reducing the network's size will serve to several social network analysis tasks, we can put forth a practical application of such reduction to a well known task which is community detection problem. Multiple approaches have focused on reducing the network's size as a preliminary step for community detection from huge networks. In order to test the effect of the reduction strategy, the clustering quality known as the modularity [7] can be used which is defined as:

$$Q = \frac{1}{2m} \sum_{i,j} (A_{i,j} - \frac{k_i k_j}{2m}) \delta(c_i, c_j) \quad (4)$$

where m is the number of edges in the graph G, A represents its adjacency matrix, k_i denotes the degree of a node i, c_i is the community to which the node i is assigned and the function $\delta(c_i, c_j)$ indicates whether nodes i and j are members of the same community.

A literature review allowed us to distinguish two main families of approaches for reducing the network's size that reflect how the reduced version is generated: graph coarsening and graph sampling. Li-Chun Zang [34] introduced in a recent work a survey on graph sampling as a representation of relevant units of a given graph. The paper talks about sampling from different areas including real graphs, bipartite graphs and conventional graphs. However, there is a lack of a clear categorization of graph sampling approaches. In fact, the author did not illustrate the variety of graph sampling techniques nor their advantages and limits. A recent survey on graph coarsening [33] was proposed with the aim to take a broad look into coarsening techniques. The authors started by showing the several techniques of graph coarsening and its applications in scientific computing with a clear categorization. Then, the emergence of graph coarsening in machine learning, which is of great interest nowadays, was discussed. However, only graph coarsening is presented in the paper for reducing graphs' size, although we distinguish many techniques for reduction. As for our work, the main concern is is to investigate the existing approaches and to talk about their advantages and limits. In this detailed survey, we divide the existing models into two main categories, graph sampling and graph coarsening models.

3   Graph Sampling

One of the well known methods for reducing the graph's size is that of graph sampling. The main idea of sampling is to find a representative pattern from the original network while maintaining its properties. Choosing a subset of nodes or edges from the initial graph is the simplest way to create the sample [2, 3]. We distinguish three popular techniques for this family of approaches: Node Sampling, Edge Sampling and Traversal Based Sampling. Figure 1 describes the three sampling techniques for selecting six nodes from the original graph.

3.1   Node Sampling

The aim of node sampling technique is to select a set of k nodes then to retain links between them. The choice of the k nodes can be done randomly. In this direction, Leskovec et al. [5] have developed the well known approach RN (Random Node). It starts by an uniform choice of a subset of nodes and the sample is created on the basis of the selected nodes and edges connecting them. Even its simplicity, RN may create samples with isolated nodes. To overcome this limit, manifold heuristics have been suggested such that RPN (Random PageRank sampling) and RDN (Random Degree Node) [5]. Indeed, the probability of a node being selected is proportional to its PageRank value for RPN and its degree for RDN. Based on degree distribution, Zhu et al. [12] proposed two sampling strategies. The first one called NS-d (Node Degree-Distribution Sampling) tends to create three nodes clusters for high degree nodes, medium degree nodes and low degree nodes using the k-Means algorithm. As for the second strategy, it uses the CountingSort algorithm to sort high degree nodes and then to put them into the reduced graph using a specified sample fraction. Cai et al. [22] proposed two variant algorithms of the existing UNI (uniform sampling) model. In fact, UNI attributes an uniform sampling probability to each node while ignoring the original network's structure. To overcome the inefficiency of UNI and further improve connectivity, the main purpose of the proposed algorithms is to study nodes distribution and connections between them. The first model called AdpUNI divides the userID space into several intervals to make an adaptive change for sampling probabilities. As for the second algorithm AdpUNI+N, it exploits the neighborhood of nodes to obtain a more representative version of the original graph. Based on contextual structures, Zhou et al. [21] have transformed vertices into vectors. Then, in order to have a reduced graph with high quality clusters, nodes are selected from the vectorized space and a sample that maintains graph connectivity is created.

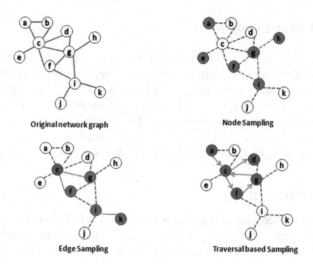

Fig. 1. Graph sampling techniques

3.2 Edge Sampling

Edge sampling is the second technique of sampling. It is based on the idea of choosing a subset of edges at random then including their end nodes in the reduced graph [5]. Random Edge sampling (RE) is the simplest method in this line [3]. During edges selection, only the random chosen ones are added to the sample. A key difference of TIES model (Totally Inducted Edge Sampling) [20] is that, after choosing an initial set of edges and their end nodes, it adds other edges that exist in the original graph among any of this sampled nodes. Wang et al. [9] proposed two algorithms to treat complex networks having a self-similarity structure. In fact, the authors have investigated relations between edges and their neighborhood to choose only those caused by self similarity. DGS (Distributed Graph Sampling) [2] is a recent model designed to sample large scale networks choosing only important edges. Based on the degree centrality, a new measure qualifying the network edges called Edge Importance was proposed. For distribution authors used the MapReduce paradigm [4]. The proposed method demonstrated its efficiency to preserve the original network's structure compared to well known approaches. Similar to DGS, Yanagiya et al. [35] tried to find important edges. The proposed model starts by converting the original graph into a line graph in order to represent edges and connections between them. Then, based on edge smoothness principle, important one are selected to create the sample.

3.3 Traversal Based Sampling

The last family of approaches is traversal based sampling also known as sampling by exploration. Indeed, sampling strategy starts with an initial set of nodes, then it expands the sample according to some observations. Forest Fire (FF) [5] is inspired by spreading fire in the woods. It begins by picking an initial seed node, then, it burns its outgoing links based on a forward burning probability. The end nodes of the selected edges constitute the next seed set and the process is repeated until reaching the required sample size. RWS (Random Walk Sampling) [15] is one of the most commonly used methods in the literature. It starts by picking an initial seed node at random, then it traverses the graph randomly to move to a random neighbor. Although RW was proved to be simple, in some cases, it gets stuck in an isolated component of the graph. A fast and recent variant called CNARW (Common Neighbor Aware Random Walk) is proposed in order to speed up the convergence of the RW [11]. The basic idea of CNARW is to consider common neighbors of the recent visited nodes to choose next step ones. The proposed scheme reduces RWS cost. DRaWS [16] is another improvement of the random walk model. This work aims to estimate the degree distribution and clique structures while reducing computational costs. In fact, for random walk paths, DRaWS exploits the many-to-one formation between nodes and a clique and the one-to-many formation between one node and many nodes in a clique. DRaWS demonstrates an efficiency in both maintaining the graph's structure and reducing the computational costs. Rank Degree (RD) [13,15] is a deterministic graph exploration model. Initially, a set of nodes is selected at random. Then, for each node, Rank Degree selects top-k highest degree neighbors and the new nodes are added to the seed set. The process iterates until reaching the desired sample size.

Node Sampling (NS) and Edge Sampling (ES) are very simple. This is because samples are created by selecting a subset of nodes (for NS) or a subset of edges (for ES) randomly or based on some measures such that the degree, the MinCut, etc. However, in some cases we can not apply NS or ES directly due to some constraints like the space [3]. In this case, Traversal Based Sampling (TBS) becomes more suitable for the simple reason that it starts by a few number of seed nodes and then tries to expand it while traversing the network. It is worth noting that the three sampling techniques are not totally different. In fact, some traversal based models are used for NS or ES, for example RWS (Random Walk Sampling) results in uniform edge distribution.

4   Graph Coarsening

Graph coarsening, also known as multilevel approach, is another line of research for reducing the network's size. It is a widely used technique for the resolution of large scale classification problems. The principle aim of coarsening is to convert the original network, level by level, into smaller ones. In other terms, starting from the original graph G, a series of decreasing graphs' size is created. At each level i, the construction of G_i is based on the graph G_{i-1} generated at the previous level [2]. In fact, vertices and edges are collapsed to form the new graph. The choice of vertices or edges to be collapsed can be decided according to several heuristics [1]. RM (Random Matching) [18] is a first implementation of the multilevel approach. It starts by the selection of a non-contracted node randomly.

Then, it matches the selected node with one of its uncontracted neighbors. As a result, the two nodes are merged into one super-node and the algorithm updates the graph's weight. The procedure is stopped when all vertices are marked. In the same direction, Hendrickson and Leland [18] presented an algorithm that starts by constructing a sequence of contracted graphs while adding weights on links and nodes. It is achieved using the maximal cut. Then, the last graph is partitioned by a spectral way. Finally, this grouping is projected and it is periodically improved with the local refinement algorithm Kernigham and Lin (KL) [29]. This algorithm has demonstrated good performance, but it requires a lot of memory to store the graphs during the contraction phase. In fact, another contraction strategy generating fewer intermediate graphs would be a way to reduce this problem. An improvement was introduced in [19] whose particularity lies in the phases of contraction and refinement. Indeed, a new heuristic is presented (hard-edge heuristic) as well as a faster variation of KL is used. As for the partitioning phase, this method tries to find groups of balanced size. As for Edge Matching (EM), it aims to find bisection which minimizes the cutting cost. LEM (Light Edge Mathing) [19] tends to minimize edges' weight. In effect, it selects a node u at random. After that, it chooses an uncontracted neighbor v in such a way that the weight $w(u, v)$ of the edge (u, v) is minimal. Chen et al. [27] proposed an hybrid approach based on three matching schemes. First, this approach merges two randomly chosen nodes while keeping the links between them, which guarantees first-order proximity. To guarantee second-order proximity, the second matching is based on the idea of merging nodes having common neighbors. As for the last scheme, it is designed to combine the two previous ones. Another category of contraction approaches is based on optimizing an objective function. In this regard, MCCA (Multi-level Coarsening Compact Areas) [1] is proposed with the aim to minimize the contraction rate. MCCA is designed to merge well connected nodes at every level, then, to update nodes and edges weight until a stopping criterion is met. The algorithm proceeds in a greedy way and at each iteration the number of nodes and edges is reduced. The graph coarsening schema is stopped when the contraction rate reaches a given threshold. Other works have opted to maximize the quality of partition based on the modularity function. In this light was proposed the work of LaSalle et al. [17]. The principal aim of their method is to visit nodes in a random order and to merge each one with a non visited neighbor while maximizing the modularity variation. If the node has an empty neighboring it is shown as contracted. This algorithm prevents big matching in case of power low degree distribution graph. To overcome this shortcoming, the authors developed another variant called M2M which uses a jump secondary during matching. It follows the same principle as the previous method except that if all the neighbors of a node are marked, the latter will be merged with one of its unmarked neighbors. Therefore, M2M improves the old version to generate a small size sample. The contraction has affected various axes in the analysis of social networks, in particular it has been widely used in network partitioning often known as community detection problem [28]. In this context, the Louvain algorithm [30] has been proposed by alternating between contraction and partitioning. Based on the modularity optimizer, the authors proposed to assign all the vertices to different partitions according to the optimal modularity. Once arrived at a first optimal situation, the process passes to the higher level while processing each partition as a vertex. The process continues until no improvement in modularity is possible. This method solves the problem of the modularity resolution. However, it offers a result that depends on the order of node processing. To overcome this shortcoming, a Louvain variant (with multi-level refinement) is suggested [31]. Another work called SLM (Smart Local Moving) [32] has been developed with the aim of detecting groups in large graphs while maximizing modularity. SLM gives higher modularity values than Louvain, however it requires more calculation time. All the works mentioned above consider only the network structure. However, semantic measures can be used to merge nodes or/end edges for graph coarsening [14]. Among this measures we can talk about homophily [14], which is designed to understand common interests of the network's users.

Discussion
To conclude, graph sampling was widely used. It is one of the first typical thoughts to process massive data. The performance of such strategy is indeed studied in a number of methods to prove that the initial network' s structure is preserved [2]. However, graph sampling has some limits. The first one is that it requires a prior knowledge of the global graph which is impracticable in certain cases as in decentralized social networks. Another limit is that, this approach depends on the network law degree distribution. In other terms, nodes degree has an impact in sampling the network and in some cases we can obtain a dense graph or a sample with a distribution different from that of the original one.

In the other hand, graph coarsening was widely used to process large networks but there is no theoretical proof that the initial network' s structure is preserved [2]. Another limit is that in many cases, there is a large portion of well connected components which will lead to a poor quality of graph partitioning. Graph coarsening has been proved to be more suitable for the community detection problem [17,19,28].

A summary of the different state-of-the-art approaches is highlighted in Table 1. This table presents the advantages, limits and complexity analysis of each approach. We denote by n the number of nodes of a graph G and by m the number of edges.

5   Recent Directions

In order to reduce the network's size, the approaches mentioned above take as input the structure of the entire network. In some cases, due to the distribution of networks, because of their huge sizes or decentralized controls, it is not possible to access to the whole network. In this context, we present distributed approaches as recent directions. Distributed approaches have demonstrated rapidity of implementation and ability to cope with large scale networks. For example, Zang et al. [23] used MapReduce-Spark to implement three distributed algorithms. The

Table 1. Summary on approaches for reducing networks' size

Family	Approach	Complexity	Advantages	Limits
Coarsening	RM [18]	O(m)	- Linear - Simple	- Damage the quality of partition
	Hendrickson et Leland [18]		- Good performance in a variety of graphs	- Significant memory consumption
	LEM [19]	O(m)	- Simple	- Contracted graph with high degree nodes
	Chen et al. [27]	O(n+mlog(n))	- Treat scale free networks - Conserve the proximity of first and second order	- Parameterized algorithm
	MCCA [1]	O(m+nlog(n))	- Conserve the original graph properties - Suitable for large scale networks	- Pseudo-linear time complexity
	M2M [17]	O(m+n)	- A small size contracted graph	- Unsuitable for power low degree distribution networks
	SLM [32]		- Improve the modularity of the Louvain algorithm	- Significant calculation time compared to Louvain
Sampling	RN [5]	O(n)	- Simple strategy for nodes selection	- Unstable due to its random strategy - The low degree distribution is not conserved
	RDN and RPN [5]	O(m)	- Improve the random node choice by using a specified heuristic	- RDN: significant number of high degree nodes - RPN: dense graphs
	Zhu et al. [12]	O(ktn)+O(m+n) with t the number of iterations and k the number of K-Means clusters	- Preserve the low degree distribution	- Limited by the sampling rate
	AdpUNI and AdpUNI+ [22]		- Improve the existing UNI (uniform sampling) model in terms of biased sampling	- The sampling size is manually defined
	Zhou et al. [21]		- Use contextual structures to create the sample	- Samples with high number of clusters
	RE [5]	O(m)	- Simple, random choice of edges	- The random strategy may affect the network's structure
	TIES [20]	O(m)	- Improve the random choice of RE	- Samples with a significant number of links
	Wang et al. [9]		- Low computational complexity compared to other models	- Networks with self-similarity structure
	DGS	O(m)	- Scale over huge networks using distributed edge sampling - Preserve the original network's properties	
	Yanagiya et al. [35]		- Sampling from large scale graphs	- Experimental study limited on some edge selection methods
	FF [5]		- A small size sample - Conserve structural properties	- Polynomial time complexity
	RWS [15]		- Simple, choose nodes based on random walks	- Risk of getting stuck in an isolated component
	DRaws [16]		- Improve the choice of nodes with the simple random walks strategy	- Require an additional time for shortest paths calculation
	CNARW [11]		- Speed up the convergence of random walk sampling	- Degree distribution deviation for the created samples
	RD [13,15]		- Deterministic graph exploration	- Parameterized algorithm

choice of Spark [25] is argued by the fact that this paradigm puts intermediate data into memory instead of saving it on disk, which makes it much faster. Spark offers also libraries helping to manipulate data in parallel (e.g. GraphX). Indeed, in this work the input graph is partitioned using the node cutting strategy by Graphx and is distributed over a set of machines. Each machine contains one or more partition(s). For the first distributed sampling algorithm, nodes are selected randomly from each partition until the desired sample size is reached. For the second type (distributed edge sampling), a set of links is randomly selected in the reduced sub-graph. For the last one (topology-based distributed sampling), two steps are developed. The first step is to label some nodes and the second stage is designed to sample the graph based on the assigned labels. The advantage of this work is to draw several comparisons between centralized and distributed approaches. In this direction, Gomes et al. [24] proposed distributed versions of some sampling algorithms. In fact, they implemented four distributed versions of RVS models (Random Vertex Sampling), RES (Random Edge Sampling), NS (Neighborhood Sampling) and RWS (Random Walk Sampling). Distribution is performed using the paradigm Spark. Another recent direction is that of using deep learning to formulate the graph sampling problem as a reinforcement learning process. In this direction, Wu et al. [10] have adopted a deep learning strategy to make agents able to select nodes at each time stamp. Then the sample is formed at the end of an episode by the best nodes. Yang et al. [26] have also investigated the deep reinforcement learning to sample networks using directed associative graph. The purpose of this work is to preserve the connection relation of all samples among all episodes. The proposed model demonstrated its efficiency especially for verifying the directed associative graph criteria.

6   Conclusion

In this paper, we focus on the problem of reducing social networks' size. Many current types of research on this problem are developed, and we have discussed their advantages and limits. Our main contribution is to survey the existing models for such problems and present a clear categorization of them. Recent directions for treating large-scale networks are also presented. For the time being, our immediate concern is to test some efficient models on real-world, large-scale social networks and to present an experimental study.

References

1. Rhouma, D., Ben Romdhane, L.: An efficient multilevel scheme for coarsening large scale social networks. Appl. Intell. **48**, 3557–3576 (2018)
2. Jaouadi, M., Ben, R.L.: A distributed model for sampling large scale social networks. Expert Syst. Appl. **186**, 115773 (2021)
3. Hu, P., Lau, W.C.: A survey and taxonomy of graph sampling. CoRR (2013)
4. Liao, Q., Yang, Y.: Incremental algorithm based on wedge sampling for estimating clustering coefficient with MapReduce. In: 2017 8th IEEE International Conference on Software Engineering and Service Science (ICSESS), pp. 700–703 (2017)

5. Jure, L., Christos, F.: Sampling from large graphs. In: Proceedings of the 12th ACM SIGKDD International Conference on Knowledge Discovery and Data Mining, pp. 631–636 (2006)
6. Seshadhri, C., Pinar, A., Kolda, T.: Edge sampling for computing clustering coefficients and triangle counts on large graphs. Stat. Anal. Data Min. **7**, 294–307 (2014)
7. Newman, M.E.J., Girvan, M.: Finding and evaluating community structure in networks. Phys. Rev. E Stat. Nonlinear Soft Matter Phys. **69**, 026113 (2004)
8. Wakisaka, Y., Yamashita, K., Tsugawa, S., Ohsaki, H.: On the effectiveness of random node sampling in influence maximization on unknown graph. In: 2020 IEEE 44th Annual Computers, Software, and Applications Conference (COMPSAC), pp. 613–618 (2020)
9. Wang, W., Fu, X., Lin, X.: Edge-based sampling for complex network with self-similar structure. In: 2021 IEEE Intl Conference on Parallel and Distributed Processing with Applications, Social Computing and Networking, pp. 955–962 (2021)
10. Wu, M., Zhang, Q., Gao, Y., Li, N.: Graph signal sampling with deep Q-learning. In: 2020 International Conference on Computer Information and Big Data Applications (CIBDA), pp. 450–453 (2020)
11. Wang, R., et al.: Common neighbors matter: fast random walk sampling with common neighbor awareness. IEEE Trans. Knowl. Data Eng. (2022)
12. Zhu, J., Li, H., Chen, M., Dai, Z., Zhu, M.: Enhancing stratified graph sampling algorithms based on approximate degree distribution. In: Silhavy, R. (ed.) CSOC2018 2018. AISC, vol. 764, pp. 197–207. Springer, Cham (2019). https://doi.org/10.1007/978-3-319-91189-2_20
13. Salamanos, N., Voudigari, E., Yannakoudakis, E.: Deterministic graph exploration for efficient graph sampling. Soc. Netw. Anal. Min. **7**, 1–14 (2017)
14. Khanam, K.Z., Srivastava, G., Mago, V.: The homophily principle in social network analysis: a survey. Multimed. Tools Appl. (2022)
15. Voudigari, E., Salamanos, N., Papageorgiou, T., Yannakoudakis, E.: Rank degree: an efficient algorithm for graph sampling. In: 2016 IEEE/ACM International Conference on Advances in Social Networks Analysis and Mining (ASONAM), pp. 120–129 (2016)
16. Zhang, L., Jiang, H., Wang, F., Feng, D.: DRaWS: a dual random-walk based sampling method to efficiently estimate distributions of degree and clique size over social networks. Knowl.-Based Syst. **198**, 105891 (2020)
17. LaSalle, D., Karypis, G.: Multi-threaded modularity based graph clustering using the multilevel paradigm. J. Parallel Distrib. Comput. **76**, 66–80 (2014)
18. Hendrickson, B., Leland, R.: A multi-level algorithm for partitioning graphs. Supercomputing 1995: Proceedings of the 1995 ACM/IEEE Conference on Supercomputing, p. 28 (1995)
19. Karypis, G., Kumar, V.: Multilevel k-way partitioning scheme for irregular graphs. J. Parallel Distrib. Comput. **48**, 96–129 (1998)
20. Ahmed, N., Neville, J., Kompella, R.: Network sampling via edge-based node selection with graph induction. Department of Computer Science Technical Reports (2011)
21. Zhou, Z., et al.: Context-aware sampling of large networks via graph representation learning. IEEE Trans. Vis. Comput. Graph. **27**, 1709–1719 (2021)
22. Cai, G., Lu, G., Guo, J., Ling, C., Li, R.: Fast representative sampling in large-scale online social networks. IEEE Access **8**, 77106–77119 (2020)
23. Zhang, F., Zhang, S., Lightsey, C.: Implementation and evaluation of distributed graph sampling methods with spark. Electron. Imaging 1–9 (2018)

24. Gomez, K., Täschner, M., Rostami, M.A., Rost, C., Rahm, E.: Graph sampling with distributed in-memory dataflow systems. CoRR (2019)
25. Apache Spark. Apache Spark Lightning-Fast Cluster Computing (2015). Spark.Apache.Org. Last accessed April 2022
26. Yang, D., Qin, X., Xu, X., Li, C., Wei, G.: Sample-efficient deep reinforcement learning with directed associative graph. China Commun. 18(6), 100–113 (2021)
27. Chen, H., Perozzi, B., Hu, Y., Skiena, S.: HARP: hierarchical representation learning for networks. CoRR (2017)
28. Preen, R.J., Smith, J.: Evolutionary n-level hypergraph partitioning with adaptive coarsening. IEEE Trans. Evol. Comput. 23, 962–971 (2019)
29. Kernighan, B.W., Lin, S.: An efficient heuristic procedure for partitioning graphs. Bell Syst. Tech. J. 49, 291–307 (1970)
30. Blondel, V., Guillaume, J., Lambiotte, R., Lefebvre, E.: Fast unfolding of communities in large networks. J. Stat. Mech. Theory Exp. (2008)
31. Noack, A., Rotta, R.: Multi-level algorithms for modularity clustering. In: Vahrenhold, J. (ed.) SEA 2009. LNCS, vol. 5526, pp. 257–268. Springer, Heidelberg (2009). https://doi.org/10.1007/978-3-642-02011-7_24
32. Waltman, L., van Eck, N.J.: A smart local moving algorithm for large-scale modularity-based community detection. Eur. Phys. J. B 86, 1–14 (2013)
33. Chen, J., Saad, Y., Zhang, Z.: Graph coarsening: from scientific computing to machine learning. SeMA J. 79, 187–223 (2022)
34. Zhang, L.-C.: Graph sampling: an introduction. Surv. Stat. 83, 27–37 (2021)
35. Yanagiya, K., Yamada, K., Katsuhara, Y., Takatani, T., Tanaka, Y.: Edge sampling of graphs based on edge smoothness. In: ICASSP 2022 -IEEE International Conference on Acoustics, Speech and Signal Processing (ICASSP), pp. 5932–5936 (2022)

AuCM: Course Map Data Analytics for Australian IT Programs in Higher Education

Jianing Xia, Yifu Tang$^{(\boxtimes)}$, Taige Zhao$^{(\boxtimes)}$, and Jianxin Li

School of Information Technology, Deakin University, Geelong, Australia
{xiaji,tangyif,zhaochr,jianxin.li}@deakin.edu.au

Abstract. Concept maps have emerged as an essential tool for illustrating the mutual relationships between knowledge in the domain. It provides guidance and suggestions in many educational applications such as optimal study order generating, curriculum design and course evaluation. However, there are no consistent datasets for concept map research. In this work, we aim to build a comprehensive dataset for learning concept maps. Specifically, we collect 1292 undergraduate courses in Information Technology (IT) and Computer Science (CS) from 14 Australian universities including course ID, category and prerequisite requirements. Besides, we analyze the semantic properties based on the concepts retrieved from the course description and visualize them to illustrate how our dataset could be used. To the best of our knowledge, this is the first dataset containing course information from Australian universities.

Keywords: Course map · Educational dataset · Prerequisite relation

1 Introduction

With the increasing availability of online education resources, it is significant to organize learning resources in a reasonable order. This has led to a boom in research into concept maps. Concept Map is a graphical tool to obtain relations between concepts such as dependencies, associations, co-occurrences and correlations. The educational concept map is a means of subject knowledge visualization. It focuses on the pedagogic relationship between concepts, especially prerequisite dependency. For instance, we need to have the basic knowledge of "probability density" in order to learn "mathematical expectation". Therefore, the concept "probability density" is a prerequisite concept of the concept "mathematical expectation". Educational concept map plays a critical role in lots of educational applications such as curriculum planning, teaching diagnosis, knowledge tracing and intelligent tutoring [1–4]. However, current educational concept map analysis works do not have available public datasets to support related research. To solve this problem, they generate private datasets by using

© The Author(s), under exclusive license to Springer Nature Switzerland AG 2022
W. Chen et al. (Eds.): ADMA 2022, LNAI 13725, pp. 158–172, 2022.
https://doi.org/10.1007/978-3-031-22064-7_13

concept and link extraction methods [5–8]. However, the data collection is time-consuming and the private synthesis data is tendentious to their analysis result. Therefore, it is important to provide a public well-organized dataset to generate an educational concept map.

In this work, we aim to build such a dataset named AuCM (Australian Courses Map data). In the first step, we collect the related data of undergraduate courses from universities in Australia. We browse all accessible undergraduate programs in the area of IT and CS from Sydney University, the University of Queensland, Monash University and other eleven universities. Overall, we collect 1292 courses from 14 universities with comprehensive information including course ID, title, category and course description. Besides basic information about the course, the course catalogs also provide course prerequisite information. For example, "SIT221 - Data Structures and Algorithms" is a prerequisite course for "SIT320 - Advanced Algorithms". In our organized dataset, all the information is well-formatted and easily accessible.

In summary, our main contributions of this paper are as follows:

1. We collect a dataset including course information of bachelor programs in IT&CS area and the corresponding prerequisite dependencies from 14 universities in Australia, called Australian Courses Map data (AuCM).
2. We analyze the properties of AuCM from the number of courses, curriculum design and prerequisites among different universities through comparative analysis and visualization.

The structure of this article is as follows. We review the related work on the task of concept map dataset collection in Sect. 2. In Sect. 3, we introduce the construction of AuCM dataset. In Sect. 4, we analyze the statistical properties of AuCM. Finally, we explore the concept semantics of our dataset in Sect. 5, and summarize our work in Sect. 6.

2 Related Work

Many works on concept map-related dataset collection have been done in order to generate concept maps in educational environments. Based on the data source, it can be roughly divided into university websites, online education websites and textbooks.

Dataset from University Websites. Yang et al. [9] collected a dataset with 3509 courses data and available prerequisite requirements from four universities (MIT, CMU, Caltech and Princeton) in U.S. Their dataset infer the course prerequisites by constructing concept-level directed graphs. Liang et al. [10] provided the other related dataset. Their work includes 654 courses data with prerequisite relations from 11 U.S universities in their dataset. They manually annotated 1008 pairs of concepts with prerequisite relations.

Dataset from Online Education Websites. Besides the course data collected from various websites of universities, some efforts were also devoted to building the dataset through online education websites, i.e., MOOCs. The dataset mentioned in [11] is based on video playlists from a leading MOOC platform. It collected a total of 1346 course videos from 20 courses with different domains. In addition, this dataset annotated 573 course concepts with 3504 pairs of prerequisite relations among them manually. Similarly, Roy et al. [12] built a dataset based on the subtitles of videos from a famous Mooc platform in India. They used 382 videos from 38 different playlists in computer science departments. In total, their dataset includes 345 concepts and 1455 prerequisite edges between video pairs.

Dataset from Textbooks and Others. In addition, Wang et al. [13] firstly built a dataset to construct a concept map using six textbooks: computer networking, physics, databases and so on. Then they extracted key concepts through mining Wikipedia concepts and content cosine similarity with sub-chapter topics. They also manually labeled the prerequisite relationships between key concepts. Followed by Wang et al., Huang et al. [14] used multi-source data including mathematics textbooks as well as student question logs to build a dataset. A Similar method [13] is also employed to extract key concepts and relations. Lu et al. [15] created a dataset by extracting concepts from 18 textbooks of three domains: Calculus, Data Structure, and Physics. They annotated related domain concepts (318) and prerequisite relations (1522 pairs) among them.

Compared with them, our dataset includes courses from Australian universities with updates in 2022. To our best knowledge, there are no similar datasets in other works. In addition, our dataset includes more attributes, such as compatible courses and category information, to recover more semantic relations between courses.

Prerequisite Relationship Extraction. Based on Educational Data Mining, various downstream applications have been explored to leverage educational resources [16–19]. Our work is highly related to prerequisite relationship extraction among concepts. Based on existing raw data from websites, several efforts have been devoted to extracting prerequisite relations to construct educational concept map. Yang et al. [9] created a concept graph by mapping courses to concepts and learning concept-level dependencies in order to accomplish interlingua-style transfer learning. Liang et al. [10] recovered an accurate and universally shared concept graph based on the observed course dependencies. Unlike the studies above which used course-related data, Huang et al. [5] considered student test records and the relationship between questions and various concepts in order to perform data mining algorithms for generating concept maps. Pan et al. proposed a method MOOC-RF to recover the concept prerequisites from Coursera data [11]. They built a set of features and trained a classifier to recognize prerequisite relationships between concepts in video transcripts. Huang et al. [14] proposed a framework to extract prerequisite and collaboration relationships from multi-source of education data, like textbooks, student question logs and Wikipedia.

3   The AuCM Dataset

We collect all possible courses of bachelor degree from 14 Australian universities in IT and CS. For comparative analysis, these universities are selected from five states including the Australian Capital Territory (ACT) and half of the universities are from the "Group of Eight (G8)". Our work on the AuCM dataset construction consists of two steps: data scraping and data processing. During the data scraping phase, we collect web pages from 1292 undergraduate courses. Then we extract the relevant information and organize the raw data during the data processing phase.

3.1   Data Scraping

For data scraping, we create multiple crawlers under Python 3.8 to scrape data from various university courses. First, we analyze the information structure of the target web to cover the key information. The requests library is used to send HTTP requests to the server during the crawling process. In general, our crawler downloads the pages of each undergraduate program which list the course requirements and extracts the URLs of each course using the Beautiful Soup library. Then, the crawler uses these URLs to request courses' web pages and saves the HTML files for each course. During the crawling process, crawlers may trigger anti-crawler techniques on some websites of institutions. When the crawler accesses the websites of specific courses frequently, the website may refuse the next request. To solve this, we use a longer random waiting period for each visit interval ranging from 2–8 s to prevent these mechanisms from being triggered frequently.

3.2   Data Processing

After scraping the raw data, we aim to extract the useful data during data processing. We first analyze the page structure of the downloaded HTML file and use Beautiful Soup to extract critical information from it. Since different universities have their own HTML structure, we program various crawlers to get useful information. Generally, attributes of a course include course ID, title, category, course description and relations between courses. During the data processing, we may encounter an error called "404 Not Found". This is mainly because some courses are no longer offered by the university. In this case, we need to remove it from our URL list and retrieve other possible course information in the IT&CS bachelor's program.

We create an Excel file for each IT&CS undergraduate program of different universities and write course information into each row. Then the data can be readily turned into the specific data format by using existing python packages, e.g., the Pandas package. The CIAU22 dataset is generated after data processing, which includes 6 attributes. These attributes together describe the features of a course. The specific attributes are listed and explained in Table 1.

4 Statistical Analysis of AuCM

In this section, we conduct statistical analysis about the number of courses, curriculum design for IT&CS bachelor degree programs, core curriculum and courses with prerequisites.

Table 1. Overview of course feature

#	Attribute	Description	Example
1	CourseID	Unique code of a specific course	COMP1600
2	Title	Course name	Foundations of computing
3	Category	Course category information such as core units or a specific major/minor	Compulsory courses of Software information systems
4	Prerequisites	Course(s) that would have been completed before enrolling in a certain course	COMP2303 or CSSE2310
5	Incompatible	You can't enrol in a particular course if you have completed any of these incompatible units	None
6	Summary	Description of course content	Statistical modeling is a set of techniques for analyzing data in order to obtain insight into real-world issues. This unit's goal is to...

Table 2. IT&CS courses of universities in NSW/ACT

#	G8 or Not	University	#Courses	#Courses with prerequisites
1	G8	Sydney University, USYD	126	52
2		The Australian National University, ANU	116	109
3	Not G8	University of Technology Sydney, UTS	71	62
4		Australian Catholic University, ACU	25	16

4.1 Analysis of the Number of Courses

Our dataset AuCM focuses on all the courses required to get a Bachelor degree in IT (BIT) and/or CS (BCS). Students are generally required to complete the core courses, specialized courses, and elective courses in order to earn a bachelor's degree in IT/CS. We obtain all available course information of core units and courses of optional majors or minors from the websites of BIT&BCS programs. The core units, major or minor information corresponds to the category feature of Table 1. For comparative analysis, we select 7 universities from the "Group of Eight (G8)" and 7 non-G8 universities from 5 states and ACT. For example, Sydney University and the Australian National University which belong to G8 in New South Wales (NSW)/ACT are included. University of Technology Sydney and Australian Catholic University (non-G8) in NSW are also contained.

The statistics of IT&CS courses from these NSW/ACT universities are listed in Table 2, which covers the total number of courses and the number of courses with prerequisite requirements. Data comparison in Table 2 shows that universities belonging to G8 generally offer significantly more courses than non-G8 universities in BIT&BCS programs. Like courses of USYD and ANU are more than 100 while UTS offers 71 courses and ACU only offers an example study plan which has 25 courses in total. This phenomenon is also evident in the other three states (Queensland, Western Australia and South Australia, shorted for QLD, WA and SA), as shown in Table 3.

Table 3. IT&CS courses of universities in QLD, WA, SA

#	G8 or not	University	#Courses	#Courses with prerequisites
1	G8	The University of Queensland, UQ	160	148
2	Not G8	Queensland University of Technology, QUT	85	63
3	G8	University of Western Australia, UWA	131	111
4	Not G8	Edith Cowan University, ECU	50	26
5	G8	The University of Adelaide, ADELAIDE	85	68
6	Not G8	University of South Australia, UN	69	50

It is obvious that the G8 universities of Queensland (UQ), Western Australia (UWA) and Adelaide (ADELAIDE) have more courses than their respective non-G8 universities in each state. In fact, the course number of UWA is more than 2 times that of ECU, with 131 and 50 courses respectively.

There is an exception to this phenomenon in the state of Victoria (VIC), Deakin University, not belonging to G8, offers more courses than G8 universities - Monash University and the University of Melbourne, with a total of 126 courses, the most in VIC. The IT&CS course statistics of universities in VIC can be seen in Table 4.

Table 4. IT&CS courses of universities in VIC

#	G8 or not	University	#Courses	#Courses with prerequisites
1	G8	Monash University, MONASH	119	44
2		University of Melbourne, UofMELB	70	59
3	Not G8	Deakin University, DEAKIN	126	81
4		Victoria University, VU	60	45

Figure 1 provides a comprehensive view of IT&CS courses from 14 universities. It can also be seen that G8 universities (green bars) generally have more total courses than non-G8 universities (blue bars). The average number of courses of these G8 universities is 115 which corresponds to 69 of non-G8 universities. This may be attributed to the fact that most non-G8 universities don't offer the bachelor of CS or CS as a major is included in the bachelor of IT. Among them, UQ has the largest total number of IT&CS courses, with 160 courses and 8 majors to be selected, While the smallest one belongs to ACU which has 25 courses as also mentioned in Table 1.

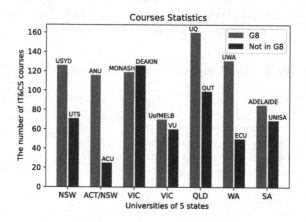

Fig. 1. #Courses vs. universities in 5 states/ACT of Australia.

4.2   Analysis of Curriculum Design

The total number of courses indicates the scope of courses in the IT/CS field. We also find that 12 of the 14 universities offer an explicit plan for IT/CS bachelor's degrees. The other 2 universities are the universities of Melbourne and Western Australia, which don't offer a bachelor's degree in IT, like the University of Melbourne places IT&CS courses into a bachelor of design, so our dataset only covers major courses in this field. Most universities that offer a bachelor's degree in IT&CS provide a rich curriculum plan. For instance, there are many curriculum options to choose from in UQ. In general, students have to complete core courses, courses for plan options and elective courses in UQ. Core units are compulsory which lay a foundation for the study area. In addition, students can choose plan from many options like single major, singer minor, major plus minor, two majors, or even no major; Elective courses are divided into program electives and general electives, for the area breadth and general purpose respectively. Our dataset doesn't cover elective courses as the majority of them are not in the IT&CS domain, like general electives in UQ are courses that can be selected from any other undergraduate course list.

We can get an intuitive impression of the distribution of course categories in UQ from Fig. 2 and 3. Notably, the majority of courses belong to various majors and minors in both pie charts, with 36 courses and 55.4% in the BIT program

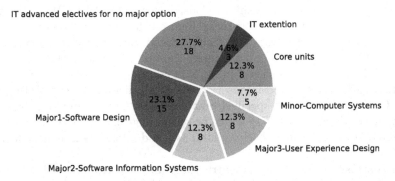

Fig. 2. Distribution of course categories for BIT in UQ. There are 3 majors and one minor, accounting for 55.4% of the courses.

and 50 courses and 52.6% in the BCS program. Similarly, most courses in the curriculum of other universities are also major-related courses. Like Monash University has 7 majors for students to choose from in the BIT program. Deakin University has a total of 6 majors and 13 minors in the BIT and BCS programs.

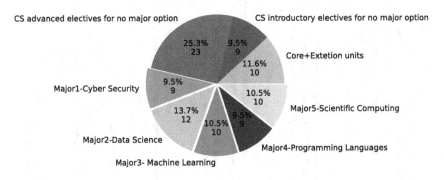

Fig. 3. Distribution of course categories for BCS in UQ. 5 majors account for 52.6% of the total number of courses.

4.3   Analysis of Core Curriculum

We also compare and analyze the core curriculum of BIT/BCS programs at different universities. Out of Australia's 14 universities, 12 offer bachelor's degrees in IT and 5 universities provide CS bachelor degrees for undergraduates to pursue. 7 of the 12 universities offering the Bachelor of IT list their core units.

Other universities make these important foundation courses compulsory instead of referring to them as "core units." Consistently, UTS, VU, UWQ and ADE-LAIDE universities all offer 8 core units. Figure 4 shows the different core courses from these schools. Notably, programming-related courses appear in the core curriculum of each university. Similar courses can also be found in the core units of the other 3 universities, as shown in Fig. 5. In addition, ANU sets the "Structured Programming" course as a compulsory course, ACU lists the course of "Programming Concepts" as an "IT specified unit", and the course requirements for the first year at UNISA and ECU include programming-related courses. Among the core modules, other courses such as databases and information management are well-liked.

In contrast, courses of "Data Structures and Algorithms" and "Computer System" are more common in core units of CS field. The core courses for the BCS at USYD, UQ, Deakin, Adelaide, and Monash Universities are displayed in Fig. 6, algorithm-related courses are shown in orange, and computer systems courses are indicated in blue. As can be seen from the figure, the CS undergraduate core courses of the 5 schools all include courses related to "Data Structure and Algorithm" and "Computer System". In addition to these common courses, many universities also offer programming courses as part of their core curriculum.

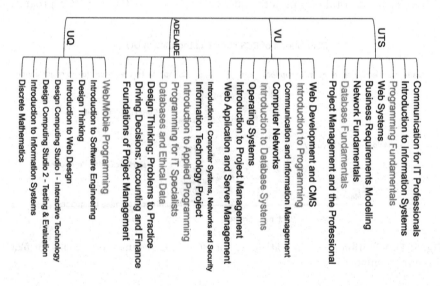

Fig. 4. Core units of BIT program from UTS, VU, UQ and ADELAIDE universities, programming-related courses are shown in blue, and database courses are indicated in green. (Color figure online)

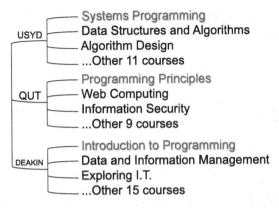

Fig. 5. Core units of BIT program from USYD, QUT and DEAKIN universities with programming-related courses are shown in blue. (Color figure online)

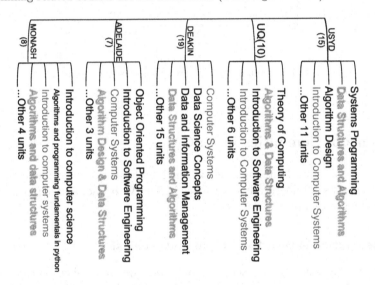

Fig. 6. Core courses for the BCS program at USYD, UQ, DEAKIN, ADELAIDE, and MONASH Universities, algorithm-related courses are shown in orange, and computer systems courses are indicated in blue. (Color figure online)

4.4   Analysis of Prerequisites

Another observation of dataset AuCM is that a majority of courses have at least one prerequisite course. Prerequisite requirement represents the sequential orders between courses. In total from the 14 universities, there are 811 unique IT&CS undergraduate courses, 511 courses with prerequisite requirements, and 1475 pairs of courses with prerequisite relations. In addition, the average number of prerequisite links per course is 1.82. Figure 7 shows the proportion of the number of courses with prerequisites at each university. The red line is drawn by a linear regression algorithm, indicating that the proportion of prerequisite courses is around 75%. The largest proportion belongs to ANU with 94.8% of courses having prerequisite requirements, while the smallest one refers to Monash University with 37.0% of courses having prerequisite dependencies.

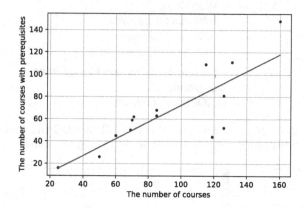

Fig. 7. #Courses with prerequisites vs. #courses.

5   Concept Semantics in AuCM

In this section, we evaluate and analyze the latent semantic features included in diverse concepts of undergraduate courses in AuCM dataset and explore concept map learning based on the dataset.

5.1   Semantic Feature Extraction and Analysis

To recover the concept-level semantic features, we first collect concepts from Wikipedia and process course descriptions by employing various traditional NLP tools, such as stop word removal, sentence segmentation, and lemmatization. Then we match the Wikipedia concept appearance with the course description. By using a pre-trained tokenizer from the BERT model, we tokenize the course description and retrieved concepts respectively in order to capture semantic characteristics. Furthermore, we use t-SNE to project these high-dimensional features into a two-dimensional map for visual analysis.

By drawing concept word clouds and scatter plots for concept/course description based on the projected 2D features, we can see that the focus areas of concepts and courses in BIT/BCS programs vary from school to school. We compare the word clouds of the course concepts between the universities of "Group of 8 (G8)" and non-G8 which are shown in Fig. 8.

As the word clouds show, common concepts such as data structures and operating systems are popular with high frequency in the course descriptions

Fig. 8. Word clouds of concepts extracted from the course descriptions of G8 (left) and non-G8 (right) universities.

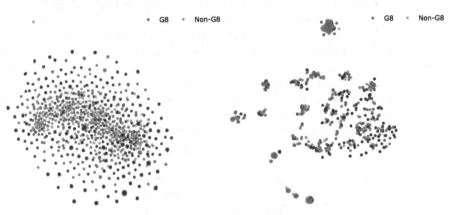

Fig. 9. Scatter plots of projected features extracted from course descriptions/concepts in the courses from G8 and non-G8 universities.

of various universities, and the difference is that advanced topics like "artificial intelligence" and "neural network" are frequently included in the course offerings at G8 universities, while basic topics of "operating systems" and "software development" are more frequently covered in the courses offered by universities of non-G8. This may indicate that G8 universities pay more attention to develop students' capability in depth, whereas Non-G8 universities emphasize application and fundamentals. Another comparison of curriculum design between G8 schools and non-G8 institutions is shown in Fig. 9. The scatter plot on the left gives the distribution of words in the course descriptions, where the blue dots are the words described in the courses of G8 universities, and the red dots belong to the non-G8 school courses. It is clear that the blue dots are more scattered and span a larger area, indicating that the G8 course descriptions employs more semantically rich words and covers a broader diversity. The scatter plot on the right, which displays the distribution of concepts, also demonstrates this. Compared to relatively dispersed blue points, the red points tend to cluster in some conceptual areas, showing that non-G8 courses may pay more attention to the particular concepts rather than breadth.

5.2   Concept Map Learning

In addition to conceptual semantic comparison among domestic universities in Australia, our dataset is primarily utilized to create concept maps. Concept map construction intends to extract structured information from unstructured text and represent it as a graph, where concepts constitute the vertices and prerequisite dependencies make up the links. An example of the concept map is shown in Fig. 10, where the vertices come from the extracted concepts, and the links, represented by blue dotted lines, indicate the relationships between concepts based on a simple assumption: the prerequisite relationship between two courses means that the concepts contained in these courses have links. It is observed that the vertex "data type" at the center of the concept graph has many connections with other vertices, indicating that the "data type" constitutes a prerequisite for many concepts. Our AuCM dataset can facilitate relevant research on concept map learning by complementing the course information from Australian universities since the present benchmark dataset is primarily from the United States, like Liang et al. [10] and Yang et al. [9] both constructed concept maps based on datasets collected from the US universities.

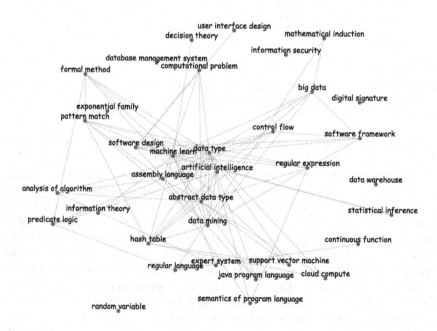

Fig. 10. An example of concept map.

Dataset AuCM also allows for validation of state-of-the-art methods on concept map construction in the Australian data setting and a comparative investigation of concept maps based on AuCM and other datasets. Other downstream applications such as course recommendation and knowledge tracing can also be supported by our dataset.

6   Conclusion

In this work, we build a dataset about Courses Map Data of Australian Universities named AuCM. It contains 1292 undergraduate courses of bachelor degree in the area of IT or CS from 14 Australian universities. The usability of our dataset is demonstrated by statistical and semantic studies. Our statistical analysis shows the superiority of the number of courses offered by G8 universities. By semantic analysis of concepts/course descriptions, we find the variances in course design in terms of semantic richness and focus areas among G8 and non-G8 universities. Future work would be to construct concept maps on AuCM to validate existing concept map generation method. Another direction would be to perform a more comprehensive model comparison and evaluation by bringing other possible variations of models to learn a concept graph.

Acknowledgements. This research was supported by a grant from Australian Research Council Linkage Project with Grant Number LP180100750.

References

1. Hirashima, T., Yamasaki, K., Fukuda, H., Funaoi, H.: Framework of kit-build concept map for automatic diagnosis and its preliminary use. Res. Pract. Technol. Enhanc. Learn. **10**(1), 1–21 (2015). https://doi.org/10.1186/s41039-015-0018-9
2. Boguski, R.R., Cury, D., Gava, T.: TOM: an intelligent tutor for the construction of knowledge represented in concept maps. In: IEEE Frontiers in Education Conference, FIE 2019, Cincinnati, OH, USA, pp. 1–7 (2019)
3. Song, X., Li, J., Tang, Y., Zhao, T., Chen, Y., Guan, Z.: JKT: a joint graph convolutional network based deep knowledge tracing. Inf. Sci. **580**, 510–523 (2021)
4. Song, X., Li, J., Lei, Q., Zhao, W., Chen, Y., Mian, A.: Bi-CLKT: Bi-graph contrastive learning based knowledge tracing. Knowl. Based Syst. **241**, 108274 (2022)
5. Huang, X., Yang, K., Lawrence, V.B.: Classification-based approach to concept map generation in adaptive learning. In: 15th IEEE International Conference on Advanced Learning Technologies, pp. 19–23 (2015)
6. Anwar, M.M., Liu, C., Li, J.: Discovering and tracking query oriented active online social groups in dynamic information network. World Wide Web **22**(4), 1819–1854 (2019)
7. Yang, C., Zhang, J., Wang, H., Li, B., Han, J.: Neural concept map generation for effective document classification with interpretable structured summarization. In: ACM SIGIR Conference on Research and Development in Information Retrieval, China, 25–30 July 2020, pp. 1629–1632. ACM (2020)
8. Lu, J., Dong, X., Yang, C.J.: Weakly supervised concept map generation through task-guided graph translation. CoRR, abs/2110.15720 (2021)

9. Yang, Y., Liu, H., Carbonell, J., Ma, W.: Concept graph learning from educational data. In: Proceedings of the Eighth ACM International Conference on Web Search and Data Mining, WSDM 2015, Shanghai, China, 2–6 February 2015, pp. 159–168 (2015)

10. Liang, C., Ye, J., Wu, Z., Pursel, B., Giles, C.L.: Recovering concept prerequisite relations from university course dependencies. In: Proceedings of the Thirty-First AAAI Conference on Artificial Intelligence, 4–9 February 2017, San Francisco, California, USA, pp. 4786–4791 (2017)

11. Pan, L., Li, C., Li, J., Tang, J.: Prerequisite relation learning for concepts in moocs. In: Proceedings of the 55th Annual Meeting of the Association for Computational Linguistics, pp. 1447–1456. Association for Computational Linguistics (2017)

12. Roy, S., Madhyastha, M., Lawrence, S., Rajan, V.: Inferring concept prerequisite relations from online educational resources. In: The Thirty-Third AAAI Conference on Artificial Intelligence, pp. 9589–9594. AAAI Press (2019)

13. Wang, S., et al.: Using prerequisites to extract concept maps from textbooks. In: Proceedings of the 25th ACM International on Conference on Information and Knowledge Management, pp. 317–326. ACM (2016)

14. Xiaoqing, H., et al.: Constructing educational concept maps with multiple relationships from multi-source data. In: IEEE International Conference on Data Mining, Beijing, China, 8–11 November 2019, pp. 1108–1113 (2019)

15. Weiming, L., Zhou, Y., Yu, J., Jia, C.: Concept extraction and prerequisite relation learning from educational data. In: The Thirty-Third AAAI Conference on Artificial Intelligence, pp. 9678–9685. AAAI Press (2019)

16. Hu, Y., Li, H., Zhou, Z., Li, H.: A new intelligent learning diagnosis method constructed based on concept map. Int. J. Pattern Recognit. Artif. Intell. 35(7), 2159023:1–2159023:18 (2021)

17. Kong, X., Xia, F., Li, J., Hou, M., Li, M., Xiang, Y.: A shared bus profiling scheme for smart cities based on heterogeneous mobile crowdsourced data. IEEE Trans. Ind. Inform. 16(2), 1436–1444 (2020)

18. Wang, X., Chai, L., Qiang, X., Yang, Y., Li, J., Wang, J., Chai, Y.: Efficient subgraph matching on large RDF graphs using mapreduce. Data Sci. Eng. 4(1), 24–43 (2019)

19. Al Hasan Haldar, N., Li, J., Reynolds, M., Sellis, T., Yu, J.X.: Location prediction in large-scale social networks: an in-depth benchmarking study. VLDB J. 28(5), 623–648 (2019). https://doi.org/10.1007/s00778-019-00553-0

Profit Maximization Using Social Networks in Two-Phase Setting

Poonam Sharma and Suman Banerjee[✉]

Department of Computer Science and Engineering, Indian Institute of Technology
Jammu, Jammu and Kashmir 181221, India
{2021rcs1023,suman.banerjee}@iitjammu.ac.in

Abstract. Now-a-days, *Online Social Networks* have been predominantly used by commercial houses for viral marketing where the goal is to maximize profit. In this paper, we study the problem of Profit Maximization in the two-phase setting. The input to the problem is a *social network* where the users are associated with a cost and benefit value, and a fixed amount of budget splitted into two parts. Here, the cost and the benefit associated with a node signify its incentive demand and the amount of benefit that can be earned by influencing that user, respectively. The goal of this problem is to find out the optimal seed sets for both phases such that the aggregated profit at the end of the diffusion process is maximized. First, we develop a mathematical model based on the *Independent Cascade Model* of diffusion that captures the aggregated profit in an *expected* sense. Subsequently, we show that selection of an optimal seed set for the first phase even considering the optimal seed set for the second phase can be selected efficiently, is an NP-Hard Problem. Next, we propose two solution methodologies, namely the *single greedy* and the *double greedy* approach for our problem that works based on marginal gain computation. A detailed analysis of both methodologies has been done. Experimentation with real-world datasets demonstrate the effectiveness and efficiency of the proposed approaches. From the experiments, we observe that the proposed solution approaches leads to more profit, and in some cases the single greedy approach leads to up to 23% improvement compared to its single-phase counterpart.

1 Introduction

In recent times, Online Social Networks have emerged as an additional dimension of human life. Among many one of the important phenomena of online social networks is the *diffusion of information* using which information propagates from one part of the network to the other [5]. How we can make use of this influence in the context of viral marketing such that the profit can be maximized remains an active area of research [4,12].

The work of Dr. Suman Banerjee is supported by the Start Up Grant provided by the Indian Institute of Technology Jammu, India (Grant No.: SG100047).

Problem Background. Commercial houses use online social networks for viral marketing purposes [3]. The goal here is to identify a small set of influential users to activate initially that leads to maximum influence (in turn profit). These initially active nodes are called *seed nodes*. In particular, given a social network and positive integer k the problem is to choose a subset of k nodes to maximize the influence in the network [6]. This problem remains an active area of research in the domain of social network analysis [1].

Motivation. The traditional seed set selection methodologies for the influence and profit maximization problem consider that all the seed nodes will be deployed at one go before the diffusion process starts. However, some recent studies show that instead of one go if we split the budget into two or more parts and conduct the diffusion process in two or more rounds, then the number of influenced nodes increases [2,13]. Naturally, a question arises, does the same thing happen even for the Profit Maximization Problem. In this paper, we elaborately investigate this question.

Our Contributions. Our goal is to choose optimal seed sets for both the phases such that the aggregated profit is maximized. In particular, we make the following contributions in this paper:

- We study the PROFIT MAXIMIZATION PROBLEM using social networks in the two-phase setting. To the best of our knowledge, this is the first study on profit maximization in this direction.
- We develop a mathematical model for this problem that captures the expected profit at the end of the diffusion process. Subsequently, we show that selecting an optimal seed set for the first phase even considering the optimal seed set for the second phase can be selected efficiently is an NP-Hard Problem.
- We propose two algorithms, namely the single greedy and the double greedy approach along with their detailed analysis and both of them work based on marginal profit gain computation.
- Finally, we conduct an extensive set of experiments with real-world datasets to show that the proposed methodologies can lead to more amount of profit than the baseline methods.

Organization of the Paper. The rest of the paper is organized as follows. In Sect. 2, we describe the background and define the problem formally. Section 3 contains the proposed solution methodologies. Section 4 contains the experimental evaluation of the solution approaches. Finally, Sect. 5 concludes our study.

2   Background and Problem Definition

We represent the input social network by a simple (un)directed, and edge-weighted graph denoted by $G(V, E, \mathcal{P})$. Here, $V(G) = \{u_1, u_2, \ldots, u_n\}$ are the set of n users and $E(G) = \{e_1, e_2, \ldots, e_m\}$ are the set of m social ties. $\mathcal{P}$ denotes the edge weight function that maps each edge to its corresponding influence probability; i.e.; $\mathcal{P} : E(G) \longrightarrow (0, 1]$. For any edge $(uv) \in E(G)$, $\mathcal{P}_{uv}$ denotes the influence probability of the user u on v. If $(uv) \notin E(G)$ then $\mathcal{P}_{uv} = 0$. Each user of the network is associated with a cost and benefit value that are characterized by the cost and the benefit function denoted as C and b. Hence, $C : V(G) \longrightarrow \mathbb{Z}^+$ and $b : V(G) \longrightarrow \mathbb{Z}^+$. For any $u \in V(G)$, let $C(u)$ and $b(u)$ denote the cost and benefit associated with the user u.

To conduct the diffusion process in the network, a subset of the users are chosen as a seed user and they are considered to be influenced at time step $t = 0$. Now, the information is diffused in the network based on some rules. In this paper, we consider that the information in the network is diffused by the rule of the Independent Cascade Model (ICM). The node is either 'uninfluenced' or 'influenced'. Every influenced node activates its inactive neighbor in discrete time step t and an influenced node can't change to its previous uninfluenced state. The diffusion process ends when no more node activation is possible. The diffusion process can be expressed as a *live graph* and they are 2^m many where m denotes the number of edges in G. We denote these graphs as $L(G) = \{\mathcal{G}_1, \mathcal{G}_2, \dots, \mathcal{G}_{2^m}\}$. More about live graphs can be found in [6]. Next, we state the influence of a seed set in Definition 1.

Definition 1 (Influence of a Seed Set). *Assume that $\mathcal{S} \subseteq V(G)$ is the seed set. Now, at the end of the diffusion process starting from $\mathcal{S}$, the number of nodes that are influenced is called the influence of $\mathcal{S}$. We denote this by $\sigma(\mathcal{S})$ where $\sigma()$ is the social influence function that maps each subset of the user to their corresponding influence value, i.e., $\sigma : 2^{V(G)} \longrightarrow \mathbb{R}_0$ with the condition $\sigma(\emptyset) = 0$.*

Definition 2 (Social Influence Maximization Problem). *Given a social network $G(V, E, \mathcal{P})$, and a positive integer k the goal of the social influence maximization problem is to choose a subset of k nodes $\mathcal{S} \subseteq V(G)$ such that their initial activation leads to maximum number of influenced nodes. Mathematically, this problem can be stated as follows:*

$$\mathcal{S}^{OPT} = \underset{\mathcal{S} \subseteq V(G) \ and \ |\mathcal{S}| \leq k}{argmax} \ \sigma(\mathcal{S}) \tag{1}$$

Here, $\mathcal{S}^{OPT}$ denotes an optimal seed set of size k.

Definition 3 (Benefit Earned by a Seed Set). *Given a social network $G(V, E, \mathcal{P})$ and a seed set $\mathcal{S}$ we denote the benefit obtained by $\mathcal{S}$ as $\beta(\mathcal{S})$ and defined in terms of expectation over the set of all possible live graphs. Mathematically, this can be defined using Eq. 2.*

$$\beta(\mathcal{S}) = \mathbb{E} \left[\sum_{v \in I_{\mathcal{G}_i}(\mathcal{S}) \cap V(G)} b(v) \right] \tag{2}$$

Here, $I_{\mathcal{G}_i}(\mathcal{S})$ denotes the set of influenced nodes from the seed set $\mathcal{S}$ in the i-th live graph. $C(\mathcal{S})$ denotes the total cost of the seed set, i.e., $C(\mathcal{S}) = \sum_{v \in \mathcal{S}} C(v)$.

The expectation mentioned in Definition 2 is taken over the probability distribution of the benefit values in different live graphs. For a seed set $\mathcal{S}$, we denote its profit by $\phi(\mathcal{S})$ and can be defined using Eq. 3.

$$\phi(\mathcal{S}) = \beta(\mathcal{S}) - C(\mathcal{S}) \tag{3}$$

Definition 4 (Profit Maximization Problem). *Given a social network $G(V, E, \mathcal{P})$ where users of the network are associated with cost and benefit value and a fixed amount of budget $\mathcal{B}$ is given. The goal is to choose a subset of the nodes $\mathcal{S} \subseteq V(G)$ such that the profit is maximized. Mathematically, this problem can be stated as follows:*

$$\mathcal{S}^{OPT} = \underset{\mathcal{S} \subseteq V(G) \text{ and } C(\mathcal{S}) \leq \mathcal{B}}{argmax} \phi(\mathcal{S}) \tag{4}$$

Definition 5 (Two-Phase Profit Maximization Problem). *Given a social network $G(V, E, \mathcal{P})$ where users of the network are associated with cost and benefit value and budget for two phases $\mathcal{B}_1$ and $\mathcal{B}_2$, our goal is to choose seed nodes $\mathcal{S}_1$ and $\mathcal{S}_2$ to maximize the profit at the end of second phase such that $C(\mathcal{S}_1) \leq \mathcal{B}_1$ and $C(\mathcal{S}_2) \leq \mathcal{B}_2$.*

3   Mathematical Model and Solution Methodologies

Mathematical Model. Let, a live graph $\mathcal{G} \in L(G)$ with its generation probability $P(\mathcal{G})$ is destined to occur and $\mathcal{S}_1$ be the seed set for the Phase I. The diffusion process starts on a live graph $\mathcal{G}$ with seed set $\mathcal{S}_1$ by the rule of IC Model and observe this diffusion process till time step d. Then, we will have the information regarding which nodes are influenced and which are not. We call this as the partial observation till time step d and denoted as Y. So, at the end of time step d, we have already activated nodes denoted by A_Y and newly activated nodes (at time step d) R_Y. These two sets A_Y and R_Y are determined from the partial observation Y.

Now, as we have the partial observation Y, for a subset of the edges of the live graph $\mathcal{G}$ we are sure whether they have appeared or not and based on that we can update the generation probability $P(\frac{\mathcal{G}}{Y})$. Now, the second phase needs to begin, and assume that $\mathcal{S}_2^{OPT(Y,\mathcal{B}_2)}$ denotes the optimal seed set for Phase II when the partial observation Y and the budget is $\mathcal{B}_2$. At the time step d, we deploy the nodes in the set $\mathcal{S}_2^{OPT(Y,\mathcal{B}_2)}$ along with the nodes in R_Y both of them together will act as seed set for Phase II. Now, our goal is to calculate the expected profit that can be earned in Phase II. Now, it is important to observe that in Phase II , the nodes from which the profit can be earned will be the subset from $V(G) \setminus A_Y$. So, for the given partial observation Y (hence, newly activated nodes R_Y) along with an optimal seed set for Phase II, i.e., $\mathcal{S}_2^{OPT(Y,\mathcal{B}_2)}$ will be equals to $\sum_{\mathcal{G} \in L(G)} P(\frac{\mathcal{G}}{Y})[\phi^{V(\mathcal{G}) \setminus A_Y}(R_Y \cup \mathcal{S}_2^{OPT(Y,\mathcal{B}_2)})]$. Here, $\phi^{V(\mathcal{G}) \setminus A_Y}(\mathcal{S})$ denotes the profit earned by the seed set $\mathcal{S}$ from the graph $V(G) \setminus A_Y$.

We can observe that given a live graph $\mathcal{G}$, seed set of Phase I; i.e.; $\mathcal{S}_1$, and the time step d, we can get the partial observation Y. Hence, $S_2^{OPT(Y,\mathcal{B}_2)}$ can be written as $S_2^{OPT(X,\mathcal{S}_1,d,\mathcal{B}_2)}$. Now, to develop an objective function where the decision variable will be the seed set for Phase I, we assume that given the partial observation Y, we will select an optimal seed set for Phase II. It is important to observe that at the starting of Phase I, the partial observation Y is not known. Let our objective function be $\mathbb{F}(\mathcal{S}_1, d, \mathcal{B}_2)$ as the expected profit with respect to all possible occurrences Y. Assuming that d and $\mathcal{B}_2$ are already given, so we can write $\mathbb{F}(\mathcal{S}_1, d, \mathcal{B}_2)$ as $f(\mathcal{S}_1)$. In the following derivation, by $\mathcal{S}_2'$ we denote the set $R_Y \cup S_2^{OPT(Y,\mathcal{B}_2)}$.

$$
\begin{aligned}
f(\mathcal{S}_1) &= \sum_Y P(Y)\left\{\left\{\sum_{\mathcal{G}} P(\tfrac{\mathcal{G}}{Y})\phi(A_Y) + \sum_{\mathcal{G}} P(\tfrac{\mathcal{G}}{Y})[\phi^{V(\mathcal{G})\backslash A_Y}(\mathcal{S}_2')]\right\}\right\} \\
&= \sum_Y P(Y)\sum_{\mathcal{G}} P(\tfrac{\mathcal{G}}{Y})\left\{\left\{\phi(A_Y) + [\phi^{V(\mathcal{G})\backslash A_Y}(S_2^{OPT(X,S_1,d,\mathcal{B}_2)})]\right\}\right\} \\
&= \sum_Y P(Y)\sum_{\mathcal{G}} P(\tfrac{\mathcal{G}}{Y})\left\{\phi^{\mathcal{G}}(S_1 \cup S_2^{OPT(X,S_1,d,\mathcal{B}_2)})\right\} \\
&\left\{\because \sum_Y P(Y)\sum_{\mathcal{G}} P(\tfrac{\mathcal{G}}{Y}) = \sum_Y\sum_{\mathcal{G}}\frac{P(\mathcal{G},Y)}{P(Y)}P(Y)\right. \\
&= \left.\sum_Y\sum_{\mathcal{G}}P(\mathcal{G},Y) = \sum_{\mathcal{G}}\sum_Y P(\mathcal{G},Y) = \sum_{\mathcal{G}}P(\mathcal{G})\right\}
\end{aligned}
$$

The objective function formulated of our Two-Phase Profit Maximization Problem is as follows:

$$
\therefore f(\mathcal{S}_1) = \sum_{\mathcal{G}} P(\mathcal{G})\phi^{\mathcal{G}}(S_1 \cup S_2^{OPT(X,\mathcal{S}_1,d,\mathcal{B}_2)}) \tag{5}
$$

Now, it is important to observe that the developed model considers the optimal seed set selection in Phase II, which is itself an NP-Hard problem. So, Theorem 1 holds.

Theorem 1. *Finding the optimal seed set $\mathcal{S}_1$ that maximizes $f(\mathcal{S}_1)$ as mentioned in Eq. 5 is NP-Hard.*

Solution Methodologies. In this section, we describe two solution methodologies, namely Single Greedy and Double Greedy for the Two-Phase Profit Maximization Problem and both of them are based on Marginal Profit Gain which is stated in Definition 6.

Definition 6 (Marginal Profit Gain). *Given a social network $G(V, E, \mathcal{P})$, a seed set S, and a node $u \in V(G) \setminus S$ we denote the Marginal Profit Gain for the node u with respect to the seed set S as $\phi_u(S)$ and it is defined as the difference of profit earned when u is added to S and when u is not in S. The profit function*

$\phi()$ *may be non-monotone as well. As we are considering the 'gain', for any seed set S and node $u \in V(G) \setminus S$, $\phi_u(S)$ is defined only when $\phi_u(S) > \phi(S)$. Mathematically, this can be defined using Eq. 6.*

$$\phi_u(S) = \phi(S \cup \{u\}) - \phi(S) \quad such\ that\ \phi_u(S) > \phi(S) \tag{6}$$

Single Greedy Approach. Now, we describe the single greedy algorithm in two-phase setting. We have the following inputs: the social network $G(V, E, \mathcal{P})$, Budgets for both the phases $\mathcal{B}_1$ and $\mathcal{B}_2$, and the duration of Phase I which is d. Now, in the first phase, until the budget $\mathcal{B}_1$ is exhausted, we iteratively select seed node based on the marginal profit gain. In each iteration, for every non seed node u, we compute the marginal profit gain to its cost ratio and the node maximizes this quantity is found out. If the marginal profit gain of this node is strictly positive then it is included in the seed set of Phase I. It may so happen that the allocated budget for the first phase has not been exhausted totally. If so, the remaining budget of Phase I is added to Phase II. Now, we conduct diffusion process based on the IC Model starting from the seed set $\mathcal{S}_1$ till d-th time step. Thus, at the end of first Phase we have A_Y and R_Y as the influence of $\mathcal{S}_1$ at time step d. Now, we begin Phase II with the updated budget. The process of seed set selection is quite similar to the first phase with one difference. During the first phase we are dealing with the entire social network, and hence, while computing the marginal profit gain we consider the whole network. However, for the second phase we deal with the network obtained by deleting the already activated nodes from the original network. Accordingly, during the second phase while computing the marginal profit gain we consider the remaining network. Algorithm 1 shows the pseudocode of the proposed approach.

Next, we analyze this algorithm to understand its time and space requirement. Let, $\mathcal{C}_{min}$ denotes the minimum cost among all the nodes; i.e.; $\mathcal{C}_{min} = \min_{u \in V(G)} C(u)$. So it is easy to observe that the maximum number of nodes that can be selected as seed in Phase I will be of $\mathcal{O}(\frac{\mathcal{B}_1}{\mathcal{C}_{min}})$ and also these many iterations are required. In each iteration, the main computation involved is the marginal profit gain. It is easy to observe that in every iteration the number of nodes for which the marginal profit gain needs to be computed is of $\mathcal{O}(n)$. Now, for a given seed node computing the marginal gain is equivalent to traversing the graph and this takes $\mathcal{O}(m+n)$ time. In the worst case, the size of $\mathcal{S}_1$ can be of $\mathcal{O}(n)$. Hence, for one marginal gain computation time requirement is of $\mathcal{O}(n \cdot (m + n))$. As, there are $\mathcal{O}(n)$ many marginal profit gain computation, hence time requirement for this purpose is of $\mathcal{O}(n^2 \cdot (m + n))$. Now, choosing the node that causes the maximum marginal profit gain that takes $\mathcal{O}(n)$ time. So, the time requirement for the Phase I is of $\mathcal{O}(\frac{\mathcal{B}_1}{\mathcal{C}_{min}} \cdot n^2 \cdot (m + n))$. Now, before starting Phase II, we need to delete the already activated nodes in Phase I from the Graph. Now, the number of already activated nodes are of $\mathcal{O}(n)$. In the worst case, they may be incident with $\mathcal{O}(n^2)$ many edges. So, deleting these vertices from the graph requires $\mathcal{O}(n^2)$ time. The analysis of the second phase will remain the same except one difference. As in Phase II, the graph has been reduced by deleting the already activated nodes so $\mathcal{C}_{min}$ may not be minimum cost of the nodes in

---

Algorithm 1: Simple Greedy Algorithm for Two-Phase Profit Maximization Problem

---

    Data: G, B_1, B_2, d

    Result: S

1   Initialize $S \leftarrow \emptyset$;

2   **Seed set selection for First Phase;**

3   Initialize $S_1 \leftarrow \emptyset$;

4   **while** *TRUE* **do**

5   Find $u' \leftarrow argmax_{u \in V(G) \setminus S_1} \frac{\phi(S_1 \cup \{u\}) - \phi(S_1)}{C(u)}$;

6   **if** $(\phi_{u'}(S_1)) \leq 0$ **then**

7 | Break;

8   **end**

9   **if** $C(u') \leq B_1$ **then**

10 | $S_1 \longleftarrow S_1 \cup \{u'\}$; $B_1 \longleftarrow B_1 - C(u')$;

11   **end**

12   **end**

13   Return S_1;

14   From the partial observation in G using seed set S_1 at time step d, we have recently activated nodes, R_Y and already activated nodes A_Y;

15   **Seed set selection for Second Phase;**

16   Initialize $S_2 \leftarrow \emptyset$;   Update $B_2 \leftarrow B_2 + B_1$;

17   **while** *TRUE* **do**

18   Find $v' \leftarrow argmax_{v \in V(G) \setminus A_Y} \frac{\phi_{V(G) \setminus A_Y}(S_2 \cup \{v\}) - \phi_{V(G) \setminus A_Y}(S_2)}{C(v)}$;

19   **if** $(\phi_{v'}(S_2)) \leq 0$ **then**

20 | Break;

21   **end**

22   **if** $C(v') \leq B_2$ **then**

23 | $S_2 \longleftarrow S_2 \cup \{v'\}$; $B_2 \longleftarrow B_2 - C(v')$;

24   **end**

25   **end**

26   Return S_2; $S \longleftarrow S_1 \cup S_2$;   Return S;

Phase II. Let, it be C'_{min} and this means $C'_{min} = \min\limits_{u \in V(G) \setminus A_Y} C(u)$. However, the remaining computations remains the same. So, the time requirement for Phase II will be of $\mathcal{O}(\frac{B_2}{C'_{min}} \cdot n^2 \cdot (m+n))$. Now, summing everything up, the total time requirement of the single greedy approach will be of $\mathcal{O}((\frac{B_1}{C_{min}} + \frac{B_2}{C'_{min}}) \cdot n^2 \cdot (m+n))$. Now, the extra space consumed by this method is to store the seed sets for both the phases which can be of $\mathcal{O}(n)$. Hence, Theorem 1 holds.

Theorem 2. *The time and space requirement of Single Greedy Approach is of* $\mathcal{O}((\frac{B_1}{C_{min}} + \frac{B_2}{C'_{min}}) \cdot n^2 \cdot (m+n))$ *and* $\mathcal{O}(n)$, *respectively.*

It is easy to convince that two-phase setting is a generalization of single-phase setting and it has been mentioned in [14] that the single greedy approach in one-phase setting does not lead to any constant factor approximation guarantee. Next, we describe the double greedy approach.

Double Greedy Approach. First phase of this method goes like this. We initialize two sets S_1 and T_1. The first one is with $\emptyset$ and the second one is with $V(G)$. Now, for every node $u \in V(G)$, we compute two measures r_u^+ and r_u^- which are mentioned in Eq. 7 and 8, respectively.

$$r_u^+ \leftarrow \frac{\phi(S_1 \cup \{u\}) - \phi(S_1)}{C(u)} \quad (7) \qquad r_u^- \leftarrow \frac{\phi(T_1 \setminus \{u\}) - \phi(T_1)}{C(u)} \quad (8)$$

Now, if $r_u^+ \geq r_u^-$ and the cost of the current nodes is less than the available budget, then the set S_1 is updated as $S_1 \cup \{u\}$ and the budget B_1 is updated as $B_1 - C(u)$, though T_1 remains the same. If the budget is not sufficient or if $r_u^+ < r_u^-$, T_1 is reduced by deleting the current node and in that case S_1 remains the same. After repeating these steps we obtain the seed set for Phase I; i.e.; S_1. Now, we conduct the diffusion process and observe upto the time step d and this obtain the the already activated nodes and recently activated nodes. If there is any unutilized budget of Phase I, that has been added to the budget of Phase II. For the seed set selection of Phase II, we repeat the same process however on the reduced graph; i.e.; the graph obtained by deleting the already activated nodes in the first phase. Now, we proceed to describe the analysis of the double greedy approach.

The analysis is quite similar to the single greedy approach in two-phase setting. As stated previously, computing one marginal profit gain computation requires $\mathcal{O}(m + n)$ time. It is important to observe that in this method we are performing two marginal profit gain computations per node. Hence, the time requirement for the seed set selection of the first phase is of $\mathcal{O}(n \cdot (m + n))$. Other than the marginal gain computations, all the remaining statements from Line 7 to 17 will take $\mathcal{O}(1)$ time. Now, in the worst case the size of S_1, R_Y, and A_Y can be of $\mathcal{O}(n)$. So, there can be $\mathcal{O}(n^2)$ many edges associated with the vertices of A_Y. Hence, deleting the set A_Y leads to the modification of the $\mathcal{O}(n^2)$ many adjacency matrix entries of the input social network. Thus, performing the deletion step after Phase I requires $\mathcal{O}(n^2)$ time. Like Phase I, it is easy to observe that the time requirement for seed set selection in Phase II will be of $\mathcal{O}(n \cdot (m+n))$. Hence, the total time requirement for the double greedy approach in two-phase setting will be of $\mathcal{O}(n \cdot (m + n) + n^2) = \mathcal{O}(n(m + n))$. The extra space consumed by this algorithm is to store the sets S_1, T_1, S_2, T_2, and S. In the worst case, all of them will consume $\mathcal{O}(n)$ space. Hence, Theorem 3 holds.

Theorem 3. *Running time and the space requirement of the double greedy approach in the two-phase setting is of $\mathcal{O}(n(m + n))$ and $\mathcal{O}(n)$, respectively.*

4   Experimental Evaluation

In this section, we describe the experimental evaluation of the proposed solution approaches. First, we mention the datasets that we have used.

Algorithm 2: Double Greedy Algorithm for Two-Phase Profit Maximization Problem

---

Data: G, B_1, B_2, d

Result: S

1 Initialize $S \leftarrow \emptyset$;

2 **Seed set selection for First Phase**;

3 Initialize $S_1 \leftarrow \emptyset$, $T_1 \leftarrow V(G)$;

4 **for** *All* $u \in V(G)$ **do**

5 $\quad$ $r_u^+ \leftarrow \frac{\phi(S_1 \cup \{u\}) - \phi(S_1)}{C(u)}$; $\quad$ $r_u^- \leftarrow -\frac{\phi(T_1 \setminus \{u\}) - \phi(T_1)}{C(u)}$;

6 $\quad$ **if** $r_u^+ \geq r_u^-$ **then**

7 $\quad\quad$ **if** $C(u) \leq B_1$ **then**

8 $\quad\quad\quad$ $S_1 \leftarrow S_1 \cup \{u\}$; T_1 *remains same*; $B_1 \longleftarrow B_1 - C(u)$;

9 $\quad\quad$ **else**

10 $\quad\quad\quad$ $T_1 \leftarrow T_1 \setminus \{u\}$; S_1 *remains same*;

11 $\quad\quad$ **end**

12 $\quad$ **else**

13 $\quad\quad$ $T_1 \leftarrow T_1 \setminus \{u\}$; S_1 *remains same*;

14 $\quad$ **end**

15 **end**

16 Return $S_1 (= T_1)$;

17 From the partial observation in G using seed set S_1 at time step d, we have recently activated nodes, R_Y and already activated nodes A_Y;

18 **Seed set selection for Second Phase**;

19 Initialize $S_2 \leftarrow \emptyset$, $T_2 \leftarrow V(G) \setminus A_Y$; $\quad$ Update $B_2 \leftarrow B_2 + B_1$;

20 **for** *All* $v \in V(G) \setminus A_Y$ **do**

21 $\quad$ $r_v^+ \leftarrow \frac{\phi_{V(G) \setminus A_Y}(S_1 \cup \{v\}) - \phi_{V(G) \setminus A_Y}(S_1)}{C(v)}$;

22 $\quad$ $r_v^- \leftarrow -\frac{\phi_{V(G) \setminus A_Y}(T_1 \setminus \{v\}) - \phi_{V(G) \setminus A_Y}(T_1)}{C(v)}$;

23 $\quad$ **if** $r_v^+ \geq r_v^-$ **then**

24 $\quad\quad$ **if** $C(v) \leq B_2$ **then**

25 $\quad\quad\quad$ $S_2 \leftarrow S_2 \cup \{v\}$; T_2 *remains same*; $\quad B_2 \longleftarrow B_2 - C(v)$;

26 $\quad\quad$ **else**

27 $\quad\quad\quad$ $T_2 \leftarrow T_2 \setminus \{v\}$; S_2 *remains same*;

28 $\quad\quad$ **end**

29 $\quad$ **else**

30 $\quad\quad$ $T_2 \leftarrow T_2 \setminus \{v\}$; S_2 *remains same*;

31 $\quad$ **end**

32 **end**

33 Return $S_2 (= T_2)$; $\quad S \longleftarrow S_1 \cup S_2$; $\quad$ Return S;

Datasets. We have used three datasets for our experiments namely, *email-Eucore* [11,15], *soc-sign-bitcoin-alpha* [7,8] and *wiki-Vote* [9,10]. They have been downloaded from https://snap.stanford.edu/data/ and listed in Table 1.

Experimental Setup. In our experimental setup Independent Cascade Model is used for diffusion process. The influence probability $\mathcal{P}_{uv}$ is set to 0.01. The cost setting of a $u \in V(G)$ is $C(u) \longrightarrow [50, 100]$ and benefit setting is $b(u) \longrightarrow$

Table 1. Basic Statistics of the Datasets

Dataset Name	Type of Graph	Number of Nodes	Number of Edges	Maximum Degree	Average Degree
email-Eu-core	Undirected	1,005	16,706	347	33.25
soc-sign-bitcoin-alpha	Directed	3,783	24,186	888	12.79
wiki-Vote	Directed	7,115	103,689	1167	29.15

[800, 1000]. Hence, a cost and benefit is associated for every node u in the graph. These settings are followed for all these three datasets specified above. We have evaluated datasets for Budget = {500, 1000, 1500, 2000, 2500}. Each budget is split into a ratio of 60% at time step $d = 3$. The budgets $\mathcal{B}_1$ and $\mathcal{B}_2$ are consumed in first phase and second phase, respectively. The first phase of experiment is run for 100 iterations which generates 100 sets of recently active nodes. Now, the second phase runs for 100 times for all 100 sets of recently active nodes. The maximum of the profits from the 100 recently active nodes is the profit earned from the second phase. We compare the performance of the proposed solution approaches with the following methods, namely Random, High Degree, Clustering Coefficient and Single Discount baseline algorithms are compared as per the experimental setup elaborated above.

Results and Discussions. Now, we describe the experimental results. Figure 1 shows the budget vs. cardinality of the seed set plots selected by proposed as well as baseline methods over different budget values for different datasets. From this figure, we observe that in most of the problem instances the number of seed nodes selected by the single greedy algorithm is more than the other methods. As an example, for the Email-Eu-Core Dataset when the budget value is 2500, among the baseline methods the cardinality of the seed sets selected by all of them is 33. However, in the same setting the number of seed nodes selected by the single greedy approach is 36. This observation is consistent even for the other two datasets. For the Soc-Sign-Bitcoin-Alpha Dataset when the budget value is 2000, among the baseline methods the maximum number of seed nodes selected by both Random and Single Discount Heuristics and the number is 29. However, the same for the single greedy approach is 31. For the Wiki-Vote Dataset, for the budget value 2500, among the baseline methods the maximum number of seed nodes selected by Random and the number is 34 whereas the same by single greedy algorithm is 39.

Figure 2 shows the budget vs. difference between profit in two-phase and one-phase for different datasets. From the figure, we observe that for most of the budget values the difference in the earned profit remains positive where as for most of the baseline methods this difference is negative. As an example, for the 'email-Eu-core' dataset when the budget value is 2500, for the single greedy algorithm the earned profit in single-phase and two-phase is 56354.35 and 69676.21, respectively. Hence, the gain in two-phase is 13321.86 which is approximately 24%. In general, we observe that when the budget value increases (upto 2000) this gain increases. However, if we increase the budget further the gain decreases. This is due to the fact when the budget exceeds a threshold,

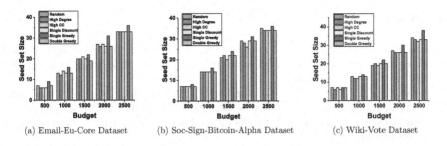

Fig. 1. Budget Vs. Cardinality of the Seed Set Plots in two-phase setting for all the datasets

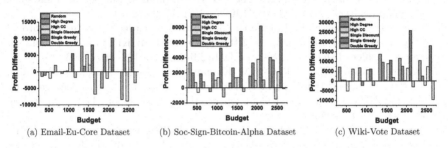

Fig. 2. Budget Vs. Differences in Profit in single and two-phase for all the datasets

then the earned benefit does not increase much. However, the cost of the seed set increases as the budget increases. Consequently, the profit does not increase even if the budget increases.

Another important observation, we make from Fig. 2 is that the performance of the single greedy and double greedy algorithm is complementary, meaning for the budget values when the single greedy approach does not perform well in those budget values the double greedy approach. As an example, for the Email-Eu-Core dataset for the budget value 500, the gain in case of the single greedy approach is negative where as the same for the double greedy approach is positive. For the other budget values the scenario is just the reverse.

5   Conclusion and Future Direction

In this paper, we have studied the problem of profit maximization in two-phase setting using online social networks. For this problem, first we develop a mathematical model for this problem based on IC Model of diffusion that captures the profit in an expected sense. Subsequently, we propose two solution methodologies namely single greedy and double greedy. Experimental analysis with real-world datasets show that effectiveness of the proposed solution methodologies. Now, in this study we have not taken care of how the total budget $\mathcal{B}$ is divided into $\mathcal{B}_1$ and $\mathcal{B}_2$ which we pose for future study.

References

1. Banerjee, S., Jenamani, M., Pratihar, D.K.: A survey on influence maximization in a social network. Knowl. Inf. Syst. **62**(9), 3417–3455 (2020). https://doi.org/10.1007/s10115-020-01461-4
2. Dhamal, S., Prabuchandran, K., Narahari, Y.: Information diffusion in social networks in two phases. IEEE Trans. Network Sci. Eng. **3**(4), 197–210 (2016)
3. Domingos, P.: Mining social networks for viral marketing. IEEE Intell. Syst. **20**(1), 80–82 (2005)
4. Gao, C., Gu, S., Yu, J., Du, H., Wu, W.: Adaptive seeding for profit maximization in social networks. J. Global Optim. **82**(2), 413–432 (2022)
5. Guille, A., Hacid, H., Favre, C., Zighed, D.A.: Information diffusion in online social networks: a survey. ACM SIGMOD Rec. **42**(2), 17–28 (2013)
6. Kempe, D., Kleinberg, J., Tardos, É.: Maximizing the spread of influence through a social network. In: Proceedings of the Ninth ACM SIGKDD International Conference on Knowledge Discovery and Data Mining, pp. 137–146 (2003)
7. Kumar, S., Hooi, B., Makhija, D., Kumar, M., Faloutsos, C., Subrahmanian, V.: Rev2: fraudulent user prediction in rating platforms. In: Proceedings of the Eleventh ACM International Conference on Web Search and Data Mining, pp. 333–341. ACM (2018)
8. Kumar, S., Spezzano, F., Subrahmanian, V., Faloutsos, C.: Edge weight prediction in weighted signed networks. In: Data Mining (ICDM), 2016 IEEE 16th International Conference on, pp. 221–230. IEEE (2016)
9. Leskovec, J., Huttenlocher, D., Kleinberg, J.: Predicting positive and negative links in online social networks. In: Proceedings of the 19th International Conference on World Wide Web, pp. 641–650 (2010)
10. Leskovec, J., Huttenlocher, D., Kleinberg, J.: Signed networks in social media. In: Proceedings of the SIGCHI Conference on Human Factors in Computing Systems, pp. 1361–1370 (2010)
11. Leskovec, J., Kleinberg, J., Faloutsos, C.: Graph evolution: densification and shrinking diameters. ACM Trans. Knowl. Discov. From Data (TKDD) **1**(1), 2-es (2007)
12. Lu, W., Lakshmanan, L.V.: Profit maximization over social networks. In: 2012 IEEE 12th International Conference on Data Mining, pp. 479–488. IEEE (2012)
13. Sun, L., Huang, W., Yu, P.S., Chen, W.: Multi-round influence maximization. In: Proceedings of the 24th ACM SIGKDD International Conference on Knowledge Discovery & Data Mining, pp. 2249–2258 (2018)
14. Tang, J., Tang, X., Yuan, J.: Profit maximization for viral marketing in online social networks: algorithms and analysis. IEEE Trans. Knowl. Data Eng. **30**(6), 1095–1108 (2017)
15. Yin, H., Benson, A.R., Leskovec, J., Gleich, D.F.: Local higher-order graph clustering. In: Proceedings of the 23rd ACM SIGKDD International Conference on Knowledge Discovery and Data Mining, pp. 555–564 (2017)

On-Device Application

SESA: Fast Trajectory Compression Method Using Sub-trajectories Segmented by Stay Areas

Shota Iiyama[1,3]([✉]), Tetsuya Oda[2,3][iD], and Masaharu Hirota[1,3][iD]

[1] Graduate School of Informatics, Okayama University of Science,
1-1 Ridaicho, Kita-ku, Okayama, Japan
i21im01cs@ous.jp, hirota@ous.ac.jp
[2] Department of Information and Computer Engineering,
Okayama University of Science, 1-1 Ridaicho, Kita-ku, Okayama, Japan
oda@ous.ac.jp
[3] Department of Information Science, Okayama University of Science,
1-1 Ridaicho, Kita-ku, Okayama, Japan

Abstract. The increase in trajectory data is associated with problems, such as increased storage costs and difficulty in analysis. One solution to these problems is trajectory compression. It reduces the data size by removing redundant positioning points from the original trajectory data. This study proposes SQUISH-E(μ) with Stay Areas (SESA), a fast batch compression method for trajectory data based on the stay area. A stay area is where the user stays for a certain period, such as the waiting time for a traffic light or bus. Moreover, SESA segments the trajectory into sub-trajectories using stay areas and applies SQUISH-E(μ) to each sub-trajectory. The experiments with trajectory datasets show that SESA achieves a compression time reduction of approximately 65% with little change in trajectory feature retention and approximately the same compression rate as the direct application of SQUISH-E(μ).

Keywords: Trajectory compression · GPS · Batch compression

1 Introduction

Devices with global positioning systems (GPS), such as smartphones, drones, and car navigation systems, are widespread. These devices generate an enormous amount of trajectory data that record an object's activities. Such trajectory data can be used to track the movements of people, vehicles, and other objects. Therefore, trajectory data are important sources of information for analyzing human behavior patterns [11,13] and next location prediction [2,12].

However, the vast amount of trajectory data causes problems. First, communicating trajectory data increases the amount of communication in the network. Second, storing a large amount of trajectory data incurs enormous storage costs. Third, because several positioning points in the trajectory data are redundant,

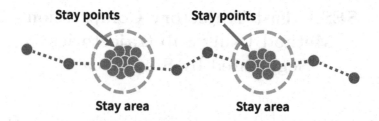

Fig. 1. Stay points and stay areas in a trajectory data.

the analysis requires extra time, which is essentially unnecessary. One solution to these problems is trajectory compression. Trajectory compression is a method that reduces the data size by removing redundant positioning points from the original trajectory data.

Several trajectory compression methods have been proposed. These include the Douglas-Peucker (DP) algorithm [3] and Spatial QUalIty Simplification Heuristic (SQUISH) [8]. However, such trajectory compression methods preserve the shape of the trajectory or movement speed of the object; nonetheless, they do not preserve the information on the area where the user stays. In this study, we refer to such an area as a stay area. A stay area provides valuable information, such as analyzing the area where a user stays and the stay of the associated traffic flow. Therefore, it is desirable to preserve the stay area of the original trajectory data when applying the trajectory compression method.

Here, we define the stay area and point in the trajectory data. Figure 1 illustrates trajectory data containing stay areas and points. The purple-dashed circles represent the stay areas, and the red points surrounded by circles represent the stay points. Stay points are positioning points (such as a bus stop or traffic lights) recorded when a user remains at a particular location for a certain period. Therefore, the stay area is defined by a set of consecutive stay points. Because the first and last stay points are sufficient to represent the information of a user's stay time and location, we consider the remaining points as redundant points in this study. In this study, we also refer to the first and last points of the stay area and positioning points where the shape of the trajectory and speed of movement have changed significantly as the feature points.

Even compression results from the existing trajectory compression methods that do not have a specific process for extracting the stay area may accidentally preserve the stay areas. However, the compression results do not necessarily preserve the first and last points of the stay area. This indicates that the compression results may not preserve information regarding the exact stay time of the user at that location. In addition, several current trajectory compression methods change the degree of compression by adjusting the parameters. Consequently, it may be possible to adjust these methods to produce compression results within a stay area. For example, when setting the parameters that decrease the compression rate, the compression results obtained by these methods may preserve both the first and last points of the stay area. However, because the compression result preserves the stay points other than the first and last points of the stay area and other redundant points, its compression rate of the compression result is low. Furthermore, when setting the parameters that increase the compression rate, the compression method considers almost positioning points in the trajectory data as redundant points and removes them. This possibly removes the stay points. These facts make it difficult for the existing methods to compress trajectory data with a high compression rate, while preserving the stay areas.

In this study, we propose a batch method of SQUISH-E(μ) with Stay Areas (SESA) for fast trajectory compression. Furthermore, SESA uses Spatial QUal-Ity Simplification Heuristic - Extended (μ) (SQUISH-E(μ)) [9], which preserves the trajectory shape and moving speed, and Stay Area (SA) to extract stay areas based on the method for extracting stay points proposed in [14]. The SESA applies SQUISH-E(μ) to the sub-trajectories segmented by the stay area extracted by the SA. Consequently, the compression time of the SESA is fast, and the compression rate is high, while ensuring that the stay area in the trajectory data is preserved.

The contributions of this research are summarized as follows.

1. We propose a batch compression method SESA extended from SQUISH-E(μ). The SESA reliably preserves the stay area of trajectory data, because SESA applies SQUISH-E(μ) to sub-trajectories segmented by stay areas.
2. SESA can compress most trajectory data approximately faster than applying original SQUISH-E(μ) with almost no change in the compression rate of the locus and the preservation rate of the feature amount.

The rest of this study is organized as follows. Section 2 describes the previous compression methods related to our method. Section 3 describes the proposed method in detail. In Sect. 4, we describe an experimental evaluation of the proposed method. Section 5 provides a summary of the study.

2   Related Work

There are two types of trajectory compression methods: batch and online methods.

Batch compression methods compress the trajectory data using all the positioning points. Therefore, the batch compression method can be applied only to the trajectory data after positioning is complete. The advantage of batch compression methods is their high performance. This is because they use all positioning point information for compression. The DP [3] preserves the shapes of trajectories. Additionally, DP recursively computes the compression points and segments a trajectory based on the perpendicular Euclidean distance (PED). The top-down time-ratio (TD-TR) [7] is an extension of the DP. It uses the synchronized Euclidean distance (SED), which considers the timestamp of the positioning points. We elaborate the SED in Sect. 3.3. This method preserves the trajectory shape and speed of movement. However, the compression time of TD-TR is slower than that of DP because SED requires more computational

Fig. 2. Overview of SESA.

time than the PED. The top-down time-ratio Reduce (TD-TR Reduce) [5] is an extension of the TD-TR to reduce the compression time. This method extracts feature points from the trajectory data with movement direction and speed have changed and applies TD-TR to the extracted points. The method proposed in [4] preserves the contour of the trajectory data (particularly turning corners and U-turns), considering the distance, angle, and velocity of the trajectory data. Furthermore, SQUISH-E(μ) [9] is a trajectory compression method that uses a priority queue. This method adds positioning points to the priority queue and removes points with a low priority. The remaining points in the priority queue are regarded as the compression results. The method proposed in [6] to compress a trajectory while preserving feature points applies SQUISH-E(μ) and the method for extracting stay points proposed in [14] to trajectory data in parallel.

On the other hand, online compression methods compress trajectory data using only the acquired positioning and few other points. Therefore, an online compression method can be used during the positioning process. Online compression methods compress data sequentially; thus, their performance regarding the compression rate and preservation of the original trajectory features is often lower than that of batch processing. Dead-Reckoning [10] is a fast trajectory compression method that uses the speed of an object to predict the subsequent moving point and its distance from the actual positioning point to determine whether it is a compression point. Critical Point (CP) [1] is a trajectory compression method that focuses on the direction of movement. Regarding this method, the compression point is the location where the direction of movement of the user has changed.

The SESA of our proposed method compresses each sub-trajectory after the SA segments the trajectory data. Therefore, SESA can select any compression method, batch, or online. In this study, we use one of the primary methods for trajectory compression: SQUISH-E(μ).

3   Proposed Method

Figure 2 presents an overview of SESA. A trajectory is defined as $T = \{p_1, p_2, \cdots, p_n\}$, and the i-th positioning point $p_i = (x_i, y_i, t_i)$ represents the longitude, latitude, and timestamp, respectively. First, we apply SA to extract the stay areas from the trajectory T. Second, we segment the trajectory into sub-trajectories using the first and last stay points of each stay area. Third, we apply SQUISH-E(μ) to each sub-trajectory to extract the compression points from them. Finally, we integrate the compression points extracted from each sub-trajectory and obtain the compression result T'.

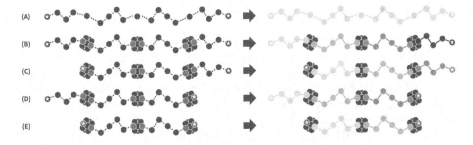

Fig. 3. Segmentation process.

3.1 Extraction of Stay Area

In this section, we explain the process of extracting stay areas from trajectory data using SA. The SESA uses SA to extract the stay areas from trajectory data T.

The SA defines the positioning points in the trajectory data that satisfy all of the following conditions as stay points:

$$a < i \leq b$$
$$Distance(p_a, p_i) \leq T_{distance}$$
$$Distance(p_a, p_{b+1}) > T_{distance} \tag{1}$$
$$Interval(p_a, p_b) \geq T_{time}$$

$Distance(p_a, p_i)$ is the Euclidean distance between positioning points p_a and p_i. $T_{distance}$ is the threshold distance between the two positioning points. $Interval(p_a, p_b)$ is the difference in the timestamps between positioning points p_a and p_b. T_{time} is the threshold of the difference in the timestamps between the two positioning points.

After extracting the stay points, we regard each consecutive set of stay points in the trajectory data as stay areas. Subsequently, we remove all stay points, excluding each extracted first and last point of the stay area. We used each extract stay area to segment the trajectory into sub-trajectories.

3.2 Segmentation

Here, we describe the segmentation process of a trajectory to sub-trajectories. We segment the trajectory into sub-trajectories by the stay area extracted by SA, as shown in Fig. 3. In the figure, the segmentation process has five cases depending on the locations of extracted stay area. The red points are the first point and the last point of stay area. The points marked with a star are the first point p_1 and the last point p_n of the trajectory. Also, we apply SQUISH-E(μ) described in the next section to the segmented sub-trajectories.

3.3   SQUISH-E(μ)

Next, we describe the SQUISH-E(μ). SQUISH-E(μ) is a trajectory compression method that uses priority queues. Priority is the degree of change in the shape of the trajectory data when a positioning point p_i is dequeued from the priority queue. In SQUISH-E(μ), the shape differences in the trajectory before and after the dequeuing do not exceed the threshold μ. Therefore, this method preserves the shape and moving speed of the trajectory data.

Moreover, SQUISH-E(μ) comprises the following two steps.

step 1: Enqueue all the positioning points of a trajectory into the priority queue Q.

step 2: Dequeue the low-priority points from Q.

Step 1 enqueues the positioning point p_i of the trajectory to the priority queue Q. Here, we denote the priority of p_i as $priority(p_i)$. When enqueuing p_i to Q, the method initializes $priority(p_i) \leftarrow \infty$ and $\pi(p_i) \leftarrow 0$. The variable $\pi(p_i)$ is used to maintain the neighborhood's maximum priority of the removed points in the subsequent step. When enqueuing a positioning point other than the first point $(i \geq 3)$, we change the priority $priority(p_{i-1})$ as follows:

$$priority(p_{i-1}) = SED(p_{i-1}, pred(p_{i-1}), succ(p_{i-1})), \tag{2}$$

where $pred(p_i)$ and $succ(p_i)$ are p_i's closest predecessor and successor among the points in Q, respectively. The SED is the Euclidean distance between the positioning $p_i = (x_i, y_i, t_i)$ and pseudo $p_i' = (x_i', y_i', t_i)$ points. The pseudo-point p_i' is the point approximated on the line between the two positioning points $\{p_s(x_s, y_s, t_s) | 1 \leq s < i\}$ and $\{p_e(x_e, y_e, t_e) | i < e \leq n\}$. In this study, we calculate the SED between p_i and p_i' based on p_s and p_e, and we denote it as $SED(p_i, p_s, p_e)$. Moreover, $SED(p_i, p_s, p_e)$ is calculated as follows:

$$(x_i', y_i') = \left(x_s + \frac{t_i - t_s}{t_e - t_s}(x_e - x_s), y_s + \frac{t_i - t_s}{t_e - t_s}(y_e - y_s) \right)$$
$$SED(p_i, p_s, p_e) = \sqrt{(x_i - x_i')^2 + (y_i - y_i')^2} \tag{3}$$

Hence, the priorities of the first p_f^Q and last p_l^Q points of Q are $priority(p_f^Q) \leftarrow \infty, priority(p_l^Q) \leftarrow \infty$, which prevent them from being removed from Q. After inquiring about all the positioning points to Q, the method moves to Step 2.

In step 2, the method dequeues the positioning points with low-priority values in Q and compresses the trajectory using the following procedure.

1. We find the lowest priority positioning point p_j in Q.
2. We update $\pi(pred(p_j))$ and $\pi(succ(p_j))$ based on $priority(p_j)$. If $\pi(pred(p_j)) < priority(p_j)$, we set $\pi(pred(p_j)) \leftarrow priority(p_j)$. Furthermore, if $\pi(succ(p_j)) < priority(p_j)$, then $\pi(succ(p_j)) \leftarrow priority(p_j)$.

3. We dequeue p_j from Q and update the priorities of $pred(p_j)$ and $succ(p_j)$. Here, we let p_k be either $pred(p_j)$ or $succ(p_j)$. If p_k is the first or last point in Q, then the priority is not updated. Otherwise, $priority(p_k) = \pi(p_k) + SED(p_k, pred(p_k), succ(p_k))$.
4. We repeat the process until the minimum priority value in Q is larger than μ.

After the completion of Step 2, the points in Q are the compression results of SQUISH-E(μ).

3.4   Integration

The last process of SESA is the integration of the compression results from the sub-trajectories. We integrate the compression points of each sub-trajectory extracted using SQUISH-E(μ). Their integration result is the compression result T' obtained by the SESA. Considering (C), (D), and (E) in Fig. 3, T' also includes the first and last points of the trajectory as described in Sect. 3.1. Consequently, the SESA can reliably preserve the compression points of SQUISH-E(μ) and the stay areas.

4   Experiment

In this section, we evaluate the performance of the SESA of our proposed method and the conventional SQUISH-E(μ).

4.1   Experimental Conditions

We used the Microsoft GeoLife dataset [15–17] as the experimental data. The total number of trajectories was 18,670. They were recorded using transportation such as walking, cars, and trains. The mean of positioning points for each trajectory were approximately 1,295. The mean sampling rates of the trajectories were approximately 7 s.

We used three evaluation criteria: the SED error, compression rate, and compression time. The SED error evaluates the difference in the trajectory shape before and after the trajectory compression. The smaller the SED error value is, the better the compressed trajectory data that preserve the shape of the original trajectory data. The compression rate evaluates the percentage of the positioning points removed using the compression method. The higher the value of the compression rate is, the smaller the data size of the compressed trajectory. The compression time evaluates the time required for the trajectory compression method to compress the data. The smaller the compression time is, the faster the method compresses the trajectory data.

We evaluated each SED threshold of SESA and SQUISH-E(μ), varying from 10 m to 100 m for each 10 m. We adjusted the SA thresholds of the SESA, setting the thresholds for the stay distance and time to 15 m and 30 s, respectively.

4.2   Experiment Results

Figure 4, 5, and 6 show the evaluation results of the SESA of the proposed method and SQUISH-E(μ) with three evaluation criteria. The vertical and horizontal axes represent each criterion and SED threshold, respectively. The red and blue lines represent the performances of SESA and SQUISH-E(μ), respectively.

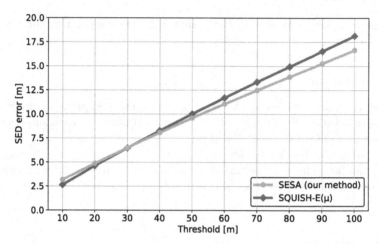

Fig. 4. SED Error.

SED Error. Figure 4 shows the SED errors of SESA and SQUISH-E(μ). When the SED threshold is less than 30 m, the SESA has more significant SED errors than SQUISH-E(μ). This is because SQUISH-E(μ) preserves several redundant points in the stay area. Moreover, the shape of the compression result is similar to that of the original trajectory, whereas the SESA preserves only the first and last points in the stay area. Therefore, SQUISH-E(μ), which has a more significant number of compression points, produces compression results closer to the shape of the original trajectory data than the SESA. However, the SESA reliably preserves the stay area without redundant stay points.

In addition, when the SED threshold is higher than 40 m, the SESA has lower SED errors than SQUISH-E(μ). Furthermore, the larger the threshold is, the more significant the difference. This is because SQUISH-E(μ) removes most of the positioning points and cannot preserve the stay area, whereas the SESA can preserve the stay area. When the SED threshold of SQUISH-E(μ) is significant, the method may excessively compress the trajectory data because it deletes the stay area to be preserved in addition to the redundant points. In contrast, the compression result by SESA close to the shape of the original trajectory data because it preserves the first and last points of the stay area.

Compression Rate. Figure 5 shows the compression rates of SESA and SQUISH-E(μ). The compression rates of the SESA and SQUISH-E(μ) are approximately identical. This difference is caused by the SESA, which preserves the stay points.

When the SED threshold is less than 10 m, SESA has a higher compression rate than SQUISH-E(μ). This is because SQUISH-E(μ) preserves several stay points in each stay area, whereas SESA preserves only the first and last points in each stay area. Thus, SESA has a higher compression rate than SQUISH-E(μ).

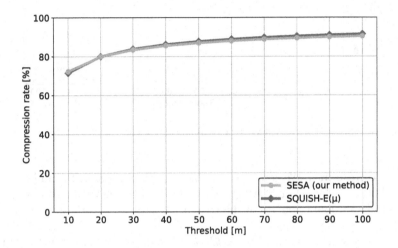

Fig. 5. Compression rate.

In addition, when the SED threshold is larger than 20 m, SESA has a lower compression rate than SQUISH-E(μ). This is because the SESA preserves the stay area, which cannot be preserved by SQUISH-E(μ). The compression points of SESA increased because the number of preserved stay areas increased.

Compression Time. Figure 6 shows the compression times of SESA and SQUISH-E(μ). Regarding all the SED thresholds, the compression time of SESA is approximately 65% less than that of SQUISH-E(μ).

This is because SESA segments the trajectory into sub-trajectories using SA. The time complexity of SQUISH-E(μ) is $\mathcal{O}(N \log N)$, which tends to have a shorter compression time because the length of the trajectory to be compressed decreases by the segmentation process. Consequently, despite the increase in the number of processes for extracting the area of stay, the compression time of the SESA is fast for most movement trajectories.

4.3 Discussion

Here, we discuss the time complexity and applicability of SESA.

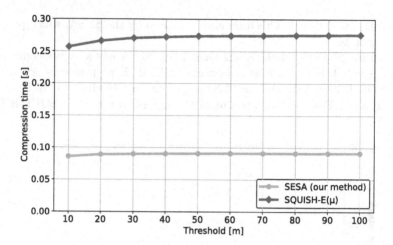

Fig. 6. Compression time.

Time Complexity. We compare the time complexities of SESA and SQUISH-E(μ). The complexity of SQUISH-E(μ) is $\mathcal{O}(N \log N)$, where N denotes the number of positioning points in the trajectory data. In addition, the time complexity of the SA is $\mathcal{O}(N)$.

The best complexity of SESA is $\mathcal{O}(N + N \log \frac{N}{|st|})$, where $|st|$ is the number of sub-trajectories in the trajectory data. Because the method applies SQUISH-E(μ) to each sub-trajectory, the compression time is fastest when the number of positioning points in each sub-trajectory is equal.

Subsequently, the worst case is that the SA does not find the stay area in the trajectory data. The time complexity of SA $\mathcal{O}(N)$ is added to SQUISH-E(μ). Consequently, the worst-case time complexity of SESA is $\mathcal{O}(N + N \log N)$. In addition, when the number of positioning points of the sub-trajectories segmented by SA is non-uniform, the time complexity of SESA approaches $\mathcal{O}(N + N \log N)$. Therefore, the time complexity of the SESA varies between $\mathcal{O}(N + N \log \frac{N}{|st|})$ and $\mathcal{O}(N + N \log N)$, depending on the location and number of extracted sub-trajectories in the trajectory.

Applicability. Because SESA segments the trajectory data by the stay areas into sub-trajectories, the compression time of SESA depends on the number of stay areas in the trajectory data. Therefore, to investigate the possibility of the SESA working effectively, we count the number of stay areas in the trajectory data. Figure 7 shows a scatter plot of the number of stay areas in 1,000 randomly sampled trajectory data of our dataset. Furthermore, the average number of stay areas in all trajectory data are 13.22. This indicates that most of the trajectory data may contain multiple stay areas. This is because, the trajectory data may include stay areas owing to events such as waiting at a traffic light or stopping at a stoplight during the user or object movement. Therefore, the SESA can

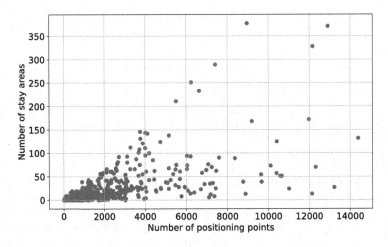

Fig. 7. The number of stay areas in each trajectory data.

potentially compress the trajectory data at high speeds although this depends on the location of stay areas in the trajectory data.

Furthermore, based on the compression time in Fig. 6, SESA is faster than SQUISH-E(μ) throughout the SED threshold. Therefore, SESA is a fast trajectory compression method.

5   Conclusion and Future Work

In this study, we have presented SESA, a fast trajectory compression method that is an extension of SQUISH-E(μ). The SESA creates sub-trajectories based on the user stay area extracted from the trajectory data and applies SQUISH-E(μ) to each sub-trajectory. In evaluation experiments, we compared SQUISH-E(μ) to SESA based on several evaluation criteria. The results show that SESA requires approximately one-third of the compression time of SQUISH-E(μ) with a slight change in the feature retention performance.

We plan to extend this study in the following directions. We will improve the SESA for a faster compression time. In addition, because SESA is faster with a more significant number of trajectory segments, we consider using feature points other than the stay area to extract the points segmented from the trajectory.

Acknowledgements. This work was supported by JSPS KAKENHI Grant Number JP19K20418 and 20K12081.

References

1. Barbeau, S., et al.: Dynamic management of real-time location data on GPS-enabled mobile phones. In: The Second International Conference on Mobile Ubiquitous Computing, Systems, Services and Technologies, pp. 343–348 (2008)

2. Chen, M., Zuo, Y., Jia, X., Liu, Y., Yu, X., Zheng, K.: CEM: a convolutional embedding model for predicting next locations. IEEE Trans. Intell. Transp. Syst. **22**(6), 3349–3358 (2021)

3. Douglas, D.H., Peucker, T.K.: Algorithms for the reduction of the number of points required to represent a digitized line or its caricature. Cartographica: Int. J. Geogr. Inf. Geovisualization **10**(2), 112–122 (1973)

4. Feng, S., Chen, L., Ma, M., Yang, A.: A turning contour maintaining method of trajectory data compression. IOP Conf. Ser.: Earth Environ. Sci. **513**(1), 012058 (2020)

5. Hansuddhisuntorn, K., Horanont, T.: Improvement of TD-TR algorithm for simplifying GPS trajectory data. In: 2019 First International Conference on Smart Technology Urban Development, pp. 1–6 (2019)

6. Iiyama, S., Oda, T., Hirota, M.: An algorithm for GPS trajectory compression preserving stay points. In: Advances in Internet. Data & Web Technologies, pp. 102–113. Springer, Cham (2022). https://doi.org/10.1007/978-3-030-95903-6_12

7. Meratnia, N., de By, R.A.: Spatiotemporal compression techniques for moving point objects. In: Bertino, E., et al. (eds.) EDBT 2004. LNCS, vol. 2992, pp. 765–782. Springer, Heidelberg (2004). https://doi.org/10.1007/978-3-540-24741-8_44

8. Muckell, J., Hwang, J.H., Patil, V., Lawson, C.T., Ping, F., Ravi, S.S.: SQUISH: an online approach for GPS trajectory compression. In: Proceedings of the 2nd International Conference on Computing for Geospatial Research & Applications, pp. 1–8 (2011)

9. Muckell, J., Olsen, P.W., Hwang, J.H., Lawson, C.T., Ravi, S.S.: Compression of trajectory data: a comprehensive evaluation and new approach. GeoInformatica **18**(3), 435–460 (2014). https://doi.org/10.1007/s10707-013-0184-0

10. Trajcevski, G., Cao, H., Scheuermanny, P., Wolfsonz, O., Vaccaro, D.: On-line data reduction and the quality of history in moving objects databases. In: Proceedings of the 5th ACM International Workshop on Data Engineering for Wireless and Mobile Access, pp. 19–26. Association for Computing Machinery (2006)

11. Yang, L., Wu, L., Liu, Y., Kang, C.: Quantifying tourist behavior patterns by travel motifs and geo-tagged photos from flickr. ISPRS Int. J. Geo-Inf. **6**(11), 345 (2017)

12. Yao, D., Zhang, C., Huang, J., Bi, J.: Serm: a recurrent model for next location prediction in semantic trajectories. In: Proceedings of the 2017 ACM on Conference on Information and Knowledge Management, pp. 2411–2414 (2017)

13. Yuan, Y., Medel, M.: Characterizing international travel behavior from geotagged photos: a case study of Flickr. PLoS ONE **11**(5), 1–18 (2016)

14. Zheng, V.W., Zheng, Y., Xie, X., Yang, Q.: Collaborative location and activity recommendations with GPS history data. In: Proceedings of the 19th International Conference on World Wide Web, pp. 1029–1038. Association for Computing Machinery (2010)

15. Zheng, Y., Li, Q., Chen, Y., Xie, X., Ma, W.Y.: Understanding mobility based on GPS data. In: Proceedings of the 10th International Conference on Ubiquitous Computing, pp. 312–321. Association for Computing Machinery (2008)

16. Zheng, Y., Xie, X., Ma, W.: GeoLife: a collaborative social networking service among user, location and trajectory. IEEE Data Eng. Bull. **33**(2), 32–39 (2010)

17. Zheng, Y., Zhang, L., Xie, X., Ma, W.Y.: Mining interesting locations and travel sequences from GPS trajectories. In: Proceedings of the 18th International Conference on World Wide Web, pp. 791–800. Association for Computing Machinery (2009)

Android Malware Detection Based on Stacking and Multi-feature Fusion

Qin Zhaowei, Xie Nannan[⊠], and Asiedu Collins Gyamfi

Changchun University of Science and Technology, Changchun, Jilin, China
xienn@cust.edu.cn

Abstract. The rapid development of Mobile Internet technology has caused an increasing number of Internet users. However, this technology has its advantages and disadvantages, and malware such as information leakage, Trojan horses, and push advertisements are hidden in smart terminals. Smartphones are now important tools for people to communicate with each other. Android dominates the current smartphone operating system. To effectively detect malware, reducing the detection rate and improving its detection efficiency have always been historical and challenging topics. Focusing on the problems of slow detection rate, low efficiency of malware detection as well as unsatisfactory detection rate of a single kind of feature, a framework of Android malware detection combined with multiple features, MS-MalDetect, is proposed. It mainly consists of three parts: multi-feature fusion which includes permissions, opcodes, application APIs, and hardware information, feature selection which combines Information Gain and Chi-square test to reduce the dimension of features, and Stacking which constructs Machine Learning classifiers that integrates multiple models to detect malware. Experiments show the proposed MS-MalDetect achieves 96.55% and 98.56% detection accuracies in different datasets and gets better performance than the compared related works.

Keywords: Malware detection · Multi-feature fusion · Feature selection · Stacking

1 Introduction

Mobile Internet is one of the core driving forces and important trends in the development of information technology in the past decade. The rapid development of colorful mobile Internet applications is changing social life in the information age. However, the phenomenon of the growth and outbreak of mobile phone malware has gradually entered the public's field of vision and has gradually become an urgent concern and problem to be solved. Statcounter global [1] stats pointed out that the Android operating system is generally on the rise in global mobile operating system market shares, which has increased from 0.66% in January 2009 to around 70% in 2022. In Kaspersky's report on the evolution of mobile malware in 2021 [2], it found 3,464,756 malicious installation packages in its mobile terminal products. Due to a large number of Android malware, the fast update speed, and the continuous emergence of new types of malware, the phenomenon of potential security risks in the Mobile Internet has gradually entered the public's field of vision. To effectively detect malware, reduce detection time and improve detection efficiency has become more challenging.

Research on Android malware detection involves mainly two aspects. One aspect is the selection of feature kinds for classification, including permissions, API calls, communication between components, bytecodes, and so on. The other is the detection of malware using different kinds of algorithms or a combination of algorithms combined with different machine learning methods such as KNN, SVM, or CNN. Research in these two aspects also corresponds to two typical challenges: the first is how to find the most representative and meaningful features of the object to ensure high efficiency and low cost, and the second is how to obtain a more stable and robust classification model [3]. The purpose of these studies is to improve the efficiency, predictability, and accuracy of Android malware detection systems with the hope that these methods are effective and operational in practical applications.

In this work, we use 5 kinds of static features to characterize applications and combine these static features with machine learning algorithms to achieve Android malware detection. Permissions, API, Dalvik opcodes, Intent, and Hardware are used as the extracted classification features with Information Gain and Chi-square test feature selection methods to optimize the feature set, and then Stacking is used to further construct a classifier to detect the malicious applications. The main contributions of this paper are as follows:

(1) A multiple feature fusion Android malware detection framework MS-MalDetect is presented. The framework can be divided into three parts, including extracting multiple features to construct feature sets, constructing optimal feature subsets, and implementing a two-layer classifier to detect malware.
(2) Feature dimension reduction. In the feature preprocessing, because the high-dimensional features affect training efficiency, a two-step feature selection method combining Information Gain and Chi-square test is employed to select the optimal feature subset.
(3) Design and implementation of the detection model. The stacking algorithm is used to combine five classification prediction models to generate a predicted model with higher accuracy.

2   Related Work

Although Google has been trying to solve the security problems of the system due to the increasing number of Android malware yet the security crisis exists. In 2009, Shabtai et al. [4] delivered a survey report on how to use classifiers on static features to detect malware. The traditional malware detection methods include static and dynamic techniques, which extract the features and detect the malware through machine learning or pattern recognition.

2.1   Features of Android Malware Detection

Permission. In the Android platform, the relevant descriptions and functions of an application are related to its permission list. Android [5] defines 206 official permissions, such as "android.permission. ACCEPT_HANDOVER", which allows the calling application to continue the call started in another application, which can be used by all developers.

Using permissions for Android malware detection and analysis has become one of the most widely used methods in recent years [6]. Zhou et al. [7] first proposed to use permission behavior and then applied heuristic filtering to detect unknown Android malware. Zhu et al. [8] presented Privilege Feature Selection, which uses Android permissions and greatly reduces the time cost of detection without reducing the accuracy. To identify the importance of Android permissions, Rathore et al. [9] proposed a detection system with a set of 16 permissions to build a malware detection engine. Dharmalingam et al. [10] proposed a permission ranking system that identifies privileges uniquely into malware and benign applications by calculating the contribution of each privilege.

API. API (Application Program Interface) is a pre-defined function in a system or program that meets certain functional requirements and common operation conventions. It can be regarded as a public abstraction of function behavior and object action and can provide data sharing for various platforms. It is found that the calling frequency and sequence of malware APIs are very different from those of benign software [11].

There are some researchers [12–14] that proposed the detection of malware by looking for the features of API calls in the system. However, the number of APIs is too large, the API extraction is complicated, and the difference in the frequency of API usage between malicious and normal programs are not considered. The LSCDroid proposed by Wang et al. [15] uses local sensitive API calls sequences as features to detect malware and represent different malicious behavior patterns. Pektas et al. [16] used an API call graph to represent all possible execution paths of malware at runtime, and introduced deep neural networks for efficient training and testing. These studies use API calls from different perspectives as a feature for malware detection.

Considering the dimension of features in this work, we only focus on "Restricted API" and "Suspicious API". "Restricted API" refers to a permission-protected API call, and "Suspicious API" refers to several artificially specified "suspicious" API calls, including functions related to encryption and decryption, functions that send HTTP requests.

Dalvik Opcode. The running of Android application needs to compile and execute the Dalvik bytecode in the dex file through virtual machine. Under the Dalvik virtual machine, every time the application runs, it needs to convert the bytecode into machine code through the JIT compiler. The Android runtime adopts the mechanism of AOT, which compiles the bytecode into machine code that can be directly executed and saves it locally when the Android application is installed.

The idea of representing applications employing the opcode operator was proposed by Moskovitch et al. [17] Jose et al. [18] use Dalvik opcodes as classification features through reverse engineering combined with machine learning algorithms to classify applications enabling malware detection. Zhang et al. [19] proposed a new method based on graphs to construct the Dalvik opcode graph, then applied graph theory and information theory to analyze its global topological properties, representing several selected global topological features to malware, and searched for the same programs like theirs. Pektas and Acarman [20] used the method of features extracted from instruction call graphs to represent malware using instruction call sequences.

Intent and Hardware Features. An Intent is a messaging object which can request actions from other Android components. It can be said that Intent is an important medium for communication between various application components. Feizollah et al. [21] evaluated the effectiveness of explicit and implicit intents for Android apps as features for identifying malicious apps. Experiment shows that the Intent of Android applications contains rich features which can be well characterized by malicious applications, and it also shows that Intent features should be combined with other features to achieve better detection results.

When an Android application runs on a phone, it depends on the hardware of the device. If a mobile phone lacks certain hardware, it will not function properly. The hardware characteristics of the application can also express the characteristics of the application.

Multi-feature Fusion. Traditional detection techniques mainly focus on analyzing a single kind of feature, which cannot give full play to the detection performance. To improve the accuracy of detection, some researchers use mixed features to detect malicious software. Focusing on the malicious behavior of malware, Liu et al. [22] used static analysis and dynamic analysis methods to extract various features, and proposed a feature fusion method based on the Simhash algorithm which reduced the original large feature dimension to a relatively small size. The small dimension ensures the accuracy of classification while improving efficiency. Ding et al. [23] represented malware with two behaviors, function call graphs and opcode sequences, and combined the proposed deep function fusion model to improve the classifier performance. To optimize the detection of malware variants, Mai et al. [24] extracted a variety of effective features from disassembled files and grayscale maps of malware maps through information gain and deep convolutional neural networks, and constructed fusion features composed of different types of features.

2.2 Feature Selection

Irrelevant features may increase the classification time and performance of the whole system. The purpose of feature selection is to select more necessary or relevant features to improve the accuracy of classification. By reducing unimportant features, the classification time is accelerated and the performance is improved [25].

Feature selection was began to be studied at 1960s [26]. By the 1970s, researchers were concerned about the correlation independence between features [27]. By the 1990s, traditional feature selection methods had great difficulties in solving various problems in machine learning based on high-dimensional features and large-scale data [28]. To adapt to the new environment, a new method of feature selection that can adapt to massive data and higher operating efficiency is urgently needed.

According to the correlations with the classification algorithm, feature selection techniques are generally divided into three kinds, Embedded, Filter, and Wrapper. The embedded model generally completes the selection of features during the model training process. The filter model uses a method independent of the classification model, which first performs feature selection and then builds the classification model, such as

Correlation Coefficient, Information Gain, and Mutual Information [29]. The wrapper model regards the target classifier as a black box and concerns the performance of the classifier as the evaluation criterion of the feature subset, then it selects the candidate feature subset through a classification algorithm. The typical method SVM-RFE is an algorithm based on backward search, proposed by Guyon et al. [30], which used the classification performance of SVM as the evaluation criteria for feature subsets.

2.3 Stacking Technique

In the past few decades, a lot of research has been implemented in machine learning to overcome the shortcomings of traditional single-model-based learning methods, the performance of a single classifier is improved by combining multiple classifiers generated by one or more learning algorithms and gradually forming three mainstream ensemble learning algorithms, Bagging, Boosting and Stacking [31]. Ensemble algorithms, which combine multiple individual classifiers, generally have better generalization performance than a single model [32].

Stacking uses meta-classifier models to learn how to combine base classifiers. Learning how to best combine predictions of two or more base machine learning algorithms is a complex strategy for ensemble models. For different models, it is important to observe the data sets from different angles. Different algorithms observe data from different data space and data structure perspectives. Stacking combines the strengths of multiple models to produce better results, it use the original data to train base classifier, and then use the output of the first layer and the original labels as new data to train the secondary classifier [33]. There are some related researches on Stacking algorithm in Android malware detection [6, 34, 35].

3 Framework and Implementation of Android Malware

To integrate a variety of different features, a multi-feature based Android malware detection framework, MS-MalDetect, is proposed, as shown in Fig. 1. The framework consists of four modules:

(1) Feature extraction. This module decompiles the obtained Android application dataset and extracts static features such as permissions, Dalvik opcodes, API, Intent, and hardware information. Then it uses a combination of multiple features to build a feature dataset of Android application samples.

(2) Feature preprocessing. The feature samples that failed to be extracted are removed, and the features extracted from the samples are digitized to construct a feature matrix, which is the basis for subsequent feature selection and classification tasks.

(3) Two-level feature selection. Feature fusion is performed on the above features, and the top 1500 features with information gain are constructed as a candidate feature set SIG. SelectKBest algorithm is further used to optimize the feature set to remove unnecessary redundant features.

(4) Malware detection. A machine learning classifier based on Stacking algorithm is constructed to determine whether a sample is a malicious program.

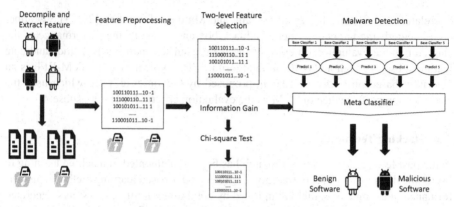

Fig. 1. Framework of MS-MalDetect

3.1 Feature Extraction and Preprocessing

The features are extracted by a python-based analysis tool, Androguard [36] from the AndroidManifest file, classes.dex file, and Smali code file respectively. The obtained features are preprocessed. The first is to remove special data and delete some samples that cannot be extracted to reduce their impact on the classification results. The features extracted from all samples are sorted by category, the features in the same category are arranged in lexicographical order to construct a feature set. The occurrence of a particular feature in the sample is marked as "1", otherwise it is marked as "0". In this way, a sample can be represented as a vector consisting of "1" and "0", thereby further representing the training sample set and the test sample set as a sample matrix. Subsequent feature selection and classification will be based on such a matrix.

3.2 Two-Level Feature Selection

The First Step with Information Gain. The initial feature set is initially processed according to the Information Gain, and a candidate feature set is constructed. As an important indicator of feature selection, Information Gain can measure the size of information contained in a feature, indicating how much information a feature can bring to the entire classification system. The more information it brings, the more important this feature is. The corresponding amount of Information Gain is also greater. Features with higher Information Gain are selected in the candidate set.

Entropy and conditional entropy are the basis of Information Gain. Information entropy represents the complexity of random variables, while conditional entropy represents the complexity of random variables under certain conditions.

Assume that the numbers of malicious samples and benign samples are N_m and N_b. For a feature F, the frequency of its appearance in malicious samples is F_m, the frequency of appearance in benign samples is F_b, the frequency of not appearing in malicious samples is F_{nm}, and the probability of not appearing in benign samples is F_{nb}.

Then the sample set information entropy is:

$$Entropy(s) = -\left(\frac{N_m}{N_m + N_b} \log\left(\frac{N_m}{N_m + N_b}\right) + \frac{N_b}{N_m + N_b} \log\left(\frac{N_b}{N_m + N_b}\right)\right) \quad (1)$$

The Information Gain of F of the feature is:

$$InfoGain = Entropy(s)$$
$$+ \frac{F_m + F_b}{N_m + N_b}\left(\frac{F_m}{F_m + F_b} \log\left(\frac{F_m}{F_m + F_b}\right) + \frac{F_b}{F_m + F_b} \log\left(\frac{F_b}{F_m + F_b}\right)\right)$$
$$+ \frac{F_{nm} + F_{nb}}{N_m + N_b}\left(\frac{F_{nm}}{F_{nm} + F_{nb}} \log\left(\frac{F_{nm}}{F_{nm} + F_{nb}}\right) + \frac{F_{nb}}{F_{nm} + F_{nb}} \log\left(\frac{F_{nb}}{F_{nm} + F_{nb}}\right)\right) \quad (2)$$

The feature candidate set S_{IG} is constructed according to the top 1500 features with information gain. To obtain a more effective high-quality feature set, the Chi-square test is further combined to select the optimal K features, and finally to constitute a feature subset S_{best}.

The Second Step with Chi-Square. The Chi-square test is a widely used hypothesis testing method. It can test whether two categorical variables are independent and related, and determine whether the null hypothesis is established by observing the deviation between the actual value and the theoretical value. It is mainly to compare two or more sample rates and the correlation analysis of two categorical variables. The Chi-square test is a hypothesis testing method based on the X^2 distribution. X^2 is used to describe the independence of two events or the degree to which the actual observed value deviates from the expected value. The larger the value of X^2, the greater the deviation between the actual observed value and the expected value, and the weaker the mutual independence of the two events.

Table 1. Feature matrix

Feature\category	Malicious	Not malicious	Total
Fi	Q_m	Q_b	$Q_m + Q_b$
Not Fi	Q_{nm}	Q_{nb}	$Q_{nm} + Q_{nb}$
Total	$Q_m + Q_{nm}$	$Q_b + Q_{nb}$	N

To describe the variance of a feature, a feature matrix is shown in Table 1. Among them, F represents the features, F_i represents a certain feature, Q_m and Q_b represent the number of times this feature appears in malware and benign samples respectively. Q_{nm} and Q_{nb} represent the number of times this feature does not appear in malware and benign sample, respectively. The global variable N represents the total number of malicious and benign samples $N_m + N_b$. The Chi-square formula is:

$$CHI(F_i, y) = x^2(F_i, y) = \frac{N(Q_m Q_{nb} - Q_b Q_{nm})^2}{(Q_m + Q_b)(Q_m + Q_{nm})(Q_b + Q_{nb})(Q_{nm} + Q_{nb})} \quad (3)$$

The larger the CHI value, the higher the correlation between the two variables. For the feature variables x_1, x_2, ..., x_n, and the categorical variable y, we only need to calculate CHI(x_1, y), CHI(x_2, y), ..., CHI(x_n, y), and sort the features according to the value of CHI from large to small, then select the threshold, and leave the features larger than the threshold, the features smaller than the threshold are removed. In this way, a set of feature subsets are selected to train the classifier, and then the performance of the classifier is evaluated.

3.3   Malware Detection Based on Stacking Structure

The malware detection framework MS-MalDetect based on Stacking uses the integration function of Stacking after constructing the feature subset of the sample. It combines the results of various classifiers to finally realize Android malware detection. This framework is set up in two layers, as shown in Fig. 2.

As shown in Fig. 2, the framework includes two layers, the base classifier layer and the meta-classifier layer.

(1)   The base classifier is used to train the data and compile the prediction results of various classification models. Stacking's base classifiers pursue "quasi-but-different", that is, each base classifier should be independent and have a high classification accuracy, and there should be differences between different base classifiers. We choose SVM, KNN, LGBM, CatBoost, and RF as the first layer basic model. The base layer takes the training set as input, and each base classifier in it will output a feature vector. These vectors are combined into a new feature matrix as the output of this layer. The specific process of the five basic classifiers in the first layer is shown in Fig. 3.

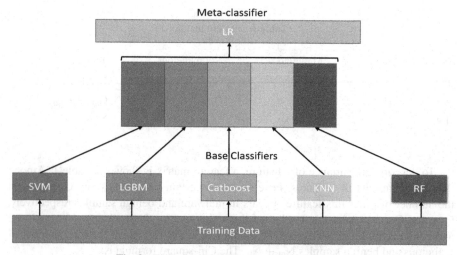

Fig. 2.   The overall framework of MS-MalDetect

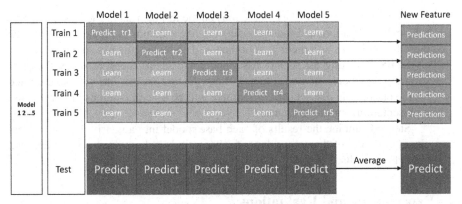

Fig. 3. The specific workflow of the first-layer base learner

In this work, the training set and the test set are divided according to the ratio of 7:3. Using 5-fold cross-validation, the training set is divided equally into five parts, four pieces of data are used to predict each base learner model, and one piece of data is used to predict labels. And stack each prediction result. After five-fold cross-validation, each prediction result is stacked in rows to form a column of feature vectors, which are used as new features for the training set. At the same time, in each fold, the overall test set was also predicted with one result, and then the prediction results of the five test sets obtained from the five-fold cross-validation were averaged. This average is then used as the new feature of the test data. The above describes the specific process of one of the base classifiers, which is the same as the process of the other base classifiers. Each base classifier generates one such new feature. The number of columns of the final generated feature matrix is the number of base classifiers, the number of rows is the number of training data as well as test data, and this feature matrix is the input to the second layer of meta-classifiers.

(2) The meta-classifier, which is used to learn how to best combine the predictions of the base model to obtain the final prediction result. This layer is usually set as a classifier model, and its input data is the output result of the previous base model. We use the Logistic Regression model to effectively prevents overfitting. The second layer model no longer uses the original samples for training, but only relies on the output of the first layer.

This article uses the Stacking framework of 5-fold cross-validation, and the overall process includes the following steps:

Step1: Preparing the training dataset. The type and quantity of the base classification model are determined, and the training set is divided into five non-intersecting parts, marked as train1 to train5.

Step2: Selecting a base classification model. Starting from train1 as the prediction set, we use the other four modeling to predict train1 and keep the results.

Step3: Using train2 as the prediction set, and continue in this way until all five copies are predicted. Then the results are superimposed to obtain prediction data for the entire train data set in the first base model.

Step4: Each model makes predictions on the test data set in each round of five-fold cross-validation, and finally retains the five columns of results, then average these five columns as a prediction of the test data by the first base model data.

Step5: Selecting the second base model and repeat Step2-Step4. If there are several base models, several columns of new feature expressions will be generated for the entire train dataset. Likewise, there will be several columns of new feature representations for the test dataset.

Step6: Combine the results of each base model into a matrix, let the original classifier model of the second layer further train based on them, and get the final prediction results.

4 Experiments and Evaluations

In this section, the preliminary work of the experiments is briefly introduced, and the classification results of the proposed method are discussed through four sets of experiments. The parameter settings of Information Gain and Chi-square test are discussed to determine the dimension of final feature subsets. In the detection part, the proposed detection method is compared with related classifiers and the classification results are evaluated.

4.1 Data Set and Experimental Environment

This experiment will use two datasets: (1) CIC-AndMal2017 [37], which contains 426 malicious samples and 1700 benign samples. (2) CICMalDroid 2020 [38], which contains 13204 malicious samples and 4039 benign samples. Both are collected and created by the Canadian Institute for Cybersecurity. And they contain various types of malicious samples, including advertisements, banking maliciousness, SMS maliciousness, risk, extortion, scaring, and so on. The largest benign sample is 49.9M, and the smallest is 12.4K. The maximum malicious sample is 107M and the minimum is 271 bytes.

All experiments are performed on Windows 10 64-bit operating system, the processor model is Intel(R) Core(TM) i7-10750H, and the memory is 16G. The programming tool is Visual Studio Code1.62.0, the python version is 3.8.3, and the libraries used are Numpy, Pandas, and Skfeature.

4.2 Experimental Process and Results

This section contained 4 experiments which include feature extraction and feature vector construction, first-layer feature selection with Information Gain; second-layer feature selection using the Chi-square test, and the classification using the proposed stacking framework. Finally, the proposed MS-MalDetect for malware detection is compared with other works.

(1) Feature Extraction and Feature Vector Construction. We integrated 5 kinds of features and extracted 7666 and 14600 features from the two datasets. The distribution of various feature classes is shown in Table 2.

Table 2. Extracted feature quantity distribution

	CIC-AndMal2017	CICMalDroid2020
Permission	133	154
API	310	405
Dalvik	2372	2401
Hardware	59	107
Intent	4792	11533
Total	**7666**	**14600**

(2) First-layer Feature Selection Based on Information Gain. The Information Gain between each feature and the category label is calculated and the feature candidate set S_{IG} is constructed. To determine the appropriate dimension of Information Gain, we conduct experiments on two datasets separately. The Information Gain was calculated for each feature and the appropriate number of features were extracted and ranked. Thus make settings of 1000 to 1600 with an interval of 100, and the feature ranking files are obtained. Through the experiments on the two data sets, the indicators obtained are shown in Table 3.

Table 3. Comparison of the different feature dimensions by IG

Data set/number of candidate feature subsets		Accuracy	F1-score	Precision	Recall
CIC-AndMal2017	1000	95.14%	88.21%	92.80%	84.06%
	1100	95.14%	88.03%	94.21%	82.61%
	1200	95.61%	89.39%	93.65%	85.51%
	1300	95.92%	90.15%	94.44%	86.23%
	1400	95.77%	89.81%	93.70%	86.23%
	1500	**96.08%**	**90.57%**	**94.49%**	**86.96%**
	1600	96.08%	90.64%	93.80%	87.68%
CICMaDroid2020	1000	98.53%	99.02%	99.01%	99.03%
	1100	98.49%	98.99%	98.98%	99.00%
	1200	98.58%	99.05%	99.06%	99.03%
	1300	98.45%	98.96%	98.98%	98.95%
	1400	98.49%	98.99%	98.87%	99.12%
	1500	**98.66%**	**99.10%**	**99.06%**	**99.15%**
	1600	98.60%	99.06%	99.06%	99.09%

It can be seen from Table 3 that in the two datasets, with the increase of feature numbers, the accuracy fluctuates but the overall trend is on the rise. When the dimension of features is 1500, it can achieve the highest overall 96.08% and 98.66% accuracy, while F1-score, Precision, and Recall can all achieve almost the best results. Therefore, 1500 features will be selected in the initially constructed feature subset.

(3) Second-level Feature Selection Based on Chi-square test. To determine the number of optimal feature subsets, the training set and the test set are divided according to the ratio of 7 to 3, the training data is trained by the method of 5-fold cross-validation, and the top K features most relevant to the label are selected according to the given selector to set different K values to train our classification model. In this experiment, the number of final feature subsets S_{best} is determined to be 800 to 1400 with an interval of 100, and the number of final feature subsets is determined through subsequent experiments. Through the experiments on the two data sets, the indicators obtained are shown in Table 4.

In CIC-AndMal2017, these metrics increase as the feature dimension increases. When k is equal to 1000, it achieves 96.55% accuracy, and then as the value of k increases, the detection effect is declined. In the data set CICMalDroid2020, the three indicators of accuracy, F1-score, and precision, all achieved the best results when k is 1000. Therefore, we determine the value of k as 1000 for further work.

(4) Comparison with other classifiers. The parameter settings for each base-level classifier in the stacking model are shown in Table 5. Using the previously constructed feature sets as training and testing samples, we compared the designed stacking framework classifier with SVM, KNN, LGBM, CatBoost, and RF. The results are shown in Figs. 4 and 5.

Table 4. Comparison of the different feature dimensions by Chi-square

Data set/Number of optimal feature subsets		Accuracy	F1-score	Precision	Recall
CIC-AndMal2017	800	95.45%	89.06%	92.91%	85.51%
	900	96.08%	90.49%	95.20%	86.23%
	1000	**96.55%**	**91.60%**	**96.77%**	**86.96%**
	1100	95.92%	90.15%	94.44%	86.23%
	1200	96.08%	90.57%	94.49%	86.96%
	1300	96.08%	90.57%	94.49%	86.96%
	1400	95.92%	90.23%	93.75%	86.96%
CICMaDroid2020	800	98.51%	99.01%	98.78%	99.23%
	900	98.53%	99.02%	98.89%	99.15%
	1000	**98.56%**	**99.03%**	**98.98%**	**99.09%**
	1100	98.53%	99.02%	98.89%	99.15%

(continued)

Table 4. (*continued*)

Data set/Number of optimal feature subsets		Accuracy	F1-score	Precision	Recall
	1200	98.51%	99.01%	98.87%	99.15%
	1300	98.56%	99.03%	98.92%	99.15%
	1400	98.53%	99.02%	98.89%	99.15%

Table 5. Parameters of the basic model of staking

Classifier	The main parameters
SVM	dual = False,C = 0.01,max_iter = 5000
LGBM	boosting_type = 'gbdt',objective = 'binary',learning_rate = 0.035, max_depth = 8,n_estimators = 500,num_leaves = 128, min_child_samples=20,min_child_weight=0.001
CatBoost	depth= 11, iterations=600, l2_leaf_reg=9, learning_rate=0.034
KNN	n_neighbors = 6 , weights='distance', leaf_size= 10 , p = 1
RF	criterion = 'entropy',min_samples_leaf = 2,n_estimators = 400, max_depth = 19,random_state=42

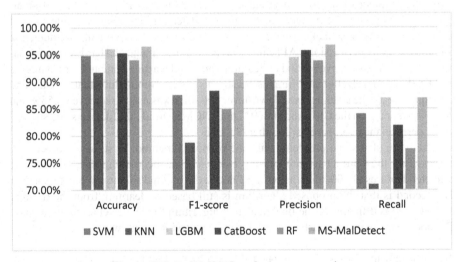

Fig. 4. CIC-AndMal2017 performance comparison

As shown in Figs. 4 and 5, the experimental framework based on the Stacking structure designed by us can detect malware better than a single classifier model. The accuracy can reach 96.55% and 98.56% respectively on the two data sets. This also proves the effectiveness and feasibility of the proposed MS-MalDetect method.

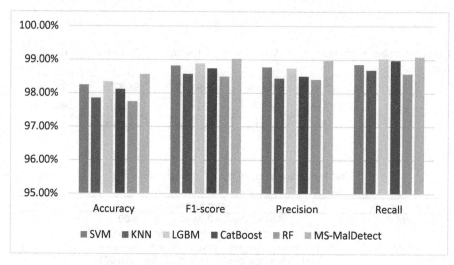

Fig. 5. CICMalDroid2020 performance comparison

5   Conclusion

Since the Android platform is popular on smartphones and intelligent terminals, the increasing number of malware is resulting in serious threats to the security of smart devices. Considering the problem of insufficient detection accuracy of single kind of features, we propose a stacking-based multi-feature fusion Android malware detection framework, using permissions, API calls, Dalvik opcodes, Intent, and hardware information as features. To remove redundant features and avoid wasting computing resources, a two-level feature selection algorithm is constructed by combining Information Gain and Chi-square test. The stacking-based two-layer classifier was used to achieve 96.55% and 98.56% accuracy on the CIC-AndMal2017 and CICMalDroid 2020 datasets and showed better performance compared with other classifiers.

There are also some problems with the proposed method. First, in the process of constructing the optimal feature subset, it depends on the training data set, and the extracted features of 1000 dimensions may not be the optimal solution in other data sets; The second is that the proposed algorithm is still based on learning from the training dataset, so it is difficult for the final detection algorithm to identify emerging malware. The above two points are further problems to be solved in future work.

Acknowledgment. This work was supported in part by the 13th Five-Year Science and Technology Research Project of the Education Department of Jilin Province under Grant No. JJKH20200794KJ, the Innovation Fund of Changchun University of Science and Technology under Grant No. XJJLG-2018-09.

References

1. Statcount. https://gs.statcounter.com/os-market-share/mobile/worldwide. Accessed 15 Oct 2021

2. Kaspersky. https://securelist.com/mobile-malware-evolution-2021/105876/. Accessed 30 Nov 2021
3. Guimaraes, R.R., Passos, L.A., Holanda Filho, R., et al.: Intelligent network security monitoring based on optimum-path forest clustering. IEEE Netw. **33**(2), 126–131 (2019)
4. Shabtai, A., Moskovitch, R., Elovici, Y., et al.: Detection of malicious code by applying machine learning classifiers on static features: a state-of-the-art survey. Inf. Secur. Tech. Rep. **14**(1), 16–29 (2009)
5. Google Developers. https://developer.android.google.cn/guide/topics/permissions/overview. Accessed 15 Aug 2021
6. Zhu, H., Li, Y., Li, R., et al.: SEDMDroid: an enhanced stacking ensemble framework for Android malware detection. IEEE Trans. Netw. Sci. Eng. **8**(2), 984–994 (2020)
7. Zhou, Y., Wang, Z., Zhou, W., et al.: Hey, you, get off of my market: detecting malicious apps in official and alternative Android markets. In: 19th Annual Network & Distributed System Security Symposium, pp. 50–52. ISOC, San Diego (2012)
8. Zhu, D., Xi, T.: Permission-based feature scaling method for lightweight Android malware detection. In: Douligeris, C., Karagiannis, D., Apostolou, D. (eds.) KSEM 2019. LNCS (LNAI), vol. 11775, pp. 714–725. Springer, Cham (2019). https://doi.org/10.1007/978-3-030-29551-6_63
9. Rathore, H., Sahay, S.K., Rajvanshi, R., Sewak, M.: Identification of significant permissions for efficient Android malware detection. In: Gao, H., Durán Barroso, R.J., Shanchen, P., Li, R. (eds.) BROADNETS 2020. LNICSSITE, vol. 355, pp. 33–52. Springer, Cham (2021). https://doi.org/10.1007/978-3-030-68737-3_3
10. Dharmalingam, V.P., Palanisamy, V.: A novel permission ranking system for Android malware detection—The permission grader. J. Ambient Intell. Humaniz. Comput. **12**(5), 5071–5081 (2020). https://doi.org/10.1007/s12652-020-01957-5
11. Peiravian, N., Zhu, X.: Machine learning for Android malware detection using permission and API calls. In: 2013 IEEE 25th International Conference on Tools with Artificial Intelligence, pp. 300–305. IEEE (2013)
12. Aafer, Y., Du, W., Yin, H.: DroidAPIMiner: mining API-level features for robust malware detection in Android. In: Zia, T., Zomaya, A., Varadharajan, V., Mao, M. (eds.) SecureComm 2013. LNICSSITE, vol. 127, pp. 86–103. Springer, Cham (2013). https://doi.org/10.1007/978-3-319-04283-1_6
13. Zhu, J., Wu, Z., Zhi, G., et al.: API sequences based malware detection for Android. In: 2015 IEEE 12th International Conference on Ubiquitous Intelligence and Computing and 2015 IEEE 12th International Conference on Autonomic and Trusted Computing and 2015 IEEE 15th International Conference on Scalable Computing and Communications and Its Associated Workshops (UIC-ATC-ScalCom), pp. 673–676. IEEE Computer Society, Beijing (2015)
14. Li, Y., Shen, T., Sun, X., Pan, X., Mao, B.: Detection, classification and characterization of Android malware using API data dependency. In: Thuraisingham, B., Wang, X.F., Yegneswaran, V. (eds.) SecureComm 2015. LNICSSITE, vol. 164, pp. 23–40. Springer, Cham (2015). https://doi.org/10.1007/978-3-319-28865-9_2
15. Wang, W., Wei, J., et al.: LSCDroid: malware detection based on local sensitive API invocation sequences. IEEE Trans. Reliab. **69**(1), 174–187 (2019)
16. Pektaş, A., Acarman, T.: Deep learning for effective Android malware detection using API call graph embeddings. Soft. Comput. **24**(2), 1027–1043 (2019). https://doi.org/10.1007/s00500-019-03940-5
17. Moskovitch, R., et al.: Unknown malcode detection using OPCODE representation. In: Ortiz-Arroyo, D., Larsen, H.L., Zeng, D.D., Hicks, D., Wagner, G. (eds.) EuroISI 2008. LNCS, vol. 5376, pp. 204–215. Springer, Heidelberg (2008). https://doi.org/10.1007/978-3-540-89900-6_21

18. Puerta, J., Sanz, B., Santos, I., et al.: Using Dalvik opcodes for malware detection on Android. Logic J. IGPL **25**(6), 938–948 (2017)
19. Zhang, J., Qin, Z., Zhang, K., et al.: Dalvik opcode graph based Android malware variants detection using global topology features. IEEE Access **6**, 51964–51974 (2018)
20. Pektaş, A., Acarman, T.: Learning to detect Android malware via opcode sequences. Neurocomputing **396**, 599–608 (2020)
21. Feizollah, A., Anuar, N.B., Salleh, R., et al.: AndroDialysis: analysis of Android intent effectiveness in malware detection. Comput. Secur. **65**, 121–134 (2017)
22. Liu, X., Dong, X., Lei, Q.: Android malware detection based on multi-features. In: Proceedings of the 8th International Conference on Communication and Network Security, pp. 69–73. Association for Computing Machinery, Qingdao (2018)
23. Ding, Y., Hu, J., Xu, W., et al.: A deep feature fusion method for Android malware detection. In: 2019 International Conference on Machine Learning and Cybernetics (ICMLC), pp.1–6. IEEE, Kobe (2019)
24. Mai, J., Cao, C., Wu, Q.: Malware variants detection based on feature fusion. In: Cheng, J., Tang, X., Liu, X. (eds.) CSS 2020. LNCS, vol. 12653, pp. 67–77. Springer, Cham (2021). https://doi.org/10.1007/978-3-030-73671-2_7
25. Dhal, P., Azad, C.: A comprehensive survey on feature selection in the various fields of machine learning. Appl. Intell. **9**, 1–39 (2021). https://doi.org/10.1007/s10489-021-02550-9
26. Lewis, P.: The characteristic selection problem in recognition systems. IRE Trans. Inf. Theory **8**(2), 171–178 (1962)
27. Cover, T.M.: The best two independent measurements are not the two best. IEEE Trans. Syst. Man Cybern. SMC-**4**(1), 116–117 (1974)
28. Jain, A., Zongker, D.: Feature selection: evaluation, application, and small sample performance. IEEE Trans. Pattern Anal. Mach. Intell. **19**(2), 153–158 (1997)
29. Guyon, I., Elisseeff, A.: An introduction to variable and feature selection. J. Mach. Learn. Res. **3**(Mar), 1157–1182 (2003)
30. Guyon, I., Weston, J., Barnhill, S., et al.: Gene selection for cancer classification using support vector machines. Mach. Learn. **46**(1), 389–422 (2002)
31. Wolpert, D.H.: Stacked generalization. Neural Netw. **5**(2), 241–259 (2017)
32. Zhang, C., Lim, P., Qin, A.K., et al.: Multiobjective deep belief networks ensemble for remaining useful life estimation in prognostics. IEEE Trans. Neural Netw. Learn. Syst. **28**(10), 2306–2318 (2017)
33. Li, Y., Yang, Z., Chen, X., et al.: A stacking model using URL and HTML features for phishing webpage detection. Future Gener. Comput. Syst. **94**, 27–39 (2019)
34. Xue, Z., Niu, W., Ren, X., et al.: A stacking-based classification approach to Android malware using host-level encrypted traffic. J. Phys.: Conf. Ser. 012049 (2021)
35. Shafin, S.S., Ahmed, M.M., Pranto, M.A., et al.: Detection of Android malware using tree-based ensemble stacking model. In: 2021 IEEE Asia-Pacific Conference on Computer Science and Data Engineering (CSDE), pp. 1–6. IEEE, Brisbane (2021)
36. Junaid, M., Liu, D., Kung, D.: Dexteroid: detecting malicious behaviors in Android apps using reverse-engineered life cycle models. Comput. Secur. **59**(2), 92–117 (2016)

37. Lashkari, A.H., Kadir, A.F.A., Taheri, L., et al.: Toward developing a systematic approach to generate benchmark Android malware datasets and classification. In: 2018 International Carnahan Conference on Security Technology (ICCST), pp. 1–7. IEEE, Holiday Inn Montreal Centreville, Canada (2018)
38. Mahdavifar, S., Kadir, A.F.A., Fatemi, R., et al.: Dynamic Android malware category classification using semi-supervised deep learning. In: 2020 IEEE International Conference on Dependable, Autonomic and Secure Computing, International Conference on Pervasive Intelligence and Computing, International Conference on Cloud and Big Data Computing, International Conference on Cyber Science and Technology Congress (DASC/PiCom/CBDCom/CyberSciTech), pp. 515–522. IEEE, Calgary, AB, Canada (2020)

Influential Billboard Slot Selection Using Pruned Submodularity Graph

Dildar Ali, Suman Banerjee$^{(\boxtimes)}$, and Yamuna Prasad

Department of Computer Science and Engineering, Indian Institute of Technology
Jammu, Jammu & Kashmir 181221, India
{2021rcs2009,suman.banerjee,yamuna.prasad}@iitjammu.ac.in

Abstract. Billboard Advertisement has emerged as an effective out-of-home advertisement technique and adopted by many commercial houses. In this case, the billboards are owned by some companies and they are provided to the commercial houses slot-wise on a payment basis. Now, given the database of billboards along with their slot information which k slots should be chosen to maximize the influence. Formally, we call this problem as the INFLUENTIAL BILLBOARD SLOT SELECTION Problem. In this paper, we pose this problem as a combinatorial optimization problem. Under the 'triggering model of influence', the influence function is non-negative, monotone, and submodular. However, as the incremental greedy approach for submodular function maximization does not scale well along with the size of the problem instances, there is a need to develop efficient solution methodologies for this problem.

In this paper, we apply the pruned submodularity graph-based pruning technique for solving this problem. The proposed approach is divided into three phases, namely, preprocessing, pruning, and selection. We analyze the proposed solution approach for its performance guarantee and computational complexity. We conduct an extensive set of experiments with real-world datasets and compare the performance of the proposed solution approach with many baseline methods. We observe that the proposed one leads to more amount of influence compared to all the baseline methods within reasonable computational time.

Keywords: Submodular function · Pruned submodularity graph · Out-of-home advertisement · Trajectory database

1 Introduction

Creating and maximizing influence among the customers is one of the main goals of a commercial house. For this purpose, they spend around 7–10% of their annual revenue. Now, how to make use of this budget effectively remains an active area of research. There are several ways through which advertisement

The work of Dr. Suman Banerjee is supported by the Start Up Research Grant provided by Indian Institute of Technology Jammu, India (Grant No.: SG100047).

W. Chen et al. (Eds.): ADMA 2022, LNAI 13725, pp. 216–230, 2022.
https://doi.org/10.1007/978-3-031-22064-7_17

can be done such as social media, television channels, and many more. However, billboard advertisement has emerged as an effective approach for out-of-home advertisement as it provides more return on investment compared to other advertisement techniques. In recent times the billboards are digital and they are allocated time slot-wise to the commercial houses based on their payments.

Problem Background. In this advertisement technique, often different commercial houses select a limited number of billboard slots with the hope that the advertisement content will be observed by many people and a significant number of them will be influenced towards the product. This may increase the sales and revenue earned from the product. Now, due to the budget constraint, only a very small number of billboard slots can be affordable for the E-Commerce house. So, it's a prominent question that for a given value of $k \in \mathbb{Z}^+$, which of the billboard slots should be chosen for posting the advertisement content. Recently, this problem has been studied by many researchers and different kinds of solution approaches have been proposed [6].

Problem Definition. Now-a-days due to the advancement of wireless devices and mobile internet, capturing the location information of moving objects become easier. This leads to the availability of many trajectory datasets in different repositories and they are being used to solve many real-life problems including route recommendation [1,4], driving behavior prediction [7], and many more. As mentioned previously, these trajectory datasets are also used to place billboards effectively. Consider for any city a trajectory database $\mathcal{D}$ is available. This database contains the location information of people along with the corresponding time stamps. Locations can be of different kinds e.g.; 'Mall', 'Beach', 'Metro Station' and so on. Digital billboards are placed over those places and these billboards can be hired by different E-Commerce houses to show their advertisement content. Now, due to financial constraints, only very limited number of billboard slots can be hired. It's a natural research question among the available n slots which k of them (where $k << n$) should be chosen such that the influence is maximized. This is the problem that we have worked out in this paper.

Related Work. There are several studies in the literature related to billboard advertisements. Zahradka et al. [8] did a case study analysis of the cost of billboard advertising in the different regions of the Czech Republic. Zhang et al. [9] studied the trajectory-driven influential billboard placement problem where a set of billboards along with its location and cost is given. The goal here is to choose a subset of the billboards within the budget that influence the largest number of trajectories. Wang et al. [6] studied the problem of the Targeted Outdoor Advertising Recommendation (TOAR) problem considering user profiles and advertisement topics. Their main contribution was a targeted influence model that characterizes the advertising influence spread along with the user mobility. Based on the divide and conquer approach they developed two solution strategies. Implementation with real-world datasets shows the efficiency and effectiveness of the proposed solution approaches. Also, there are few studies in the context of the billboard advertisement, that consider the minimization of regret that causes due to providing influence to the advertiser. Zhang et al. [10] studied the problem of regret minimization and proposed several solution methodologies. Experimentation with real-world datasets showed the efficiency of the approaches.

Our Contributions. In all the existing studies, the problem that has been considered is to identify influential locations for placing billboards. However, as now-a-days the billboards are digital, and they are allocated slot wise. So, it is important to consider these issues. In this paper we have made the following contributions:

- We formulate the Influential Billboard Slot Selection Problem as a discrete optimization problem and showed that this problem is NP-Hard and hard to approximate within a constant factor.
- We propose a Pruned Submodularity Graph-based solution approach to solve this problem with its detailed analysis and illustration with a problem instance.
- We conduct an extensive set of experiments with real-world billboard and trajectory datasets and compare the performance of the proposed algorithm with the existing solution approaches.

Organization of the Paper. The rest of the paper is organized as follows. Section 2 describes the background and defines the problem formally. Section 3 describes the proposed solution approach. The experimental evaluations of the proposed solution approach have been described in Sect. 4. Section 5 concludes our study and gives future research directions.

2   Preliminaries and Problem Definition

In this section, we describe the background and define the problem formally. Consider there are m billboards $\mathcal{B} = \{b_1, b_2, \ldots, b_m\}$ and each one of them is running for the interval $[T_1, T_2]$ and assume that $T = T_2 - T_1$. Also, assume that all the billboards are allocated slot-wise for display advertisement, and the duration of each slot is fixed and it is denoted by Δ. These billboards are placed at different locations (e.g., street junctions, shopping malls, airports. metro stations, etc.) of a city. If some person u_j comes close to any billboard b_i at time t and at that time if the advertisement content for some commercial house is running on that billboard then u_j will be influenced towards the item with the probability $Pr(b_i, u_j)$. In this study, we assume that this value is known. However, the standard way of computing these values has also been described in the literature [9]. For any positive integer n, by $[n]$ we denote the set $\{1, 2, \ldots, n\}$. We denote any arbitrary billboard slot as a tuple containing two items: the first one is billboard id and the second one is the starting and ending time of a slot. Let, $\mathbb{BS}$ denotes the set of all billboard slots; i.e.; $\mathbb{BS} = \{(b_i, [t_j, t_j + \Delta]) : i \in [k] \text{ and } j \in \{1, \Delta + 1, 2\Delta + 1, \ldots, \frac{T}{\Delta} + 1\}\}$. Now, if $|\mathbb{BS}| = n$ then it is easy to observe that $n = k \cdot \frac{T}{\Delta}$.

A trajectory database contains the location information of different persons and this is stated in Definition 1.

Definition 1 (Trajectory Database). *A trajectory database $\mathcal{D}$ is a collection of tuples of the following form :$< u_{id},$ `loc`, `time_stamp` $>$. The description of each attribute are given below:*

- u_{id}: *This is the unique identification of a people.*
- `loc`: *This is the location information of the people u_{id}.*
- `time`: *This attribute stores the time information*

If there is a tuple $<$ u_{126}, `Chihago_Airport`, $[1800, 2000]$ $>$ in the trajectory database $\mathcal{D}$ then it signifies that the people with its unique identification number u_{126} was at the place `Chicago_Airport` from time stamp 1800 to 2000.

One important point to highlight here is that in Definition 1 we have listed only the required attributes. However, in real-world datasets, we may have some more attributes as well; such as `trip_id`, `vehicle_id`, and many more. The set of unique user_ids that are present in the trajectory database $\mathcal{D}$ is $\mathcal{U} = \{u_1, u_2, \ldots, u_n\}$. Next, we describe the Billboard Database.

Definition 2 (Billboard Database). *The billboard database $\mathbb{B}$ contains the tuples of the following form: $<$ b_{id}, `loc`, `cost` $>$. The meaning of each attribute has been given below:*

- b_{id}: *This attribute stores the unique ids of the billboards.*
- `loc`: *This attribute stores the location information of the billboard.*
- `cost`: *This attribute stores the cost that needs to be paid by the E-Commerce house for renting one billboard.*

If there is a tuple $<$ b_{245}, `Chihago_Airport`, $6 >$ in the billboard database $\mathbb{B}$, then it signifies that the billboard whose unique id is b_{245} is placed at the location `Chicago_Airport` and \$6 needs to be paid by the E-Commerce house for renting one billboard slot.

Now, assume that the people $u_i \in \mathcal{U}$, is at the location `Chicago_Airport` for the duration $[t_i, t_i']$ and for the duration $[t_j, t_j']$ at the billboard b_i an advertisement of the brand XYZ. If $[t_j, t_j'] \cap [t_i, t_i'] \neq \emptyset$, then we can hope that the people u_i will look into the advertisement and he will be influenced with the probability $Pr(b_j, u_i)$. In this study, we assume that for all $b_j \in \mathbb{BS}$, and people $u_i \in \mathcal{U}$, we have these probability values. Next, we state the notion of the influence for a given subset of billboard slots in Definition 3.

Definition 3 (Influence of Billboard Slots). *Given a subset of billboard slots $\mathcal{S} \subseteq \mathbb{BS}$, we denote its influence by $I(\mathcal{S})$ and defined it as the sum of the influence probabilities of the individual people. Mathematically, this is characterized by Eq. 1.*

$$I(\mathcal{S}) = \sum_{u_i \in \mathcal{U}} [1 - \prod_{b_j \in \mathbb{BS}} (1 - Pr(b_j, u_i))] \tag{1}$$

It can be easily observed that the influence function I is a set function that maps each possible subsets of billboard slots to their respective influence; i.e.; $I : 2^{\mathbb{BS}} \longrightarrow \mathbb{R}_0^+$ where $I(\emptyset) = 0$. We list out the properties of the influence function in Lemma 1. Due to space limitations, we are unable to provide its proof which will come in a subsequent journal version of this paper.

Lemma 1. *The influence function $I()$ follows non-negativity, monotonicity, and submodularity property.*

Now, it is important to observe that as the selection of billboard slots is involved with money, hence a limited number of them can be chosen. In commercial campaigns, the goal will be to maximize the influence. Then the question arises which k billboard slots should be chosen to maximize the influence. This problem has been referred to as the INFLUENTIAL BILLBOARD SLOT SELECTION Problem which is stated in Definition 4.

Definition 4 (Influential Billboard Slot Selection Problem). *Given a trajectory database $\mathcal{D}$ and its corresponding billboard database $\mathbb{B}$, the problem of Influential Billboard Slot Selection asks to choose a subset of k billboard slots such that the influence as stated in Eq. 1 is maximized. Mathematically, this problem can be stated as follows:*

$$\mathcal{S}^{OPT} = \underset{\mathcal{S} \subseteq \mathbb{BS}, |\mathcal{S}| = k}{argmax} \; I(\mathcal{S}) \qquad (2)$$

From the algorithmic point of view, the INFLUENTIAL BILLBOARD SLOT SELECTION PROBLEM can be given as the following text box.

INFLUENTIAL BILLBOARD SLOT SELECTION PROBLEM

Input: The Set of Billboard Slots $\mathbb{BS}$, The Influence Function $I()$, The Trajectory Database $\mathcal{D}$, The number of slots k.

Problem: Find out a set $\mathcal{S} \subseteq \mathbb{BS}$ with $|\mathcal{S}| = k$ such that $I(\mathcal{S})$ is maximized.

We denote any arbitrary instance of the Influential Billboard Slot Selection Problem by $I = (\mathbb{BS}, \mathcal{S}, k)$. First, we show that the Influential Billboard Slot Selection Problem is NP-Hard by a reduction from the Set Cover Problem.

Theorem 1. *The Influential Billboard Slot Selection Problem is NP-Hard.*

Proof. (Outline) We prove this statement by a reduction from the Hitting Set Problem. We denote an arbitrary instance of the set cover problem by $I' = (\mathcal{U}, \mathcal{X}, k')$. Here, $\mathcal{U}$ is the ground set, $\mathcal{X}$ is the collection of the subsets over the ground set. The goal here is to choose k' many elements from the ground set $\mathcal{U}$ such that every subset in $\mathcal{X}$ contains at least one element from the chosen elements. It is well known that this problem is NP-Hard [5].

Now, we provide a polynomial time reduction from the Hitting Set Problem to the Influential Billboard Slot Selection Problem. Without loss of generality, we assume that the elements of the ground set are the subset of the set of natural numbers and starting from 1. Also, for simplicity, we assume that there is only one slot in a billboard. Now, the construction is as follows. For every $i \in \mathcal{U}$, we create one location with its id as ℓ_i, one billboard with id b_i, and place the billboard at that location. For every subset $x \in \mathcal{X}$, we create one trajectory that contains the locations. We fix $k = k'$. We want to influence all the trajectories. Now, it is easy to observe that the hitting set problem instance will have a solution of size k' if and only if the influential billboard slot selection problem instance has a solution of size k. Due to space limitations, we have only given an outline of the proof.

3   Proposed Solution Approach

In this section, we describe the proposed solution approaches for this problem. Initially, we start by describing the Marginal Influence Gain of a Billboard Slot in Definition 5.

Definition 5 (Marginal Influence Gain of a Billboard Slot). *Given a subset of billboard slots $\mathcal{S} \subseteq \mathbb{BS}$ and a particular billboard $b \in \mathbb{BS} \setminus \mathcal{S}$, the marginal influence gain of the billboard slot b with respect to the billboard slots in $\mathcal{S}$ is denoted by $\Delta(b|\mathcal{S})$ and defined as the difference between the influence when the billboard slot is included with $\mathcal{S}$ and when it is not. Mathematically, this is stated in Eq. 2*

$$\Delta(b \mid \mathcal{S}) = I(\mathcal{S} \cup \{b\}) - I(\mathcal{S}) \tag{3}$$

As shown that the influence function follows the summodularity property, hence this problem reduces to the problem of submodular function maximization subject to the cardinality constraint. This can be solved in the following way. Starting with an empty set, in each iteration we pick up the slot that causes maximum marginal gain. As shown by Nemhauser et al. [2,3] this method leads to $(1 - \frac{1}{e})$ factor approximation guarantee. However, it is easy to observe that in each iteration the number of marginal gain computation is of $\mathcal{O}(n)$ where n is the total number of billboard slots which is quite large. This leads to huge computational burden for real-world problem instances. So, there is a need to develop efficient solution methodology. In this paper, we apply the pruned submodularity graph-based pruning technique to solve our problem [11]. Before describing the proposed solution methodology, first we introduce the notion of pruned submodularity graph in Definition 6 which is the heart of the proposed solution approach.

Definition 6 (Pruned Submodularity Graph). *This is a weighted, directed graph $G(V, E, w)$ where the vertex set is the set of billboard slots; between every pair of billboard slots b_i, b_j, $(b_i b_j)$ will be an edge in G. The edge weight function w maps each edge to a real number. Hence, the vertex set $V(G) = \mathbb{BS}$, the edge* set $E(G) = \{(b_i b_j) : i, j \in [p] \text{ and } i \neq j\}$, and the edge weight function w is defined as follows:

$$w(b_i, b_j) = I(b_j \mid b_i) - I(b_i \mid \mathbb{BS} \setminus b_i) \tag{4}$$

This edge weight has got practical significance. Consider an edge $(b_i b_j)$ and its edge weight $w(b_i b_j)$. It is important to observe that the quantity $I(b_j \mid b_i)$ signifies that the maximum influence that the billboard slot can contribute when the billboard slot b_i is already contained in the solution. Whereas the quantity $I(b_i \mid \mathbb{BS} \setminus b_i)$ signifies the billboard slot. As in our problem context, a vertex in the pruned submodularity graph corresponds to a billboard slot, hence in the rest of the paper we use the terminology 'vertex' and 'slot' interchangeably. Next, we state the divergence of a node of a pruned submodularity graph in Definition 7.

Definition 7 (Divergence of a Node). *On the pruned submodularity graph $G(V, E, w)$, the divergence of a node v from a set of nodes V' is defined as $w_{V'v} = \min\limits_{x \in V'} w_{xv}$.*

Next, we present the description of the proposed solution approach.

Description of the Algorithm. First, we perform the following preprocessing task. Billboard slots having their individual influence 0 is removed from the list. As the influence function is submodular, for any billboard slot the marginal gain with respect to any set will always be less than its individual influence. After that, we construct the pruned submodularity graph with the slots having non-zero individual influence at Line No. 8. Once we are done with the construction of the tree, we perform the pruning from Line No. 9 to 16. This works in the following way. In each iteration of the `while` loop we randomly sample out $r \cdot \log n$ many slots and put them in the list $\mathcal{U}$. Now, for all the remaining slots in $\mathbb{BS}$, we compute the value of $w_{\mathcal{U}b}$ as mentioned in Definition 7. From this list, we remove $(1 - \frac{1}{\sqrt{c}})$ fraction of the slots having the smallest value of $w_{\mathcal{U}b}$. Once we are done with the pruning step, we execute the incremental greedy approach to finally select the billboard slots. The described solution approach has been represented in the form of pseudo code in Algorithm 1.

Complexity Analysis. Now, we analyze our proposed solution approach to understand its time and space requirements. Initialization at Line No. 1 will take $\mathcal{O}(1)$ time. Now, for any arbitrary billboard slot $b \in \mathbb{BS}$, its individual influence computation using Equation No. 1 will take $\mathcal{O}(t)$ time where t is the number of tuples in the trajectory database. As there are n billboard slots, the time required for execution from Line No. 2 to 6 will take $\mathcal{O}(n \cdot t)$ time. After that, the initialization statements of Line No. 7 take $\mathcal{O}(1)$ time. In the worst case, it may so happen that all the billboard slots lead to non-zero influence. In that case, the pruned submodularity graph will have n vertices. So, there are $\mathcal{O}(n^2)$ billboard slot pairs for whom the edge weight needs to be calculated. From Equation No. 4, it is easy to observe that for each edge, its corresponding weight can be calculated

---

Algorithm 1: Pruned Submodularity Graph+Incremental Greedy Approach for Influential Billboard Slot Selection Problem

---
Data: $\mathbb{BS}$, $I()$, $\mathcal{D}$, r, c, and, k
Result: A subset of $\mathbb{BS}$
1 Initialize $\mathcal{S} \leftarrow \emptyset$;
2 **for** *All* $b \in \mathbb{BS}$ **do**
3   **if** $I(b) == 0$ **then**
4 | $\mathbb{BS} \leftarrow \mathbb{BS} \setminus \{b\}$;
5   **end**
6 **end**
7 Initialize $\mathcal{S}' \leftarrow \emptyset, \mathcal{U} \leftarrow \emptyset, |\mathbb{BS}| = n$;
8 Construct the pruned submodularity graph with the slots in $\mathbb{BS}$;
9 **while** $|\mathbb{BS}| > r \cdot \log n$ **do**
10   Sample out $r \cdot \log n$ slots uniformly at random from $\mathbb{BS}$ and place them in $\mathcal{U}$;
11     $\mathbb{BS} \leftarrow \mathbb{BS} \setminus \mathcal{U}, \mathcal{S}' \leftarrow \mathcal{S}' \cup \mathcal{U}$;
12   **for** *All* $b \in \mathbb{BS}$ **do**
13 | $w_{\mathcal{U}b} \leftarrow \min_{u \in \mathcal{U}} [I(b|u) - I(u|\mathbb{BS} \setminus \{u\})]$;
14   **end**
15   Remove $(1 - \frac{1}{\sqrt{c}}) \cdot |\mathbb{BS}|$ elements from $\mathbb{BS}$ having smallest value of $w_{\mathcal{U}b}$;
16 **end**
17 $\mathbb{BS} \leftarrow \mathbb{BS} \cup \mathcal{S}'$;
18 **while** $|\mathcal{S}| \neq k$ **do**
19     $u^* \leftarrow \operatorname*{argmax}_{u \in \mathbb{BS} \setminus \mathcal{S}} I(\mathcal{S} \cup \{u\}) - I(\mathcal{S})$;
20     $\mathcal{S} \leftarrow \mathcal{S} \cup \{u^*\}$;
21 **end**
22 Return $\mathcal{S}$;

---

in $\mathcal{O}(n \cdot t)$ time. Hence, the construction of the pruned submodularity graph will take $\mathcal{O}(n^2 \cdot t)$ time. It is easy to observe that the number of iterations of the while loop will be of $\mathcal{O}(\log_{\sqrt{c}} n)$. Removing $\mathcal{O}(\log n)$ many elements from $\mathbb{BS}$ uniformly at random will take $\mathcal{O}(\log n)$ time. Also, in each iteration, the cost of computing the quantity $w_{\mathcal{U}b}$ is also involved. In the set $\mathcal{U}$, there are $\mathcal{O}(r \cdot \log n) = \mathcal{O}(\log n)$ many slots. In the pruned submodularity graph every vertex is directly linked with every other vertex in the graph. Now, for every vertex $b \in \mathbb{BS}$ in the pruned submodularity graph, we need to compute the function as mentioned in Line No. 13 of Algorithm 1. Now, we can store these values while computing the weights of the edges of the pruned submodularity graph. So, for a billboard slot $b \in \mathbb{BS}$ and any $u \in \mathcal{U}$ the computation w_{bu} can be computed in $\mathcal{O}(1)$ time. As there are $\mathcal{O}(\log n)$ many elements in $\mathcal{U}$, so computing the value of $w_{\mathcal{U}b}$ will take $\mathcal{O}(\log n)$. As the number of elements in $\mathbb{BS}$ is upper bounded by $\mathcal{O}(n)$, performing the steps from Line No. 12 to 14 will take $\mathcal{O}(n \log n)$ time. Sorting these values will take $\mathcal{O}(n \log n)$ time and removing $(1 - \frac{1}{\sqrt{c}})$ fraction of the elements of $\mathbb{BS}$ will take $\mathcal{O}(n)$ time in the worst case. So, a single iteration of the while loop will take $\mathcal{O}(\log n + n \cdot \log n + n) = \mathcal{O}(n \cdot \log n)$ time. So, the execution time of the `while` loop will be of $\mathcal{O}(n \cdot \log n \cdot \log_{\sqrt{c}} n) = \mathcal{O}(n \cdot \log^2 n)$. Now, it is easy to observe that the `while` loop at Line No. 18 will iterate $\mathcal{O}(k)$ many times. The only operation involved within this `while` loop is the marginal gain computation. For any billboard slot $b \in \mathbb{BS} \setminus \mathcal{S}$, its marginal gain computation will take $\mathcal{O}(n \cdot t)$ time. So, the execution of this `while` loop will take $\mathcal{O}(k \cdot n \cdot t)$ time. Hence, the total time requirement of Algorithm 1 will be of $\mathcal{O}(n \cdot t + n^2 \cdot t + n \cdot \log^2 n + k \cdot n \cdot t)$. As $k << n$, this quantity is reduced to $\mathcal{O}(n^2 \cdot t + n \cdot \log^2 n)$.

Additional space requirement by Algorithm 1 is to store the following lists $\mathcal{S}$, $\mathcal{S}'$ and $\mathcal{U}$ which will take $\mathcal{O}(n)$, $\mathcal{O}(n)$, and $\mathcal{O}(\log n)$, respectively. Other than this storing the pruned submodularity graph will take $\mathcal{O}(n^2)$ space. One important thing to observe here is that we can store the edge weights in the adjacency matrix of the pruned submodularity graph and this can be used for the computation involved in Line No. 13. This does not require any more extra space. So, the total space requirement by Algorithm 1 is of $\mathcal{O}(n + \log n + n^2) = \mathcal{O}(n^2)$. Hence, Theorem 2 holds.

Theorem 2. *Time and space requirement by Algorithm 1 is of $\mathcal{O}(n^2 \cdot t + n \cdot \log^2 n)$ and $\mathcal{O}(n^2)$, respectively.*

Theoretical Analysis Related to the Performance. Due to the space limitation, we do not elaborate the analysis. In our analysis, we have used the result from the study in [11] and we state the result in Theorem 3.

Theorem 3. *[11] If we apply the pruned submodularity graph-based pruning technique then the size of the reduced ground set will be $|V'| = \frac{cp}{\log \sqrt{c}} \cdot k \cdot \log^2 n$ where p is the probability of sampling. With high probability (e.g.; $n^{1-qp} \cdot \log_{\sqrt{c}} n$), for all $v \in V \setminus V'$, $w_{V'v} \leq 2 \cdot w_{V*}$. Thus the incremental greedy algorithm for the submodular function maximization on the ground set leads to a solution $\mathcal{S}'$ that follows the following criteria:*

$$I(\mathcal{S}') \geq (1 - \frac{1}{e}) \cdot (I(\mathcal{S}^{OPT}) - 2k\epsilon) \tag{5}$$

where $\mathcal{S}^{OPT}$ is the set of optimal slots of size k.

One important thing to point out is that in Algorithm 1 there are two parameters r and c. The parameter c controls the shrink rate of the billboard slots (decreases at a rate of $\frac{1}{\sqrt{c}}$) whereas the parameter r controls the size of the set $\mathcal{U}$, and eventually it controls the size of the final output of the pruning method. Now, it is important what value we should choose for c and r. As mentioned in the study by Zhou et al. [11], when the value of both r and c is 8, the pruning method converges very quickly. In this study also we consider these values only.

4   Experimental Evaluation

In this section, we describe the experimental evaluation of the proposed solution methodology. Initially, we start by describing the datasets.

4.1 Datasets Used

The datasets that have been used in our study have also been used by existing studies as well [9]. The trajectory data is obtained from two real-world datasets. The first one is the TLC trip record dataset[1] for NYC and the second one is the Foursquare check-in dataset[2] Both these datasets contain the records of green taxi trips from Jan 2013 to Sep 2016 and different location types such as Mall, Beach, Airport, and so on. From these trajectory datasets, we separate the tuples corresponding to the following six locations: 'Beach', 'Mall', 'Bank', 'Bus Stop', 'Train Station', 'Airport', and create six different datasets. The basic statistics of this dataset have been listed in Table 1.

Table 1. Basis statistics of the datasets

Location type	# Rows1	# Rows2	# Billboard slots
Beach	76	575	21888
Mall	86	1186	24768
Bank	671	2232	193248
Bus stop	1056	4473	304128
Train station	288	6407	82944
Airport	313	2852	90144

4.2 Experimental Setup

Now, we describe the experimental setup involved in our experiments. The only parameter whose value needs to be set in this study is the influence probability whose value needs to be fixed. In the dataset (provided by LAMAR[3]) we have information about the panel size of the billboards. Among all the billboards, we choose the panel size of the biggest billboard and we fix the influence probability as the ratio between the size of that billboard and the size of the biggest billboard. After fixing the influence probabilities, we compute the influence for four different values of k, namely 10, 15, 20, 25. All source codes for this experiment are available in the given link[4].

4.3 Algorithms Compared

We compare the performance of the proposed solution approach with the following approaches:

- **RANDOM:** In this method, for a given value of k, we pick any k billboard slots uniformly at random. This is possibly the most simple approach and it will take $\mathcal{O}(k)$ time to execute.

---

[1] http://www.nyc.gov/html/tlc/html/about/trip_record_data.shtml.
[2] https://sites.google.com/site/yangdingqi/home.
[3] http://www.lamar.com/InventoryBrowser.
[4] https://github.com/dildariitjammu/.

- **Top-k**: In this method, for every individual billboard slot we compute their influence and sort them in descending order based on their individual influence value. From this sorted list, we pick up Top-k slots. If there are s many tuples in the trajectory dataset, then computing influence for a single billboard slot will take $\mathcal{O}(s)$ time. As there are total n many billboard slots, hence the total time requirement for influence computation is of $\mathcal{O}(n \cdot s)$. Sorting the slots based on the individual influence value will take $\mathcal{O}(n \cdot \log n)$ time. Choosing k of them will take additional $\mathcal{O}(k)$ time. Hence the total time required will be of $\mathcal{O}(n \cdot s + n \cdot \log n + k)$. As $k << n$, this has been reduced to $\mathcal{O}(n \cdot s + n \cdot \log n)$.
- **Maximum Coverage (MAX_COV)**: In this method, we calculate the individual coverage of every billboard slot. We say that a billboard slot $b \in \mathbb{BS}$ covers a tuple t in the trajectory database $\mathcal{D}$ if the recorded time present in the tuple contained in the duration of the billboard slot. After computing the coverage for every billboard slot, we sort them in descending order and pick k slots of them with the highest coverage value. For a single billboard slot, computing its coverage will take $\mathcal{O}(s)$ time. As there are n many billboard slots hence computing coverage for all the slots will take $\mathcal{O}(n \cdot s)$ time. Now, sorting the slots based on the coverage value will take $\mathcal{O}(n \cdot \log n)$ time. Now, choosing k slots from the sorted list will take $\mathcal{O}(k)$ time. As $k << n$, hence the total time requirement of this approach will be $\mathcal{O}(n \cdot s + n \cdot \log n)$.
- **Pruned Submodularity Graph+Random (P.S.Graph+RAND)**: In this method, first the pruned submodularity graph-based approach is applied to obtain the pruned set of billboard slots, and from these slots we pick k of them uniformly at random.

4.4 Goals of the Experiments

The goals of the experiments are four folded and they are mentioned below:

- **The Effectiveness of Preprocessing**: The first experimental goal of our study is to understand how effective the preprocessing step is. As mentioned previously, if we remove the billboard slots having zero influence then due to submodularity property of the influence function, it does not make any difference.
- **The Effectiveness of Pruning**: One of our goals is to study what is the percentage of the billboard slots pruned out and it is good if it is sufficiently large. In that case, we can easily apply the incremental greedy approach without much computational burden.
- **The Influence Spread**: Another goal of our study is to make a comparative study of the proposed as well as the baseline methods based on the influence spread.
- **Computational Time Requirement**: Finally, our goal is to have a comparison regarding the computational time among the proposed as well as baseline methods.

Next, we proceed to describe the experimental observations along with their explanations.

4.5 Observations with Explanation

The Effectiveness of Preprocessing. In most of the instances, we observe that a significant number of billboard slots are removed after the preprocessing phase. When the location type is 'Beach', among the 21888 many billboard slots only 98 of them are remaining after the preprocessing phase. Percentage wise more than 99% of billboard slots are removed and these observations are consistent with the other location type as well. Table 2 lists out the number and percentage wise billboard slots removed after preprocessing for different location types. Here, $\mathbb{BS}'$ denotes the set of billboard slots after preprocessing.

Table 2. Experimental results related to the Preprocessing Step

Location type	$|\mathbb{BS}|$	$|\mathbb{BS}'|$	Percentage
Beach	21888	98	99.55%
Mall	24768	155	99.37%
Bank	193248	273	99.85%
Bus stop	304128	1030	99.66%
Train station	82944	804	99.03%
Airport	90144	557	99.38%

The Effectiveness of Pruning. As mentioned previously, pruning is an important step in the proposed methodology. When the location type is 'Beach', the number of billboard slots available after the preprocessing step is 98. If we apply the pruned submodularity graph-based approach the number of slots is even reduced to 58 and the percentage-wise reduction is more than 40%. So, including both preprocessing and pruning, the proposed solution approach reduces 99.997% slots. Even for other location types also our observation is consistent that the proposed preprocessing and pruning technique reduces a significant portion of the billboard slots. Table 3 contains experimental results related to the pruning method. Here, $\mathbb{BS}''$ denotes the set of billboard slots after pruning. It is easy to observe that $\mathbb{BS}'' \subseteq \mathbb{BS}' \subseteq \mathbb{BS}$.

Table 3. Experimental results related to the Pruning Step

Location type	$|\mathbb{BS}'|$	$|\mathbb{BS}''|$	Pruning percentage	(Preprocessing+Pruning) Percentage
Beach	98	58	40.81%	99.73%
Mall	155	73	52.9%	99.7%
Bank	273	95	65.2%	99.95%
Bus stop	1030	162	84.27%	99.94%
Train station	804	147	81.71%	99.82%
Airport	557	128	77%	99.85%

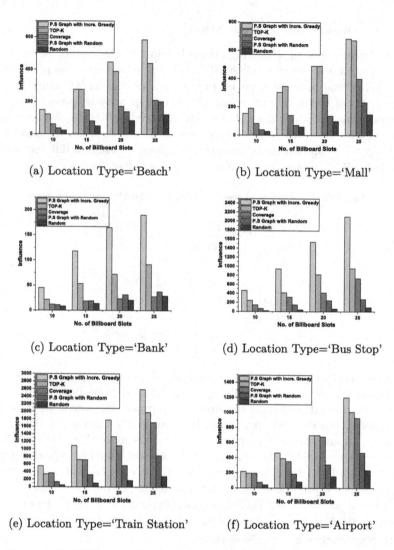

(a) Location Type='Beach'   (b) Location Type='Mall'

(c) Location Type='Bank'   (d) Location Type='Bus Stop'

(e) Location Type='Train Station'   (f) Location Type='Airport'

Fig. 1. No. of billboard slots vs. Influence plots for all the location types

The Influence Spread. Now, we describe our comparative study on influence spread. From the Fig. 1 we observe that the billboard slots selected by the proposed solution approach leads to the maximum influence compared to the baseline methods. As an example, when the value of k is 25, the influence due to the billboard slots selected by the pruned submodularity graph along with the incremental greedy approach is around 1169. However, among the baseline methods the Top-k method leads to maximum amount of influence and that is 655. So, in this dataset the proposed solution approach leads to approximately 78% more influence compared to the baseline methods. This observations are consistent with other location types as well with few exceptions when the location type is 'Mall' and the budgets are 10 and 15. In these two cases we observe that the Top-k Approach is giving more influence.

Computational Time Requirement. From the results, we observe that the computational time requirement of the proposed methodology is reasonable. As an example when the location type is 'Beach' the preprocessing and the pruning time requirement is 84 s and 110 s, respectively. After that to apply the incremental greedy approach for the final selection of billboard slots requires 7.45 min when the value of k is 10. With the increment of k, the time required for the incremental greedy approach is increasing rapidly. These observations are consistent with the other location types also. Table 4 contains the experimental results related to the computational time requirement. Due to the space limitations, we only provide the computational time required for $k = 10$. In case of the PS Graph+Greedy Approach time requirement is $x + y + z$ means, preprocessing time is x s, pruning time is y s, and the selection time is z s.

Table 4. Computational time requirement for all the algorithms for different location types (in Secs.)

Location type	PS Graph+Greedy	Top-k	Coverage	PS Graph+Random	Random
Beach	84+110+447	64	105	190	120
Mall	191+487+1554	102	209	864	432
Bank	5527+4136+14004	5544	5544	10440	6372
Bus stop	17174+42984+74160	17316	17352	61200	10440
Train station	5497+37605+57600	5652	5616	10440	7200
Airport	1625+7201+12600	1656	1656	9360	2196

5   Conclusion and Future Research Directions

In this paper, we have studied the problem of influential billboard slot selection and posed this as a discrete optimization problem. We propose an effective solution approach that works in two phases. First, we apply the pruned submodularity graph-based approach to obtain the pruned set of billboard slots, and subsequently, we apply the incremental greedy approach for finally selecting the billboard slots. We analyze the proposed solution approach to understand its time and space requirements. Also, we have analyzed the proposed methodology to obtain the performance guarantee. Our future work on this problem will remain concentrated on developing efficient solution methodologies for this problem.

References

1. Dai, J., Yang, B., Guo, C., Ding, Z.: Personalized route recommendation using big trajectory data. In: 2015 IEEE 31st International Conference on Data Engineering, pp. 543–554. IEEE (2015)

2. Fisher, M.L., Nemhauser, G.L., Wolsey, L.A.: An analysis of approximations for maximizing submodular set functions-II. In: Balinski, M.L., Hoffman, A.J. (eds.) Polyhedral Combinatorics, pp. 73–87. Springer, Heidelberg (1978). https://doi.org/10.1007/BFb0121195

3. Nemhauser, G.L., Wolsey, L.A., Fisher, M.L.: An analysis of approximations for maximizing submodular set functions-I. Math. Program. **14**(1), 265–294 (1978)

4. Qu, B., Yang, W., Cui, G., Wang, X.: Profitable taxi travel route recommendation based on big taxi trajectory data. IEEE Trans. Intell. Transp. Syst. **21**(2), 653–668 (2019)

5. Vazirani, V.V.: Approximation Algorithms, vol. 1. Springer, Heidelberg (2001)

6. Wang, L., et al.: Data-driven targeted advertising recommendation system for outdoor billboard. ACM Trans. Intell. Syst. Technol. (TIST) **13**(2), 1–23 (2022)

7. Xue, Q., Wang, K., Lu, J.J., Liu, Y.: Rapid driving style recognition in car-following using machine learning and vehicle trajectory data. J. Adv. Transp. **2019** (2019)

8. Zahrádka, J., Machová, V., Kučera, J.: What is the price of outdoor advertising: a case study of the Czech Republic? Ad Alta J. Interdisc. Res. (2021)

9. Zhang, P., Bao, Z., Li, Y., Li, G., Zhang, Y., Peng, Z.: Towards an optimal outdoor advertising placement: when a budget constraint meets moving trajectories. ACM Trans. Knowl. Discovery Data (TKDD) **14**(5), 1–32 (2020)

10. Zhang, Y., Li, Y., Bao, Z., Zheng, B., Jagadish, H.: Minimizing the regret of an influence provider. In: Proceedings of the 2021 International Conference on Management of Data, pp. 2115–2127 (2021)

11. Zhou, T., Ouyang, H., Bilmes, J., Chang, Y., Guestrin, C.: Scaling submodular maximization via pruned submodularity graphs. In: Artificial Intelligence and Statistics, pp. 316–324. PMLR (2017)

Quantifying Association Between Street-Level Urban Features and Crime Distribution Around Manhattan Subway Entrances

Nanxi Su[1]([✉]) [iD], Waishan Qiu[2] [iD], Wenjing Li[3] [iD], and Dan Luo[4] [iD]

[1] Columbia University, New York, NY 10027, USA
ns3315@columbia.edu
[2] Cornell University, Ithaca, NY 14850, USA
[3] The University of Tokyo, Kashiwa-shi, Chiba 277-8568, Japan
[4] The University of Queensland, Saint Lucia, QLD 4072, Australia

Abstract. The Manhattan subway system serves 39% of its commuters as an essential public transit option; however, its annual ridership dropped by 3.48% from 2015 to 2018. This study hypothesizes that ground-level urban-design quality relates to passengers' perceived safety and actual crime rates, subsequently affecting metro ridership. Current literature lacks intensive investigations into how the intertwined physical features and subjective perceptions of micro-scale street environments around subway stations correlate with crime frequencies. It sets out to quantify the correlations between crime reports and urban design quality within the ¼-mile buffer zone of Manhattan subway entrances with the application of Street View Imagery (SVI) and the artificial intelligence of computer vision (CV) and machine learning (ML). Key findings are 1) subjectively and objectively measured urban design quality from SVIs improve explanations of crime. 2) higher perceived safety does not necessarily link with lower crime risks. 3) parks as a point of interest (POI) serve as a crime deterrent. This study has significant implications for urban design and transportation policies and provides references for other urban areas to facilitate safer public transit services and systems by enhancing built environments.

Keywords: Urban design quality · Subway entrance crime · Street View Imagery (SVI) · Objective measure · Human perception

1 Introduction

Decades ago, criminologists pointed out that subway stations and bus stops have ideal settings and opportunities for crimes [1]. Existing literature investigating the correlations between built environment features and urban crime activities often relies on land-use and socio-demographic features, especially in city-scale research. Subway entrances extend the whole subway system underground onto the street-level urban network. They are attached to various neighborhood characteristics (e.g., land use, surveillance management strength, demographics, and socioeconomics). Empirical studies have shown that

© The Author(s), under exclusive license to Springer Nature Switzerland AG 2022
W. Chen et al. (Eds.): ADMA 2022, LNAI 13725, pp. 231–243, 2022.
https://doi.org/10.1007/978-3-031-22064-7_18

the physical appearance of the street-level environment affects human perceptions and behaviors. For example, blocks with visually densely packed buildings, narrower streets [2], and fewer exposed sky [3] are subjectively safer. Perceptions resulting from such visual measurement would influence the popularity [4] and the walkability of a neighborhood [5]. The quality of physical street environments around subway entrances also interacts with the surrounding neighborhood characteristics. For instance, the visual safety of a neighborhood is positively correlated with its average income [6]; neighborhood management failures, such as vacant buildings, broken windows, trash, and abandoned vehicles, are all linked to crimes [7]; subway ridership positively reflects housing prices [8]. With the fast development of deep learning technology and accessible data resources such as Street View Images (SVI), emerging studies have incorporated eye-level streetscape characteristics in examining criminal activities; however, current research lacks specific settings for crime scenarios. Does subjective and objective urban design quality around subway entrances help explain crime behaviors aside from socioeconomic factors? Which urban design qualities are associated with less crime around subway entrances? As part of the whole experience of taking the subway, this study hypothesizes that street-level urban design quality and perceived safety of the streets relate to public transit crime behaviors, which will subsequently alter metro ridership. To fill the gap and test this hypothesis, the paper sets out to extend the conventional factors affecting crime by adding subjective and objective measures of the street environment around metro entrances to comprehensively quantify the association between urban design quality and on-street crime.

2   Literature Review

2.1   Key Dimensions in Assessing Crimes Around Subway Stations

Crime activities and subway stations are highly correlated. Subway stations, with their unique settings of steady streams of massive crowds during certain times of the day, are referred to as "generators" and "attractors" of crimes by previous studies, for it is ideal for offenders to target vulnerable victims on a routine pattern. From sociological views, a rich body of literature has discussed the links between crime and structural defects in society or family. They argue that social disorganizations or inequalities, such as distinctions between neighborhood characteristics in household income, residential stability, employment status, educational level, ethnic heterogeneity, etc., result in crime when people try to achieve their goals valued by the masses. Brantingham and Brantingham [1] claim that urban criminological settings and fear-creating situations are unintentionally created to support regular operations and to meet the satisfaction of people's daily lives. From urban design and architectural standpoints, Jacobs [9] stated that neighborhood life could be revitalized and crime could be reduced through urban planning and urban design; natural surveillance such as night shops, restaurants, pubs, bars, and buildings oriented to the streets, etc. could serve as 'eyes on the street' to diminish crimes and lessen feelings of insecurity. Later Newman [10] indicates in his "defensible space theory" that spaces with territorial markers, surveillance management signals, and ownership indicators deter criminal activity. A growing amount of investigative research emerged based on a similar idea; for instance, Xu et al. [11] found that improved street lights can be a possible solution to reduce crime in a high-crime area. Yang et al. [16] measured the environmental factors affecting offenders' decision-making using virtual reality.

2.2 SVI, CV, and ML for Street Measures

As SVI data become widely accessible, it has improved the efficiency and accuracy of street audits over the traditional in-person field survey. In addition, with the application of CV segmentation models, features can be effortlessly extracted as view indexes by computer for further analysis. Street view features were widely used to study street-level activities and human perceptions. Zhou et al. [13] extracted streetscape features such as traffic signs, terrains, and building out of SVIs to test the associations between the micro-built-environment and drug-related activities. The largest and urban scale crowd-sourcing human perceptions project of Place Pulse has made outstanding contributions in scene understanding algorithms, innovating the way of the field audit. Place Pulse 1.0 [14] ranked around 4,000 geo-coded SVIs from major cities in the United States and Austria, with inputs from nearly 8,000 unique participants from 91 countries of their evaluations on perceptions of safety, class, and uniqueness. In total, around 209,000 pairwise votes were collected. Significant spatial and quantitative correlations between urban perceptions (safety and class) and homicide in NYC at the zip code level were also found after controlling for the effects of income, population, area, and age. Place Pulse 2.0 [15] further upgraded this project on more perceptual dimensions (beautiful, depressing, wealthy, boring, lively, and safe), by expanding the geographic ranges and volumes of the SVI sample collection, and the number of participants and final pairwise votes were raised around ten times. Based on the abundant database of human-perceived urban design qualities, researchers further excavated the possibility of human perception predictions using SVI features, and various model algorithms were tested and compared. Without considering urban design factors, generic features and compositions of the images were explored, and predictive power of 0.57 R^2 was yielded using Place Pulse 1.0 [16]. Besides, with training the dataset of Place Pulse 2.0, the six subjective perception scores were predicted and applied to describe cityscapes in new urban regions. Also worth mentioning are two recent studies that applied the Place Pulse dataset in conjunction with SVI human perception predictions in the drug crime [13] and housing price [17].

3 Data and Methods

3.1 Hotspots of Crime Around Subway Stations in Manhattan

More than 39% of commuters in Manhattan choose the subway as a daily transit method, and it is the most popular one among other mode options. However, according to the MTA statistics, from 2015 to 2018 (the year before the COVID-19 breakout), the annual ridership of the subway in Manhattan continuously declined by 3.48%. An empirical study shows that over-dense crime reports of a place can negatively impact its ridership [18]. According to historical crime statistics of 2018, complaints reported to the NYPD tend to happen more frequently around subway entrances (Fig. 1), with a density of 192 count/km^2 compared to the average crime density of 119 count/km^2 in Manhattan. Safe mobility and stable metro ridership are indispensable attributes of a healthy urban society. Because of the steady drop in ridership, MTA faces unprecedented budget shortfalls in the coming years, making it a huge burden to maintain the service. It is important for researchers, city planners, and decision-makers to give more explicit urban design guidelines on designing safer subway entrances.

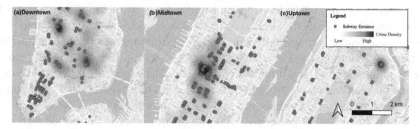

Fig. 1. Manhattan's 2018 total on-street crime reports density maps across (a) downtown, (b) midtown, and (c) uptown. Evident overlaps of crime hotspots and subway station clusters exist, which indicate high correlations between the two.

3.2 Analytical Framework

To investigate the key questions, this study implements the framework of Fig. 2. The critical factors in assessing crime issues are categorized into built environment and neighborhood socioeconomic characteristics. First, the urban design quality on the street level within the ¼-mile buffer area of subway entrances in Manhattan borough on both objective and subjective aspects are evaluated using SVI, CV, and ML. Second, the remaining factors of the built environment are collected (including land use and POI) within ¼ mile buffer area of subway entrances, along with neighborhood socioeconomics (including demographics, residential stability, and structural defects) where the entrances located in the Neighborhood Tabulation Areas (NTA) level by applying Geographic Information System (GIS) and the dataset of American Community Survey (ACS) of the period from 2015 to 2019. Third, the counts of NYPD crime reports within the same ¼-mile buffer area of each subway entrance in Manhattan are aggregated. Fourth, using the data above, this study builds two Ordinary Least Squares (OLS) regression models, one including urban design quality variables and one without, to compare the performance of the variables in explaining the crime behaviors around the subway entrances.

3.3 Data for Constructing Variables

Independent Variables

According to the official count of the MTA, there are 151 subway stations and 775 subway entrances in total in the Manhattan borough. A ¼-mile buffer zone around each subway entrance as a sphere of influence is created, and the street centerline segments

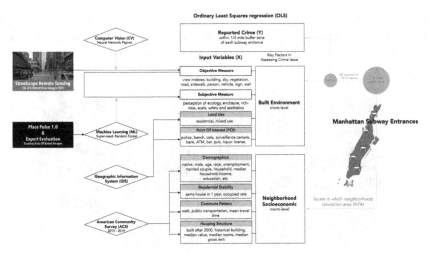

Fig. 2. Analytical framework

[19] falling into the joined area of the buffers are considered as sampling streets to pull out SVIs. The sampling points are extracted from the street segments at an interval of 150 feet. The standard block in Manhattan is about 264 feet by 900 feet, so the interval could ensure around 15 images to be downloaded for a walkable block distance around the entrances. The sampling process is conducted in QGIS. Each sampling point's SVI could be downloaded by giving Google Street View (GSV) API unique coordinates. To get consistent sampling results, the camera settings are kept consistent (pitch = 0, fov = 90, the heading is parallel with the street) and image resolution (600 pixels × 400 pixels). In total, 10,860 valid SVIs are downloaded. The individual visual index is an important indicator representing the percentage that the pixels of a specific visual element takes up the total pixels of an SVI. For example, tree, building, and road are typical view indexes that describe the percentages of the tree, building, and road pixels in an image. To simulate a pedestrian's perspectives when walking down the street, the index of each visual feature of the SVIs is calculated and measured by the function below:

$$VIfeature = \frac{\sum_{i=1}^{n} Pixelfeature}{\sum_{i=1}^{m} Pixeltotal} = \frac{1}{n}\sum_{i=1}^{n} Pixelfeature, feature \in \{tree, \ sky, \ etc.\}$$

(1)

To automatically and efficiently collect such information, Pyramid Scene Parsing Network (PSPNet) is adopted. PSPNet works on a pixel level and can classify and distinguish different visual elements from each other in the SVIs, and the percentage of each feature is calculated. Visual elements of the street environment can form people's feelings when walking in a neighborhood. Following previous studies, this study trains perceived safety, enclosure, complexity, imageability, and human scale based on the datasets of Place Pulse 1.0 [14] and expert panel evaluations of SVIs [20]. Random forest (RF) is adopted for the prediction process, and both PSPNet and Mask R-CNN features were used for training the prediction model. Figure 3 shows examples of 4 groups of SVIs with PSPNet and Mask R-CNN segmentation results and their respective perception scores on five dimensions ranging from 0 to 10 around different subway entrances in Manhattan. Both evaluated objective and subjective qualities were aggregated back to the entrance by summarizing mean values within the buffer zones.

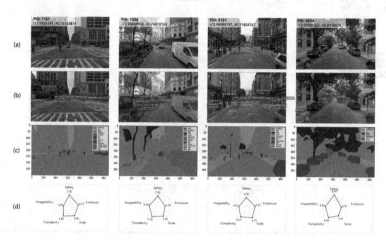

Fig. 3. Samples of (a) original SVIs, (b) instance segmentation results, (c) semantic segmentation results, (d) predicted subjective urban design quality scores. The radar charts scores from 0 to 10 inside out.

The land use data of Manhattan is acquired from MapPLUTO [21]. POI information (parks, surveillance cameras, nightclubs, and cafes) is collected from Geofabrik [22]. The liquor license data are gathered from New York State Liquor Authority Mapping Project [23]. In QGIS, the percentage of the residential land use and the mixed land use within the ¼-mile buffer area for each subway entrance is calculated and aggregated along with the POI counts. From NYC Planning Population Fact Finder [24], the American Community Survey (ACS) five-year estimates at Neighborhood Tabulation Areas (NTA) levels (from 2015 to 2019) are collected. Demographics (native, young age, and race diversity), residential stability (same house), and structural defects (male divorced, poverty population, and education below high school) are taken as independent variables to build the models.

Dependent Variable

To minimize the impact of COVID-19 on crime behaviors and the ridership drop, the historically reported crime data in the Manhattan borough for the whole year of 2018 are acquired from the publicly accessible data source of CompStat 2.0 [25]. Each crime record is documented with where it was completed or attempted (longitude and latitude information), a description of the offense, and the law category code. As our study focuses on the ground-level built environment around the subway stations, among the 114,084 reports, 25,667 pieces of records of on-street crimes are filtered for this analysis. Then the crime incidents are overlapped with the ¼-mile buffer around each subway station in QGIS and the counts are aggregated as the dependent variable (Table 1).

Table 1. Independent variable description (N = 766)

Variables	Description	Mean	Std. Dev	Minimum	Maximum
SUBJECTIVE					
Q1_Safety	Perceived safety	6.73	0.30	5.60	7.71
Q2_Enclosure	Perceived enclosure	5.40	0.28	3.96	5.97
Q3_Complexity	Perceived complexity	5.72	0.39	3.90	6.32
Q4_Scale	Perceived human scale	5.49	0.33	4.00	6.01
Q5_Imagebility	Perceived imageability	4.78	0.52	3.58	5.78
OBJECTIVE					
Person	Number of persons	10.0	4.1	2.3	23.5
Vehicle	Total number of bicycles, motorcycles, cars, trucks and buses	16.6	3.7	5.8	26.2
Stop Sign	Number of stop signs	0.1	0.1	0	0.5
Sky	Sky view index	6.635%	3.745%	0.779%	23.571%
Tree	Tree view index	7.804%	5.461%	0.001%	37.286%
Wall	Wall view index	0.969%	1.504%	0.016%	8.082%
Fence	Fence view index	0.579%	0.521%	0.036%	4.651%
Signboard	Signboard view index	0.297%	0.173%	0.014%	1.009%
Railing	Railing view index	0.120%	0.152%	0%	1.311%
Earth	Earth view index	0.077%	0.158%	0%	1.170%
Window Panel	Window panel view index	0.012%	0.034%	0%	0.227%
Sidewalk	Sidewalk view index	4.453%	1.778%	0.601%	12.753%
Road	Road view index	26.559%	3.631%	11.575%	31.977%
Building	Building view index	39.554%	10.065%	5.211%	58.984%
LAND USE					
Residential	Ratio of family building areas	13.81%	11.82%	0%	47.68%
POI					
POI_Park	Number of cafes	0.02	0.14	0	1
POI_Camera	Number of surveillance cameras	1.97	8.24	0	65
POI_Nightclub	Number of nightclubs	0.08	0.29	0	2
DEMOGRAPHICS					
Native	Native rate	72.00%	7.78%	44.30%	84.00%

(*continued*)

Table 1. (*continued*)

Variables	Description	Mean	Std. Dev	Minimum	Maximum
Young Age	Rate of population age 15 to 24 years	10.73%	4.39%	5.00%	33.10%
Race Diversity	Shannon diversity index of race compositions	102.32%	16.71%	65.74%	140.82%
Male Divorced	Divorce rate male 25 years and above	6.98%	1.75%	3.10%	9.70%
Poverty Population	Rate of population below poverty level	11.74%	6.90%	4.40%	35.60%
Same House	Rate of population resident in the same house 1 year ago	82.16%	4.70%	73.70%	91.80%

4 Findings and Discussion

4.1 Regression Results

During the process of building OLS regression models, Q4_Scale, Q5_Imageability, Building, Mixed Land Use, POI_Cafe, POI_Liquor, Population, Unemployment, and Less Education are removed due to either high VIF-Value (>10) or insignificant P-Value (>0.1), as they are not performing properly in describing the crime rates of any kind. By adding urban design quality attributes, the baseline model is improved in R^2 with 0.270, yielding an overall R^2 of 0.644. Regarding urban design quality variables, both objective measures and subjective measures are found to be significantly correlated with total crime. In general, the results of objectively measured features are largely consistent with the literature, however, some of the subjective perceptions are found inconsistent with general beliefs, which are discussed below.

4.2 Urban Design Quality Matters

Biased Perceived Safety and Street Usage
Perceived safety is positively correlated with crime counts, contrary to the general belief in urban design and criminology [13, 14]. In other words, based on the perceived safety prediction results, subway entrances with more crime counts around tend to have higher safety scores. Meanwhile, it is worth noting that the count numbers of "Person" and "Vehicle" are significantly and positively related to crime counts. The count of "Signboard," usually captured at commercial streets' SVIs; view index of "Window Panel" indicating denser use; and "Surveillance Cameras," suggesting higher criminal behavior potentials, are all found to be positively correlated to crime. On the one hand, unlike the statement that street eyes protect the neighborhood from crimes in the encounter hypothesis [9].

This could be explained by the fact that the population density, work density, and pedestrian and metro users in those perceived safer places are all higher, such that there are more potential victims and more crime opportunities on busier streets [1]. Our controversial finding is, to some extent, consistent with Zhang et al. [26], who explained that unemployment rate, visitor counts throughout the day, and use density are associated with biased perceived safety, making a place be assumed more dangerous or overestimated safe, especially for outsiders. Human perceptions are sophisticated. People of different ages, genders, educational backgrounds, family relations, lifestyles, personalities, or experiences of being offended could have a completely different judgment of perceived safety. This calls for a more detailed evaluation system for future perceived safety surveys and predictions – where the crowd orientation should be specific and determined.

Perceived Enclosure and Territory Markers
View indexes of "Wall", "Fence", "Railing" and count of "Stop Sign", attributes which are considered as key territory markers for deterring offenders, are found negatively correlated with crime around subway stations. This is consistent with Newman's defensible space theory [10], however, a similar term - "perceived enclosure", is found to be positively correlated with the total crime in this study. "Enclosure" is often used in the criminology or crime prevention context to describe a precinct with different forms of barrier, such as the wall, fence, boom, and gate, etc., to restrict access, violence, and unwanted traffic (vehicular or pedestrian). In such contexts, the "enclosure" is the physical description of a neighborhood with well-equipped defensible features, and a place with low permeability tends to have low risks. The discrepancy between "perceived enclosure" and "fence and wall" in the regression result indicates that objective features of the urban environment should not be confused with subjective features by an arbitrary decision before carefully investigating or quantifying research into their interrelations (Table 2).

Perceived Complexity and Other Built Environment Factors
Subjectively measured complexity of the urban environment around subway entrances is found to be negatively and significantly correlated to crime. A higher complexity score suggests the environment's appearance has more diverse and mixed visual elements, in scales, shapes, colors, quantities, etc. Complexity may increase people's enthusiasm and encourage citizen engagement, hence promoting the neighborhood's vitality and management. On the contrary, such revitalization would have a deterrent effect on criminals' decision-making or target sourcing. Likewise, natural streetscape feature such as "Tree" is also negatively correlated with crime, while vacancy indicators of "Sky" and "Earth" perform the opposite. At the same time, the park is also correlated with a lower crime rate around metro entrances. These findings support that street afforest helps improve mental and physical health, ultimately enhancing public safety [27]. Higher residential land use may suggest lower mixed-use and lower diversity of a neighborhood, which is found to be positively correlated with crime. Noticeably, the positive effect of diversity mentioned above does not mean the more elements included, the better. For example,

Table 2. Regression results of baseline and main model.

Variables	Baseline		Model 1		
	Coef.	P-value	Coef.	P-value	VIF
SUBJECTIVE MEASURE					
Q1_Safety			38.8236	***	7.9
Q2_Enclosure			64.0056	***	4.4
Q3_Complexity			−59.9618	***	5.0
OBJECTIVE MEASURE					
Person			5.0099	***	4.2
Vehicle			2.0432	**	5.4
Stop Sign			−127.8557	***	1.8
Sky			625.6886	***	5.1
Tree			−429.3092	***	6.6
Wall			−1274.1322	***	3.2
Fence			−823.6556	**	2.0
Signboard			12580.0000	***	2.7
Railing			−6311.7973	***	1.9
Earth			3160.9658	***	1.6
Window Panel			30820.0000	***	1.4
LAND USE					
Residential	−1.2396	***	0.8301	***	3.2
POI					
POI_Park	−29.3245	***	−29.3514	**	1.2
POI_Camera	1.8547	***	1.1637	***	1.3
POI_Nightclub	47.4612	***	27.2020	***	1.2
DEMOGRAPHICS					
Native	231.8710	***	257.6062	***	3.3
Young Age	−463.0810	***	−322.7878	***	4.4
Race Diversity	−99.7221	***	−94.9332	***	2.8
RESIDENTIAL STABILITY					
Same House	−604.8034	***	−371.0220	***	7.1
STRUCTURAL DEFECTS					
Male Divorced	556.1117	***	334.9005	**	3.4
Poverty Population	917.6979	***	757.2256	***	7.6

Note: p values are shown in parentheses, ***, **, and * indicate significance level of 1%, 5% and 10%, respectively.

the nightclubs around subway stations may indicate more risks, which is consistent with the fact that nightclub is often believed to be risky facilities.

Socio-economic Factors
In addition to urban design quality attributes, the socioeconomic characteristics of the neighborhood where a subway entrance locates are also noteworthy. The poverty population and male divorce rate are found positively correlate with crime, which is consistent with previous studies supporting that social instability and structural defects would trigger crime [28]. Additionally, the same house rate is negatively associated with crime. Similarly, the higher native percentage is positively correlated with crime. This might be because when group size divergence increases, conflicts are easier to arise, and crime is more likely to follow. Hate crime and racism often occur in public places, especially in the transit system, when people of different colors, backgrounds, and ethnic beliefs are brought together. Eventually, the divergent findings contrary to the general and common knowledge are highlighted. First, although many people believe that joblessness leads criminals into crime to achieve goals valued in the common belief of the society or solely to survive, our study suggests that a higher unemployment rate in a neighborhood is not relevant to the crime rate at least around subway stations. Second, this study found young age rate of a neighborhood is negatively correlated with crime.

5   Conclusion

This study leverages the current development in CV technology, visual data source of SVI, and the urban informatics of Manhattan reported crime geography to test whether urban design qualities matter in assessing public transit crime issues. It quantitatively explores the association between crime around subway stations and both objectively and subjectively measured urban design environments – they are essential factors in building a comprehensive evaluation system to support a safer and sustainable public transportation environment and crime control. By comparing the OLS model performances before and after introducing the urban design quality attribute, a notable increase in P-Value of 0.278 is established, implying that urban design quality is essential for related analysis. We recommend that future studies further investigate the interlocked associations taking into account of spatial interaction and ridership flows. This study has important implications for urban planners and policymakers in managing urban design strategies and optimizing subway transit services. It also casts light on the power and potential of CV and ML technology and SVI as effective tools to investigate urban-related issues for urban researchers.

References

1. Brantingham, P., Brantingham, P.: Criminality of place: crime generators and CrimeAttractors. Eur. J. Crim. Policy Res. **13**, 5–26 (1995)
2. Harvey, C.W.: Measuring streetscape design for livability using spatial data and methods (2014). https://scholarworks.uvm.edu/graddis/268/

3. Yin, L., Wang, Z.: Measuring visual enclosure for street walkability: using machine learning algorithms and Google Street View imagery. Appl. Geogr. **76**, 147–153 (2016). https://doi.org/10.1016/j.apgeog.2016.09.024

4. Ma, X., et al.: Measuring human perceptions of streetscapes to better inform urban renewal: a perspective of scene semantic parsing. Cities **110**, 103086 (2021). https://doi.org/10.1016/j.cities.2020.103086

5. Ewing, R., Handy, S.: Measuring the unmeasurable: urban design qualities related to walkability. J. Urban Des. **14**, 65–84 (2009). https://doi.org/10.1080/13574800802451155

6. Naik, N., Kominers, S., Raskar, R., Glaeser, E., Hidalgo, C.: Do people shape cities, or do cities shape people? The Co-evolution of physical, social, and economic change in five major U.S. Cities. SSRN Electron. J. (2015). https://doi.org/10.2139/ssrn.2698292

7. Kelling, G.L., Wilson, J.Q.: Broken Windows - The police and neighborhood safety. https://www.theatlantic.com/magazine/archive/1982/03/broken-windows/304465/

8. Chen, C.: Exploring Housing Market in Toronto, Ontario: Spatial Hedonic Modeling of Crime Rates, Subway Ridership, Dwelling Density and House Prices (2015). http://hdl.handle.net/10012/9656

9. Jacobs, J.: The death and life of great American cities (1961)

10. Newman, O.: Defensible Space: Crime Prevention Through Urban Design. Macmillan (1972)

11. Xu, Y., Fu, C., Kennedy, E., Jiang, S., Owusu-Agyemang, S.: The impact of street lights on spatial-temporal patterns of crime in Detroit, Michigan. Cities **79**, 45–52 (2018). https://doi.org/10.1016/j.cities.2018.02.021

12. Yang, J.W., Kim, D., Jung, S.: Using eye-tracking technology to measure environmental factors affecting street robbery decision-making in virtual environments. Sustainability **12** (2020). https://doi.org/10.3390/su12187419

13. Zhou, H., et al.: Using Google Street View imagery to capture micro built environment characteristics in drug places, compared with street robbery. Comput. Environ. Urban Syst. **88**, 101631 (2021). https://doi.org/10.1016/j.compenvurbsys.2021.101631

14. Salesses, P., Schechtner, K., Hidalgo, C.A.: The collaborative image of the city: mapping the inequality of urban perception. PLoS ONE **8**, e68400 (2013). https://doi.org/10.1371/journal.pone.0068400

15. Dubey, A., Naik, N., Parikh, D., Raskar, R., Hidalgo, C.A.: Deep learning the city: quantifying urban perception at a global scale. In: Leibe, B., Matas, J., Sebe, N., Welling, M. (eds.) ECCV 2016. LNCS, vol. 9905, pp. 196–212. Springer, Cham (2016). https://doi.org/10.1007/978-3-319-46448-0_12

16. Raskar, R., Naik, N., Philipoom, J., Hidalgo, C.: Streetscore – predicting the perceived safety of one million streetscapes. In: IEEE Conference on Computer Vision and Pattern Recognition Workshops. IEEE Computer Society (2015). https://doi.org/10.1109/CVPRW.2014.121

17. Kang, Y., et al.: Understanding house price appreciation using multi-source big geo-data and machine learning. Land Use Policy **111**, 104919 (2021). https://doi.org/10.1016/j.landusepol.2020.104919

18. Zhang, W.: Does compact land use trigger a rise in crime and a fall in ridership? A role for crime in the land use–travel connection. Urban Stud. **53**, 3007–3026 (2016). https://doi.org/10.1177/0042098015605222

19. NYC Open Data. https://opendata.cityofnewyork.us/

20. Qiu, W., et al.: Subjective or objective measures of street environment, which are more effective in explaining housing prices? Landsc. Urban Plan. **221**, 104358 (2022). https://doi.org/10.1016/j.landurbplan.2022.104358

21. NYC Department of City Planning. https://www1.nyc.gov/site/planning/index.page

22. GEOFABRIK Downloads. https://download.geofabrik.de/

23. New York State Liquor Authority Mapping Project. https://lamp.sla.ny.gov/

24. NYC Planning Population Fact Finder. https://popfactfinder.planning.nyc.gov/
25. New York City Police Department. https://compstat.nypdonline.org/
26. Zhang, F., Fan, Z., Kang, Y., Hu, Y., Ratti, C.: "Perception bias": deciphering a mismatch between urban crime and perception of safety. Landsc. Urban Plan. **207**, 104003 (2021). https://doi.org/10.1016/j.landurbplan.2020.104003
27. Jing, F., et al.: Assessing the impact of street-view greenery on fear of neighborhood crime in Guangzhou, China. Int. J. Environ. Res. Public. Health **18** (2021). https://doi.org/10.3390/ijerph18010311
28. Merton, R.K.: Social structure and anomie. Am. Sociol. Rev. **3**, 672–682 (1938). https://doi.org/10.2307/2084686

The Coherence and Divergence Between the Objective and Subjective Measurement of Street Perceptions for Shanghai

Qiwei Song[1]([⊠]) [iD], Meikang Li[2] [iD], Waishan Qiu[3] [iD], Wenjing Li[4] [iD], and Dan Luo[5] [iD]

[1] IBI Group, Toronto, ON M5V 3T5, Canada
qiwei.song@mail.utoronto.ca
[2] Shenzhen Technology University, Shenzhen 518118, Guangdong, China
[3] Cornell University, Ithaca, NY 14850, USA
[4] The University of Tokyo, Tokyo 113-8654, Japan
[5] University of Queensland, St. Lucia, QLD 4072, Australia

Abstract. Recent development in Street View Imagery (SVI), Computer Vision (CV) and Machine Learning (ML) has allowed scholars to quantitatively measure human perceived street characteristics and perceptions at an unprecedented scale. Prior research has measured street perceptions either objectively or subjectively. However, there is little agreement on measuring these concepts. Fewer studies have systematically investigated the coherence and divergence between objective and subjective measurements of perceptions. Large divergence between the two measurements over the same perception can lead to different and even opposite spatial implications. Furthermore, what street environment features can cause the discrepancies between objectively and subjectively measured perceptions remain unexplained. To fill the gap, five pairwise (subjectively vs objectively measured) perceptions (i.e., complexity, enclosure, greenness, imageability, and walkability) are quantified based on Street View Imagery (SVI) and compared their overlap and disparity both statistically and through spatial mapping. With further insights on what features can explain the differences in each pairwise perceptions, and urban-scale mapping of street scene perceptions, this research provides valuable guidance on the future improvement of models.

Keywords: Street view imagery · Human perceptions · Subjective and objective · Coherence and divergence · Machine learning

1 Introduction

Street provides significant public space where people gather, meet, and interact [1]. How people sense and perceive the street environment directly influences human behaviors such as walking [2]. Therefore, it is essential to maintain consistency and efficiency in evaluating perceptions of the streets. Recently, SVI provides big dataset for micro-level human-perceived street characteristics [3] and studies have taken advantage of it to map the street environment perceptions [4–6].

© The Author(s), under exclusive license to Springer Nature Switzerland AG 2022
W. Chen et al. (Eds.): ADMA 2022, LNAI 13725, pp. 244–256, 2022.
https://doi.org/10.1007/978-3-031-22064-7_19

Using SVI, scholars were able to measure street perceptions objectively or subjectively [4, 5, 7]. However, to date there is little consensus on the measurement of perceptions. Objectively measured perceptions rely on complex math formulas by recombing extracted view indices using CV to proxy perceptions such as walkability [5]. Subjectively measured perceptions are typically collected through ML predicted scores based on crowd-sourcing visual survey results [6, 8]. However, previous study revealed it may exhibits different results even measuring the same perception concept [9]. Fewer studies have systematically compared the coherence and divergence between objective and subjective measurements of perceptions. Furthermore, it is largely unknown what street features can cause the disparities in pairwise objectively measured and subjectively measured perceptions.

To bridge these knowledge gaps, using Shanghai as case study site, we collected SVI to quantify five pairwise (objective vs subjective) perceptions [9], namely the enclosure, greenness, complexity, imageability, walkability. Our contribution is three-fold. First, we provide a comprehensive and high-throughput framework integrating Artificial Intelligence (AI) and SVI data that accurately reflects objective and subjective measurements. Second, we provide pairwise comparison (statistically and spatially) between objectively and subjectively measured perceptions to identify their overlap and divergence. Third, we investigate what physical features may explain the differences between two measures and further provide suggestions to help improve the measurement framework, which hasn't been studied in previous research.

2 Literature Review

The conventional methods to measure street perceptions rely on low-throughput questionaries or surveys, while this process is labor-intensive, time-consuming and challenging to deploy over larger territories [7]. Open-source SVI data such as Google Street Views largely overcomes these limitations and was increasingly used in urban studies regarding human-perceived street environment [10, 11]. Moreover, the rapid development in CV, deep learning (DL), and ML technology have enhanced efficiency and accuracy in processing SVI data. Combining SVI and computational frameworks, many studies have proved good robustness in mapping streetscape features and human perceptions [8, 12].

On the one hand, objective measure uses CV to extract pixels of various elements as view indices from SVI to describe the street environment [13]. Apart from simple indicator such as sky view factor, recent studies developed complex mathematical formulas to proxy the human perceptions. Specifically, Ma et al. [5] formulated equations deriving from operative definitions of each perception concept [14], and measured five perceptions (i.e., greenness, openness, enclosure, walkability, imageability) by recombining the view indices of key physical elements like sidewalk and tree. On the other hand, traditional subjective measures collect opinions from surveys and panel of experts [14]. Along this line, new studies emerged to integrate SVI and crowdsourced visual surveys [15]. Extensive studies have followed this approach and successfully predicted citywide subjective perceptions using ML algorithms [6, 12]. For example, Zhang et al. [6] mapped six subjective perceptions (e.g. lively) for different Chinese cities.

Nevertheless, few studies have compared the overlap and divergence of the two perception schemes but revealed mixed results. Xu et al. [9] compared six pairwise objectively and subjectively measured perceptions and found that objective measures outperform subjective measures in explaining housing price variances for self-evident concepts such as greenness. While Song et al. [16] demonstrated that subjectively measured perceptions have a higher correlation with social inequality. However, previous research needs to better capture spatial differences and the underlining mechanism behind the discrepancy between the two frameworks.

Previous empirical studies attempted to understand the correlation between perceptions and the street features [14]. Zhang et al. [6] investigated the correlations between visual elements and six subjective perceptions (e.g., lively) and detected the negative impact of wall on all perceptions and the positive influences of natural features. Similarly, Qiu et al. [4] revealed that less typical features like signboard are important in affecting urban design perceptions. Nevertheless, prior studies only focused on the impact of visual elements on subjective perceptions. Acknowledging the difference between subjectively and objectively measured perceptions, it is crucial to further analyze and evaluate what features can potentially explain the discrepancies in measuring the same perception concept.

3   Methods and Process

3.1   Analytical Framework

With the research gap identified, this research is set to investigate the coherence and divergence systematically and spatially between subjective and objective measures of perceptions. It proposes high-throughput quantification and analysis methods with AI and big data. Furthermore, it is the first study which sheds light on the features that lead to differences of the two measurement streams (Fig. 1).

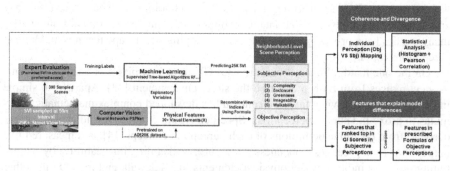

Fig. 1. Analytical framework.

3.2 Site Investigation and Data Preparation

As one of China's major financial, trade and shipping hubs, a city-wide analysis of the street perceptions in Shanghai using both measures across neighborhoods can provide meaningful comparisons and draw conclusions for the urban planning. The data are collected from (1) SVI: Baidu Street View API, and (2) shapefile of road networks: Open Street Map (OSM).

3.3 Quantifying Objective and Subjective Perception Scores

Extracting Physical Elements from SVIs. SVIs can reflect the human-centric perspectives of pedestrians or cyclists [10].We followed steps from previous studies [9] and sampled SVIs at 50 m intervals in QGIS along the road networks and requested SVI data from Baidu Street View Static API, and we retrieved 25,276 valid SVIs. Previous studies have used View Index to denote the ratio of the visual feature's total pixels to the full image [5, 9]. We applied Pyramid Scene Parsing Network (PSPNet) [17], to extract the view indices of physical elements from SVIs efficiently. The example semantic segmentation results are shown in Fig. 2. The process provided quantifiable view indices of 33 types of physical street elements from the dataset.

Fig. 2. Example of CV parsing raw SVI inputs and results.

Calculating Objective Perception Scores. Objective perceptions are calculated based on their operative definitions [14]. Following previous objective measurement framework [5], we calculated each perception using complex equations (Table 1) by recombining view indices. The results is then normalized to a 0–1 scale (worst to best) and obtained objectively measured perception scores.

Table 1. Measurements of objective perceptions.

Perceptions	Qualitative definition	Objective score equations
1. Complexity	The visual richness of a place [14]	$O1\_Cmplx_i =$ $$\frac{VI_{persn}+VI_{signb}+VI_{strlgh}+VI_{tree}+VI_{chair}+VI_{windwp}}{VI_{bldg}+VI_{road}} \quad (1.1)$$

(*continued*)

Table 1. (*continued*)

Perceptions	Qualitative definition	Objective score equations
2. Enclosure	The degree to which streets are visually defined [14]	$O2\_Encls_i = \frac{VI_{bldg}+VI_{tree}}{VI_{road}+VI_{sidewlk}+VI_{earth}+VI_{grass}}$ (1.2)
3. Greenness	Visual urban greenery [5]	$O3\_Green_i = VI_{tree}$ (1.3)
4. Imageability	The quality of a place that makes it distinct [14]	$O4\_Imgbl_i = VI_{bldg} + VI_{skycrp} + VI_{signb}$ (1.4)
5. Walkability	The psychological walking experience [5]	$O5\_Walkb_i = \frac{VI_{sidewlk}+VI_{fence}}{VI_{road}}$ (1.5)

Notes: VI_{tree}, $VI_{sidewlk}$, VI_{fence}, VI_{road}, VI_{persn}, VI_{signb}, VI_{strlgt}, VI_{windwp}, VI_{skycrp}, VI_{earth}, VI_{grass}, and VI_{chair} denotes the view index of tree, sidewalk, fence, road, person, signboard, streetlight, windowpane, skyscraper, earth, grass, and chair, respectively.

Calculating Subjective Perception Scores. Following Qiu et al. [4]'s method to quantify subjective perceptions, we sampled 300 SVIs across Shanghai, covering urban center to countryside and a visual survey website was developed. Participants were shown randomly paired SVIs side by side to choose preferred image to reply to each perception. And we further adopted the Microsoft TrueSkill algorithm to convert the collected pairwise preferences into interpretable scores [18]. Since our explanatory variables encompass roughly thirty physical features, 300 samples are sufficient because scholars mentioned ten times the number of variables could attain reasonable results [8]. We split 300 SVIs by 80% for training and 20% for testing. The five perceptions of 300 SVIs are used as training labels and the view indices are used as independent variables for prediction. Five tree-based algorithms are selected for prediction, the balance performance of R-squared (R2) and Mean Absolute Error (MAE) were used to judge the results (Table 2).

Overall, Gradient Boosting (GB) had best performances (lowest MAE) in predicting four qualities. And Random Forest (RF) performed the best in predicting 'complexity'. The five models had R2 values (0.41–0.51) which explain around half of the variance, and they partially or entirely outperformed previous research outcomes by Ito & Biljecki [19] and Naik et al. [8]. And MAEs range from 1.2 to 1.51, indicating that the prediction errors would not offset fitted value away from true scores in the 0–10 scale. Results revealed that people exhibit more similarities in evaluating complexity, greenness and imageability perceptions. The best-performed model is selected for each perception to predict subjective scores for the entire SVIs.

Verifying Perception Scores. Zhang et al. [6] reported high correlations in 'beautiful-wealthy' and 'depressing-safe', presenting multicollinearity issues. We applied Pearson correlation analysis to the five perceptions within each framework. We found that

Table 2. Performance of ML algorithms.

Model criterion	S1_ Cmplx		S2_Encls		S3_Green		S4_Imblt		S5_ Walkb	
	R^2	MAE	R^2	MAE	R^2	MAE	R^2	MAE	R^2	MAE
Random Forest (RF)	0.49*	1.21*	0.43	1.55	0.41	1.43	0.29	1.73	0.46	1.36
Decision Tree (DT)	0.08	2.14	0.26	2.29	0.12	1.96	0.05	2.36	0.13	1.94
Voting Selection (VS)	0.31	1.60	0.35	1.60	0.35	1.53	0.26	1.78	0.36	1.55
Gradient Boosting (GB)	0.14	2.01	0.41*	1.52*	0.49*	1.39*	0.51*	1.62*	0.48*	1.33*
ADA Boost (ADAB)	0.32	1.63	0.41	1.52	0.31	1.57	0.20	1.84	0.48	1.33

Notes: S1_Cmplx, S2_Encls, S3_Green, S4_Imblt, S5_Walkb represents Complexity, Enclosure, Greenness, Imageability and Walkability, respectively. And * denotes the best performance model

within the subjective perceptions, enclosure-complexity, walkability-complexity, and walkability-enclosure indicated relatively high (between ± 0.50 and ± 1) degree correlations, other pairs showed moderate correlations (between ± 0.30 and ± 0.5). Comparing to subjectively measured perceptions, objective perceptions in general reveal low or moderate correlations except one pair (greenness-imageability). This indicates that choosing objective perceptions help reduce the multi-collinearity issues.

3.4 Coherence and Divergence of the Subjective and Objective Perceptions

The descriptive statistics of perceptions are listed in Appendix. Their coherence and divergence are further examined statistically using histogram (Fig. 3). First, scores measured from both strands were close to normal distribution. Second, only imageability revealed more coherence in the mean value, variance, and data distribution. Though we discovered some overlap for enclosure and walkability, they have different variances. Third, complexity and greenness have the most evident differences. Lastly, all subjective scores are larger in mean value than objective counterparts. The overall low median values of objective perceptions manifested that simply recombining view indices might not comprehensively capture all indicators of visual experience, for example, the psychological emotions may contribute to how people sense a space [4]. Accounting for the overlap and divergence between the two measurement approaches, the neighborhood perceptions is represented by the average of scores withing 1km radius of each downloaded housing property [4, 7], and mapped subjective and objective perceptions using natural breaks (Jenks) to examine the spatial distribution and within-perception heterogeneity pattern of each perception.

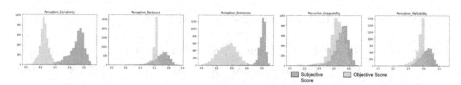

Fig. 3. Histograms of subjective and objective scores for five pairwise perceptions.

3.5 Features that Cause Differences Between Objective and Subjective Measures

Although pairwise perceptions exhibited similarities in subjective and objective scores, it exhibited more disagreements both statistically and spatially. This led to another unanswered question: what features in the streetscape can potentially lead to the divergence between subjective and objective perceptions?

Statistical inferences from prior studies revealed that visual elements have a different weighting in predicting subjective perceptions [6, 9]. On the one hand, for subjective perceptions, we chose tree-based algorithm models during prediction process as they can calculate Gini Importance (GI) score, which represents the importance of each explanatory variable [20] that contributes to the perception scores. We applied Tree-Based Regressor in the Python Scikit-learn package to calculate GI scores, ranking each physical feature in its impact. On the other hand, objectively measured perceptions are formula-derived by nature as framework was developed based on operative definitions [5, 14]. In essence, each perception is influenced by features prescribed in its formula. Hence, by comparing ranked physical elements from GI scores of subjective scores with features prescribed in their pairwise objective perception formula, we could identify elements that explain the disparity between the two measures for each perception.

4   Results and Discussion

4.1 Spatial Mismatch Between Subjective and Objective Perceptions

We mapped the distribution of subjective and objective scores of five perceptions in Shanghai (Fig. 4). For greenness, both streams have a strong consensus in identifying areas of bad or low quality. Regarding walkability and enclosure, higher subjective scores concentrate in sub-city centers in Lu Wan, Xu Hui, Zha Bei, Jing'An, Hong Kou and downtown Pu Dong. While higher objective scores scatter across districts, we see more areas are with high walkability and enclosure than their subject counterparts.

Both statistically and spatially, this research sheds new lights on the coherence and divergence of the two measurement frameworks. The mapping of each perception using both measures shows that street perceptions distribute unevenly. Subjective scores spatially exhibit a more uneven pattern. It is observed that districts with more urban and sub-city center areas show high complexity, enclosure, imageability and walkability while exhibit relatively lower greenness. This finding aligns well with our understanding that typically in densely populated downtown neighborhoods, the street interface is more complex (higher complexity), has more towers (higher enclosure), with its distinct identity (higher imageability) and is more walking-friendly (higher walkability).

However, due to the narrow street profile, trees are typically not lushly planted (lower greenness).

When making pairwise comparisons across measures, enclosure, greenness, and walkability seem to exhibit closer within-perception heterogeneity patterns. However, we find a significant discrepancy on pairwise perceptions regarding complexity and imageability. For example, most lands in downtown Pu Dong are rated as low quality in objective complexity, while subjective score shows them as high quality.

4.2 Key Urban Features for Variances Between Two Models

Figure 5 illustrates the feature importance of each subjectively measured score. Following prior research [4], the GI score was applied to assess the importance of various features contributing to the subjective scores, and we compared them with the prescriptive formulas of objective scores. This enabled us to derive preliminary assumptions on elements that can explain the differences between the two measurement frameworks for each perception.

Regarding the Complexity, besides sky, the top ten important features exhibit very close GI values in explaining subjective scores. For objective perception, complexity is affected by various elements such as the signboard. The low median value of objective complexity suggest it can further be improved by calculating the diversity of elements in the scene. The subjective measure of Enclosure seems to reveal people sense the enclosure predominantly by feeling the overall ratio of sky with its relationship with other vertical or horizontal features, thus the sky can be further added into the objective formula. Objective Greenness is solely dependent on 'tree'. Its scores show huge discrepancy from subjectively perceived greenness, which is determined jointly by trees, buildings, roads, cars, plants, earth, sky and wall. We speculate that greenness depends not only on quantity but on quality [21] and its structural composition of different types of greenery [22]. These assumptions need further verification in the future. Imageability represents the memorable quality of the site. The objective imageability clearly neglected the quality of softscape. For example, in Shanghai the lush row of London Plane trees along Heng Shan Road located in the French Concession Zone render the uniqueness of the neighborhood because of its cultural identity. This assumption is supported by examining the GI scores, of which the tree and grass rank 2nd and 3rd, respectively. The objective Walkability would benefit from incorporating additional elements such as the presence of cars and planting. However, regarding GI scores for subjective perceptions, some features that are important in previous research, such as street furniture and streetlights, were not among the top ten ranked GI scores. Current ML framework uses view indices of elements as explaining variables, this might cause some bias as the street furniture functions as street amenity and improves the walkability, but the limited pixel ratios of the furniture can mislead to a weak correlation. In the future it can be improved by counting object instances or judging its presence [19].

5   Conclusion

This study collected five urban design perceptions (i.e., complexity, enclosure, greenness, imageability, and walkability) using subjective and objective measures in Shanghai. It is

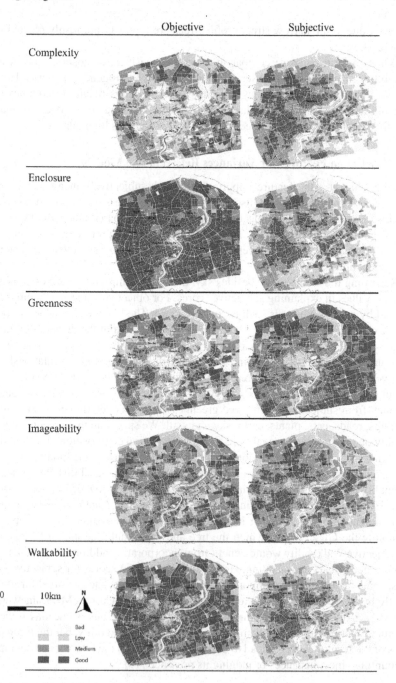

Fig. 4. Spatial distribution of subjective and objective perceptions. (Color figure online)

pioneering research that systematically compared their overlap and divergence between

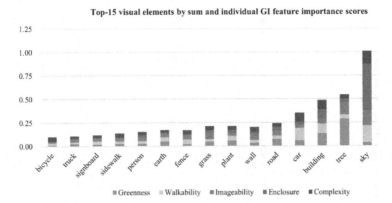

Fig. 5. Top 15 physical elements for each subjective score and their sum Gini importance.

the two streams. First, we identified both similarities and discrepancies between subjective and objective pairwise perceptions, they are reflected by statistical differences in variances and mean values. In general, subjective scores exhibit higher mean values than their objective counterparts. While objective perceptions can potentially help reduce the multi-collinearity issues identified in subjective perceptions. Second, we identified the disparity regarding within-perception heterogeneity across pairwise perceptions. It revealed clear spatial mismatch regarding the distribution pattern when mapping pairwise perceptions. It is observed in the findings that straightforward perceptions exhibit more similarities for the pairwise measurement, while complex qualities (e.g., imageability) show more discrepancies and even demonstrate contrary within-perception heterogeneity patterns. Third, it provides preliminary inferences on physical features that lead to the discrepancies of the two measures and provide advice on future model improvement. Finally, it overall provides a high-throughput measurement and comparison framework for subjective and objective perceptions, which can be applied to studies of other cities in the future as long as SVI data is available.

There are also limitations to the study. First, for subjective measure, the training samples were based on experts' preference selections on 300 images, the dataset can be further expanded. The prediction accuracy can be further improved by adding low-level features as explanatory variables [19]. Second, when comparing the pairwise concept, individual perceptions can be jointly analyzed by using other statistical models such as Principle Component Analysis to derive more interpretations [23]. Finally, our preliminary conclusions on improving the measurement models can be tested in future work.

Appendix

General descriptive statistics of perceptions.

Neighbourhood attributes		Count	Mean	Std. Dev.	Min	Max	Data source
Subjective streetscape attributes							
S1_CMPLX	Subjectively perceived complexity	40,159	0.6	0.0	0.5	0.9	Predicted by ML models from Baidu SVIs
S2_ENCLS	Subjectively perceived enclosure	40,159	0.7	0.1	0.3	0.9	
S3_GREEN	Subjectively perceived greenness	40,159	0.8	0.0	0.4	0.9	
S4_IMBLT	Subjectively perceived imageability	40,159	0.7	0.1	0.3	0.9	
S5_WALKB	Subjectively perceived walkability	40,159	0.6	0.1	0.4	0.8	
Objective streetscape attributes							
O1_CMPLX	Objectively calculated complexity	40,159	0.3	0.1	0.0	0.6	Recombined selected physical feature view indices
O2_ENCLS	Objectively calculated enclosure	40,159	0.6	0.0	0.1	0.7	
O3_GREEN	Objectively calculated greenness	40,159	0.4	0.1	0.0	0.8	
O4_IMBLT	Objectively calculated imageability	40,159	0.6	0.1	0.0	0.9	
O5_WALKB	Objectively calculated walkability	40,159	0.6	0.1	0.2	0.7	

References

1. Mehta, V.: Lively streets: determining environmental characteristics to support social behavior. J. Plan. Educ. Res. **27**, 165–187 (2007). https://doi.org/10.1177/0739456X0730 7947
2. Salazar Miranda, A., Fan, Z., Duarte, F., Ratti, C.: Desirable streets: using deviations in pedestrian trajectories to measure the value of the built environment. Comput. Environ. Urban Syst. **86**, 101563 (2021). https://doi.org/10.1016/j.compenvurbsys.2020.101563
3. Rundle, A.G., Bader, M.D.M., Richards, C.A., Neckerman, K.M., Teitler, J.O.: Using google street view to audit neighborhood environments. Am. J. Prev. Med. **40**, 94–100 (2011). https://doi.org/10.1016/j.amepre.2010.09.034
4. Qiu, W., et al.: Subjective or objective measures of street environment, which are more effective in explaining housing prices? Landsc. Urban Plan. **221**, 104358 (2022). https://doi.org/10.1016/j.landurbplan.2022.104358
5. Ma, X., et al.: Measuring human perceptions of streetscapes to better inform urban renewal: a perspective of scene semantic parsing. Cities **110**, 103086 (2021). https://doi.org/10.1016/j.cities.2020.103086
6. Zhang, F., et al.: Measuring human perceptions of a large-scale urban region using machine learning. Landsc. Urban Plan. **180**, 148–160 (2018). https://doi.org/10.1016/j.landurbplan.2018.08.020
7. Zhou, H., He, S., Cai, Y., Wang, M., Su, S.: Social inequalities in neighborhood visual walkability: using street view imagery and deep learning technologies to facilitate healthy city planning. Sustain. Cities Soc. **50**, 101605 (2019). https://doi.org/10.1016/j.scs.2019.101605
8. Naik, N., Philipoom, J., Raskar, R., Hidalgo, C.: Streetscore – predicting the perceived safety of one million streetscapes. In: 2014 IEEE Conference on Computer Vision and Pattern Recognition Workshops, pp. 793–799 (2014). https://doi.org/10.1109/CVPRW.2014.121
9. Xu, X., et al.: Associations between street-view perceptions and housing prices: subjective vs. objective measures using computer vision and machine learning techniques. Remote Sens. **14**, 891 (2022). https://doi.org/10.3390/rs14040891
10. Biljecki, F., Ito, K.: Street view imagery in urban analytics and GIS: a review. Landsc. Urban Plan. **215**, 104217 (2021). https://doi.org/10.1016/j.landurbplan.2021.104217
11. Song, Q., Liu, Y., Qiu, W., Liu, R., Li, M.: Investigating the impact of perceived micro-level neighborhood characteristics on housing prices in Shanghai. Land. **11**, 2002 (2022). https://doi.org/10.3390/land11112002
12. Dubey, A., Naik, N., Parikh, D., Raskar, R., Hidalgo, C.A.: Deep learning the city: quantifying urban perception at a global scale. In: Leibe, B., Matas, J., Sebe, N., Welling, M. (eds.) ECCV 2016. LNCS, vol. 9905, pp. 196–212. Springer, Cham (2016). https://doi.org/10.1007/978-3-319-46448-0_12
13. Li, X., Zhang, C., Li, W., Ricard, R., Meng, Q., Zhang, W.: Assessing street-level urban greenery using google street view and a modified green view index. Urban For. Urban Green. **14**, 675–685 (2015). https://doi.org/10.1016/j.ufug.2015.06.006
14. Ewing, R., Handy, S.: Measuring the unmeasurable: urban design qualities related to walkability. J. Urban Des. **14**, 65–84 (2009). https://doi.org/10.1080/13574800802451155
15. Salesses, P., Schechtner, K., Hidalgo, C.A.: The collaborative image of the city: mapping the inequality of urban perception. PLoS ONE **8**, e68400 (2013). https://doi.org/10.1371/journal.pone.0068400
16. Song, Q., Li, W., Li, M., Qiu, W.: Social Inequalities in Neighborhood-Level Streetscape Perceptions in Shanghai: The Coherence and Divergence between the Objective and Subjective Measurements (2022). https://papers.ssrn.com/abstract=4179127. https://doi.org/10.2139/ssrn.4179127

17. Zhao, H., Shi, J., Qi, X., Wang, X., Jia, J.: Pyramid scene parsing network. In: Proceedings of the 2017 IEEE Conference on Computer Vision and Pattern Recognition (CVPR), Honolulu, HI, USA, 21–26 June 2017, pp. 6230–6239 (2017)
18. Minka, T., Cleven, R., Zaykov, Y.: TrueSkill 2: an improved Bayesian skill rating system (2018)
19. Ito, K., Biljecki, F.: Assessing bikeability with street view imagery and computer vision. Transp. Res. Part C Emerg. Technol. **132**, 103371 (2021). https://doi.org/10.1016/j.trc.2021.103371
20. Nembrini, S., König, I.R., Wright, M.N.: The revival of the Gini importance? Bioinformatics **34**, 3711–3718 (2018). https://doi.org/10.1093/bioinformatics/bty373
21. Wang, R., Feng, Z., Pearce, J., Yao, Y., Li, X., Liu, Y.: The distribution of greenspace quantity and quality and their association with neighbourhood socioeconomic conditions in Guangzhou, China: a new approach using deep learning method and street view images. Sustain. Cities Soc. **66**, 102664 (2021). https://doi.org/10.1016/j.scs.2020.102664
22. Li, X., Zhang, C., Li, W., Kuzovkina, Y.A.: Environmental inequities in terms of different types of urban greenery in Hartford, Connecticut. Urban For. Urban Green. **18**, 163–172 (2016). https://doi.org/10.1016/j.ufug.2016.06.002
23. Kang, Y., Zhang, F., Gao, S., Peng, W., Ratti, C.: Human settlement value assessment from a place perspective: considering human dynamics and perceptions in house price modeling. Cities **118**, 103333 (2021). https://doi.org/10.1016/j.cities.2021.103333

Other Application

A Comparative Study of Question Answering over Knowledge Bases

Khiem Vinh Tran[1,2], Hao Phu Phan[4], Khang Nguyen Duc Quach[3],
Ngan Luu-Thuy Nguyen[1,2], Jun Jo[3], and Thanh Tam Nguyen[3(✉)]

[1] University of Information Technology, Ho Chi Minh City, Vietnam
{khiemtv,ngannlt}@uit.edu.vn
[2] Vietnam National University, Ho Chi Minh City, Vietnam
[3] Griffith University, Gold Coast, Australia
j.jo@griffith.edu.au, thanhtamlhp@gmail.com
[4] HUTECH University, Ho Chi Minh City, Vietnam

Abstract. Question answering over knowledge bases (KBQA) has become a popular approach to help users extract information from knowledge bases. Although several systems exist, choosing one suitable for a particular application scenario is difficult. In this article, we provide a comparative study of six representative KBQA systems on eight benchmark datasets. In that, we study various question types, properties, languages, and domains to provide insights on where existing systems struggle. On top of that, we propose an advanced mapping algorithm to aid existing models in achieving superior results. Moreover, we also develop a multilingual corpus COVID-KGQA, which encourages COVID-19 research and multilingualism for the diversity of future AI. Finally, we discuss the key findings and their implications as well as performance guidelines and some future improvements. Our source code is available at https://github.com/tamlhp/kbqa.

Keywords: Question answering · Knowledge base · Query processing

1 Introduction

Question Answering (QA) is a long-standing discipline within the field of natural language processing (NLP), which is concerned with providing answers to questions posed in natural language on data sources, and also draws on techniques from linguistics, database processing, and information retrieval [18]. One important type of data sources is knowledge bases, also known as knowledge graphs, which have been automatically constructed from web data and have become a key asset for search engines and many applications [17]. Finding answers for a question in a KB, on the other hand, is not always straightforward. The user needs to have a thorough understanding of the KB as well as a structured query language in order to express their queries in a structured manner that can be utilized to locate matches in the KB.

© The Author(s), under exclusive license to Springer Nature Switzerland AG 2022
W. Chen et al. (Eds.): ADMA 2022, LNAI 13725, pp. 259–274, 2022.
https://doi.org/10.1007/978-3-031-22064-7_20

In order to address this issue, a significant number of QA systems that allow users to express their information requirements using natural language have been created. Additionally, factoid question answering has two main approaches, information retrieval (IR) based QA and knowledge-based QA. In the first approach, many research have been established in recent years with the advancement of machine reading comprehension (MRC) task. The second approach is question answering over knowledge base (KBQA) with precision of question answering over knowledge graph (KG). Many studies have been conducted with KG such as [8] and recently research related to KG [4,6,9,12].

However, understanding the performance implications of these techniques is a challenging problem. While each of them has distinct performance characteristics, their performance evaluations often lack diversity in domains, query languages, natural languages and datasets. First, many developed datasets contain questions expressed in plain language but answers are specific to a KB format such as DBPedia. Additionally, the number of questions in each of the existing benchmarks is substantially different, rendering it difficult for users to select a practical guideline. Second, there is a lack diversity in terms of query languages. Another issue is the application domain. Most KBQA systems assume the generalization over different domains, but only a few domain-specific datasets are evaluated [2].

More precisely, the salient contributions of our benchmark are highlighted as follows:

- **Reproducible benchmarking:** We present the first large-scale replicable benchmarking framework for comparing KBQA techniques. Additionally, it can be applied to novel methods proposed in future studies. Our findings are reliable and reproducible, and the source code is publicly available at https://github.com/tamlhp/kbqa.
- **Surprising findings:** While some of our experimental findings confirm the state-of-the-art, we demonstrate the surprising superiority of methods using both SPARQL and SQL query languages. Additionally, we discovered that there are differences in their effectiveness when compared together.
- **New KGQA dataset:** We present COVID-KGQA, a new knowledge graph question answering benchmarking corpus that includes the most questions about COVID-19. COVID-KGQA includes more than 1000 questions with different types.
- **Performance guideline.** We provide valuable performance guidelines to academics working on KBQA in terms of benchmarks and QA systems selection. Alternatively, we suggest some future attempts to improve the performance.

The remainder is organized as follows. We discuss in Sect. 2 the problem setting, our benchmarking procedure, and the most representative approaches in KBQA. Section 3 introduces the setup used for our benchmark, including the analysis of KBs, datasets, metrics, and evaluation procedures. Section 4 reports the experimental results. A discussion of practical guidelines and conclusion are provided in Sect. 5.

2 Methodology

2.1 Problem Setting

Question Answering over Knowledge Bases (KBQA) is the term used for the task of retrieving the answer from executing query matching with natural language questions over a knowledge base as follow. Formally, we can define the task of KBQA as follow. Let KB be a knowledge base, Q is a question, q is the matching query and A is an answer extracted by matching query q executed from given question Q over KB. According to [1], the set of all possible answers can be in the form as follows. The first is the union of the power set P of entities E and literals L in KB: $P(E \cup L)$. The second is the number of outcomes set for all potential functions of aggregation $F : P(E \cup L) \mapsto \mathbb{R}$. The third is a Boolean set of answers for yes/no questions. Figure 1 illustrates the general process of question answering over knowledge bases.

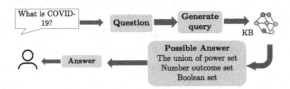

Fig. 1. Question answering over knowledge bases.

2.2 KBQA Approaches

In recent years, the number of off-the-shelf methods for KBQA in a variety of applications has increased. It is fascinating to compare and evaluate them so that users can make informed choices. KBQA can be divided into four primary strategies: (i) *embedding-based* - converts a query into a logic form, which is then executed against KBs to discover the appropriate responses; (ii) *subgraph matching* - constructs the query subgraph using a semantic tree; (iii) *template-based* - converts user utterances into structured questions through the use of semantic parsing; and (iv) *context-based*. This will be followed by a discussion of the concept underlying these approaches and some representative systems.

Embedding Methods. Embedding techniques take advantages of semantic and syntactic information included in a question as well as a database schema to produce a SQL logic form (parsing tree).

TREQS. The underlying concept of Translate-Edit Model for Question-to-SQL (TREQS) [16] is to convert healthcare-related questions posed by physicians into database queries, which are then used to obtain the response from patient medical records. Given that questions may be linked to a single table or to many tables, and that keywords in the questions may not be correct owing to the fact that the questions are written in healthcare language, using the language generation method, this model is able to address the problems of general-purpose applications. Using a language generation model, this translate-edit model produces a query draft, which is subsequently edited in accordance with the table schema.

TREQS++. TREQS++ or TREQS+Recover is the additional method with the added step as described below. *Recover Condition Values with Table Content:* This step is necessary because the use of the translate-edit model may not provide complete confidence that all of the queries will be executed since the condition values may not be precise in certain cases. With the goal of retrieving the precise condition values from the projected ones, this model employs a condition value recovery method in order to address this issue.

Subgraph Matching Methods. Subgraph matching constructs the query subgraph using a semantic tree, while others deviate significantly from this approach by building the subgraph from the entity. A natural language phrase may have many interpretations, each of which corresponds to a different set of semantic elements in the knowledge graph. After the semantic tree is located, the semantic relationships must be extracted from it before the semantic query graph is constructed (Fig. 2).

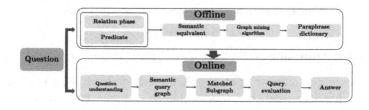

Fig. 2. Overview of gAnswer model

gAnswer. gAnswer [5] is the most advanced subgraph matching method to convert natural language questions into query graphs that include semantic information. This model responds to natural language queries using a graph data-driven, offline and online solution. Using a graph mining technique, the semantic equivalence of relation terms and predicates is determined during the online phase. The discovered semantic equivalence is then included into a vocabulary of paraphrased phrases. The online part includes the question comprehension and assessment stages. During the question comprehension phase, a semantic query graph is constructed to capture the user's purpose by extracting semantic relations from the dependency tree of the natural language question using the previously constructed paraphrase dictionary. Then, a subgraph of the knowledge graph is chosen that fits the semantic query graph through subgraph isomorphism. The final result is determined by the chosen subgraph during the query assessment phase.

Template-Based Methods. The usage of templates is critical in question answering (QA) over knowledge graphs (KGs), where user utterances are converted into structured questions via the use of semantic parsing . Using templates has the advantage of being traceable, and this may be used to create explanations for the users, so that they can understand why they get certain responses (Fig. 3).

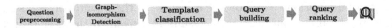

Fig. 3. Overview of TeBaQA model

TeBaQA. TeBaQA [15] is the most advanced template-based technique. First, all questions undergo (1) *Preprocessing* to eliminate semantically unnecessary terms and provide a meaningful collection of n-grams. By evaluating the underlying graph structure for graph isomorphisms, (2) *the Graph-Isomorphism Detection and Template Classification* phase trains a classifier based on a natural language question and a SPARQL query using the training sets. The key assumption is that structurally comparable SPARQL searches correspond to questions with similar syntax. A query is categorised into a sorted list of SPARQL templates at runtime. During (3) *Information Extraction*, TeBaQA collects all relevant information from the question, such as entities, relations, and classes, and identifies the response type according on a set of KG-independent indexes. The retrieved information is entered into the top templates, the SPARQL query type is selected, and query modifiers are applied during the (4) *Query Building* step. The conducted SPARQL queries are compared to the predicted response type. The following (5) *Ranking* is determined by a mix of all facts, the natural language inquiry, and the returning responses.

Context-Based Methods. These methods attempt to comprehend questions from various perspectives, such as question analysis, classification of questions, answer path, context of answers, and type of answers.

QAsparql. QAsparql [7] is a model using five steps as listed below, to translate questions to SPARQL queries. Figure 4 shows an overview of QAsparql model.

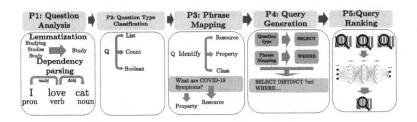

Fig. 4. Overview of QAsparql model

At each stage, each individual software component separately solves a related job. First, the question analysis component processes the incoming question based only on syntactic characteristics. The question's phrases are then mapped to relevant resources and attributes in the underlying RDF knowledge network once the question's type has been determined. On the basis of the mapped resources and attributes, many SPARQL queries are built. A ranking model based on Tree-structured Long Short-Term Memory (Tree-LSTM) is used to order potential questions based on the closeness of their syntactic and semantic structure to the input query. Finally, results are given to the user by running the produced query against the knowledge network underpinning the system.

QAsparql* (Our proposal). While carrying out a resource and property mapping task (Phase 3 as described in Sect. 2.2) utilizing EARL [3], we find that there is a significant difference between the two methods, as described in more detail below. In empirical evaluation, we have discovered that new algorithm produces superior results to previous one.

These methods are utilised to determine the number of hops and connections for each candidate node. This information is then sent to a classifier, which ranks and scores the features. After the creation of three distinct types of objects (Entity, Golden item, and the URI of each item), the next step is to convert them to a list. Consequently, we now have three separate listings. The List of Entities, Golden Item, and URI are examples of lists. The primary purpose of this method is to identify which entity URI corresponds to the golden item URI. The new algorithm performs the same function as the old algorithm, but with a different approach. During the first phase, it determines which items in the list correspond to the golden item and moves those that do not into a separate list known as the "not found list." It will then conduct a second search for entities, this time in surface form, to ensure that no entity is overlooked. After that, it will delete any non-included items and return a list of the items.

2.3 Summary

In summary, we use six representative systems - TREQS, TREQS++, gAnswer, TeBaQA, QAsparql and QAsparql* across domains, natural languages, and query languages. Table 1 compares the key characteristics of each technique studied in this benchmark.

Table 1. Characteristics comparison between KBQA techniques.

Technique	Features	Query language	Paradigm	Number of steps	Domain	Natural language
TREQS	Deep learning	SQL, SPARQL	Embedding	5	Single	Single
TREQS++	Deep learning	SQL, SPARQL	Embedding	6	Single	Single
gAnswer	Hand-crafted	SPARQL	Subgraph matching	4	Multiple	Multiple
TeBaQA	Hand-crafted	SPARQL	Template-based	5	Multiple	Multiple
QAsparql	Hybrid	SPARQL	Context-based	5	Multiple	Multiple
QAsparql*	Hybrid	SPARQL	Context-based	5	Multiple	Multiple

3 Experimental Setup

Datasets. We construct a sizable collection of benchmarking datasets.

Multi-Domain Datasets. Multi-domain attempts to improve performance by distributing it over several domains, and has been successfully used in a variety of areas. *Generic* are data gathered from a wide range of sources, including mathematics, physics, computer science, and a variety of other fields. LC-QUAD [13] is Large-Scale Complex Question Answering Dataset. It consists of 5000 questions and answers pair with the intended SPARQL queries over knowledge base in DBPedia. *Biomedical* is data about human health. A large-scale healthcare Question-to-SQL dataset, MIMICSQL [16], was created by utilizing the publicly available real world Medical Information Mart for Intensive Care III (MIMIC III) dataset to generate 10,000 Question-to-SQL pairs. MIMICSQL* [10] is the modified version of MIMICSQL which improves on the disadvantages of MIMIC-SQL as their tables are unnormalized and simpler than the tables used in actual hospitals. MIMICSPARQL is an SPARQL-based version of MIMICSQL*.

Multilingual Datasets. Multilingual datasets are those that support multiple languages. With multilingual datasets, the performance of a model can be evaluated across a variety of languages, and the differences between each language can be determined. Beginning in 2011 and lasting until 2020, Question Answering over Linked Data (QALD) is a series of evaluation campaigns. The most recent version is QALD-9 [14], which contains 408 questions compiled and selected from previous tests and is available in eleven languages.

Our New Benchmarking data (COVID-KGQA). In addition to the provided datasets, we create a bilingual dataset about COVID-19 to increase the diversity of domains and languages. Using the most recent version of DBPedia as a starting point, we compiled a corpus of more than one thousand question-answer pairs in two languages.

Measurements. We use the following evaluation metrics.

F1-Score. Let Q denotes the number of questions in benchmark , G is the number of answers processed by the system for given question Q, and A is the number of corrected answers extracted by executing query q from question Q over knowledge graph KG. F1 is defined as $F1 = \frac{1+\beta^2}{\beta^2*(1/P+1/R)}$ where $P = \frac{|A|}{|G|}$ and $R = \frac{|A|}{|Q|}$. β weighs whether precision or recall is more important, and equal emphasis is given to both when $\beta = 1$.

Execution Accuracy. Execution accuracy (Acc_{EX}) is computed as the accuracy of the response retrieved through SQL/SPARQL [16], $Acc_{EX} = N_{EX}/N$, where N represents the total number of Question-SQL pairings in the MIMIC-SQL database and N_{EX} represents the number of created SQL queries with the potential to provide accurate responses. Execution accuracy may also take into account questions formulated with invalid SQL queries but yielding valid query results. Therefore, we employ a different metric, Logical form accuracy, which eliminates the disadvantage of execution accuracy in the execution.

Logical Form Accuracy. The Logical form accuracy (Acc_{LF}) is used for if a string match exists between a produced SQL query and a ground truth query [10]. In order to compute Acc_{LF}, we must compare the produced SQL/SPARQL with the true SQL/SPARQL, token by token. That is, $Acc_{LF} = N_{LF}/N$ where N_{LF} counts how many requests are completely matched to the ground truth query.

Computation Time. Another measure for KBQA methods is the amount of time required to complete a task.

4   Results

4.1   End-to-end Comparison

We provide empirical finding and examined the overall performance of knowledge bases in terms of answering questions on three biomedical datasets, four commonly used datasets, and one new dataset as part of this investigation. TREQS and TREQS++ are the most accurate among the top performers shown in Table 2 on the first three biomedical datasets, with the highest accuracy on the fourth. TREQS++ achieves a better result on MIMICSQL, but its overall performance on the development and test datasets is inferior to that of TREQS on MIMICSQL*. Due to the design of TREQS++, which is to Recover Condition Values with Table Content, it is currently unable to perform with a knowledge graph and therefore cannot operate with MIMICSPARQL. The figure Fig. 5a depicts the visualisation of our results on the first three biomedical datasets, while the figure Fig. 6 illustrates the average time required to answer 100 random questions. With the TREQS technique, Acc_{EX} and Acc_{LF} perform better on the development set than on the test set in the majority of instances. However, with TREQS, the outcomes are different; Acc_{LF} performs better on the test set.

Table 2. End-to-end comparison on biomedical datasets

Dataset	Method	Acc_{EX}	Acc_{LF}	Time (s)
MIMICSQL	TREQS (Dev)	0.543	0.345	0.161
	TREQS++ (Dev)	**0.626**	**0.43**	
	TREQS (Test)	0.469	0.354	
	TREQS++ (Test)	**0.533**	**0.404**	
MIMICSQL*	TREQS (Dev)	**0.636**	**0.506**	0.311
	TREQS++ (Dev)	0.626	0.43	
	TREQS (Test)	**0.563**	**0.524**	
	TREQS++ (Test)	0.533	0.404	
MIMICSPARQL	TREQS (Dev)	**0.822**	0.580	0.25
	TREQS (Test)	0.698	**0.641**	

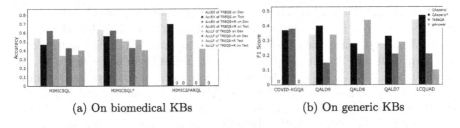

Fig. 5. Accuracy comparison

Table 3 compares the four systems and their scores for the four metrics - Precision, Recall, F1 and Time used in our paper with five datasets (four generic dataset and one additional biomedical dataset) and Fig. 5b shows visualisation of them in terms of F1 . Our system has been chosen alongside two new systems that are up to date and scheduled to be launched in 2021. These systems performed outstandingly in the previous comparison [7]. The unique aspect of this discovery is that altering the entity and resource algorithm, as stated in Sect. 2.2, improves performance over and beyond the default method used before. This is referred to as QAsparql* in order to differentiate it from the default. gAnswer is our next chosen system, which is state of the art in the QALD-9 challenge and is performed well on QALD-7, QALD-8, and LC-QUAD.

In terms of datasets, we select four existing datasets in addition to a large number of additional datasets that are split into the LC-QUAD and QALD series, which are the most frequently used datasets in the DBPedia. We choose QALD versions 7 to 9 from the QALD series since the target KG of these datasets are from version 2016, which is suited with a system that is nearly as current as the QALD series, and this will help these systems perform better. Our COVID-KGQA uses the latest version of DBPedia.

Table 3. Performance comparison on multilingual datasets (time in s)

Dataset	QAsparql				QAsparql*				TeBaQA				gAnswer			
	P	R	F1	Time	P	R	F1	Time	P	R	F1	Time	P	R	F1	Time
COVID-KGQA (Our)	0.00	0.00	0.00	**0.26**	0.29	0.52	0.37	**2.69**	**0.38**	**0.38**	**0.38**	13.2	0.00	0.00	0.00	40.99
QALD-9 (Train)	0.4	0.45	0.42	5.89	0.33	0.71	0.45	32.37	0.17	0.18	0.16	8.5	0.39	0.43	0.39	13.03
QALD-9 (Test)	0.32	0.36	0.34	9.56	0.3	0.6	0.4	27.45	0.16	0.15	0.15	8.15	0.33	0.37	0.34	13.30
QALD-8 (Train)	0.35	0.49	0.41	5.16	0.31	0.67	0.42	7.9	0.21	0.22	0.21	7.52	**0.45**	**0.51**	**0.46**	12.93
QALD-8 (Test)	**0.57**	0.44	0.5	3.56	0.2	0.44	0.28	7.35	0.21	0.22	0.21	8	0.42	**0.54**	0.44	**12.07**
QALD-7 (Train)	0.33	0.55	0.42	6.42	**0.38**	0.77	**0.51**	5.69	0.2	0.21	0.2	7.92	0.43	0.5	0.44	13.09
QALD-7 (Test)	0.21	0.44	0.28	1.74	0.24	0.5	0.33	7.3	0.22	0.23	0.21	**7.72**	**0.29**	0.35	0.29	13.04
LC-QUAD (Test)	0.34	**0.62**	**0.44**	5.41	0.31	**0.98**	0.47	13.75	0.21	0.22	0.21	16.45	0.09	0.11	0.1	60.45

For the COVID-KGQA dataset, we evaluated the effectiveness of the existing system with our new benchmark corpus in COVID-19 using four distinct systems. gAnswer is deployed via gStore with version 2016–10 of DBPedia as the guidance for gAnswer systems, whereas we conduct experiments with the most recent version of DBPedia for other systems. In this version, the knowledge graph contains no COVID-19-related information. Consequently, gAnswer achieves the worst performance in our corpus for both F1 and Time. On the other hand, TeBaQA and QAsparql have the best performance; while TeBaQA outperforms QAsparql 0.01 in F1, QAsparql outperforms TeBaQA in some fields, as discussed in Sect. 4.3. TeBaQA has the best performance on our new benchmarking dataset, whereas this system performs poorly on other datasets with a performance range of only 0.15 to 0.21. Possible causes include the fact that QALD is a more difficult benchmark with a large number of questions requiring complex queries with more than two triples to answer. A deeper examination revealed that QALD-9 questions frequently required sophisticated templates that were either absent from the training questions or provided very limited assistance. The template-based approach in TeBaQA is inferior to the Context-based and Subgraph matching techniques, with the exception of LC-QUAD. QAsparql*, our new finding approach, achieves the best performance in almost all corpora, with the exception of COVID-KGQA (with a length of less than approximately 0.01) and QALD-8. This demonstrates that enhancing the algorithm for entity matching can improve the system's performance.

4.2   Running Time

Figure 6 displays the experimental results. In the first three datasets, 100 questions are selected at random and their time performance is compared. In comparison to two other corpora, MIMICSPARQL with a knowledge graph approach achieves the best results overall. Compared to MIMICSPARQL, MIMICSQL* queries are nearly twice as long. A single JOIN in SQL requires 11 tokens (including the operators "=" and "."), whereas a single hop in SPARQL requires only three tokens (i.e., subject, predicate, and object). An additional distinction between SQL and SPARQL is that the model must comprehend the hierarchy between a table and its columns and the relationships between tables. SQL and SPARQL differ syntactically due to this inherent distinction between relational tables and a knowledge graph in terms of how information is linked (joining multiple tables versus hopping across triples). Consequently, the graph-based method performs significantly better than the table-based method. [10]. We compare the execution times of four systems across five additional datasets. We evaluate every question in the test set of five corpora. gAnswer reaches its highest point in COVID-KGQA and becomes the fastest time overall. The explanation is summarised as follows: gStore is used to install gAnswer with the default knowledge graph version set to 2016–10. However, our dataset is about COVID-19, which was introduced in 2019, so gAnswer takes longer to query but returns no results.

4.3   Influence of Question Taxonomy

Over the course of this investigation, the impacts of question type on question answering are investigated using knowledge graph methods. From these experiments, we conduct our studies on test set of LCQUAD and COVID-KGQA

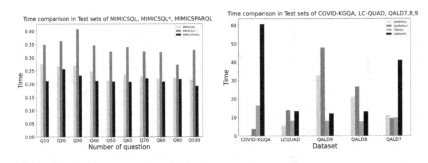

Fig. 6. End to end time (s) comparison

because the quantity of question of them is the highest and almost the same (1000 questions). Figure 7 depicts the results of MIMIC datasets, and Fig. 8, Fig. 9, which are conducted with four systems in COVID-KGQA and LCQUAD, as stated in Sect. 4.3. According to [11], the Wh-question taxonomy comprises six types: 'Who', 'What', 'When', 'Where', 'Why', and 'How'. Additionally, throughout the data analysis process, we discover that the dataset contains other types such as 'Whose' and 'Whom'. On the other hand, certain inquiries are not Wh-questions, such as 'Yes/No' questions, which contain phrases starting with the words, i.e., 'Can', 'May', 'Am', 'Is', 'Are', 'Was', 'Were', 'Will', 'Do', and some request questions, such as 'Count', 'Provide', 'Give', 'Tell', 'Specify'. As a result, we undertake this experiment using eleven distinct sorts of questions. For the MIMICSQL dataset, Fig. 7a depicts the Acc_{EX}, Acc_{LF}, and Time of TREQS and TREQS++ for eleven question categories, with regard to each sort of query. The plot depicts the Acc_{EX}, Acc_{EX} with TREQS++, and Time for each question type across the whole test set in this dataset. It is apparent that TREQS++ outperformed TREQS for all question categories on this dataset in Acc_{EX} when using the Question-to-SQL techniques TREQS and TREQS++. Acc_{LF} outperforms all other variables except What, How, and Request. Similarly, MIMIC-SQL* and MIMICSPARQL results are depicted in Fig. 7b and c. Regarding the F1 in COVID-KGQA and LCQUAD results, these are illustrated in Fig. 8 and Fig. 9. With respect to specific question categories, TeBaQA surpassed both of them. For COVID-KGQA, QAsparql* outperforms in the 4 categories of questions, including 'What', 'When', 'Where', and 'Which', whereas TeBaQA hits a high in the category of 'How' (Figs. 10 and 11).

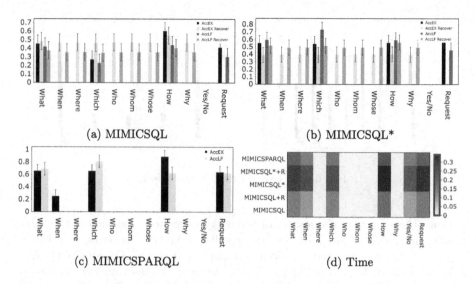

(a) MIMICSQL

(b) MIMICSQL*

(c) MIMICSPARQL

(d) Time

Fig. 7. Comparison on MIMICSQL, MIMICSQL*, MIMICSPARQL with the influence of question type, the underscores represent zeroes

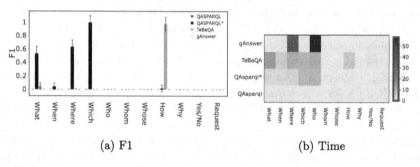

(a) F1

(b) Time

Fig. 8. Comparison on COVID-KGQA with the influence of question type, the underscores represent zeroes

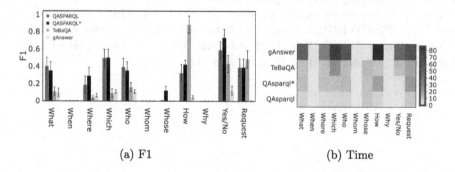

(a) F1

(b) Time

Fig. 9. Comparison on LCQUAD with the influence of question type, the underscores represent zeroes

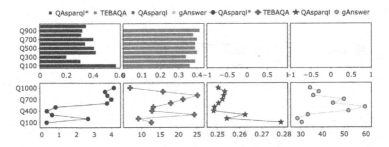

Fig. 10. Comparison on COVID-KGQA with the influence of question type from test set

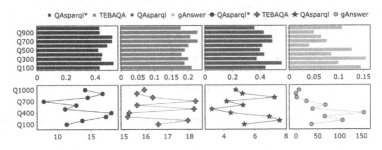

Fig. 11. F1 and Time comparison on LCQUAD with the influence of question quantity from test set

4.4 Effects of Quantity of Questions

The effects of next characteristic, the number of questions, are then examined. Figure 12 depicts the results of an experiment in which the quantity factor of the question is changed from each 100 question segments to 1000 questions of each dataset. The accuracy of the test set improves significantly as the number of questions increases from 200 to 700, decreases from 800 to 900, and in the majority of instances Acc_{EX} outperforms Acc_{LF}. The performance of both metrics typically peaks between the 500th and 700th question. Regarding Time in MIMICSQL, MIMICSQL*, and MIMICSPARQL, as depicted in Fig. 12, MIMICSQL and MIMICSQL* are quite comparable in terms of overall performance on the Test sets. The primary difference is that in MIMICSPARQL, the time for both questions is initially extremely short, but steadily increases until the 1000th question, where it reaches a maximum.

As shown in Figs. 8 and 9, F1 in COVID-KGQA and LCQUAD. Regarding context-based techniques, it is evident that QAsparql* outperformed all other methods, with all questions scoring over 0.4 and 300 questions scoring over 0.5, and that it was surpassed by QAsparql* in terms of overall performance. In contrast, the gAnswer technique, which employs Subgraph Matching, had the lowest F1 scores of any technique. QAsparql* was superior to all other techniques and procedures. In terms of time, QAsparql achieved the highest performance, followed by QAsparql* and TeBaQA; however, gAnswer once again achieved the lowest performance in addition to F1 score.

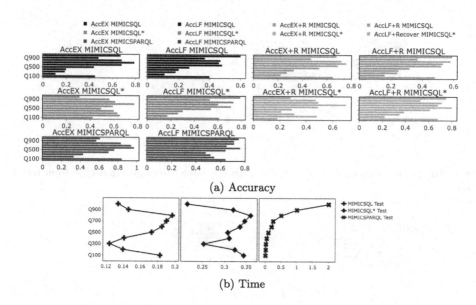

(a) Accuracy

(b) Time

Fig. 12. Effects of #questions to Accuracy and Time on domain-specific datasets

5   Conclusion

Performance Guidelines. To assist end users in selecting a suitable solution for a certain application need, we offer the following set of recommendations, which are based on our experimental findings.

- In general, context-based methods, such as QAsparql and QAsparql*, are the most effective overall for both F1 and Time. The effectiveness of QAsparql decreased with question type. Despite its low computation time, TeBaQA struggles with question types and performs best with multiple criteria.
- In terms of Acc_{EX} and Acc_{LF}, TREQS++ outperforms MIMICSQL in nearly all question types. TREQS++ is ineffective in MIMICSPARQL, whereas TREQS is effective at Acc_{EX} in How and Request questions. Regarding Time, QAsparql is the first runner up in nearly all kinds of questions, while QAsparql* is the second runner up in a negligible number of questions.
- Regarding the effect of the number of questions on the performance of COVID-KGQA, we recommend avoiding the use of gAnswer and QAsparql, as these algorithms are inadequate in this regard. In terms of diversity, QAsparql* and TeBaQA are better.
- Regarding multi-domain and multi-language, we advise avoiding the use of gAnswer and QAsparql, as these algorithms are ineffective in this regard. In terms of variety, QAsparql* and TeBaQA are both outstanding alternatives.

The findings are also summarized in Table 4, where the best, the second-best and the worst techniques are shown for each percategory.

Table 4. Performance guideline

Category	Winner	1st Runner-up	Worst
Overall	QAsparql*	QAsparql*	gAnswer
Question type	QAsparql*	QAsparql	gAnswer
Question quantity	QAsparql*	QAsparql	gAnswer
Acc_{EX}	TREQS++	TREQS	TREQS
Acc_{LF}	TREQS++	TREQS	TREQS
Multi-domain	QAsparql*	QAsparql*	gAnswer
Multi-language	QAsparql*	QAsparql*	gAnswer
Cross-Query Language	TREQS	TREQS	TREQS++

Summary. We compared KBQA systems from a variety of perspectives and discussed the ramifications of our findings. Experiments were conducted on a wide range of topics, including simple subgraph matching, context-based, deep learning models, and others. The results ultimately demonstrated the superiority of deep learning models when combined with a context-based approach, which produced the best results. In addition, as a result of this work, we create the COVID-19 large-scale KGQA dataset, which consisted of over a thousand questions.

Acknowledgement. This research is funded by University of Information Technology, Vietnam National University HoChiMinh City under grant number D1-2022-25.

References

1. Chakraborty, N., Lukovnikov, D., Maheshwari, G., Trivedi, P., et al.: Introduction to neural network-based question answering over knowledge graphs. WIREs DMKD **11**(3), e1389 (2021)
2. Costa, J.O., Kulkarni, A.: Leveraging knowledge graph for open-domain question answering. In: WI, pp. 389–394 (2018)
3. Dubey, M., Banerjee, D., Chaudhuri, D., Lehmann, J.: EARL: joint entity and relation linking for question answering over knowledge graphs. In: ISWC, pp. 108–126 (2018)
4. Gao, Y., Tian, X., Zhou, J., Zheng, B., Li, H., Zhu, Z.: Knowledge graph embedding based on quaternion transformation and convolutional neural network. In: ADMA, pp. 128–136 (2022)
5. Hu, S., Zou, L., Yu, J.X., Wang, H., Zhao, D.: Answering natural language questions by subgraph matching over knowledge graphs. TKDE **30**(5), 824–837 (2018)
6. Hung, N.Q.V., Tam, N.T., Tran, L.N., Aberer, K.: An evaluation of aggregation techniques in crowdsourcing. In: WISE, pp. 1–15 (2013)
7. Liang, S., Stockinger, K., de Farias, T.M., Anisimova, M., Gil, M.: Querying knowledge graphs in natural language. J. Big Data **8**(1), 1–23 (2021). https://doi.org/10.1186/s40537-020-00383-w

8. Ma, J., Zhong, M., Wen, J., Chen, W., Zhou, X., Li, X.: RecKGC: integrating recommendation with knowledge graph completion. In: ADMA, pp. 250–265 (2019)

9. Nguyen, T.T., et al.: Monitoring agriculture areas with satellite images and deep learning. Appl. Soft Comput. **95**, 106565 (2020)

10. Park, J., Cho, Y., Lee, H., Choo, J., Choi, E.: A knowledge graph-based question answering with electronic health records. In: MLHC, vol. 149, pp. 1–17 (2021)

11. Pomerantz, J.: A linguistic analysis of question taxonomies: research articles. J. Assoc. Inf. Sci. Technol. **56**(7), 715–728 (2005)

12. Toan, N.T., Cong, P.T., Hung, N.Q.V., Jo, J.: A deep learning approach for early wildfire detection from hyperspectral satellite images. In: RiTA, pp. 38–45 (2019)

13. Trivedi, P., Maheshwari, G., Dubey, M., Lehmann, J.: LC-QuAD: a corpus for complex question answering over knowledge graphs. In: ISWC, pp. 210–218 (2017)

14. Usbeck, R., Gusmita, R.H., Ngomo, A-C.N., Saleem, M.: 9th challenge on question answering over linked data (QALD-9). In: ISWC, pp. 58–64 (2018)

15. Vollmers, D., Jalota, R., Moussallem, D., Topiwala, H., Ngomo, A.C.N., Usbeck, R.: Knowledge graph question answering using graph-pattern isomorphism. arXiv preprint arXiv:2103.06752 (2021)

16. Wang, P., et al.: Text-to-SQL generation for question answering on electronic medical records. In: WWW, pp. 350–361 (2020)

17. Weikum, G.: Knowledge graphs 2021: a data odyssey. PVLDB **14**(12), 3233–3238 (2021)

18. Zheng, Y., et al.: Quality prediction of newly proposed questions in CQA by leveraging weakly supervised learning. In: ADMA, pp. 655–667 (2017)

A Deep Learning Framework for Removing Bias from Single-Photon Emission Computerized Tomography

Josh Jia-Ching Ying[1](✉), Wan-Ju Yang[1], Ji Zhang[2], Yu-Ching Ni[3,4], Chia-Yu Lin[3], Fan-Pin Tseng[3], and Xiaohui Tao[2]

[1] Department of Management Information Systems, National Chung Hsing University, Taichung 402, Taiwan
jashying@gmail.com
[2] School of Mathematics, Physics and Computing, The University of Southern Queensland, Toowoomba, Australia
[3] Health Physics Division, Institute of Nuclear Energy Research, Atomic Energy Council, Taoyuan 325, Taiwan
[4] Department of Biomedical Engineering and Environmental Sciences, National Tsing-Hua University, Hsinchu 300, Taiwan

Abstract. After being photographed by medical equipment, noise in the unprocessed medical image is removed through manual processing and correction to create a proper medical image. However, manually processing medical images takes a long time. Suppose the current medical images are used with artificial intelligence to predict the type and severity of the disease. In that case, patients can be prioritized based on the predicted results, reducing the probability of patients most in need of care not getting timely treatment and increasing the efficiency of visits. Most experts use deep learning image feature segmentation to learn all the features in the image. However, some features in the image are not needed. These unwanted image features will affect subsequent training, which we call "biased information." In the process of training image features through artificial intelligence, biased information may overpower the more important image features in the target learning task, resulting in poor training results. Therefore, instead of learning all the features in the image, we should only learn what we need. This paper uses the architecture of biomedical image segmentation convolutional neural network combined with principal component analysis to extract the main feature weights in the image data and determine whether the feature is something we want to learn. If not, the feature is deleted, which prevents it from affecting subsequent training. The feature vector we need is associated with the first principal component. After learning the results, we can verify its accuracy through the image classification model. It is found that after biased information is removed, the classification effect is reduced, and the accuracy of disease classification has increased significantly from less than 35% to more than 60%.

Keywords: Deep learning · Bias correction · Image segmentation

© The Author(s), under exclusive license to Springer Nature Switzerland AG 2022
W. Chen et al. (Eds.): ADMA 2022, LNAI 13725, pp. 275–289, 2022.
https://doi.org/10.1007/978-3-031-22064-7_21

1 Introduction

Medical imaging plays an important role in the diagnosis and treatment of diseases. The most common medical images in the medical field today are tomography (CT) and magnetic resonance (MRI) [1, 2], which are mainly used to monitor the course of diseases and help with medical diagnosis. MRI is a type of nuclear medical imaging. Nuclear medical imaging [3] is one of the few examination tools with high sensitivity and can reflect abnormalities in the brain at an early stage, which can assist in the diagnosis of dementia (Alzheimer's disease) and other brain diseases in the elderly. However, current domestic nuclear medical imaging mainly relies on doctors interpreting the image with the naked eye. If enhanced images can be provided to highlight areas in the brain that are functioning abnormally, doctors will be able to more objectively and efficiently interpret results and relieve the workload caused by the surge in the amount of medical data. In addition, to improve the efficiency of diagnosis, we can use artificial intelligence to predict the classification of brain diseases [4–6], which can not only prioritize patients with more severe diseases with earlier visits but also make the diagnosis more accurate by helping doctors with the interpretation. When generating MRI images, random noise is generated during operation, by the equipment, or by the environment, which in turn affects the accuracy of subsequent image analysis and prediction. Manjón and Coupe [7] believes that noise removal is one of the most fundamental steps in image preprocessing, combining the latest development of deep learning architecture with classic noise removal. Vaishali et al. [8–10] provide various noise removal techniques and believe that the technique used should depend on the type of image it is suitable for. Hong et al. [11] believes that Rician noise may be accidentally injected during the image acquisition process [12], interfering with diagnostic and treatment. If the squared Euclidean distance is used to train the neural network for noise reduction, some image information may be lost during training in the deep convolutional neural network. Using perceptual loss squared to train the Euclidean distance of the network [13] helps solve the aforementioned problems. This convolutional network surpasses the state-of-rician MRI noise reduction method, produces satisfactory image results, and enables image segmentation to have more refined features and obtain better quality noise-reduced brain MR images. Chauhan and Choi [14] provide several noise reduction filters based on fuzzy logic to remove noise from brain images.

Denoising the structure of 3D MRI is a key step in medical image analysis. Many characteristic methods and impressive algorithms have been proposed in the past. Ran, et al. [15] introduced a residual-based MRI denoising method with an encoder-decoder Wasserstein Generative Adversarial Network (RED-WGAN). The residual auto-encoder and the deconvolution operation are combined into the generative network; Liu et al. [16] proposed a new multisurface approximation-based FCM with the interval membership method for brain MRI with simultaneous bias correction and segmentation. Intensity unevenness is one of the common artifacts in image processing. Khosravanian et al. [17] proposed a new region-based level set method for segmenting images with uneven intensity and applied it to brain tumor segmentation in MRI scans. Bouhrara et al. [18] introduced a new high-performance nonlocal noise filter to reduce MR image sets composed of multispectral images. In addition, principal component analysis (PCA) can also be used to remove noise [19–21]. Combined with other related noise removal technology, PCA can reduce the dimensionality of the information in the image and extract the part of the whole image from the lower-dimensional image to remove noise. This improves the predictability of pixel intensities in the reconstructed image. It is a great challenge to correct the artifacts produced in brain magnetic resonance imaging, especially when the bias field and noise are important. It adversely affects the performance of the image processing algorithm because the artifacts make image segmentation more difficult. A new and more effective denoising model and bias correction formula have been developed based on regression analysis and the Monte Carlo simulation [22].

To summarize, we find that most of the current research and development goals are on how to remove the noise generated during the operation or how to improve the clarity of the image so that the doctor can more accurately determine the size of the brain lesions. A more accurate classification can be obtained with image segmentation, screening out the areas with important characteristics and training only on those areas. However, the use of the above experimental method model cannot improve the classification accuracy of our disease dataset. We presume that in addition to general noise in the image, some feature information may affect classification, which cannot be learned by manual labeling or deletion. In feature selection, if the deep learning model learns all the features in the image, some specific features cannot be learned, which makes it impossible to determine which features are biased information directly. Therefore, to analyze the bias for solving this problem with feature regions, in this paper, we mainly experiment using biomedical image segmentation models to find and remove biased information in the single-photon emission computerized tomography (SPECT) images with PCA before disease prediction and classification, and hope to improve the accuracy of brain disease classification.

The rest of this paper is organized as follows. We first briefly review the related work in Sect. 2. The details of our proposed framework is given in Sect. 3. We present the experimental evaluation results of our proposed approach in Sect. 4 and finally present our conclusions and future work in Sect. 5.

2   Related Works

With regards to our experimental process, we divided the relevant research needed to be implemented into the following four parts for discussion and conducted a number of studies and speculations:

2.1   Segmentation of Brain MRI Image

Image segmentation is one of the most important tasks in medical image analysis and is usually the most critical step in many clinical applications. It is usually used to measure and visualize the anatomical structure of the brain, analyze brain changes, display diseased areas, and can be applied to surgical planning. Akkus et al. and Despotović et al. [23, 24] aim to develop a brain MRI segmentation method based on deep learning and explores the most commonly used brain MRI segmentation methods. It discusses the various performances, speed, and characteristics of deep learning methods and compares their learning effects' differences, abilities, advantages, and disadvantages. In addition, Akkus et al. and Despotović et al. [23, 24] provide different brain MRI preprocessing steps, such as removing nonbrain tissues and image enhancement.

The precise segmentation of infant brain tissue is very challenging. As the image quality is relatively poor, the contrast between the brain's white matter and gray matter is less obvious because the volume is mapped to a relatively small brain image. If you zoom in, the image quality will lower, and the size of the brain in the image will be different from the real brain, affecting the operation and making it difficult to segment. Dolz et al. [25] uses a semidense fully convolutional neural network (CNN) to solve this problem and is the first 3D CNN integration for annotation in images. Dolz et al. [25] also studies the effect of image multimodal fusion on the performance of deep architecture. In the MICCAI iSEG-2017 Challenge Open Data Competition, Dolz et al. [25] got good results. Moeskops and Pluim [26] use a combination of an expanded three-plane CNN and a nonexpanded 3D CNN. The above methods are valuable references for the smaller image sizes in our dataset.

Glioma is one of the most common and aggressive primary malignant brain tumors. Accurate segmentation is very important because it helps monitor the pathological changes of glioma and regularly evaluate and observe the effect of treatment. Chen, L., Wu et al. [27] developed a 3D CNN to segment gliomas automatically. The main difficulty in implementing a segmentation model lies in gliomas' location, structure, and shape between different dimensional spaces. To accurately classify each voxel, multi-scale context information is captured by extracting features from the receptive field. Billot et al. [28] proposed a deep learning strategy that, unlike anything before, can perform semantic segmentation on the MRI brain scan without preprocessing, additional training, or fine-tuning for the new model. Moeskops [29] used adversarial training methods to improve CNN-based brain MRI segmentation. It also contains an additional loss function, which is used to stimulate the network to generate a segmentation method that is difficult to distinguish from manual segmentation. From all the above segmentation methods, we can see that there are many types of brain diseases with different image segmentation model architectures. Because brain structure is very complex and vital, everyone's goal is to segment the brain lesions as detailed as possible.

2.2 Brain Image Segmentation Based on U-Net Architecture

The most famous biomedical image segmentation method is U-Net [30]. It is currently one of the most commonly used methods in the field of biomedical image segmentation research, such as brain and liver image segmentations. The U-Net network is very simple. The first half is for feature extraction, and the second half is for upsampling. The main goal is to restore the feature map of the learning image to the original image. The infrastructure is a CNN. Siddique et al. [31] reviewed many developments and breakthroughs in the U-Net architecture, provided their own observations, and discussed recent trends. They also discussed the innovative ideas obtained in deep learning and how these tools promote U-Net.

Most of the current medical image segmentation architectures are based on the U-Net architecture. In the past few years, many different versions of U-Net have been derived or optimized from the original. Kolařík et al. [32] proposed a method for using modern deep learning: a fully automatic method for 3D segmentation of brain tissue in MRI scans using a densely connected layer of 3D Dense-U-Net neural network architecture. Compared with many previous methods, accurate segmentation can be performed without any preprocessing of the input image. It can also be easily applied to the U-Net network as a segmentation algorithm to enhance its results. Zhang et al. [33] showed that the U-Net network could be deeper, thereby improving the performance of segmentation tasks. Zhang et al. [33] integrated the Inception-Res module and the densely connected convolution module into the U-Net architecture.

Zhou et al. [34] proposed U-Net++, which gave U-Net new neural network semantics and instance segmentation architecture, effectively reducing the unknown network depth and the collection of U-Nets with different depths. Some shortcomings in U-Net are improved, and some advantages are improved, such as partial use of deep shared encoders and simultaneous learning and supervision, highly flexible feature fusion, and sped up inference. Rehman et al. [35] proposed a two-dimensional brain tumor segmentation image segmentation method called BU-Net. It uses remaining extended skip (RES) and wide context in conjunction with custom loss in the baseline to find more diverse features by increasing the effective receptive field. Markov Random Field (MRF) coding has a simple label distribution. Although it is not as flexible as U-Net, it is not prone to overfitting. Dinsdale et al. [36] Combining U-Net and MRF by calculating the product of the distribution achieved unprecedented success in semantic segmentation tasks. It can be shown on 3D neuroimaging data that this novel network improves the generalization ability of samples outside the distribution. We can see the diversified U-Net architecture models from above, which shows how important U-Net is in biomedical segmentation. Thus, it is worth researching which U-Net image segmentation model should be used for our proposal.

2.3 Disease Classification Prediction Based on CNN Architecture

The key to the final decision of neurologists and radiologists on the diagnosis of brain diseases lies in the evaluation of magnetic resonance imaging. The manual evaluation process is time-consuming and requires domain expertise to avoid human error. To overcome this problem, we use deep learning methods. CNN is one of the most classic network architectures in classifying diseases. It has the ability of automatic feature learning, which has become a great advantage for classifying schizophrenia because it can eliminate the subjective spatial features related to the selection. Al-Masni [43] tried to implement various CNN models in MRI or PET classification tasks to predict Alzheimer's disease. At the same time, they tried to adjust various parameters to summarize its characteristics and changed CNN, from low-level to high-level features, for extraction. By increasing the filter size and moving from a lower stride, different reception areas to a higher stage are represented.

Under the theoretical background of past research, the CNN architecture gradually developed from 2D to 3D. Khagi and Kwon [44] implemented various CNN models in nuclear medicine for functional image classification tasks for predicting different 3D convolutional neural network architectures for Alzheimer's disease. CNN plays an important role in fine-tuning and analyzing the brain. Hu et al. [45] also built a 3D CNN model using different architectures and compared their performance to classify patients with mental illness between healthy people. More studies have been applying deep learning algorithms to neuro-imaging data in recent years. Advances in deep learning for image classification have so far provided a powerful framework through automatic feature generation and direct analysis. Korolev et al. [4] uses this feature to implement a powerful framework of 3D CNN architecture to classify disease prediction.

Subsequent CNN also created many classic and well-known neural network architectures. Bhanumathi and Sangeetha [46] introduced different neural network architecture classification models based on CNN, such as Alex Net, Vgg Net, Google Net [47]. Other technologies are used in the process of pretraining and feature analysis of brain tumor images. Using the above-mentioned references, the main experimental method of this paper will use the image segmentation framework of the previous trend to extract the image features, use the PCA method to find the biased information, and view the before and after changes of the image after biased information is removed. To judge whether the removed feature information is biased information, we use the characteristics of the referenced CNN architecture to build our own model and apply it to our dataset to predict the classification of diseases and to verify the comparison of the results after the learning model experiment. As this stage is just to prove the previous hypothetical method, we can use the above related research papers as the reference basis for our model.

3   Our Proposed Method

We propose our new model for removing bias through the above references to verify that our hypothesis is correct. The following implementation process is divided into three major components: feature extraction, feature analysis, and image classification. There is an extended version of U-Net called 3D U-Net [48] regarding feature extraction, which is also a type of deep learning technology. The main difference lies in the evolution from the original U-Net's 2D to 3D images. In this paper, for medical image bias correction, we use 3D U-Net. Although the nuclear medicine brain function image data is mainly 2D images, it is constructed as a three-dimensional 3D image concept. If 2D images are used for learning, some characteristic information across different images cannot be trained, so it is necessary to use the 3D U-Net model framework for model training by adding the feature information of the z-axis. 3D U-Net yields very good results in biomedical image segmentation. Compared with other deep neural networks, 3D U-Net uses fewer feature dimensions. It can be seen from the structure diagram that U-Net is divided into three parts: encoder, bottleneck, and decoder. The network model is built by the interaction between these three parts. Different from previous research that uses 3D U-Ne, in this paper, the difference is that we will change the bottleneck, extract it and use the PCA method to find the location of the biased information, and then make corrections.

First, the main purpose of the encoder is to perform feature extraction. The architecture will have two consecutive layers of convolution before reducing the dimensionality, which can make it easier for the neural network to capture accurate feature information. After the bottleneck is obtained through the previous encoder and the image feature map is obtained, we will first temporarily store this feature map in preparation for feature analysis. As for feature analysis, this paper first performs general statistical software analysis on the vectors in the temporarily stored bottleneck obtained earlier. As the information provided by the dataset only has two categories: hospital and disease, for the time being, only these two categories can be analyzed. After statistical analysis, we found that the imaging features of nuclear medicine brain function are relatively irrelevant to the disease, and most of the results indicate significant differences between hospitals.

For instance, the following is the nuclear medicine brain function imaging data of four hospitals collected by the Nuclear Energy Committee of the Taiwan Executive Yuan. The U-Net extracts the bottleneck. As the feature dimension in the image is very large, we used PCA analysis for dimensionality reduction and only took the top three principal component features for statistical analysis. Features 1–3 represent the three largest principal component features. To compare whether there is a correlation between the hospital and the disease for the features in the image, we use a multifactor analysis of variance. With the analysis, we find that in Table 1 below, the significance of each feature in the image of each different hospital is less than 0.01, indicating that there is a significant relationship with the image feature.

Table 1. Multivariate analysis of variance (MANOVA) result with hospital

Source	Dependent variable	Type III sum of squares	df	Mean square	F	Sig	Partial eta squared
Hospital	Feature 1	21223.957	3	7074.652	11.177	<.001	.080
	Feature 2	8517.580	3	2839.193	8.680	<.001	.063
	Feature 3	26951.327	3	8983.776	11.177	<.001	.080

Table 2. Multivariate analysis of variance (MANOVA) result with disease.

Source	Dependent variable	Type III sum of squares	df	Mean square	F	Sig	Partial eta squared
Disease	Feature 1	3096.784	3	1032.261	3.026	0.030	.023
	Feature 2	3262.230	3	1087.410	4.938	0.002	.037
	Feature 3	1340.060	3	446.687	0.653	0.582	.005

In Table 2, we clearly see that only the significance of the second feature is slightly related to the disease, while the other features have no significant relationship. Therefore, it can be concluded that the disease information does not have a significant relationship with the image features. From the above, we can find that the pictures of different diseases are, from the computer's point of view, not different at all. Still, the pictures of different hospitals have obvious differences. It can be seen that the difference in the characteristics of the images lies in the possibility of nuclear medicine brain function imaging between different hospitals. Because of manual processing, the image characteristics of the output image are different, which indirectly causes the natural disease characteristics in the original image to be hidden.

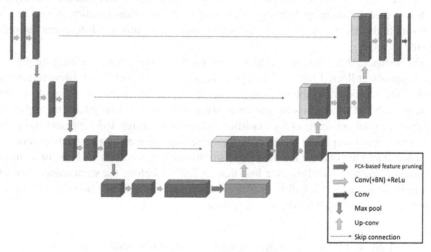

Fig. 1. 3D U-Net + PCA architecture.

Therefore, the PCA method can help study the distribution of image feature information and directly interspersed and used in the 3D U-Net architecture, as shown in Fig. 1. In the orange block, we process the feature map through PCA to analyze the important features of the image, and there is no need to take it out for analysis to facilitate subsequent image classification models for classification. According to the results of the statistical analysis of the image features obtained previously, because the hospital feature information is significant in the image, we first assume that the first principal component is not the disease feature but the biased information of the hospital feature. Therefore, we are going to remove this principal component feature. After deleting it, data is put back into the 3D U-Net architecture to determine whether the results have changed. In the process of deleting the biased information, to avoid directly deleting the principal component features and not being able to find the original feature parameter positions in the subsequent restoration of the nuclear medicine brain function image, which causes the problem of image variation, we set the value of the biased information to 0 as an alternative method of deletion.

In the PCA analysis process, an important part of this is that although we put the architecture of this stage in the 3D U-Net model, if the PCA process is also used for gradient training, it may cause some unnecessary features to be learned in the process of PCA feature analysis. This increases the complexity of the model, resulting in a poor learning effect in the process of restoring the image. Therefore, we adopt the pruning method. The purpose is to learn only the processed images and then quickly improve the training effect to restore the images, reduce the risk of overfitting, and thereby reduce the generalization error.

The image is converted into a vector, which is used for further classification. However, in image segmentation, we need to convert the feature map into a vector and reconstruct the image from this vector. This is a huge task because it is more difficult to convert a vector into an image than the other way around. The whole concept of U-Net originates from this issue. Going back to the U-Net architecture, we prepare to restore the bottleneck processed by PCA to the original medical image and use the decoder on the right half to generate an MRI image that retains the original features. Each layer is matched with a skip connection to capture the original image details and to improve the reduction of MRI images. The feature map at the end of the neural network is small so that you can see the details but cannot read the full map, and the feature map at the front end covers a large range, but the features are rough. The advantages of these two methods are that both are used for learning, so space consumed by the GPU can be reduced. As the two dimensions are different, they must be combined after deconvolution. Finally, the process of this model is completed, and the restored image is highly similar to the original MRI image.

As the main research purpose of this paper is to use the PCA method to remove the biased information in the image, the establishment of the classification model only needs to be able to verify that the theoretical direction of the implementation is correct. Thus, the image classification model structure used is relatively basic. We decided to adopt the basic CNN architecture as our classification model by referring to the above-mentioned related documents. The image classification also uses 3D images throughout. It can be seen from Fig. 1 that the three-layer convolution plus maxpooling is used, and the conversion of the intermediate dimension depends on Flatten. Finally, the model will return to the fully connected layer for classification. The convolutional layer is a multidimensional array. The input of the fully connected layer is usually one-dimensional. To avoid overfitting, we will add a dropout layer in specific places.

As the data image set we used is 3D, the order of the image dimension array is z-axis, y-axis, and x-axis. To make the following implementation more intuitive, the transpose matrix action is used to convert the general x-axis to z-axis. When observing the original image again, it is found that there is still a lot of black invalid information around the image. To reduce the training of this area, we have to cut it, and when doing deconvolution, compared with the image of the same layer, the size of the two images will be slightly different. As a result, skip connections will cause problems. Therefore, the size of all images is adjusted from $91 \times 90 \times 91$ to $88 \times 104 \times 88$, leaving the more important brain structure, which will be less affected by the surrounding invalid features during restoration. The influence of information enhances the learning effect.

In the process of training image segmentation, the main purpose is only to perform feature extraction and correction analysis. The final output image is to be handed over to the doctor for final interpretation and diagnosis. Therefore, the original image can be restored as much as possible to narrow the gap between the images before and after the output. We naturally use mean square error as the loss function for the difference. The equation of the loss function is shown below. y_i is the true voxel value of the original image, $\hat{y}_i$ is the predicted voxel value of the image restored after processing, N is the total number of voxels in the image, and i is the voxel value in the image.

$$L(y, \hat{y}) = \frac{1}{N} \sum_{i=1}^{N} (y_i - \hat{y}_i)^2 \qquad (1)$$

The dataset used in this article is an application from the Nuclear Energy Committee of the Taiwan Executive Yuan. The dataset contains medical images of brain diseases from four different hospitals. The brain diseases collected are Alzheimer's disease, normal (NC), Vascular Dementia (VaD), and Dementia with Lewy Bodies (DLB_PDD). The symptoms are mainly related to dementia. One of the features of the dataset is that it contains images of normal people's brains. Most people who take MRI images in hospitals only need it for those with brain diseases. To study the structure of brain lesions, we especially look for some people who are willing to cooperate. In this dataset, the image size is $91 \times 90 \times 91$, and the total number is 390 3D nuclear medical brain function images, divided into 80% training set and 20% validation set. For the sake of convenience, no additional preprocessing is needed. We are currently cooperating with the Institute of Nuclear Research, so the entire experiment process only uses the corrected image as our dataset.

4 Experiments

After the image segmentation model and PCA removes the biased information, the output image is compared with the original image using computer vision. From Fig. 2(a), we can intuitively find that the yellow area is more obvious in the original correction image processed by the radiologist; the yellow area of the output image 2(b) after the deviation feature correction is reduced, and the brightness is also reduced. Although we don't know the real content of the area, we can guess that this part may be the related area after artificial processing to make the target area more prominent, which caused the natural features in the original image to be masked. The diagnosis made by the doctor using just the naked eye may not make much difference. Still, for the deep learning model, because the process is to analyze each voxel in the image, small differences accumulate to form a huge biased information, which is an important reason why the model cannot learn effectively (Fig. 3).

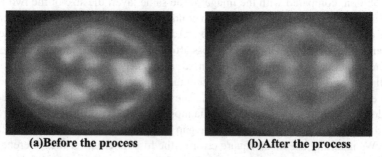

|(a)Before the process|(b)After the process|

Fig. 2. Result of our proposed 3D U-Net + PCA architecture.

Comparing the images, we cannot clearly know whether the removed image features are correct, so we use the CNN architecture to classify diseases and hospitals. After various tests, we can see from the table below that the accuracy rate of hospital classification in the original image is as high as 97.4%, while the disease classification

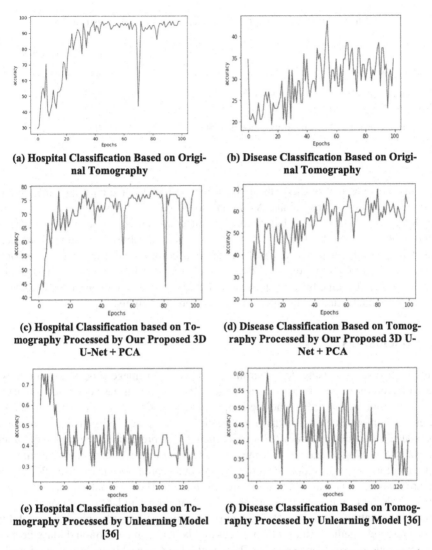

(a) Hospital Classification Based on Original Tomography

(b) Disease Classification Based on Original Tomography

(c) Hospital Classification based on Tomography Processed by Our Proposed 3D U-Net + PCA

(d) Disease Classification Based on Tomography Processed by Our Proposed 3D U-Net + PCA

(e) Hospital Classification based on Tomography Processed by Unlearning Model [36]

(f) Disease Classification Based on Tomography Processed by Unlearning Model [36]

Fig. 3. Accuracy of hospital classification and disease classification under various epoch

is only 34.6%. Therefore, we believe that most of the nuclear medicine brain function imaging features in this dataset are hospital features. After deleting this feature in the image through the above model segmentation, the results show that we can see that the original hospital classification has dropped to 78.2%. In comparison, the disease classification has increased to 63.1%, an increase of at least 30%. From the above-mentioned deep learning model, we can see that even if the biased information in the image is removed, the overall brain image structure in the output image does not have too many errors, which can also assist the doctor in diagnosis (Table 3).

Table 3. Accuracy of hospital classification and disease classification

	Hospital classification	Disease classification
Original tomography	97.436%	34.615%
Tomography processed by our proposed 3D U-Net + PCA architecture	78.205%	63.158%
Tomography processed by unlearning model [36]	35%	40%

This indicates that our hypothesis is correct. It has become an important issue to solve the problem of image bias. As no one so far has used this method to solve this problem, only a more typical basic model architecture is used for verification. We can refer to the other advanced model architectures mentioned above for training, and there is a chance to achieve better results. In addition, according to this dataset, we can only obtain information on two features. If we can obtain other information, such as gender and age, it may become biased information that affects the classification of diseases. It may be possible to analyze the image features more thoroughly.

5 Conclusions

This paper discusses using deep learning frameworks in image processing to remove biased feature information in medical images. The original disease classification accuracy is inferior (about 45%), and the hospital classification accuracy is higher (97%). After our implementation, the disease classification accuracy (65%) is better than the hospital classification accuracy (61%). Compared with previous studies, most of which are about removing image noise or methods to extract image features more accurately, this paper touches on an emerging topic. In similar research, no one has used PCA to do feature analysis and remove biased feature information. There are only similar concepts. [49] regards the scanning instrument as biased information, similar to the hospital characteristic information in our dataset. The main goal of this paper is to explore the reliability of our research method. Rather than discussing better segmentation and classification learning methods, no one has used the PCA method to remove biased information, and it is worthy of discussion. Using a more basic model architecture, in the future, we can use a more accurate and suitable U-Net model or other image segmentation architecture for the dataset and collect more image category information for training. The method can perhaps lead to better learning results. If combined with a more accurate classification architecture model, the accuracy of classifying diseases may be increased. Therefore, the direction of future research is twofold: we can continue to find how to remove the remaining biased information more carefully so that the classification results can be improved or analyze the biased information between multimodal images.

Acknowledgements. This research was funded by National Science and Technology Council grant number MOST 108-3111-Y-042A-117, 111-1401-01-27-01, 111-2221-E-005-086.

References

1. González-Villà, S., Oliver, A., Valverde, S., Wang, L., Zwiggelaar, R., Lladó, X.: A review on brain structures segmentation in magnetic resonance imaging. Artif. Intell. Med. **73**, 45–69 (2016)
2. Vijayalaxmi, Fatahi, M., Speck, O.: Magnetic resonance imaging (MRI): a review of genetic damage investigations. Mutat. Res. Mutat. Res. **764**, 51–63 (2015)
3. Visvikis, D., Le Rest, C.C., Jaouen, V., Hatt, M.: Artificial intelligence, machine (deep) learning and radio (geno) mics: definitions and nuclear medicine imaging applications. Eur. J. Nucl. Med. Mol. Imaging **46**(13), 2630–2637 (2019)
4. Korolev, S., Safiullin, A., Belyaev, M., Dodonova, Y.: Residual and plain convolutional neural networks for 3D brain MRI classification. In: 2017 IEEE 14th International Symposium on Biomedical Imaging (ISBI 2017), pp. 835–838. IEEE, April 2017
5. Fung, Y.R., Guan, Z., Kumar, R., Wu, J.Y., Fiterau, M.: Alzheimer's disease brain MRI classification: challenges and insights. arXiv preprint arXiv:1906.04231 (2019)
6. Nazir, M., Wahid, F., Ali Khan, S.: A simple and intelligent approach for brain MRI classification. J. Intell. Fuzzy Syst. **28**(3), 1127–1135 (2015)
7. Manjón, J.V., Coupe, P.: MRI denoising using deep learning. In: Bai, W., Sanroma, G., Wu, G., Munsell, B.C., Zhan, Y., Coupé, P. (eds.) Patch-MI 2018. LNCS, vol. 11075, pp. 12–19. Springer, Cham (2018). https://doi.org/10.1007/978-3-030-00500-9_2
8. Vaishali, S., Rao, K.K., Rao, G.S.: A review on noise reduction methods for brain MRI images. In: 2015 International Conference on Signal Processing and Communication Engineering Systems, pp. 363–365. IEEE, January 2015
9. Yousuf, M.A., Nobi, M.N.: A new method to remove noise in magnetic resonance and ultrasound images. J. Sci. Res. **3**(1), 81 (2011)
10. Doty, F.D., Entzminger, G., Kulkarni, J., Pamarthy, K., Staab, J.P.: Radio frequency coil technology for small-animal MRI. NMR Biomed.: Int. J. Devot. Dev. Appl. Magn. Reson. In vivo **20**(3), 304–325 (2007)
11. Hong, D., Huang, C., Yang, C., Li, J., Qian, Y., Cai, C.: FFA-DMRI: a network based on feature fusion and attention mechanism for brain MRI denoising. Front. Neurosci. **14**, 934 (2020)
12. Coupé, P., Manjón, J.V., Gedamu, E., Arnold, D., Robles, M., Collins, D.L.: Robust Rician noise estimation for MR images. Med. Image Anal. **14**(4), 483–493 (2010)
13. Panda, A., Naskar, R., Rajbans, S., Pal, S.: A 3D wide residual network with perceptual loss for brain MRI image denoising. In: 2019 10th International Conference on Computing, Communication and Networking Technologies (ICCCNT), pp. 1–7. IEEE, July 2019
14. Chauhan, N., Choi, B.J.: Denoising approaches using fuzzy logic and convolutional autoencoders for human brain MRI image. Int. J. Fuzzy Logic Intell. Syst. **19**(3), 135–139 (2019)
15. Ran, M., et al.: Denoising of 3D magnetic resonance images using a residual encoder–decoder Wasserstein generative adversarial network. Med. Image Anal. **55**, 165–180 (2019)
16. Liu, Z., Bai, X., Liu, H., Zhang, Y.: Multiple-surface-approximation-based FCM with interval memberships for bias correction and segmentation of brain MRI. IEEE Trans. Fuzzy Syst. **28**(9), 2093–2106 (2019)
17. Khosravanian, A., Rahmanimanesh, M., Keshavarzi, P., Mozaffari, S.: A level set method based on domain transformation and bias correction for MRI brain tumor segmentation. J. Neurosci. Methods **352**, 109091 (2021)
18. Bouhrara, M., Bonny, J.M., Ashinsky, B.G., Maring, M.C., Spencer, R.G.: Noise estimation and reduction in magnetic resonance imaging using a new multispectral nonlocal maximum-likelihood filter. IEEE Trans. Med. Imaging **36**(1), 181–193 (2016)

19. Zhang, Y., Liu, J., Li, M., Guo, Z.: Joint image denoising using adaptive principal component analysis and self-similarity. Inf. Sci. **259**, 128–141 (2014)

20. Manjón, J.V., Coupé, P., Buades, A.: MRI noise estimation and denoising using non-local PCA. Med. Image Anal. **22**(1), 35–47 (2015)

21. Bazin, P.L., Alkemade, A., van der Zwaag, W., Caan, M., Mulder, M., Forstmann, B.U.: Denoising high-field multi-dimensional MRI with local complex PCA. Front. Neurosci. **13**, 1066 (2019)

22. Mukherjee, P.S., Qiu, P.: Efficient bias correction for magnetic resonance image denoising. Stat. Med. **32**(12), 2079–2096 (2013)

23. Akkus, Z., Galimzianova, A., Hoogi, A., Rubin, D.L., Erickson, B.J.: Deep learning for brain MRI segmentation: state of the art and future directions. J. Digit. Imaging **30**(4), 449–459 (2017)

24. Despotović, I., Goossens, B., Philips, W.: MRI segmentation of the human brain: challenges, methods, and applications. Comput. Math. Methods Med. (2015)

25. Dolz, J., Desrosiers, C., Wang, L., Yuan, J., Shen, D., Ayed, I.B.: Deep CNN ensembles and suggestive annotations for infant brain MRI segmentation. Comput. Med. Imaging Graph. **79**, 101660 (2020)

26. Moeskops, P., Pluim, J.P.: Isointense infant brain MRI segmentation with a dilated convolutional neural network. arXiv preprint arXiv:1708.02757 (2017)

27. Chen, L., Wu, Y., DSouza, A.M., Abidin, A.Z., Wismüller, A., Xu, C.: MRI tumor segmentation with densely connected 3D CNN. In: Medical Imaging 2018: Image Processing, vol. 10574, p. 105741F. International Society for Optics and Photonics, March 2018

28. Billot, B., Greve, D., Van Leemput, K., Fischl, B., Iglesias, J.E., Dalca, A.V.: A learning strategy for contrast-agnostic MRI segmentation. arXiv preprint arXiv:2003.01995 (2020)

29. Moeskops, P., Veta, M., Lafarge, M.W., Eppenhof, K.A.J., Pluim, J.P.W.: Adversarial training and dilated convolutions for brain MRI segmentation. In: Cardoso, M.J., Arbel, T., Carneiro, G., Syeda-Mahmood, T., Tavares, J.M.R.S., Moradi, M., Bradley, A., Greenspan, H., Papa, J.P., Madabhushi, A., Nascimento, J.C., Cardoso, J.S., Belagiannis, V., Lu, Z. (eds.) DLMIA/ML-CDS -2017. LNCS, vol. 10553, pp. 56–64. Springer, Cham (2017). https://doi.org/10.1007/978-3-319-67558-9_7

30. Ronneberger, O., Fischer, P., Brox, T.: U-Net: convolutional networks for biomedical image segmentation. In: Navab, N., Hornegger, J., Wells, W.M., Frangi, A.F. (eds.) MICCAI 2015. LNCS, vol. 9351, pp. 234–241. Springer, Cham (2015). https://doi.org/10.1007/978-3-319-24574-4_28

31. Siddique, N., Paheding, S., Elkin, C.P., Devabhaktuni, V.: U-Net and its variants for medical image segmentation: a review of theory and applications. IEEE Access (2021)

32. Kolařík, M., Burget, R., Uher, V., Dutta, M.K.: 3D Dense-U-Net for MRI brain tissue segmentation. In: 2018 41st International Conference on Telecommunications and Signal Processing (TSP), pp. 1–4. IEEE, July 2018

33. Zhang, Z., Wu, C., Coleman, S., Kerr, D.: DENSE-INception U-Net for medical image segmentation. Comput. Methods Programs Biomed. **192**, 105395 (2020)

34. Zhou, Z., Siddiquee, M.M.R., Tajbakhsh, N., Liang, J.: UNet++: redesigning skip connections to exploit multiscale features in image segmentation. IEEE Trans. Med. Imaging **39**(6), 1856–1867 (2019)

35. Rehman, M.U., Cho, S., Kim, J.H., Chong, K.T.: BU-Net: brain tumor segmentation using modified U-Net architecture. Electronics **9**(12), 2203 (2020)

36. Dinsdale, N.K., Jenkinson, M., Namburete, A.I.: Deep learning-based unlearning of dataset bias for MRI harmonisation and confound removal. Neuroimage **228**, 117689 (2021)

37. Pearson, K.: On lines and planes of closest fit to systems of points in space (PDF). Phil. Mag. **2**(11), 559–572 (1901). https://doi.org/10.1080/14786440109462720

38. Kumar, V., Sachdeva, J., Gupta, I., Khandelwal, N., Ahuja, C.K.: Classification of brain tumors using PCA-ANN. In: 2011 World Congress on Information and Communication Technologies, pp. 1079–1083. IEEE, December 2011

39. Ahmadi, M., Sharifi, A., Jafarian Fard, M., Soleimani, N.: Detection of brain lesion location in MRI images using convolutional neural network and robust PCA. Int. J. Neurosci. 1–12 (2021)

40. Irzan, H., Hütel, M., Semedo, C., O'Reilly, H., Sahota, M., Ourselin, S., Marlow, N., Melbourne, A.: A network-based analysis of the preterm adolescent brain using PCA and graph theory. In: Bonet-Carne, E., Hutter, J., Palombo, M., Pizzolato, M., Sepehrband, F., Zhang, F. (eds.) Computational Diffusion MRI. MV, pp. 173–181. Springer, Cham (2020). https://doi.org/10.1007/978-3-030-52893-5_15

41. Vijay, K., Selvakumar, K.: Brain fMRI clustering using interaction K-means algorithm with PCA. In: 2015 International Conference on Communications and Signal Processing (ICCSP), pp. 0909–0913. IEEE, April 2015

42. Kaya, I.E., Pehlivanlı, A.Ç., Sekizkardeş, E.G., Ibrikci, T.: PCA based clustering for brain tumor segmentation of T1w MRI images. Comput. Methods Programs Biomed. **140**, 19–28 (2017)

43. Al-Masni, M.A., Kim, D.H.: CMM-Net: contextual multi-scale multi-level network for efficient biomedical image segmentation. Sci. Rep. 11(1), 1–18 (2021)

44. Khagi, B., Kwon, G.R.: 3D CNN design for the classification of Alzheimer's disease using brain MRI and PET. IEEE Access (2020)

45. Hu, M., Sim, K., Zhou, J.H., Jiang, X., Guan, C.: Brain MRI-based 3D convolutional neural networks for classification of schizophrenia and controls. In: 2020 42nd Annual International Conference of the IEEE Engineering in Medicine & Biology Society (EMBC), pp. 1742–1745. IEEE, July 2020

46. Bhanumathi, V., Sangeetha, R.: CNN based training and classification of MRI brain images. In: 2019 5th International Conference on Advanced Computing & Communication Systems (ICACCS), pp. 129–133. IEEE, March 2019

47. Szegedy, C., et al.: Going deeper with convolutions. In: Proceedings of the IEEE Conference on Computer Vision and Pattern Recognition, pp. 1–9 (2015)

48. Çiçek, Ö., Abdulkadir, A., Lienkamp, S.S., Brox, T., Ronneberger, O.: 3D U-Net: learning dense volumetric segmentation from sparse annotation. In: Ourselin, S., Joskowicz, L., Sabuncu, M.R., Unal, G., Wells, W. (eds.) MICCAI 2016. LNCS, vol. 9901, pp. 424–432. Springer, Cham (2016). https://doi.org/10.1007/978-3-319-46723-8_49

Popularity Forecasting for Emerging Research Topics at Its Early Stage of Evolution

Yankin Chi[✉], Raymond Wong, and John Shepherd

University of New South Wales, Sydney, Australia
z5111725@ad.unsw.edu.au, {wong,jas}@cse.unsw.edu.au

Abstract. The accurate modelling and forecasting of the popularity of emerging topics can benefit researchers by allocating resources and efforts on promising research directions. While existing forecasting approaches enjoy various levels of success, most suffer from at least one of the following three limitations: a limited scope due to having to mine topic terms from only a few documents, low generalizability due to assigning arbitrary binary classifications on topics to be either "emerging" or not, or using an emerging topic or field of study's historical features as inputs to forecast its future popularity while disregarding the existing effect of a "cold start". In this paper we propose a forecasting algorithm that address all three limitations in three steps. Firstly, we leverage the field of study taxonomy present in most academic databases to obtain a neighborhood of trending fields within the discipline of the field of study of interest. Then, dynamic time warping is used to measure the similarity of each neighbour's trending pattern compared to the trending pattern of the field of study of interest. Lastly, we conduct multivariate forecasting using a LSTM model while utilizing the historical popularity scores of similar trending neighbours as input. Experimental results on 5 emerging fields of study showcases the "cold start" phenomenon as well as the proposed algorithm reducing RMSE, MAE, and MAPE by half for 4 emerging topics. This validates the claim of the limitations for existing methods and provides insight on the dependency structure of emerging topics with their historical features.

Keywords: Emerging topic · Cold start · Time series forecasting · Forecasting popularity · Dynamic time warping · Node embedding

1 Introduction

In this digital age era, increasing volumes of data are constantly being offered on the internet as services provided by various data owners [30]. A benefit is that data sharing can potentially result in collaborative intelligence and value creation. However, it can also be overwhelming for end users that do not possess the capability to process this surplus of data. As a result, many data service platforms from social media to academic online databases often provide artificial intelligence as a service (AIaaS) in addition to raw data provisioning services. Typical examples include topic modelling and emerging topic detection that is usually offered as part of a recommendation service [7, 10, 18, 21].

© The Author(s), under exclusive license to Springer Nature Switzerland AG 2022
W. Chen et al. (Eds.): ADMA 2022, LNAI 13725, pp. 290–303, 2022.
https://doi.org/10.1007/978-3-031-22064-7_22

In academic research, the problem persists with various large academic repositories constantly being updated with large amounts of publications. For researchers, it is then beneficial to be able to quickly and effectively model and forecast the popularity of various fields of study. This is because accurate predictions can maximise the value generated from funding and resources by allocating them to promising research directions [16, 22, 27].

Existing academic popularity forecasting algorithms mainly take the form of emergence detection algorithms [8, 13, 16] that suffer from three common disadvantages. The first disadvantage is an oversimplified binary criterion to determine if emergence has been achieved. Even though most approaches agree that the popularity of a topic should directly correlate with the number of related publications published within a fixed time interval [2, 6, 9, 13, 16, 20], the difference in threshold definitions show that these criterions are not generalisable across domains.

The second disadvantage is that most algorithms mine topic terms by parsing through documents, paper abstracts or utilising other NLP techniques [13, 16, 20]. This results in the scope of extraction being biased and limited to a few documents. As an alternative, many specialised academic databases such as the Microsoft Academic Graph (MAG) [26], its successor OpenAlex developed by OurResearch, and Aminer [24] are continuously updated with categories for articles as well as an evolving field of study taxonomy. Transitioning to these databases enable fields of study labels to be more consistent yet saving on the computation requirement due to not needing to filter or validate extracted topic terms.

The last disadvantage is the presence of cold start. This occurs when algorithms over rely on the field's own historical measures as multivariate inputs [16] to forecast popularity. This is crucial especially for emerging fields because the lack of sufficient data often leads to spurious or unreliable dependencies. In these cases, alternate features should be considered to obtain reliable dependencies [15].

In this paper, we expand on [9]'s popularity model using the FoS (Field of Study) score and propose a popularity forecasting algorithm. The algorithm has three unique characteristics, each of which corresponds to the three disadvantages of existing approaches. The first disadvantage is accounted for since a numerical popularity score is forecasted instead of a binary classification. Regarding the second disadvantage of a limited scope, the algorithm embraces a macroscopic exploration by integrating the field of study taxonomy of academic databases and consolidating neighbouring fields of study that share similar trending patterns using node embeddings and a modified dynamic time warping (DTW) distance. Finally, by using the historical popularity of these neighbourhood fields of study as the multivariate inputs instead of the desired field's own historical measures, the "cold start" phenomenon is also accounted for. Preliminary results show that this combined approach significantly reduces the mean absolute error (MAE), root mean squared error (RMSE), and mean absolute percentage error (MAPE) of forecasts on topics that have only been trending for a few time intervals. As such, the contribution of our paper can be summarised:

1. A novel popularity modelling algorithm that addresses the "cold start" phenomenon.
2. An effective popularity modelling algorithm that utilises the field of study taxonomy provided by most academic databases.

3. One of the first popularity modelling algorithms that incorporates DTW despite its wide usage in other domains [17].

The outline for the rest of the paper is as follows. Section 2 formulates the problem and defines the popularity. Section 3 summarises existing methodologies for emerging topic detection and modelling. In Sect. 4 a detailed description for each stage of the proposed method is provided. In Sect. 5, we present the experiment setting, results, and discussion. Lastly, Sect. 6 concludes the paper.

2   Problem Formulation

Building on the previous work of [9] with the variables defined in Table 1, let the popularity of a field of study f at time t be recursively defined as:

$$\Gamma_t(f) = \sum_{p \in P_t} \Lambda_t(p,f) + \sum_{g \in \pi(f)} \frac{\Gamma_t(g)}{|\rho(g)|} \tag{1}$$

where the score of each field of study f at time t contributed by paper $p \in P_t$ is given by:

$$\Lambda_t(p,f) = \begin{cases} 1/|F_B(p)|, & \text{if } \in F_B(p) \\ 0, & \text{otherwise} \end{cases} \tag{2}$$

Table 1. Essential variable definitions

Variable	Table column head
P_t	Set of all papers published in time t
$F_B(p)$	Set of fields of study tagged under paper $p \in P$
$\pi(x)$	Set of all children fields of study of x defined by the field of study taxonomy provided by the database
$\rho(x)$	Set of all direct parent fields of study of x
$\Gamma_t(f)$	Popularity of field of study f at time t
τ_{abs}	Absolute threshold for DTW distance
k	Number of initial node-embedding neighbours considered
q	Quantile parameter, defines τ_q
θ	Phase lag parameter

Then, if γ represents the maximum number of years traced back, the multivariate forecasting problem for a single field of study f can be set up as the following. Let matrix $X_f \in \mathbb{R}^{\gamma \times m}$ be an input time series consisting of m features for γ past time intervals and matrix $Y_f \in \mathbb{R}^{t^* \times m^*}$ be the target output time series consisting of m^* features for t^*

future time intervals. Given $\Gamma(f) \in m^*$ and mapping function ϕ such that the forecasted time series matrix $\hat{Y} = \phi(X_f)$, the goal is to learn a set of features m such that $\hat{Y}$ is an accurate representation of Y_f.

Furthermore, since a highlight of the proposed algorithm is its better compatibility with emerging fields of study, we specify years as the time interval and further define field of study f to be emerging at time t if the time series

$$\left\{\Gamma_{t-\gamma}(f), \Gamma_{t-(\gamma-1)}(f), \ldots\ldots, \Gamma_t(f)\right\}$$

is tested by the augmented Dickey–Fuller test (ADF) and does have a unit root at the 95% confidence level while

$$\left\{\Gamma_{t-\gamma}(f), \Gamma_{t-(\gamma-1)}(f), \ldots\ldots, \Gamma_{t-h-1}(f), \Gamma_{t-h}(f)\right\}$$

does not have a unit root using the ADF test at the 95% confidence level. Note h represents the maximum allowed years of trending. In this paper we set γ and h to be 18 and 6 respectively to incorporate various emerging fields from the last decade. However, also due to the relatively small value of h, both m^* and t^* cannot be too large or else ϕ would have too few inputs to produce an accurate forecast. In this paper we set both m^* and t^* to 1. This implies at time t, m^* only consists of $\Gamma_{t+1}(f)$.

3   Related Works

Topic popularity modelling first started with methods that explored emergence using manually defined heuristics. Cataldi and Schifanella [5] aggregated topic terms from tweets using Page Rank and a topic graph then allocated them a stage within a predefined "emergence cycle". Chen et al. [6] explored topic evolution through knowledge transfers between topics. These patterns then determined if a topic transits from "adjusting" to "mature" and has thus emerged. Kim et al. [14] also used Page Rank and an influence score system to construct a weighted citation network that modelled topic evolution and diffusion. Asooja et al. [2] used the tf-idf score to model popularity of topic terms extracted from conference proceedings using linear and polynomial regression. Recently, various approaches using machine learning have also been adopted. Dridi et al. [8] measured the dynamics of keywords such as their popularity rankings, and up-rankings overtime using temporal word embeddings to predict emerging topics. Xu et al. [29] aggregated the growth, coherence, scientific impact, and novelty of publications of a field into an indicator. Then, a multi-task least squares support vector machine (MTLS-SVM) forecasted future indicator values. Jung et al. Jung et al. [13] utilised a network-based database to identify emerging topics that were assumed to possess specific structural features. Liang et al. [16] utilised an extensive set of historical measures in a multivariate setting and tested the forecasting performance of various multivariate models including naïve, autoregression, neural network autoregression, and long short-term memory (LSTM) models. Among these methods LSTM provided the best forecast. Xu et al. [28] extended the research by chaining multiple LSTMs with a graph convolutional neural network to forecast the popularity of topic terms. It can be seen that while all these methods have achieved various levels of success, no method simultaneously addresses the three disadvantages mentioned in Sect. 1.

4 Proposed Method

The core philosophy of the proposed popularity forecasting algorithm is to integrate the advantages provided by the existing academic databases. This is mainly in the form of being able to trace a paper's involved fields of study back to a structured and hierarchical field of study taxonomy. In our proposed method, there are three main steps. They are: (i) neighbourhood extraction, (ii) extraction of neighbourhood with similar trending patterns, and (iii) multivariate forecast.

At time t, given field of study of interest f; a field of study taxonomy as a directed graph $G = (V, E)$ defined by a set of field vertices V and set of edges E; the popularity time series

$$\left\{ \Gamma_{t-18}(x), \Gamma_{t-(n-1)}(x), ..., \Gamma_t(x) \right\}$$

for all $x \in V$; and the task of determining an input matrix $X_f \in \mathbb{R}^{18 \times m}$ to obtain $\hat{Y} = \left[\widehat{\Gamma_{t+1}(f)} \right]$, the proposed method produces a forecast through the following procedures:

A. Extracting a neighbourhood

The first step of the forecasting process is to extract a neighbourhood set $n_t(f)$ consisting of the top k trending fields of study that are structurally near and therefore likely to be in the same discipline as field f. One assumption present is that fields within the same discipline have a higher probability of sharing trending patterns than fields in other disciplines located further away in the taxonomy.

This step is initialised by embedding the vertices (i.e., fields of study) in G to a vector space using a node embedding algorithm to be able to cluster a local neighbourhood around field f. In this paper we choose the classic node2vec model utilising skip gram [11] due to its relatively simple and fast implementation for single-relational graphs such as G. Specifically, given $N_R(u)$ is the neighbourhood of $u \in V$ for a random walk strategy R, the following log-likelihood function is optimised

$$\mathcal{L} = \sum_{u \in V} \sum_{v \in N_R(u)} - \log \left(\frac{\exp(\mathbf{z}_u^T \mathbf{z}_n)}{\sum_{n \in V} \exp(\mathbf{z}_u^T \mathbf{z}_n)} \right) \tag{3}$$

with the approximation that the normalization is only conducted against δ random samples

$$\log \left(\frac{\exp(\mathbf{z}_u^T \mathbf{z}_n)}{\sum_{n \in V} \exp(\mathbf{z}_u^T \mathbf{z}_n)} \right) \approx \log(\sigma(\mathbf{z}_u^T \mathbf{z}_n)) - \sum_{i=1}^{\delta} \log(\sigma(\mathbf{z}_u^T \mathbf{z}_{n_i})), n_i \sim P_v \tag{4}$$

This provides a set of embeddings $f : u \to \mathbb{R}^d : f(u) = \mathbf{z}_u$.

Then, using the embeddings we extract the set V^k containing the top k fields of study whose vector cosine similarity to field f is smallest. Note that cosine similarity is preferred because it normalizes the magnitude of each vector. At this point although trending neighbours are prioritised within V^k, it could still include fields whose historical popularity is stationary. Therefore, to remove these non-trending neighbours an ADF test is conducted on $\{\Gamma_{t-18}(x), \Gamma_{t-(n-1)}(x), ..., \Gamma_t(x)\}, \forall x \in V^k$. Lastly, only fields in V^k that possess a unit root at the 95% confidence level are consolidated into the neighbourhood field set $n_t(f)$ in preparation for step 2.

B. Extracting neighbours with similar trending patterns

At time t, given a set of neighbourhood fields of study $n_t(f)$ for field f, the second step is to further extract a subset $\mathbb{N}_t(f) \in n_t(f)$ of specific neighbors that share a similar trending pattern. The dynamic time warping (DTW) distance measure [4] is chosen to evaluate the extent of similarity because the evaluated sequences can be of variable lengths. In this case let $\Gamma'_t(x)$ be the normalised value of $\Gamma_t(x)$, we evaluate the DTW distance $\sigma_{f,x} = DTW\left(T_t^x, T_t^f\right)$ for all $x \in n_t(f)$ such that:

$$T_t^x = \left\{\Gamma'_{t-18}(x), \Gamma'_{t-(n-1)}(x), ..., \Gamma'_{t-\theta}(x)\right\} \tag{5}$$

$$T_t^f = \left\{\Gamma'_{t-18}(f), \Gamma'_{t-(n-1)}(f), ..., \Gamma'_t(f)\right\} \tag{6}$$

and

$$DTW\left(T_t^x, T_t^f\right) = argmin(sum(D(i,j))) \tag{7}$$

with

$$D(i,j) = \left|\Gamma'_i(x) - \Gamma'_j(f)\right| + min \begin{bmatrix} D(i-1, j-1) \\ D(i-1, j) \\ D(i, j-1) \end{bmatrix} \tag{8}$$

Notice that the DTW distance is calculated with the popularity time series of x up until time $t - \theta$ where θ is the phase lag parameter. This is deliberate since multivariate models can account for input features such as the popularity trend from $t - \theta$ to t of field x that potentially foreshadows the future popularity trend of f beyond t [1, 16]. Note to prevent prioritising fields x whose phase difference is too drastic to f, it is recommended to set $\theta \leq 2$. In this paper we set $\theta = 1$.

For $x \in n_t(f)$ to be included in $\mathbb{N}_t(f)$, its DTW distance $\sigma_{f,x}$ must be smaller than an absolute threshold τ_{abs} and a relative quantile threshold τ_q determined by taking a prespecified q^{th} quantile on $\{\sigma_{f,x}\}, x \in n_t(f)$. This is because without τ_{abs}, $\mathbb{N}_t(f)$ will never be empty even if all neighbours have large distances. Alternatively, without τ_q, when k is large, $\mathbb{N}_t(f)$ would overflow with fields of study negating the influence of various $x \in \mathbb{N}_t(f)$ that trend especially similarly to f.

C. Conducting the multivariate forecast

At time t, given a set of similar trending neighbours $\mathbb{N}_t(f)$ for field f, the last step is to define feature set m and subsequently input matrix $X_f \in \mathbb{R}^{18 \times m}$ as well as specifying mapping function ϕ to conduct the multivariate forecast.

Feature set m will consist of two components, they are field f's own normalised popularity time series $\{\Gamma'_i(f)\}$, $t - 18 \leq i \leq t$ and the normalised popularity time series $\{\Gamma'_i(x)\}$, $t - 18 \leq i \leq t$ for all $x \in \mathbb{N}_t(f)$.

Regarding ϕ, many multivariate time series models exist, including but not limited to the autoregressive integrate moving average (ARIMA), generalised autoregressive conditional heteroskedasticity model (GARCH) [3], neural network autoregressive model (NNAR) [25], and the long short term memory with sequence to sequence structure (LSTM) [23]. In this paper since the goal is to evaluate the effectiveness of similar trending neighbours making up the feature set m, we implement the LSTM model as our multivariate forecasting model due to its ability to account for temporal dependencies [12] and proven superior performance in other studies that benchmarked it against other multivariate models for emerging topic detection [16, 28].

5   Experiment Setup and Results

5.1   Dataset and Experiment Setup

The dataset chosen is the Microsoft Academic Graph (MAG) version 2020-06-19 due to its discontinued service allowing easy reproducibility of our experiment. We define t to be 2018, such that the starting year of observed popularity is 2000 and the popularity of year 2019 is forecasted. In total, forecasting will be done on 10 fields of study. Specifically, under field of study "deep learning", 5 emerging topics satisfying the definition in Sect. 2 were chosen at random. They are "Long short term memory" (LSTM), "Deep CNN" (DCNN), "Self attention" (SA), "Class action mapping" (CAM), and "Denoising auto encoder" (DAE). 5 non-emerging topics that are non-stationary and under the field of study "computer science" were also randomly chosen to benchmark the proposed method's forecasting performance on non-emerging but trending fields of study. They are "Computer science" (CS), "Drawback" (DB), "Knowledge graph" (KG), "Gadget" (Gad), and "System information" (SI). For all forecasting tasks, we set k to be 1000 such as to ensure all fields of study possesses at least 1 neighbour with similar trending patterns for the proposed method. The other parameters q and τ_{abs} were set to be 0.003 and 1.2 respectively.

5.2   Baseline

The main target baseline is a multivariate popularity forecasting algorithm under the same LSTM setting but utilising the desired field's own historical measures as input features. A comprehensive list of such features is summarised by [16]. However, due to [16] conducting forecasts for self-extracted terms, some measures are not applicable. A list of relevant historical features used as measures for field f are listed in Table 2. Additionally, a univariate LSTM is also benchmarked along the two mentioned approaches. This can provide deeper insight on the significance and extent of influence provided by these input features. All LSTM models are configured to have two bidirectional layers with 64 and 32 units each using the rectified linear unit as the activation function. The batch size is set to 15 and the epochs set to 100.

Table 2. Historical feature set

Variable	Table column head
$\Gamma'_t(f)$	Historical popularity of field f
Doc Freq (DF_t)	Normalised number of papers tagged under field f at time t
Unique Authors (UA_t)	Normalised number of unique authors that contributed to field f at time t
Unique References (UR_t)	Normalised number of unique references used by documents in field f at time t

5.3 Evaluation Metrics

The standard evaluation metrics for time series forecasts are the root mean squared error (RMSE), mean absolute error (MAE) and mean absolute percentage error (MAPE) [16, 17] calculated using the following formulas for field f when $m^* = 1$ at time t:

$$RMSE_f = \sqrt{\frac{\sum_{u=t+1}^{t+t^*}\left(\Gamma_u(f)-\widehat{\Gamma_u(f)}\right)^2}{t^*}} \tag{9}$$

$$MAE_f = \frac{\sum_{u=t+1}^{t+t^*}\left|\Gamma_u(f)-\widehat{\Gamma_u(f)}\right|}{t^*} \tag{10}$$

$$MAPE_f = \frac{\sum_{u=t+1}^{t+t^*}\left|\frac{\Gamma_u(f)-\widehat{\Gamma_u(f)}}{\Gamma_u(f)}(100)\right|}{t^*} \tag{11}$$

Note that since $t^* = 1$, the absolute value of the residuals are identical for all forecasted popularity values. This results in RMSE and MAE being equal. We repeat 10 iterations for each field's forecast and record the average RMSE/MAE and MAPE in addition to their standard deviation.

5.4 Results and Discussion

The experimental results are separated for the 5 emerging topic fields of study and the 5 non-emerging but trending fields of study. To optimise spacing we abbreviate the proposed method to "NeighB LSTM", the external feature LSTM to "Ext LSTM", and the univariate LSTM to "UniV LSTM". The best performing score is also bolded to enhance readability. For completeness, we also include the number of neighbours extracted by the proposed method along with each neighbour's measured DTW distance. Note that since $k = 1000$ and $q = 0.003$, assuming all neighbour fields have DTW distances below τ_{abs}, there will be at most 3 neighbours, corresponding to "neighbour 1", "neighbour 2", and "neighbour 3" within Tables 4 and 6. Should the neighbour count be smaller than 3 due to some neighbours having distances greater than τ_{abs}, the distance is replaced with a "-" marking their absence.

The forecast results for emerging topics shown in Tables 3 and 4 mainly reveal three observations. Firstly, the proposed method can significantly improve the accuracy and

Table 3. Results for emerging topics

		LSTM	DCNN	SA	CAM	DAE
RMSE/MAE	UniV LSTM	**20.644** ±16.537	4.295 ±0.755	41.973 ±0.062	1.629 ±0.200	4.943 ±0.307
	Ext LSTM	25.566 ±10.237	4.712 ±0.689	40.388 ±0.870	0.923 ±0.376	5.537 ±0.344
	NeighB LSTM	28.271 ±7.692	**1.024** **±0.941**	**11.712** **±2.450**	**0.899** **±0.297**	**2.609** **±0.098**
MAPE	UniV LSTM	**19.213%** **±15.391%**	18.043% ±3.172%	96.516% ±0.142%	41.696% ±5.129%	402.549% ±24.977%
	Ext LSTM	23.794% ±9.527%	19.795% ±2.895%	92.871% ±2.000%	23.626% ±9.633%	450.922% ±28.053%
	NeighB LSTM	26.312% ±7.159%	**4.301%** **±3.955%**	**26.932%** **±5.634%**	**23.025%** **±7.600%**	**212.473%** **±7.986%**

Table 4. NeighB DTW distances (emerging)

	LSTM	DCNN	SA	CAM	DAE
Neighbour 1	0.666	0.671	0.818	0.502	0.864
Neighbour 2	0.710	0.734	0.626	0.447	0.952
Neighbour 3	0.670	0.743	0.870	0.488	1.085

precision of popularity forecasts for emerging topics. This is shown by the RSME/MAE and MAPE for NeighB LSTM being lower than both its competitors in 4 out of 5 topics by at least 2.60%, at most 76.16%, and on average 49.24% across the 4 fields. This highlights that trending patterns within emerging subtopics of a discipline can potentially be very similar, and thus, be successfully utilised in a multivariate setting to influence the predicted growth of a particular field of study.

The second observation is that the "cold start" phenomenon can be significant. From Table 3, it is obvious that Ext LSTM struggles to infer the popularity growth trend for emerging topics. This is evident by the fact that it achieves the highest RMSE/MAE and MAPE for 3 out of 5 fields, even being surpassed by the univariate LSTM. This is significant because it showcases that at a field of study's early stage of evolution, inferred dependencies between the field's popularity growth and its historical features can not only be unhelpful but also misleading. In this case, this misleading effect may be exacerbated due to fields LSTM, DCNN, and DAE trending relatively smoothly, potentially benefiting the univariate approach. Nonetheless, this showcases the importance to account for the "cold start" phenomenon in future forecasting tasks. Note that in this paper, we defined a field of study to be emerging based on testing the field's historical popularity time series for a unit root. However, it will be interesting, as further study, to investigate the varying dependency structure of a field of study with its historical features over time (Fig. 1).

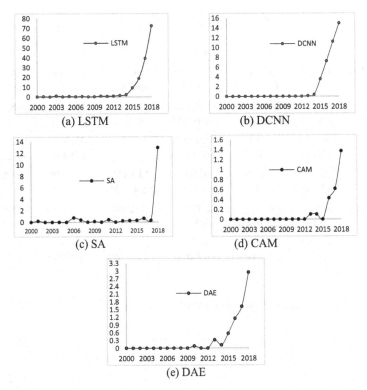

Fig. 1. History popularity of 5 emerging topics

The third observation is the DTW distance of the neighbours of a field of study does not necessarily indicate the performance gained compared to other methods. Particularly, while field DCNN and CAM both have three neighbours with relatively small DTW distances, the error reduction for DCNN is much more significant than CAM. This is likely due to DCNN's historical popularity from 2013 to 2018 trending relatively smoothly while CAM's popularity being volatile from years 2012 to 2018 as seen from Fig. 2(b) and Fig. 2(d). This shows that the underlying feature dimensions needed to generate accurate forecasts for a volatile series is far more complex than the feature list proposed by [16] and is likely to vary with different domains. Therefore, referencing similar volatility trends from neighbours once again proves to be the superior option due it not being constrained to a fix set of features. As time goes, the proposed method is also expected to improve since the backlog of popularity trend patterns only increases as these academic databases update overtime (Table 5).

The forecast results for non-emerging topics shown in Tables 4 and 6 allow deeper insight on the relative compatibility of the proposed method on fields at various stages of their trending popularity. Specifically, for non-emerging fields of study, the advantage of the proposed method quickly diminishes. This can be seen from the RMSE/MAE and MAPE values of NeighB being only slightly better in some cases and mostly worse than Ext LSTM in 3 of the 5 fields. There are two probable explanations.

Table 5. Results for non-emerging topics

		KG	CS	DB	Gad	SI
RMSE/MAE	UniV LSTM	90.574 ±8.585	9301.222 ±591.022	**9.020** **±0.566**	40.406 ±1.613	56.134 ±21.824
	Ext LSTM	**50.864** **±8.165**	**3765.102** **±1491.671**	11.648 ±2.022	41.357 ±1.443	**18.520** **±11.646**
	NeighB LSTM	87.030 ±5.189	6094.895 ±1013.960	9.884 ±0.437	**32.363** **±0.495**	19.181 ±21.608
MAPE	UniV LSTM	34.874% ±3.306%	5.925% ±0.377%	**11.845%** **±0.743%**	44.970% ±1.795%	33.569% ±13.051%
	Ext LSTM	**19.584%** **±3.144%**	**2.399%** **±0.950%**	15.296% ±2.656%	46.029% ±1.606%	**11.075%** **±6.965%**
	NeighB LSTM	33.509% ±1.998%	3.883% ±0.646%	12.979% ±0.574%	**36.019%** **±0.551%**	11.470% ±12.922%

Table 6. NeighB DTW distances (non-emerging)

	KG	CS	DB	Gad	SI
Neighbour 1	0.859	0.428	0.724	0.824	0.720
Neighbour 2	0.793	0.467	0.653	0.809	0.727
Neighbour 3	0.576	–	0.700	0.706	–

The first possibility is because NeighB has neighbours with relatively low DTW distances, overtime, a trending field's historical features stabilises to better reflect the growth rate of the field's popularity. This explanation is naturally evident when considering the fact that many previous studies have had success utilising a field's historical measures as multivariate features, as well as Ext LSTM's performance for non-emerging fields. In this case, NeighB can still be compatible, however, the similarity criteria would need to be tighter with k increasing and τ_{abs} decreasing to capture neighbours with better resembling trending patterns.

Another possible explanation is NeighB loses its effectiveness because trending fields of study have more complicated time-specific dependencies. For example, if a field of study f has children fields x and y whose trending patterns are both similar with their only difference being x trending 5 years ago, the node2vec implementation of the proposed method treats them both equally. This should not be the case since intuitively field f's popularity is more likely to be influenced by field y whose popularity is trending within the same period. This phenomenon is also more likely to impact non-emerging fields of study due to emerging fields being unlikely to possess sub/children fields of study. In this case, potential solutions include providing time-specific weights to neighbour fields during the node embedding process. Alternatively, time specific taxonomies can also be utilised to differentiate the taxonomy's varying graph structure at different time periods.

These graph structures can then be account for using other node embeddings such as struc2vec [19].

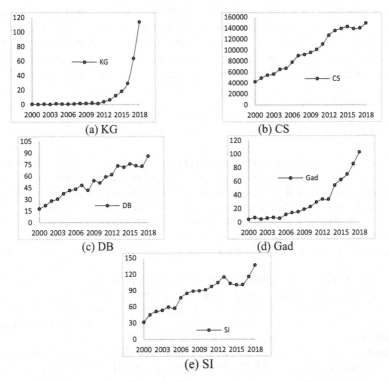

Fig. 2. History popularity of 5 non-emerging topics

6 Conclusion

In this paper, we proposed a field of study popularity forecasting algorithm. By using a field of study taxonomy, a macroscopic exploration of similar trending neighbour fields of study can be obtained. Using the historical popularity of these similar trending neighbours as multivariate input features, we provide a solution for the "cold start" phenomenon that is ignored by most existing popularity forecasting approaches. Through extensive testing using 10 fields of study within MAG, we have shown that our approach outperforms existing approaches utilising historical features as multivariate inputs on emerging fields of study. However, through the experiment, we also show that these features potentially stabilize overtime to better reflect the popularity growth of that particular field. As such, our algorithm serves as a complement to existing methods especially at a field's early stage of evolution.

Future research for the proposed algorithm mainly includes two directions. Firstly, the current definition of an "emerging field" is not robust. This is because a popularity series initially with a unit root can still reject the null hypotheses when rerunning the ADF test with the inclusion of data for additional time periods. This is slightly accounted for with two tests being run in the proposed method, however, a more consistent emerging definition should be explored. Secondly, as mentioned, the time-specific dependencies of each field of study can be accounted for during the discovery of suitable neighbour field of studies. This allows the time period of a neighbour's popularity trend to also be considered. Lastly, the current work can also be extended by testing the proposed approach on non-academic fields such as forecasting the popularity of social media entities.

References

1. Aboagye-Sarfo, P., Mai, Q., Sanfilippo, F.M., Preen, D.B., Stewart, L.M., Fatovich, D.M.: A comparison of multivariate and univariate time series approaches to modelling and forecasting emergency department demand in Western Australia. J. Biomed. Inform. **57**, 62–73 (2015)
2. Asooja, K., Bordea, G., Vulcu, G., Buitelaar, P.: Forecasting emerging trends from scientific literature. In: Proceedings of the Tenth International Conference on Language Resources and Evaluation, LREC 2016, pp. 417–420, May 2016
3. Bauwens, L., Laurent, S., Rombouts, J.V.: Multivariate GARCH models: a survey. J. Appl. Economet. **21**(1), 79–109 (2006)
4. Berndt, D.J., Clifford, J.: Using dynamic time warping to find patterns in time series. In: Proceedings of KDD Workshop, vol. 10, no. 16, pp. 359–370, July 1994
5. Cataldi, M., Di Caro, L., Schifanella, C.: Emerging topic detection on twitter based on temporal and social terms evaluation. In: Proceedings of the Tenth International Workshop on Multimedia Data Mining, pp. 1–10, July 2010
6. Chen, B., Tsutsui, S., Ding, Y., Ma, F.: Understanding the topic evolution in a scientific domain: An exploratory study for the field of information retrieval. J. Informet. **11**(4), 1175–1189 (2017)
7. Chu, V.W., Wong, R.K., Chi, C.H., Chen, F.: Extreme topic model for market eAlert service. In: Proceedings of the 2018 IEEE International Conference on Services Computing (SCC), pp. 145–152. IEEE, July 2018
8. Dridi, A., Gaber, M.M., Azad, R.M.A., Bhogal, J.: Leap2Trend: a temporal word embedding approach for instant detection of emerging scientific trends. IEEE Access **7**, 176414–176428 (2019)
9. Effendy, S., Yap, R.H.: Analysing trends in computer science research: a preliminary study using the Microsoft academic graph. In: Proceedings of the 26th International Conference on World Wide Web Companion, pp. 1245–1250, April 2017
10. Gao, Z., et al.: SeCo-LDA: mining service co-occurrence topics for recommendation. In: Proceedings of the 2016 IEEE International Conference on Web Services (ICWS), pp. 25–32. IEEE, June 2016
11. Grover, A., Leskovec, J.: node2vec: scalable feature learning for networks. In: Proceedings of the 22nd ACM SIGKDD International Conference on Knowledge Discovery and Data Mining, pp. 855–864, August 2016
12. Hochreiter, S., Schmidhuber, J.: Long short-term memory. Neural Comput. **9**(8), 1735–1780 (1997)
13. Jung, S., Datta, R., Segev, A.: Identification and prediction of emerging topics through their relationships to existing topics. In: Proceedings of the 2020 IEEE International Conference on Big Data (Big Data), pp. 5078–5087. IEEE, December 2020

14. Kim, M., Baek, I., Song, M.: Topic diffusion analysis of a weighted citation network in biomedical literature. J. Am. Soc. Inf. Sci. **69**(2), 329–342 (2018)
15. Lam, X.N., Vu, T., Le, T.D., Duong, A.D.: Addressing cold-start problem in recommendation systems. In: Proceedings of the 2nd International Conference on Ubiquitous Information Management and Communication, pp. 208–211, January 2008
16. Liang, Z., Mao, J., Lu, K., Ba, Z., Li, G.: Combining deep neural network and bibliometric indicator for emerging research topic prediction. Inf. Process. Manag. **58**(5), 102611 (2021)
17. Malarya, A., Ragunathan, K., Kamaraj, M.B., Vijayarajan, V.: Emerging trends demand forecast using dynamic time warping. In: Proceedings of the 2021 IEEE 22nd International Conference on Information Reuse and Integration for Data Science (IRI), August 2021
18. Qiu, Y., Chen, Y., Jiao, L., Huang, S.: RTA: real time actionable events detection as a service. In: Proceedings of the 2016 IEEE International Conference on Web Services, June 2016
19. Ribeiro, L.F., Saverese, P.H., Figueiredo, D.R.: struc2vec: learning node representations from structural identity. In: Proceedings of the 23rd ACM SIGKDD International Conference on Knowledge Discovery and Data Mining, pp. 385–394, August 2017
20. Saeed, Z., Abbasi, R.A., Razzak, I., Maqbool, O., Sadaf, A., Xu, G.: Enhanced heartbeat graph for emerging event detection on Twitter using time series networks. Expert Syst. Appl. **136**, 115–132 (2019)
21. Shi, M., Liu, J., Zhou, D., Tang, M., Xie, F., Zhang, T.: A probabilistic topic model for mashup tag recommendation. In: Proceedings of the 2016 IEEE International Conference on Web Services (ICWS), pp. 444–451. IEEE, June 2016
22. Small, H., Boyack, K.W., Klavans, R.: Identifying emerging topics in science and technology. Res. Policy **43**(8), 1450–1467 (2014)
23. Sutskever, I., Vinyals, O., Le, Q.V.: Sequence to sequence learning with neural networks. Adv. Neural Inf. Process. Syst. **27** (2014)
24. Tang, J.: AMiner: toward understanding big scholar data. In: Proceedings of the Ninth ACM International Conference on Web Search and Data Mining, p. 467, February 2016
25. Taskaya-Temizel, T., Casey, M.C.: A comparative study of autoregressive neural network hybrids. Neural Netw. **18**(5–6), 781–789 (2005)
26. Wang, K., Shen, Z., Huang, C., Wu, C.H., Dong, Y., Kanakia, A.: Microsoft academic graph: when experts are not enough. Quant. Sci. Stud. **1**(1), 396–413 (2020)
27. Wang, Q.: A bibliometric model for identifying emerging research topics. J. Am. Soc. Inf. Sci. **69**(2), 290–304 (2018)
28. Xu, M., Du, J., Xue, Z., Guan, Z., Kou, F., Shi, L.: A scientific research topic trend prediction model based on multi-LSTM and graph convolutional network. Int. J. Intell. Syst. (2022)
29. Xu, S., Hao, L., An, X., Yang, G., Wang, F.: Emerging research topics detection with multiple machine learning models. J. Informet. **13**(4), 100983 (2019)
30. Zheng, Z., Zhu, J., Lyu, M.R.: Service-generated big data and big data-as-a-service: an overview. In: Proceedings of the 2013 IEEE International Congress on Big Data, June 2013

Positive Unlabeled Learning by Sample Selection and Prototype Refinement

Zhuowei Wang[1,2(✉)] and Guodong Long[1]

[1] Australian Artificial Intelligence Institute, Faculty of Engineering and Information Technology, University of Technology Sydney, Sydney, Australia
zhuowei.wang@student.uts.edu.au, guodong.long@uts.edu.au
[2] Space and Astronomy, CSIRO, Marshfield, Australia
zhuowei.wang@csiro.au

Abstract. Positive and Unlabeled learning (PU learning) learns a binary classifier on training data with only positive and unlabeled instances. Recent cost-sensitive methods tackled this problem by designing unbiased loss functions and achieved state-of-the-art performance. However, we observe that the model suffers from overfitting at the late training stage caused by regarding unlabeled positive samples as negative ones. This motivates us to propose PUSP, a novel framework that leverages sample selection and prototype refinement to tackle PU learning problem. We first carefully select reliable samples based on the time consistency of the model output and spatial consistency of different views of image contents. Then we assign labels to those unselected samples based on their feature similarities with the prototypes of the selected ones. We conduct extensive experiments to show the effectiveness of our method on over different datasets and modalities.

Keywords: PU learning · Sample selection · Prototype learning

1 Introduction

In tradition binary classification problem, the classifier is learned by taking full advantage of both positive and negative samples. *Positive and Unlabeled* (PU) *learning* aims at training the classifier with only positive and unlabeled data, which is more practical in real-world setting where one certain class of binary classification might be difficult to be collected or annotated. PU learning has drawn huge attention from both academia and industry recently since it can be applied to various scenarios, including disease diagnosis [5], deceptive review detection [33], recommendation system [34], and text classification [9].

PU learning has been widely studies for decades and there are two main stream strategies based on how to handle unlabeled data, which contains *two-step* (TS) approach and *cost-sensitive* (CS) approach. The TS approach [3,44] first selects a portion of negative samples using various heuristic strategies, and then trains the classifier with original positive data and selects negative data. However, the selected negative data might be inaccurate, which may deteriorate the performance of the model. Therefore, the CS approach treats unlabeled data

W. Chen et al. (Eds.): ADMA 2022, LNAI 13725, pp. 304–318, 2022.
https://doi.org/10.1007/978-3-031-22064-7_23

directly as negative data but assigns them with different reduced importance weights. The estimation bias of the objective is corrected by leveraging well-engineered unbiased risk estimation [16,28], which refines the classification risk to an equivalent form depending only on PU data. The most representative method, called nnPU [16], proposed the non-negative PU learning method to overcome the serious overfitting problem of unbiased risk estimation, achieving state-of-the-art performance.

Although CS approaches have shown great potentials in PU learning, treating all unlabeled samples as corrupted negatives may lead to performance deterioration caused by the model overfitting to hard unlabeled instances. To demonstrate this issue, we conducted a preliminary experiment on CIFAR-10 [17] where the positive probability of each unlabeled sample predicted by the nnPU model is plotted by histogram. We observe that at the late stage of training (Fig. 1 (b)), the distribution of the positive and negative becomes less polarzied and more negative samples are wrongly regarded as positive ones. Based on this phenomenon, we conclude that CS methods may result in unreliable supervision by treating all unlabeled data as weighted negative samples.

Motivated by the above observations, we proposed a novel sample selection and prototype learning method called PUSP, which carefully selects the most reliable positive and negative samples in the unlabeled set, and assigns labels to the rest unchosen samples based on the similarity between them and the prototypes of the reliable samples. More specifically speaking, our sample selection criterion considers both the dynamics of the confidence of model on each sample and output consistency of different views of images. We use an exponential moving average (EMA) of the model to predict the positive probability of each sample with multiple data augmentations, which overcomes the instability of instantaneous model prediction without extra inference cost. To capture more class-relevant information and semantically meaningful knowledge from the selected high-confidence samples, we propose to use class-wise prototype for label assignment. Concretely, we label the remaining unchosen samples in the unlabeled set by calculating the feature similarity between them and the prototypes (average vector) of the chosen samples. We evaluate our method on multiple bench-marked datasets (including three computer vision datasets and two natural language processing datasets) and show that our method outperforms various types of PU learning approaches.

2   Related Work

In this section, we briefly review several related aspects on which our method builds: PU learning, semi-supervised learning, and noisy label learning.

PU Learning. What makes PU learning more difficult than standard binary classification is that we have only access to a set of unlabeled samples sampled from the marginal density $P(x)$ instead of labeled negative data. Several methods have been designed to address these problems using two steps. [23] finds out reliable negative data, and applies the Expectation-Maximization algorithm to train the classifier. [22] leverages the Rocchio classifier to identify negative samples, and then runs an SVM classifier. In the deep learning age, various risk estimators have been proposed to avoid tuning parameters. [16] proposes the non-negative PU method to overcome the serious overfitting problem of unbiased risk estimation. Recently, [12] utilizes generated adversarial networks to generate both positive and negative samples. [4] proposes a variational approach that does not require explicit computation of the class prior. [5] proposes a novel Self-PU learning framework that integrates PU learning and self-training.

Noisy Label Learning. Similar to the IR approach, if unlabeled samples are treated as wrongly labeled negative data, PU learning can be transferred to a noisy label learning. Sample selection methods regards samples with small loss as "clean" and trains the model only on selected clean samples. [14] selects small-loss samples and uses them to train the network. [10] leverages two networks to choose the next batch of data for each other for training. Another line of works is noise transition estimation [11,30]. [30] estimates the noise transition matrix and trains the network with two different loss corrections. [11] proposes a loss correction technique which requires a small portion of trusted samples. Some other interesting and promising directions for NLL include meta-learning [21, 35] based and semi-supervised learning [39] based approaches. Recently, some works [40,42] non-trivially extend NLL to federated setting [27].

Semi-supervised Learning. Semi-supervised learning (SSL) deals with learning with partially labeled data, where the amount of unlabeled data is usually much larger than that of the labeled data. With positive and unlabeled available for training, the problem setup for PU learning may seem similar to that of SSL. [18] applies consistency between the output of the current network and the exponential moving average (EMA) of the output from the past epochs. [38] proposes to update the network on a mini-batch level using an EMA of model parameter values. [2] introduces a holistic approach that well combines *MixUp* [45], entropy minimization, and consistency regularization.

3   Problem Setting

In this section, we briefly review the risk estimators used in two classic cost-sensitive approaches and the sample misclassification problem caused by these methods during the training process.

3.1   Review of Cost-Sensitive Approaches

Problem Setting. Let X and Y be the input and output domains. Consider a binary classification problem, $Y \in \{1, -1\}$. The training set consist of a positive set $\mathbb{D}_p$ and an unlabeled set $\mathbb{D}_u$. $\mathbb{D}_p$ contains n_p samples x_p sampled from $P(x|Y = 1)$. $\mathbb{D}_u$ contains n_u samples x_u sampled from $P(x)$. We denote the class prior probability for positive and negative data as $\pi_p = P(Y = +1)$ and $\pi_n = P(Y = -1)$. We follow [16] to assume that π is known throughout the paper. In real-world practice, if π is not known in advanced, it can be estimated from the given PU data [32].

Expected Risk Estimators. Let $g(\cdot;\theta) : \mathbb{R}^d \to \mathbb{R}$ be the model, θ be its parameters, and $\ell : \mathbb{R} \times \{+1, -1\} \to \mathbb{R}$ be loss function. For PN learning, the risk can be written as:

$$R_{\mathrm{pn}}(g) = \pi_p \mathbb{E}_{X \sim p(x|Y=1)}[\ell(g(X;\theta))] + \pi_n \mathbb{E}_{X \sim p(x|Y=-1)}[\ell(-g(X;\theta))]. \quad (1)$$

And $R_{\mathrm{pn}}(g)$ can be approximated by

$$\widehat{R}_{\mathrm{pn}}(g) = \frac{\pi_p}{n_{\mathrm{p}}} \sum_{i=1}^{n_{\mathrm{p}}} \ell(g(x_i;\theta)) + \frac{\pi_n}{n_{\mathrm{n}}} \sum_{i=1}^{n_{\mathrm{n}}} \ell(-g(x_i;\theta)), \quad (2)$$

where n_{n} is the number of N data. However, there are no labeled negative data in PU learning. Thus, we replace the negative loss on the right of Eq. (1) by

$$\pi_n \mathbb{E}_{X \sim p(x|Y=-1)}[\ell(-g(X;\theta))] = \mathbb{E}_{X \sim p(x)}[\ell(-g(X;\theta))] - \pi_p \mathbb{E}_{X \sim p(x|Y=1)}[\ell(-g(X;\theta))]. \quad (3)$$

$R_{\mathrm{pu}}(g)$ can be approximated by

$$\widehat{R}_{\mathrm{uPU}}(g) = \frac{\pi_p}{n_{\mathrm{p}}} \sum_{i=1}^{n_{\mathrm{p}}} \ell(g(x_i;\theta)) + \left(\frac{1}{n_{\mathrm{u}}} \sum_{i=1}^{n_{\mathrm{u}}} \ell(-g(x_i;\theta)) - \frac{\pi_p}{n_{\mathrm{p}}} \sum_{i=1}^{n_{\mathrm{p}}} \ell(-g(x_i;\theta)) \right) \quad (4)$$

which is the uPU method [7,8]. [16] points out that the value in the second parenthesis in Eq. (4) would turn negative under more flexible models and proposes nnPU:

$$\widehat{R}_{\mathrm{nnPU}}(g) = \frac{\pi_p}{n_{\mathrm{p}}} \sum_{i=1}^{n_{\mathrm{p}}} \ell(g(x_i;\theta)) + \max \left(0, \frac{1}{n_{\mathrm{u}}} \sum_{i=1}^{n_{\mathrm{u}}} \ell(-g(x_i;\theta)) - \frac{\pi_p}{n_{\mathrm{p}}} \sum_{i=1}^{n_{\mathrm{p}}} \ell(-g(x_i;\theta)) \right). \quad (5)$$

In nnPU, the risk estimator ignores negative estimates of risk by leveraging a max term (the value in the second parenthesis in Eq. (3.1)).

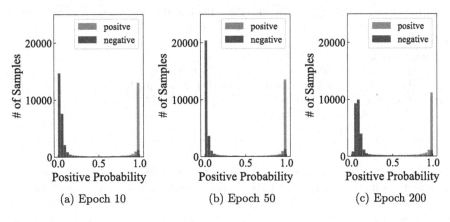

(a) Epoch 10   (b) Epoch 50   (c) Epoch 200

Fig. 1. The nnPU algorithm's positive probability histogram of all unlabeled data (20000 postive samples and 30000 negative) on CIFAR-10. Given the model θ_t and each unlabeled sample x_u^i, we calculate its probability of being positive at epoch t.

3.2   Overfitting of Cost-Sensitive Methods

[1] shows that deep neural networks tend to fit frequent or easy patterns first and difficult patterns later. This memorization effect also applies to cost-sensitive approaches that use more flexible models in PU learning. The risk estimator is trained using deep models and assigns reduced weights to all the unlabeled data. This might cause severe overfitting at the end of the training phase as the model reweights hard unlabeled samples incorrectly. This can also be explained from a noisy label learning perspective. Since all unlabeled samples are considered as negative ones, the positive samples are equivalent to samples with noisy label, which may degrade the performance of cost-sensitive methods.

In Fig. 1, we conduct a preliminary experiment by training nnPU [16] on CIFAR-10 to demonstrate the overfitting problem of cost-sensitive method. At the end of each epoch t, we can compute the output $g(x_u^i; \theta_t)$ for each sample x_u^i in the unlabeled set $\mathbb{D}_u$ and then calculate the probability of x_u^i being positive using a monotonic function (e.g., sigmoid function) f: $p_t(x) = f(g(x_u^i; \theta_t))$. Figure 1 demonstrates the histogram of epoch 10, 50, and 200 for nnPU, where purple denotes true positive data and orange denotes true negative data. From epoch 10 to epoch 50, the model makes some progress in distinguishing positive and negative samples as the distribution of Fig. 1 (b) looks more polarized. However, as the training proceeds, the positive probabilities of most negative samples are increasing, and some of them get surprisingly high positive probabilities, being wrongly classified as positive samples. We can conclude that the distinguishing ability of nnPU peaks at a certain epoch in the middle of the training process but tends to overfit by misclassifying negative data as positive data.

4   Proposed Method

Algorithm 1. PUSP

Require: positive set $\mathbb{D}_p$, unlabeled set $\mathbb{D}_u$, the initial model θ_0, number of epoch T,
    the rate of selected positive data μ, the rate of selected negative data δ.
Ensure: binary classifier θ_T.
  1: **for** $t \leftarrow 1$ to T **do**
  2:    Estimate the confidence $c_t(i)$ of $x_u^i \in \mathbb{D}_u$ using Eq. 8
  3:    Divide $\mathbb{D}_u$ into $\mathbb{X}_+$, $\mathbb{X}_-$, and $\mathbb{X}_u$ using Eq. 9 and Eq. 10
  4:    Calculate the prototypes C_{+1} and C_{-1} using Eq.11
  5:    Assign label to $x_u^i \in \mathbb{X}_u$ using Eq.12
  6:    Update θ using $\nabla_\theta L_{CE}$

In this section, we introduced our method, a simple and effective framework to select confident samples and label unconfident samples for supervision. We design a pipeline to pick up trustworthy samples (both positive and negative) to the labeled set and leave samples with high uncertainty to the remaining unlabeled set. And then we assign labels to these unconfident samples based on the similarity between their deep feature and the prototypes of positive and negative classes.

4.1 Confidence-based Selection

To deal with the overfitting problem of CS method, we carefully select a portion of confident samples in the unlabeled set and assign them hard labels instead of reweighting all the unlabeled samples in previous CS methods. A key challenge in this step is to design a reliable criterion to select confident samples and distinguish them from uncertain ones, so all the confident data can be fully exploited to guide the labeling process of uncertain samples. Inspired by confidence score in noisy label learning [29], we can figure out the confidence of the model prediction on sample x. Given the current model $g(\cdot; \theta_t)$ and an unlabeled example x_u^i, we can calculate the probability of x_u^i being positive at the current epoch t:

$$p_t(Y = +1 \mid x_u^i)) = f(g(x_u^i; \theta_t)), \qquad (6)$$

where f is a monotonic function (e.g., sigmoid function). If the value $p_t(x_u^i)$ is near 1/0, the model is confident that x_u^i is positive/negative. In contrast, samples with values near 0.5 are difficult for the model to classify. However, the confidence calculated by Eq. 6 is based on instantaneous feedback from the model from a single view of the image content x_u^i, which does not consider the training history nor different views by augmentations of each samples. The instantaneous confidence of each sample can be inconstant and change dramatically between consecutive epochs, which is caused by the randomness of stochastic gradient descent. So instantaneous confidence is not stable to represent how confident of the model to each sample in future epochs. This results in a large difference in the confident samples selected between two epochs. Hence, we leverage an exponential moving average (EMA) of the model parameters to take into account the time consistency of the model confidence between epoch t and $t-1$. And we leverage m data augmentations to consider the spatial consistency over different views of samples x_u^i. We define the confidence $c_t(i)$ of sample x_u^i at epoch t as:

$$c_t(i) \triangleq \frac{1}{m} \sum_{j=1}^{m} f\left(g(\mathcal{A}_s(x_u^i); \bar{\theta}_t)\right), \qquad (7)$$

$$\bar{\theta}_t \triangleq \gamma\theta_{t-1} + (1 - \gamma), \bar{\theta}_{t-1} \qquad (8)$$

where $\bar{\theta}_t$ is EMA of the model parameters, $\mathcal{A}_s(\cdot)$ is the strong augmentation, $\gamma \in [0, 1]$ is the rate of EMA. If the model's prediction on x_u^i is confident enough, the confidence $c_t(i)$ of sample x_u^i over time and across augmentations will still be polarizedly distributed between $[0, 1]$. After calculating the confidence of sample x_u^i, we divide the original unlabeled set $\mathbb{D}_u$ into the positive set $\mathbb{X}_+$, the negative set $\mathbb{X}_-$, and the remaining unlabeled set $\mathbb{X}_u$ using two criteria. More specifically, $\mathfrak{I}_{pos}, \mathfrak{I}_{neg} : \mathcal{X} \to \{0, 1\}$ are functions mapping x_u^i to either true or false. $\mathfrak{I}_{pos}$ is defined as:

$$\mathfrak{I}_{pos}\left(x_u^i\right) = \mathbb{I}\left(\sum_{j=1}^{n_u} \mathbb{I}\left(c_t(j) > c_t(i)\right) \leq n_u\mu\right), \qquad (9)$$

where $\sum_{j=1}^{n_u} \mathbb{I}c_t(j)$ counts the number of samples in the original unlabeled set with larger confidence than $c_t(i)$. Equation 9 is true if the count is smaller than or equal the number of unlabeled samples multiply the rate of selected positive data μ. Similarly, $\mathfrak{I}_{neg}$ is defined as:

$$\mathfrak{I}_{neg}\left(x_u^i\right) = \mathbb{I}\left(\sum_{j=1}^{n_u} \mathbb{I}\left(c_t(i) > c_t(j)\right) \leq n_u\delta\right), \tag{10}$$

where δ is rate of selected negative data. For each sample x_u^i in the original unlabeled set $\mathbb{D}_u$:

- If $\mathfrak{I}_{pos}/\mathfrak{I}_{neg}$ is true/false, x_u^i belongs to $\mathbb{X}_+$, which will be labeled as positive samples.
- If $\mathfrak{I}_{pos}/\mathfrak{I}_{neg}$ is false/true, x_u^i belongs to $\mathbb{X}_-$, which will be labeled as negative samples.
- If $\mathfrak{I}_{pos}/\mathfrak{I}_{neg}$ is false/false, x_u^i belongs to remaining unlabeled set $\mathbb{X}_u$, which will remain as unlabeled samples.

Compared to previous CS methods that requires to estimate the class prior first in order to calculate the unbiased risk estimator, our method selects confident samples and assigns them hard label, treating them as standard supervised learning samples.

4.2 Similarity-based Prototypical Labeling

After we construct the positive and the negative set, we aim to obtain a correct label for each image in the remaining unlabeled set by leveraging the prototypes generated from the confident samples. Inspired by the class prototypes [35,37], we use the model $g(\cdot; \theta_t)$ at the current epoch t to extract deep features of images in the positive set $\mathbb{X}_+$ and the negative set $\mathbb{X}_-$. We regard the output before the last fully connected layer of the model as the deep features, denoted as $h(\cdot; \phi_t)$. We define a prototype to represent the positive and negative samples in $\mathbb{X}_+$ and $\mathbb{X}_-$. For each class, the prototype is the mean vector of the deep features of all the samples in the confident set, i.e.,

$$C_{+1} = \frac{1}{n_u\mu} \sum_{x_i \in \mathbb{X}_+} h(x^i; \phi_t), C_{-1} = \frac{1}{n_u\delta} \sum_{x_j \in \mathbb{X}_-} h(x^j; \phi_t), \tag{11}$$

where C_{+1} is the prototype for the positive samples, and C_{-1} is the prototype for negative samples. Then we calculate the cosine similarity between the prototypes and each sample x_u^i in the remaining unlabeled set $\mathbb{X}_u$, and simply predict the label of x_u^i as follows:

$$\hat{y} = \operatorname{argmin}_k \frac{h(x^j; \phi_t)^T C_k}{\|h(x^j; \phi_t)\|_2 \|C_k\|_2}, \quad k \in \{-1, +1\} \tag{12}$$

Prototypes are computed based on the features of samples in the confident sets, so the class-relevant information and semantic knowledge from the selected high-confidence sample can guiding the labeling process of the remaining unlabeled samples. Since all the samples are labeled, we can calculate the cross entropy loss:

$$L_{CE} = \sum_{(x,y) \in \mathbb{X}_+ \cup \mathbb{X}_- \cup \mathbb{X}_u} \log f(g(x; \theta_t)) \mathbb{I}_{y=1} + \log (1 - f(g(x; \theta_t)) \mathbb{I}_{y=-1} \quad (13)$$

5 Experiment

In this section, we experimentally investigate the effectiveness of our method. First, we introduce the benchmark datasets and experiment setup commonly used for the evaluation of PU learning. Then, we demonstrate the experimental results, including the comparison of performance, parameter study, and ablation study.

5.1 Datasets

We evaluate the proposed method on four widely-used benchmark datasets for PU learning. The datasets include three computer vision datasets (MNIST [20], F-MNIST [41], and CIFAR-10 [17]) and two natural language processing datasets (20NEWS [19] and IMDb [26]). The detailed descriptions are given as follows:

- **MNIST.** MNIST is a 28 × 28 gray-scale image dataset of handwritten digits. It has a training set of 60,000 examples and a test set of 10,000 examples. Following [16], we set odd numbers (1, 3, 5, 7, and 9) as the positive class and even numbers (0, 2, 4, 6, and 8) as the negative class. In this case, the class prior probability is $\pi_p = 0.49$.
- **F-MNIST.** F-MNIST is a dataset of 28 × 28 gray-scale images of 10 categories of clothing, which consists of a training set of 60,000 examples and a test set of 10,000 examples. Following [4], we set the classes 'trouser', 'coat', and 'sneaker' as the positive class, and other classes as the negative class. In this case, the class prior probability is $\pi_p = 0.30$.
- **CIFAR-10-V.** CIFAR-10 is a 32 × 32 RGB image dataset with 50,000 training examples and 10,000 testing examples. Following [16], we define 'airplane', 'automobile', 'ship', and 'truck' as the positive class and 'bird', 'cat', 'deer', 'dog', 'frog', and 'horse' as the negative class in order to distinguish the vehicles from the animals. In this case, the class prior probability is $\pi_p = 0.40$.
- **CIFAR-10-M.** Following [13], we define 'cat', 'deer', 'dog', and 'horse' as the positive class and 'bird', 'frog', 'airplane', 'automobile', 'ship', and 'truck' as the negative class in order to distinguish the mammals from the non-mammals. In this case, the class prior probability is $\pi_p = 0.40$.
- **20NEWS.** The 20NEWS dataset is a text dataset with approximately 20,000 newsgroup documents, partitioned evenly across 20 different topics. The latest version has a predefined split of 11,314 train and 7,532 test documents. Following [16], we set 'alt.', 'comp.', 'misc.', and 'rec.' as the positive class and 'sci.', 'soc.', and 'talk.' as the negative class, such that the class prior probability is $\pi_p = 0.56$.

- **IMDb**. The IMDB Large Movie Reviews dataset is a text dataset with 50,000 highly polarized (positive and negative) movie reviews from the Internet Movie Database, which are split evenly into train and test sets. We set positive reviews as positive samples and negative reviews as negative samples. In this case, the class prior probability is $\pi_p = 0.50$.

5.2 Experimental Setup

In this section, we introduce the settings of our experiments, including comparing algorithms, network architecture, and computing infrastructures.

Comparing Algorithms. We compare our method with five state-of-the-art algorithms in PU learning: (1) **uPU** [8], which proposes to use the ramp loss function as a proxy to minimize 0–1 loss. (2) **nnPU** [16], which builds on uPU and formulates a new risk objective that is negatively bounded. (3) **self-PU** [5], which leverages a self-paced sample selection to choose a set of trusted samples progressively. (4) **VPU** [4], which uses a variational bound on the divergence between the true and estimated positive data distribution. (5) **PUbN(\N)** [13], which assigns more accurate weights to each example to evaluate the classification risk. (6) **PUSB** [15], which partially identifies the binary classifier by learning a scoring function. (7) **PULNS** [25], which selects negative sample by reinforcement learning [43].

Table 1. **Test accuracy** (%) of PUSP and baselines across different datasets. Num. Pos. means the number of positive (labeled) samples used for training. The number in the parenthesis is the standard deviation for five repeated runs.

Dataset	MNIST		F-MNIST		CIFAR-10-V		CIFAR-10-M		20NEWS	IMDb
Num. Pos	1000	3000	1000	3000	1000	3000	1000	3000	800	1000
uPU	51.64	59.42	73.79	77.06	60.21	65.97	51.91	54.60	48.43	65.01
	(±0.16)	(±0.63)	(±0.48)	(±0.24)	(±0.09)	(±0.71)	(±0.55)	(±0.11)	(±0.36)	(±0.79)
nnPU	81.02	84.10	87.51	88.46	86.73	88.97	77.08	78.94	87.34	82.46
	(±0.81)	(±0.58)	(±0.45)	(±0.31)	(±0.56)	(±0.31)	(±0.89)	(±0.54)	(±0.32)	(±1.04)
self-PU	90.27	92.59	90.72	91.83	89.67	90.86	77.69	79.05	84.54	81.79
	(±0.80)	(±0.31)	(±0.56)	(±0.51)	(±0.47)	(±0.82)	(±0.14)	(±0.12)	(±0.44)	(±0.60)
VPU	91.54	93.17	89.81	91.06	89.47	90.76	79.10	80.59	85.36	82.71
	(±0.19)	(±0.23)	(±0.39)	(±0.52)	(±0.54)	(±0.21)	(±0.73)	(±0.66)	(±0.27)	(±0.29)
PUbN(\N)	92.83	93.57	91.40	92.18	89.23	90.07	78.50	79.67	86.61	83.07
	(±0.24)	(±0.12)	(±0.37)	(±0.16)	(±0.17)	(±0.17)	(±0.52)	(±0.61)	(±0.61)	(±0.72)
PUSB	89.09	92.55	90.57	92.09	87.90	89.63	78.66	79.61	86.57	82.69
	(±0.34)	(±0.70)	(±0.12)	(±0.76)	(±0.45)	(±0.33)	(±0.96)	(±0.52)	(±0.19)	(±0.60)
PULNS	92.79	94.05	91.56	92.88	90.63	91.09	80.37	81.08	89.29	83.70
	(±1.07)	(±0.67)	(±1.14)	(±0.87)	(±0.69)	(±0.54)	(±0.58)	(±1.09)	(±1.26)	(±0.82)
PUSP (ours)	**95.63**	**96.52**	**92.64**	**94.90**	**91.93**	**93.61**	**81.09**	**82.40**	**90.62**	**85.06**
	(±0.24)	(±0.49)	(±0.20)	(±0.56)	(±0.27)	(±0.56)	(±0.33)	(±0.17)	(±0.46)	(±0.42)

Implementation Details. For MNIST, we follow the same setting of [16]. We use a 6-layer fully-connected *multilayer perceptron* (MLP) with ReLU. The final output $(p(x \mid Y = +1))$ is produced by a Sigmoid function. For F-MNIST, the setting is similar to MNIST, except that we forward the one-dimensional vector to a LeNet-5 network [20]. For CIFAR-10, we use a 13-layer CNN with ReLU following [5]. For 20NEWS, we preprocess the raw text data into 9216-dimensional feature vectors using ELMo [31] following [13]. We use a 3-layer fully connected neural network. For IMDb, we use a bidirectional LSTM network with 64 hidden dimensions and an MLP layer. For all methods, we use Adam optimizer with a cosine annealing learning rate scheduler. The batch size is set to 64. τ_1 and τ_2 are set to 0.25. For computer vision tasks, we use RandAugment [6] for the strong augmentations.

Table 2. Ablation study (%) of PUSP variants with one component removed/changed in all datasets.

Dataset	MNIST	F-MNIST	CIFAR-10-V	CIFAR-10-M	20NEWS	IMDb
Num. Pos	1000	1000	1000	1000	1000	1000
PUSP: original **PUSP**	**95.63**	**92.64**	**91.93**	**81.09**	**90.62**	**85.06**
	(± 0.24)	(± 0.20)	(± 0.27)	(± 0.33)	(± 0.46)	(± 0.42)
PUSP: no EMA	93.09	90.78	90.34	80.41	90.07	84.90
	(± 0.14)	(± 0.35)	(± 0.10)	(± 0.15)	(± 0.76)	(± 0.42)
PUSP: no augmentations	94.64	92.50	88.45	79.08	–	–
	(± 0.09)	(± 0.41)	(± 0.60)	(± 0.58)	–	–
PUSP: no prototypes	89.79	87.56	79.09	70.94	81.16	68.60
	(± 0.23)	(± 0.51)	(± 0.32)	(± 0.56)	(± 0.83)	(± 0.12)
PUSP: ssl	94.08	91.41	89.09	80.46	88.64	82.09
	(± 0.31)	(± 0.45)	(± 0.34)	(± 0.80)	(± 0.24)	(± 0.63)

5.3   Comparison with the State of the Art

We run each method five times on all datasets and reported the test accuracy with mean and standard deviations in Table 1. In most cases, our method outperforms other methods on both computer vision and natural language processing datasets, showing the superior performance among different modalities and tasks. Compared with uPU and nnPU, our method improves the performance by a large margin, indicating that our method can tackle the overfitting problem in tradition CS methods using the confident sample selection method and prototypical labeling method. The second best method among all datasets is PULNS, which shows that reinforcement learning might be a promising to solve PU learning problem.

5.4   Ablation Study

We conduct a thorough ablation study on four variants of our method, each removing/changing one component used in our method. Table 2 reports their results on the all benchmark datasets. Among them, "no EMA" removes the EMA of model parameters by setting $\gamma = 0$; "no augmentations" replaces the strong augmentations with weak augmentations with random crop and random horizontal flip; "no prototypes" removes the similarity-based prototypical labeling part and only trains the model with the confident samples; "ssl" replaces the prototypical labeling part using a semi-supervised learning method called FixMatch [36]. We come to the following conclusions: (1) The performance of both "no EMA" and "no augmentations" is worse than the original version, indicating the EMA of model parameters and the strong augmentations play important roles in consistently selecting confident samples. (2) "no prototypes" shows significant degradation on the test accuracy, indicating that simply discarding the unconfident samples may lead to the the loss information of the associated image contents. (3) "ssl" can leverage the unconfident by fitting them into a semi-supervised learning framework. However, our method still outperforms "ssl" since the two prototypes are specifically designed for PU learning and can generate better representative embeddings for positive and negative samples.

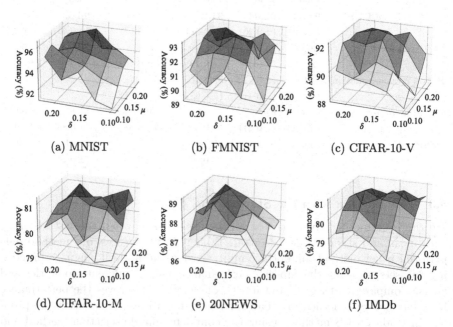

(a) MNIST   (b) FMNIST   (c) CIFAR-10-V

(d) CIFAR-10-M   (e) 20NEWS   (f) IMDb

Fig. 2. Sensitivity analysis of the ratio of selected positive samples μ and the the ratio of selected positive samples δ. A darker color indicates a higher test accuracy.

5.5 Parameter Sensitivity

We investigate the influence of two important hyperparameters regarding confident sample selection, μ which determines the ratio of positive samples in unlabeled set and δ which determines the ratio of negative samples. We set the range of μ and δ to $\{0.10, 0.14, 0.18, 0.22, 0.25\}$ in all datasets and plot the sensitivity in Fig. 2. When the μ and δ are small, the test accuracy is relatively low. This is reasonable because the lack of confident samples may result in unsatisfactory prototypes. With the growth of μ and δ, the performance improvement is significant. When the values reach 0.22, the test accuracy reaches its maximum and slightly decreases as μ and δ continue to increase. The performance trend indicates that a sufficient size of confident samples are significant to the performance since the the quality of prototypes depend on the quality and quantity of the confident samples. However, large μ and δ are harmful since more samples with low confidence will be selected. Consequently, we set the value of μ and δ to 0.22.

5.6 The Distribution of the Positive Probability

One of the advantages of our method compared to CS method is that by carefully selecting confidence samples, we can tackle the overfitting issue. In Fig. 3, we plot the histogram of the positive probability of all the training samples trained by uPU, nnPU and our method in CIFAR-10. For uPU, the positive probability of the negative samples is small while that of the positive ones is scatteredly distributed along the range of $[0, 1]$, showing that uPU can not distinguish positive and negative samples. For nnPU, the distribution of the positive probability is better than uPU since a large portion of the samples are located at the end of the range of $[0, 1]$. For our method, the distribution is more polarized than nnPU, indicating positve/negative samples are well separated.

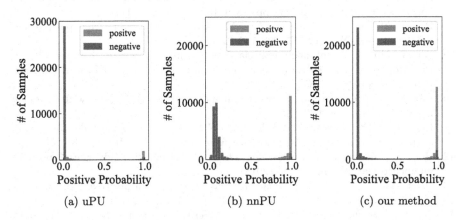

(a) uPU   (b) nnPU   (c) our method

Fig. 3. The distribution of positive probability histogram of all unlabeled data on CIFAR-10 at the last epoch.

6 Conclusion

In this paper, we propose a novel sample selection and prototype learning method to tackle PU learning. First, we carefully select both negative and positive samples from U data based on the time-consistency and spacial consistency of the model prediction from different image views. Secondly, we use prototypes to label the remaining unchosen samples in the unlabeled data. Experimental results on four benchmark datasets demonstrate the effectiveness of our method. Future work will focus on the theoretical analysis of the current method and the combination with contrastive learning [24].

References

1. Arpit, D., et al.: A closer look at memorization in deep networks. In ICML, pp. 233–242. PMLR (2017)
2. Berthelot, D., Carlini, N., Goodfellow, I., Papernot, N., Oliver, A., Raffel, C.: A holistic approach to semi-supervised learning. In NeurIPS, Mixmatch (2019)
3. Chaudhari, S., Shevade, S.: Learning from positive and unlabelled examples using maximum margin clustering. In: Huang, T., Zeng, Z., Li, C., Leung, C.S. (eds.) ICONIP 2012. LNCS, vol. 7665, pp. 465–473. Springer, Heidelberg (2012). https://doi.org/10.1007/978-3-642-34487-9_56
4. Chen, H., Liu, F., Wang, Y., Zhao, L., Wu, H.: A variational approach for learning from positive and unlabeled data. In: NeurIPS (2019)
5. Chen, X., et al.: Self-PU: self boosted and calibrated positive-unlabeled training. In: ICML, pp. 1510–1519. PMLR (2020)
6. Cubuk, E.D., Zoph, B., Shlens, J., Le, Q.V.: Randaugment: practical automated data augmentation with a reduced search space. In: Proceedings of the IEEE/CVF Conference on Computer Vision and Pattern Recognition Workshops, pp. 702–703 (2020)
7. Plessis, M.D., Niu, G., Sugiyama, M.: Convex formulation for learning from positive and unlabeled data. In: ICML, pp. 1386–1394. PMLR (2015)
8. Plessis, M.C.D., Niu, G., Sugiyama, M.: Analysis of learning from positive and unlabeled data. NeurIPS **27**, 703–711 (2014)
9. Fei, G., Liu, B.: Social media text classification under negative covariate shift. In: Proceedings of the 2015 Conference on Empirical Methods in Natural Language Processing, pp. 2347–2356 (2015)
10. Han, B., et al.: Co-teaching: robust training of deep neural networks with extremely noisy labels. In: NeurIPS (2018)
11. Hendrycks, D., Mazeika, M., Wilson, D., Gimpel, K.: Using trusted data to train deep networks on labels corrupted by severe noise. In: NeurIPS (2018)
12. Hou, M., Chaib-Draa, B., Li, C., Zhao, Q.: Generative adversarial positive-unlabelled learning. In: IJCAI (2018)
13. Hsieh, Y.-G., Niu, G., Sugiyama, M.: Classification from positive, unlabeled and biased negative data. In: ICML, pp. 2820–2829. PMLR (2019)
14. Jiang, L., Zhou, Z., Leung, T., Li, L.-J., Fei-Fei, L.: MentorNet: learning data-driven curriculum for very deep neural networks on corrupted labels. In ICML, pp. 2304–2313. PMLR (2018)
15. Kato, M., Teshima, T., Honda, J.: Learning from positive and unlabeled data with a selection bias. In: ICLR (2018)
16. Kiryo, R., Niu, G., Plessis, M.C.D., Sugiyama, M.: Positive-unlabeled learning with non-negative risk estimator. In: NeurIPS (2017)
17. Krizhevsky, A., et al.: Learning multiple layers of features from tiny images (2009)
18. Laine, S., Aila, T.: Temporal ensembling for semi-supervised learning. In: ICLR (2017)

19. Lang, K.: Newsweeder: learning to filter netnews. In: Machine Learning Proceedings 1995, pp. 331–339. Elsevier (1995)
20. LeCun, Y., Bottou, L., Bengio, Y., Haffner, P.: Gradient-based learning applied to document recognition. Proc. IEEE **86**(11), 2278–2324 (1998)
21. Li, B., Han, B., Wang, Z., Jiang, J., Long, G.: Confusable learning for large-class few-shot classification. In: Hutter, F., Kersting, K., Lijffijt, J., Valera, I. (eds.) ECML PKDD 2020. LNCS (LNAI), vol. 12458, pp. 707–723. Springer, Cham (2021). https://doi.org/10.1007/978-3-030-67661-2_42
22. Li, X., Liu, B.: Learning to classify texts using positive and unlabeled data. In: IJCAI, vol. 3, pp. 587–592. CiteSeer (2003)
23. Liu, B., Lee, W.S., Yu, P.S., Li, X.: Partially supervised classification of text documents. In ICML, vol. 2, pp. 387–394 (2002)
24. Liu, Y., Li, Z., Pan, S., Gong, C., Zhou, C., Karypis, G.: Anomaly detection on attributed networks via contrastive self-supervised learning. IEEE Trans. Neural Netw. Learn. Syst. **33**(6), 2378–2392 (2021)
25. Chuan Luo, P., Zhao, C.C., Qiao, B., Chao, D., Zhang, H., Wei, W., Cai, S., He, B., Rajmohan, S., et al.: Puls: Positive-unlabeled learning with effective negative sample selector. Proc. AAAI Conf. Artif. Intell. **35**, 8784–8792 (2021)
26. Maas, A., Daly, R.E., Pham, P.T., Huang, D., Ng, A.Y., Potts, C.: Learning word vectors for sentiment analysis. In: Proceedings of the 49th Annual Meeting of the Association for Computational Linguistics: Human Language Technologies, pp. 142–150 (2011)
27. McMahan, B., Moore, E., Ramage, D., Hampson, S., Arcas, B.A.: Communication-efficient learning of deep networks from decentralized data. In: AISTATS, pp. 1273–1282. PMLR (2017)
28. Niu, G., du Plessis, M.C., Sakai, T., Ma, Y., Sugiyama, M.: Theoretical comparisons of positive-unlabeled learning against positive-negative learning. In: NeurIPS (2016)
29. Northcutt, C., Jiang, L., Chuang, I.: Confident learning: estimating uncertainty in dataset labels. J. Artif. Intell. Res. **70**, 1373–1411 (2021)
30. Patrini, G., Rozza, A., Menon, A.K., Nock, R., Qu, L.: Making deep neural networks robust to label noise: a loss correction approach. In CVPR, pp. 1944–1952 (2017)
31. Peters, M.E., et al.: Knowledge enhanced contextual word representations. In: EMNLP-IJCNLP (2019)
32. Ramaswamy, H., Scott, C., Tewari, A.: Mixture proportion estimation via kernel embeddings of distributions. In ICML, pp. 2052–2060. PMLR (2016)
33. Ren, Y., Ji, D., Zhang, H.: Positive unlabeled learning for deceptive reviews detection. In EMNLP, pp. 488–498 (2014)
34. Schnabel, T., Swaminathan, A., Singh, A., Chandak, N., Joachims, T.: Recommendations as treatments: debiasing learning and evaluation. In: International Conference on Machine Learning, pp. 1670–1679. PMLR (2016)
35. Snell, J., Swersky, K., Zemel, R.S.: Prototypical networks for few-shot learning. In: NeurIPS (2017)
36. Sohn, K., et al.: Simplifying semi-supervised learning with consistency and confidence. In: NeurIPS, Fixmatch (2020)
37. Tan, Y., et al.: Fedproto: federated prototype learning across heterogeneous clients. In AAAI Conference on Artificial Intelligence, vol. 1, p. 3 (2022)
38. Tarvainen, A., Valpola, H.: Mean teachers are better role models: weight-averaged consistency targets improve semi-supervised deep learning results. In: NeurIPS (2017)

39. Wang, Z., et al.: SemiNLL: a framework of noisy-label learning by semi-supervised learning. in: Transactions on Machine Learning Research (2022)
40. Wang, Z., Zhou, T., Long, G., Han, B., Jiang, J.: FedNOiL: a simple two-level sampling method for federated learning with noisy labels. arXiv preprint arXiv:2205.10110 (2022)
41. Xiao, H., Rasul, K., Vollgraf, R.: Fashion-MNIST: a novel image dataset for benchmarking machine learning algorithms. arXiv preprint arXiv:1708.07747 (2017)
42. Xu, J., Chen, Z., Quek, T.Q.S., Chong, K.F.E.: FedCorr: multi-stage federated learning for label noise correction. In: Proceedings of the IEEE/CVF Conference on Computer Vision and Pattern Recognition, pp. 10184–10193 (2022)
43. Yang, Y., Jiang, J., Wang, Z., Duan, Q., Shi, Y.: BiES: adaptive policy optimization for model-based offline reinforcement learning. In: Long, G., Yu, X., Wang, S. (eds.) AI 2022. LNCS (LNAI), vol. 13151, pp. 570–581. Springer, Cham (2022). https://doi.org/10.1007/978-3-030-97546-3_46
44. Zhang, B., Zuo, W.: Reliable negative extracting based on KNN for learning from positive and unlabeled examples. J. Comput. 4(1), 94–101 (2009)
45. Zhang, H., Cisse, M., Dauphin, Y.N., Lopez-Paz, D.: mixup: beyond empirical risk minimization. In: ICLR (2018)

Handling Missing Data with Markov Boundary

Azhar Mohammed$^{(\boxtimes)}$, Dang Nguyen, Bao Duong, Melanie Nichols, and Thin Nguyen

Applied Artificial Intelligence Institute (A^{2}I^2), Deakin University, Victoria, Australia
{mohammedaz,d.nguyen,duongng,melanie.nichols,thin.nguyen}@deakin.edu.au

Abstract. In machine learning (ML) applications, high-quality data are very important to train a well-performed model that can provide robust predictions and responsible decisions. A common problem in ML applications e.g., healthcare is that the training dataset often consists of samples (or records) with missing values. As a result, the ML model cannot use such samples in its training phase. Handling missing data is thus an important and open research problem. In this paper, we propose a method to predict missing values by considering a *causal graphical model* framework. Our method exploits the Markov boundary encapsulating all necessary information about the missing variables. By utilizing the information encoded in the Markov boundary, we formulate a predictive function for each feature that has missing values to predict its missing values. Compared to existing methods, our predictive function is trained with only the features involved in the Markov boundary. To demonstrate the effectiveness of our proposed method, we compare its imputation performance with those of state-of-the-art imputation methods via a comprehensive experiment on seven real-world datasets. Our empirical results highlight that our method is significantly better than those of the baselines in terms of the imputation error thanks to its Markov information.

Keywords: Data imputation · Causal graphical model · Markov boundary

1 Introduction

Missing data can significantly impact on the overall data quality, and may adversely influence study results, interpretations, and conclusions [1]. Especially in ML applications, training a ML model with missing data is one of the biggest challenges. In supervised learning, to train classifiers effectively, the training set should be complete without missing values. However, the training samples can contain missing values or not recorded due to the measurement error during the real-time experiments. For example, electronic health records (EHR) are often used these days to collect a large amounts of digital data on a daily basis in

W. Chen et al. (Eds.): ADMA 2022, LNAI 13725, pp. 319–333, 2022.
https://doi.org/10.1007/978-3-031-22064-7_24

therapy sessions, clinics, and hospitals [3,6,46]. The data recorded is mostly in a questionnaire format. However, sometimes patients are unwilling or unable to answer some questions, or medical and nursing staff may not record all the relevant information, which can result in an incomplete dataset [4,43]. Another major problem in e-health is that the dropout rate is relatively higher, which can lead to bias and inconclusive results (i.e., findings) [4,51]. On the other hand, missing data controls the pipeline of that valid tests in many research areas. Missing data also hinders decision-making, data-mining for large scale healthcare related cases and sensor networks in internet of things devices [5]. Especially in risky policy making, the concerns are higher when missing values are responsible for triggering a raise in the bias as well as affecting the efficiency of the analysis [30]. Even few missing samples in the features can translate into a big problem especially when interpreting crucial information, as it can cause a big slump in statistical power of a study and give rise to biased estimates. This may lead to conclusions, which are biased and impact translations of findings in clinical, and medical findings [21].

With these concerns, researchers in universities (i.e., academia) as well as businesses deal with missing data in different ways. To ensure valid and generalisable results, it is necessary to clearly understand how missing data can be optimally identified, handled, and imputed. Existing methods for handling missing data can be grouped into three approaches. The first approach is to replace all the "null" values i.e., missing values with zeros called *zero-imputation* [7]. The second approach is to discard the records (samples) with missing values entirely i.e., the samples with missing values are ignored, and the analysis are conducted only on samples with the complete information [18]. The final approach is to replace missing values with a predicted value derived by either using statistical computational methods or predictive learning functions [14]. The first two approaches are simple to implement but are heavily inefficient, especially when a substantial proportion of the data is missing [20]. Furthermore, when the imputation is poorly performed, it can lead to biases [21]. Replacing missing values with zeros or deleting the records with missing values is applicable only when a small chunk of data is missing from a large dataset, as few samples will not affect the overall data quality. However, as datasets frequently contain relatively large portion of missing data, the imputation is a challenging task. Recent advancements in state-of-the-art deep learning methods have created new paths for handling the missing data. Existing imputation methods based on deep learning often use generative models e.g., autoencoders [49], to replace/impute missing values. While most of the above methods are able to complete missing values, they fail to understand the causal relationships involved with respect to each observation's other features [34].

We propose a novel method for data imputation using the Markov boundary. Instead of using all features, our method only use a minimum number of features to impute missing values. By doing this, we avoid unnecessary noises incurred by many features during the imputation process. We initially obtain a graphical causal model of the dataset to establish the direction of the flow of information in regards with each feature's cause and effect relationships. For example, in a cohort study the feature "pregnant" often carries the important information. This feature can have missing values due to the unwillingness to disclose information. To impute the missing values for the feature "pregnant", most statistical methods replace the missing values with random values or the values computed based on the feature's statistic. Some methods use the nearest distance to calculate the likelihood of the missing values. However, these methods do not consider the causal aspects to handle the missing values. As stated in the example, if the pregnancy feature has missing values, then its causal aspect i.e., data regarding the "sex" of the participant which has the most important and relevant information for imputing or predicting missing values in the pregnancy feature. Our graphical causal discovery based method allows us to eliminate unnecessary noises in the imputed values by removing features that do not carry any useful information. Thus, it reduces the biased and spurious findings the study. The main difference between an existing predictive models involving all the features and our method involving only causal features is represented in the Fig. 1.

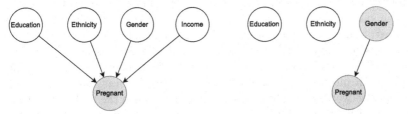

(a) Exiting method using all the features to impute missing values in "Pregnant".

(b) Our method using only causal features to impute missing values in "Pregnant".

Fig. 1. Main difference between existing methods and our method in imputing the missing values.

Our Contributions. To summarize, we make the following contributions:

1. We propose a new method for handling missing data using Markov boundary. Our method trains a predictive function to predict missing values in a feature using only its causal features i.e., the most important and relevant features.[1]
2. We develop an efficient framework to predict missing values, which consists of two main steps: (1) To obtain a causal model $\mathcal{G}$ of the dataset and (2) train a predictive function for imputing missing values.
3. We demonstrate the benefits of our method on seven real-world datasets, where our method performs better than the well-known statistical imputation methods as well as recent state-of-the-art deep learning methods.

---

[1] Source code and relevant datasets will be made publicly available upon publication.

The rest of the paper is organized as follows. In Sect. 2, we briefly outline the fundamentals of the statistical methods used in the literature for imputing missing data. In Sect. 3, we describe our proposed method involving two steps for obtaining a graphical model and to build a predictive model respectively. Our experimental settings, datasets, and results are presented in Sect. 4, where we evaluate the performance of our method and compare it with six competitive imputation methods. Finally, we conclude our work in Sect. 5.

2   Related Works

The literature related to the imputation of data focuses on presenting novel or improved data imputation methods and comparing them with existing baseline methods. Missing data has been a serious ongoing problem that many data scientists face. Missing data impacts a model's learning ability (i.e., performance) [8,19,29,31]. There are three main strategies that exist to counter this problem in the data are case deletion, imputation of missing values and learning from the non-missing data [10,26,36].

List-wise or case deletion methods directly eliminate complete information of the samples that has a missing value in other attributes. For a model to learn from missing data does not require imputation. This approach may result in serious biases if the ratio of missingness is very high [2]. Also, most machine learning models require optimization to handle missing data as they cannot be directly applied due to missingness in the dataset. Thus, many machine learning models deal with incomplete datasets initially before learning for example, in wireless sensor networks due to miscommunication between the nodes or sensors we might end with missing data [11]. They use basic and simple data imputation methods such as mean imputation or median imputation that imputes the missing data with an average value of that particular attribute/feature [52]. But with such a method, all missing values in a feature are replaced with the same value (i.e., mean).

There are other advanced methods such as kNN, decision tree, MICE (i.e., iterative method) which impute the data by using different strategies [9,27,35,38]. When there are very few missing instances most models tend to replace the missing value with either zero or the mean value of the feature. However, as the missingness increases over multiple features then methods such as kNN imputation and MICE imputation perform extremely well. They are very simple to implement and utilize the non-missing data to learn. Both approaches are accepted by most researchers for their extremely good performance. There are many modified kNN and iterative methods proposed in the past for imputing data [16,22,37,50]. The underlying concept is the same for these modified (improved) methods. For example, the kNN imputation method proposed by Garcia et al. [16] uses a feature-weighted distance metric that selects the k nearest instances considering the input attribute relevance to the target class based on the available information in non-missing data. Similarly, an extension method of MICE named Single Center Imputation from Multiple Chained Equation (SICE) combines both single and multiple imputation methods to predict the missing value using a regression method [22]. More recently, deep learning models have also been used to tackle feature imputation problems. However, these models have important limitations. GAIN and Denoising auto-encoders models (DAE) only use a single observation as input to impute the missing values in the features [17,45,49].

The mentioned methods use feature selection methods to select important features/attributes that are necessary for imputing the data in the feature which has missing values [10,13,15,47]. However, most of the methods use complete instances for imputation, and the performance is seriously limited when the attribute is randomly selected and imputed despite having independent relationships. To address this problem, we have designed a new strategy to impute missing values by considering the causal relationship between every attribute/feature such that the imputed value is more meaningful and interpretable.

3   Framework

3.1   Problem Definition

Let $\mathcal{D} \in \mathbb{R}^{n \times m}$ be a data matrix consisting of n data points and m features and a mask matrix $\mathcal{M} \in \{0,1\}^{n \times m}$ where entries of zeros represent the locations with missing values. The entry at the row i^{th} and the column j^{th} in $\mathcal{D}$ and $\mathcal{M}$ are respectively denoted as $\mathcal{D}_{ij}$ and $\mathcal{M}_{ij}$. With respect to the mask matrix, certain entries in $\mathcal{D}$ are missing i.e., the value of $\mathcal{D}_{ij}$ can only be observed if $\mathcal{M}_{ij} = 1$ and vice versa. Our task is *feature imputation*, where the goal is to predict the missing feature values, i.e., predict value for $\mathcal{D}_{ij}$ when $\mathcal{M}_{ij} = 0$.

3.2   Learning with Markov Boundary

Given a causal graphical model $\mathcal{G}$ [32] with all the edges describing the cause-effect relationships among the features (i.e., there is a directed edge from a cause to each of its effects). Graphical models generally combine qualitative and quantitative components to represent a probability distribution. Such distribution is given over a spectrum of the total number of variables. The qualitative component is represented graphically, or as a network structure of the model derived from the given data, which represents conditional independencies among all the features and determines the influence on the targeted variable. We use a traditional approach where missing data is imputed based on a prediction algorithm. However, in our method we not only use a causal graphical model $\mathcal{G}$ to establish a set of cause-effect relationships among variables, we also employ Markov boundary to capture the minimal and sufficient information (i.e., parents, children, and spouses) of the missing variables in $\mathcal{D}$ that is required for predicting the missing variables. After we obtain $\mathcal{G}$ we find a Markov boundary for each missing variable. This is in contrary with conventional machine learning methods where all variables are used to predict the missing variable. When we want to infer on an interested variable from a set of other variables, usually only a

---

Algorithm 1: The proposed algorithm.

Input: $\mathcal{D}^* = \{x_i, y_i\}_{i=1}^n$: dataset with missing values
Output: $\mathcal{D}^*$: imputed dataset without missing values

1 split $\mathcal{D}^*$ into $\mathcal{D}_c$ that contains completed records and $\mathcal{D}_i$ that contains uncompleted records;
2 learn the causal graphical model $\mathcal{G}$ by using $\mathcal{D}_c$;
3 **for** *each feature* $X_i \in \mathcal{D}^*$ **do**
4   find the Markov boundary $MB(X_i)$ from $\mathcal{G}$;
5   use kNN to impute missing values in all features of $\mathcal{D}^*$ except for X_i;
6   train a regression model $f(\cdot)$ with $MB(X_i)$ as input features and X_i as the target label;
7   use $f(\cdot)$ to predict missing values in X_i;

8 return $\mathcal{D}^*$ with no missing values;

---

subset of variables is enough for prediction, while other variables are redundant or may add biases into the predictions. This subset contains all the background information that is contained in the underlying causal graph $\mathcal{G}$ and is known as the Markov blanket.

Definition 1. (Markov blanket [33]) *Given a causal graph $\mathcal{G}$ with S being the set of variables. A subset S' of S is called a Markov blanket of a variable Y if and only if Y is conditionally independent of all other variables given S':*

$$Y \perp\!\!\!\perp S \backslash S' \mid S'$$

If, moreover, the obtained Markov blanket is minimal, such that, it cannot drop any variable without losing information (i.e., predictive performance), it is called a Markov boundary (MB) [40,41].

Definition 2. (Markov boundary [33]) *The Markov boundary of a variable Y is the Markov blanket S' where no proper subset of S' is a Markov blanket of Y. In a causal graphical model, the Markov boundary of a variable Y is composed of Y's parents, Y's children, and the parents of Y's children (i.e., Y's spouses).*

Building from the Markov boundary framework, our approach has two main steps: (1) learning $\mathcal{G}$ from an incomplete dataset and (2) training a machine learning model to predict the missing values using the MBs obtained from $\mathcal{G}$.

Learning $\mathcal{G}$ from an Incomplete Dataset. To elaborate the method further let us consider an example illustrated in Fig. 2. Let $\mathcal{D}$ be a dataset which has four features X_1, X_2, X_3 and X_4, all of which have null values. With missing values in the dataset, one major challenge that we must face is that it is generally difficult to recover the graphical model $\mathcal{G}$ over the variables if the dataset has missing values as it has incomplete or lost information.

To overcome this issue, we use the data of any standard imputation method (e.g., kNN imputation) to fill up all the missing values temporarily. More specifically, let the temporarily imputed dataset be $\mathcal{D}_t$. Then, when X_1 has missing values, for example, we replace the values of X_1 in $\mathcal{D}_t$ with the original values in $\mathcal{D}$, including the null values. Next, we filter out the incomplete dataset by

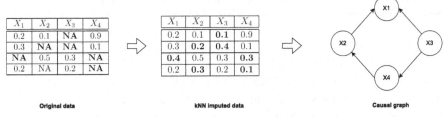

Original data   kNN imputed data   Causal graph

(a) Step one: Obtaining a graphical model (causal graph) of the example dataset

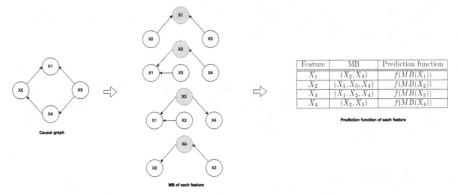

(b) Step two: Obtaining a prediction function from a causal graph of the example dataset using a Markov boundary.

Fig. 2. Demonstration of our method with steps involve in the imputation of the missing values for an example dataset.

eliminating all the rows that have missing values in X_1, so that the remaining dataset, denoted as $\mathcal{D}_c \subset \mathcal{D}$, is complete, i.e., it contains no missing values. Subsequently, with the complete dataset $\mathcal{D}_c$ and a causal discovery algorithm (e.g., PC [39]) we obtain the required graphical model $\mathcal{G}$ as shown in Fig. 2a.

Predicting Missing Values. After obtaining $\mathcal{G}$, we iteratively fill the missing values in each feature. More specifically, from the graph $\mathcal{G}$ we find the Markov boundary of the features of interest, e.g., the Markov boundary of X_1 is $MB(X_1) = \{X_2, X_3\}$ which is also the parents set of X_1 since it has no children nor spouses. Next, we split the data into train and test sets, where the training set is the complete dataset $\mathcal{D}_c$, and the test set is the incomplete dataset $\mathcal{D}_i$. In what follows, we train a machine learning model $f(\cdot)$ on dataset $\mathcal{D}_c$ with the features in the Markov boundary of X_1 being the input, and X_1 being the target variable, to predict the missing values in $\mathcal{D}_i$, i.e., $\hat{X}_1 \approx f(X_2, X_3)$. We then record both the non-missing and the imputed values of X_1 into a dataset $\mathcal{D}^*$, which stores the imputation results of all features. After predicting all the missing values in feature X_1, we repeat the process for all the features of the

Table 1. Characteristics of seven tabular datasets. We denote n: the number of samples, x: the number of features, y: target label and m_r: missing ratio (i.e., percentage).

Dataset	n	x	y	$m_r = 5\%$	$m_r = 10\%$	$m_r = 15\%$
HTRU_2	17,898	9	Class	8,054	16,108	24,162
concrete	1,030	9	Concrete strength	463	927	1,390
wine	4,898	12	Quality	2,938	5,877	8,816
cpu_small	8,192	13	Run-time	5,324	10,649	15,974
protein	45,730	10	RMSD	22,865	45,730	68,595
housing	507	14	Price	354	709	1,064
parkinsons_updrs	5,875	22	UPDRS scores	6,462	12,925	19,387

original dataset $\mathcal{D}$ that have missing values, and replace them with their respective values obtained from the predictive functions as shown in the Fig. 2b. The results for each feature are stored in the dataset $\mathcal{D}^*$. To summarize, Algorithm 1 highlights the main steps of our method.

4 Experiments

In this section we explain the experimental results where we conduct extensive numerical evaluations on seven real-world datasets to assess the performance (in terms of imputation error) of our method and compare it with six strong baselines.

4.1 Datasets

To evaluate our method, we employ seven real-world datasets where the causal connections between them are available. These datasets are commonly used for the evaluation of data imputation methods.

Table 1 shows characteristics of each dataset along with the selected treatment feature and the respective outcome.

The datasets can be briefly described as follows:

– *HTRU_2*: this data consists of pulsar candidates which were collected during high time resolution universe survey (HTRU). They originate from a rare type of Neutron star which produce radio-emission that are detectable from the earth's surface. The data set contains 16,259 examples (spurious) caused by noise, and 1,639 real pulsar samples. These samples have all been cross checked by human annotators [12].
– *Concrete*: in the stream of engineering, it is important to have accurate estimates of the quality i.e., the performance of the building materials. Estimating the strength of concrete is a challenge of particular interest in order to develop

safety guidelines governing the materials used in the construction of building, bridges, and roadways. This dataset is known as concrete comprehensive strength which is a highly non-linear function of ingredients used in making concrete and its ingredients. This dataset has nearly 1,030 examples of cement with eight features describing the components used in the mixture [48].

- *Protein*: physicochemical properties of protein tertiary structure is a dataset describing molecular structure of protein. The data is obtained from CASP 5–9. There are 45,730 decoys and size varying from 0 to 21 Armstrong. This data consists of nine features including molecular mass, non-polar residue and average deviation from standard exposed area of residue [12].

- *Housing*: this dataset contains pricing of houses in Boston, Massachusetts and surrounding suburbs during the year 1978. It has 507 samples and fourteen features with both categorical and numerical format which includes information regarding the number of bedrooms, bathrooms, square feet area and suburb's postcode which are integral part for predicting house prices [44].

- *Cpu_small*: this dataset is a collection of activity measures of a computer from a Sun Sparcstation running in the university department. The data is recorded whenever the users perform the different tasks ranging from accessing the web, editing the filed online as well as running very CPU-bound programs such that the recorded data varies from each task [44].

- *Wine:* This data is the results of an analysis (i.e., chemical) of the white wines grown in the same region of Europe (especially in Italy) but derived from three varieties of cultivars. The analysis determines the quantity of 13 constituents found in each of the three types of wines namely red, white, and sparkling. This dataset is used to determine and rank the quality of the white wine [12].

- *Parkinsons_updrs:* this dataset is created by Max Little and Athanasios Tsanas of the Oxford university, including collaborations with ten US and Intel corporation's medical centers responsible for developing monitoring device for speech signals. This dataset consists of 5,875 samples with twenty-six features that are helpful for regression method, i.e., to predict clinician's Parkinson's disease symptom score on the UPDRS scale [25].

All the datasets have no missing values. Therefore, to simulate the missing values setting, we inject the null values with respect to a given missing rate by randomly removing values in the $\mathcal{D}$. Following [24], the feature values of all the datasets are scaled uniformly to the range $[0, 1]$ before data is split into train and test.

4.2   Baselines and Evaluation Method

We compare our model against five popular imputation methods, as well as compare it with a state-of-the-art deep-learning imputation approach:

- Mean imputation (Mean) [23]: This method imputes the missing values in dataset $\mathcal{D}$ with the mean value of each feature.

- Iterative SVD (SVD) [42]: The singular value decomposition (SVD) is a method for factorizing an arbitrary non-square matrix as the product of a diagonal matrix and two orthogonal matrices, which is mainly used for approximating matrices with lower-rank components. This method imputes missing values by finding the low-rank components of the feature matrix that minimizes the reconstruction error over non-missing values.
- Spectral regularization algorithm (Spectral) [28]: Spectral regularization belongs to a class of regularization methods that are used to prevent over-fitting problems and to control the impact of noise. Soft-impute iteratively replaces the missing data with those obtained from a soft-thresholded SVD while this matrix completion model uses the nuclear norm as a regularizer.
- K-nearest neighbors (kNN) [16]: This method calculates the Euclidean distance and the weights between the non-missing samples, and the missing values in $\mathcal{D}$ are replaced with the nearest neighbors that have observed values. This method is very efficient but takes longer time than rest of the methods.
- Multivariate imputation by chained equations (MICE) [38]: It is a robust and informative method of dealing with missing data in datasets. This method runs iterative series of regression models for imputing or filling missing data. In each iteration, each targeted feature in the dataset is imputed using the other features in the dataset. These iterations will keep running until the model converge.
- Generative Adversarial Imputation Nets (GAIN) [49]: This is a state-of-the-art deep generative adversarial neural network approach. The model consists of a generator and discriminator, where generator observes few components of a real data vector and imputes the missing components, whereas the discriminator takes a completed vector and tries to differentiate between the components that are actually imputed and observed.

Additionally, for all baselines we use the best performing hyper-parameters.

Regarding the evaluation process, we compare the imputation error of our model against all the other baselines as following. Given an arbitrary complete dataset $\mathcal{D} \in \mathbb{R}^{n \times m}$, we generate randomly a mask matrix $\mathcal{M} \in \{0,1\}^{n \times m}$ with different missing rates $m_r \in \{5\%, 10\%, 15\%\}$. Next, we perform 5-fold cross-validation for each combination of method, dataset, and missing rate. For the evaluation criterion, we use the mean absolute error (MAE) over the missing entries as the performance metric:

$$MAE_{\text{imputation}} = \frac{1}{\sum_{i,j} \mathcal{M}_{ij}} \sum_{i,j}^{n,m} \mathcal{M}_{ij} \times \left| \mathcal{D}_{ij} - \hat{\mathcal{D}}_{ij} \right|$$

4.3   Results

The complete results of seven datasets of our method against all the baselines are shown in Fig. 3. Initially, when the missing rate is five percent as seen in the Sub-Fig. 3a, we can observe that our model outperforms all the baselines on all

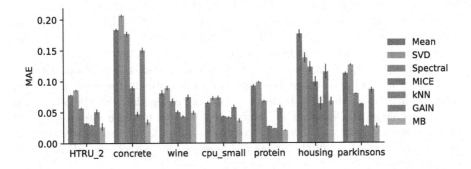

(a) Missing rate $m_r = 5\%$.

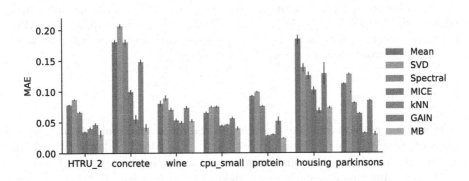

(b) Missing rate $m_r = 10\%$.

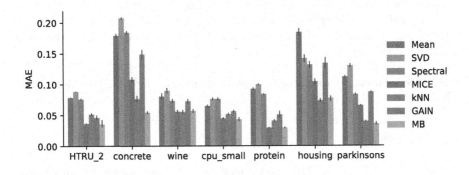

(c) Missing rate $m_r = 15\%$.

Fig. 3. Performance of our model when compared with the baselines in terms of mean absolute error. The three graphs are with respect to the injected missing rates m_r in all the seven real-world datasets.

the datasets except for the *wine quality* and the *housing* dataset. However, as the missing rate in the *wine quality* dataset increases, the MAE of the rest of the baselines also increases except for our method which remains robust to the changes in the percentage of missing rate.

When m_r is 10%, the *protein* dataset's error in all the methods slightly increase by a small percentage as shown in Sub-Fig. 3b. On the other hand, our model still remains the best performing model among the powerful statistical baselines such as MICE and kNN. Moreover, when we increase the missing rate even further, i.e., m_r to 15% as displayed in Sub-Fig. 3c, we can clearly see that despite the increase in all the errors of baselines, our model generalizes better over all datasets.

Overall, we can conclude that our model's performance is generally better compared to all the baselines in all the three missing rate settings, as shown in Fig. 3. In addition, while most statistical methods often require adjustment or optimization of hyper-parameter accordingly due to significant difference between characteristics of the data, especially the deep learning approach GAIN, our model shows significantly higher performance without the need for optimizing the hyper-parameters like epoch, learning rate, etc.

5   Conclusion

In this paper, we have introduced a new method for handling missing data using Markov boundary. Our method involves finding the Markov boundary for each feature with missing variable and learning a predictive model to predict the missing values using the information contained in the respective Markov boundary. To demonstrate the efficacy of our model, we compare it with several baselines that are often used in statistical data imputation methods. The obtained results on seven standard datasets show that our model outperforms all the competitive baselines and is more robust in constructing the original dataset. We hope our work will open a new path on handling missing data problems with Markov information such that the imputed missing value is more meaningful than just a statistical prediction.

Acknowledgement. This research is partly supported by NHMRC Ideas Grant GNT2002234.

References

1. Alamoodi, A.H., et al.: Machine learning-based imputation soft computing approach for large missing scale and non-reference data imputation. Chaos Solitons Fractals **151**, 111236 (2021)
2. Allison, P.D.: Missing data. Sage Publications (2001)
3. Armijo-Olivo, S., Warren, S., Magee, D.: Intention to treat analysis, compliance, drop-outs and how to deal with missing data in clinical research: a review. Phys. Therapy Rev. **14**(1), 36–49 (2009)

4. Blankers, M., et al.: Missing data approaches in eHealth research: simulation study and a tutorial for nonmathematically inclined researchers. J. Med. Internet Res. **12**(5), e54 (2010)
5. Chowdhury, M.H., Islam, M.K., Khan, S.I.: Imputation of missing healthcare data. In: Proceedings of the International Conference of Computer and Information Technology. IEEE (2017)
6. Cismondi, F., Fialho, A.S., Vieira, S.M., Reti, S.R., Sousa, J.M.C., Finkelstein, S.N.: Missing data in medical databases: impute, delete or classify? Artif. Intell. Med. **58**(1), 63–72 (2013)
7. De Souto, M.C.P., Jaskowiak, P.A., Costa, I.G.: Impact of missing data imputation methods on gene expression clustering and classification. BMC Bioinform. **16**(1), 1–9 (2015)
8. Rupam Deb and Alan Wee-Chung Liew: Missing value imputation for the analysis of incomplete traffic accident data. Inf. Sci. **339**, 274–289 (2016)
9. Dempster, A.P., Laird, N.M., Rubin, D.B.: Maximum likelihood from incomplete data via the EM algorithm. J. Royal Statist. Soc. **39**(1), 1–22 (1977)
10. Doquire, G., Verleysen, M.: Feature selection with missing data using mutual information estimators. Neurocomputing **90**, 3–11 (2012)
11. Jinghan, D., Minghua, H., Zhang, W.: Missing data problem in the monitoring system: a review. IEEE Sens. J. **20**(23), 13984–13998 (2020)
12. Dua, D., Graff, C.: UCI machine learning repository (2017)
13. Dzulkalnine, M.F., Sallehuddin, R.: Missing data imputation with fuzzy feature selection for diabetes dataset. SN Appl. Sci. **1**(4), 1–12 (2019). https://doi.org/10.1007/s42452-019-0383-x
14. Emmanuel, T., Maupong, T., Mpoeleng, D., Semong, T., Mphago, B., Tabona, O.: A survey on missing data in machine learning. J. Big Data **8**(1), 1–37 (2021). https://doi.org/10.1186/s40537-021-00516-9
15. García-Laencina, P.J., Sancho-Gómez, J.-L., Figueiras-Vidal, A.R.: Pattern classification with missing data: a review. Neural Comput. Appl. **19**(2), 263–282 (2010)
16. García-Laencina, P.J., Sancho-Gómez, J.-L., Figueiras-Vidal, A.R., Verleysen, M.: K nearest neighbours with mutual information for simultaneous classification and missing data imputation. Neurocomputing **72**(7–9), 1483–1493 (2009)
17. Gondara, L., Wang, K.: MIDA: multiple imputation using denoising autoencoders. In: Phung, D., Tseng, V.S., Webb, G.I., Ho, B., Ganji, M., Rashidi, L. (eds.) PAKDD 2018. LNCS (LNAI), vol. 10939, pp. 260–272. Springer, Cham (2018). https://doi.org/10.1007/978-3-319-93040-4_21
18. Haitovsky, Y.: Missing data in regression analysis. J. Roy. Stat. Soc.: Ser. B (Methodol.) **30**(1), 67–82 (1968)
19. Huang, J., et al.: Cross-validation based K nearest neighbor imputation for software quality datasets: an empirical study. J. Syst. Softw. **132**, 226–252 (2017)
20. Huisman, M.: Imputation of missing network data: some simple procedures. J. Soc. Struct. **10**(1), 1–29 (2009)
21. Jäger, S., Allhorn, A., Bießmann, F.: A benchmark for data imputation methods. Front. Big Data **4**, 693674 (2021)
22. Khan, S.I., Hoque, A.S.M.L.: SICE: an improved missing data imputation technique. J. Big Data **7**(1), 1–21 (2020)
23. Landerman, L.R., Land, K.C., Pieper, C.F.: An empirical evaluation of the predictive mean matching method for imputing missing values. Soc. Methods Res. **26**(1), 3–33 (1997)
24. Leskovec, J., Rajaraman, A., Ullman, J.D.: Mining of massive data sets. Cambridge University Press (2020)

25. Little, M., McSharry, P., Hunter, E., Spielman, J., Ramig, L.: Suitability of dyspho-nia measurements for telemonitoring of Parkinson's disease. In: Nature Preceedings (2008)

26. Luong, P., Nguyen, D., Gupta, S., Rana, S., Venkatesh, S.: Bayesian optimization with missing inputs. In: Hutter, F., Kersting, K., Lijffijt, J., Valera, I. (eds.) ECML PKDD 2020. LNCS (LNAI), vol. 12458, pp. 691–706. Springer, Cham (2021). https://doi.org/10.1007/978-3-030-67661-2_41

27. Malarvizhi, R., Thanamani, A.S.: K-nearest neighbor in missing data imputation. Int. J. Eng. Res. Develop. **5**(1), 5–7 (2012)

28. Mazumder, R., Hastie, T., Tibshirani, R.: Spectral regularization algorithms for learning large incomplete matrices. J. Mach. Learn. Res. **11**, 2287–2322 (2010)

29. Nassiri, V., Molenberghs, G., Verbeke, G., Barbosa-Breda, J.: Iterative multiple imputation: a framework to determine the number of imputed datasets. Am. Stat. **74**(2), 125–136 (2020)

30. Okafor, N.U., Delaney, D.T.: Missing data imputation on IoT sensor networks: implications for on-site sensor calibration. IEEE Sensors J. **21**(20), 22833–22845 (2021)

31. Pan, L., Li, J., et al.: K-nearest neighbor based missing data estimation algorithm in wireless sensor networks. Wirel. Sens. Netw. **2**(02), 115 (2010)

32. Pearl, J.: Causality: Models. Cambridge University Press, Reasoning and Inference (2009)

33. Pearl, J.: Probabilistic reasoning in intelligent systems: networks of plausible infer-ence. Elsevier (2014)

34. Pigott, T.D.: The Handbook of Research Synthesis and Meta-Analysis, vol. 2, chapter Handling missing data, pp. 399–416. Russell Sage Foundation (2009)

35. Rahman, G., Islam, Z.: A decision tree-based missing value imputation technique for data pre-processing. In: Proceedings of the Australasian Data Mining Confer-ence (2011)

36. Ramoni, M., Sebastiani, P.: Robust learning with missing data. Mach. Learn. **45**(2), 147–170 (2001)

37. Rani, P., Kumar, R., Jain, A.: HIOC: a hybrid imputation method to predict missing values in medical datasets. International J. Intell. Comput. Cybernetics (2021). https://doi.org/10.1108/IJICC-03-2021-0042

38. Royston, P., White, I.R.: Multiple imputation by chained equations (MICE): imple-mentation in Stata. J. Stat. Softw. **45**, 1–20 (2011)

39. Spirtes, P., Glymour, C.N., Scheines, R., Heckerman, D.: Causation, prediction, and search. MIT Press (2000)

40. Statnikov, A., Lemeir, J., Aliferis, C.F.: Algorithms for discovery of multiple Markov boundaries. J. Mach. Learn. Res. **14**(1), 499–566 (2013)

41. Teshima, T., Sugiyama, M.: Incorporating causal graphical prior knowl-edge into predictive modeling via simple data augmentation. arXiv preprint arXiv:2103.00136 (2021)

42. Troyanskaya, O., et al.: Missing value estimation methods for DNA microarrays. Bioinformatics **17**(6), 520–525 (2001)

43. Van Ginkel, J.R., Kroonenberg, P.M., Kiers, H.A.L.: Missing data in principal component analysis of questionnaire data: a comparison of methods. J. Statist. Comput. Simul. **84**(11), 2298–2315 (2014)

44. Vanschoren, J., van Rijn, J.N., Bischl, B., Torgo, L.: OpenML: networked science in machine learning. SIGKDD Explor. **15**(2), 49–60 (2013)

45. Vincent, P., Larochelle, H., Bengio, Y., Manzagol, P-A.: Extracting and composing robust features with denoising autoencoders. In: Proceedings of the International Conference on Machine Learning (2008)
46. Wells, B.J., Chagin, K.M., Nowacki, A.S., Kattan, M.W.: Strategies for handling missing data in electronic health record derived data. EGEMS **1**(3), 1035 (2013)
47. Xue, Yu., Tang, Y., Xin, X., Liang, J., Neri, F.: Multi-objective feature selection with missing data in classification. IEEE Trans. Emerg. Top. Comput. Intell. **6**(2), 355–364 (2021)
48. Yeh, I.-C.: Modeling slump flow of concrete using second-order regressions and artificial neural networks. Cem. Concr. Compos. **29**(6), 474–480 (2007)
49. Yoon, J., Jordon, J., Schaar, M.: GAIN: missing data imputation using generative adversarial nets. In: Proceedings of the International Conference on Machine Learning, pp. 5689–5698. PMLR (2018)
50. Zhang, S.: Nearest neighbor selection for iteratively kNN imputation. J. Syst. Softw. **85**(11), 2541–2552 (2012)
51. Zhang, Z., Fang, H., Wang, H.: Multiple imputation based clustering validation (MIV) for big longitudinal trial data with missing values in eHealth. J. Med. Syst. **40**(6), 1–9 (2016)
52. Zhang, Z.: Missing data imputation: focusing on single imputation. Ann. Transl. Med. **4**(1), 9 (2016)

Pattern Mining

An Efficient Method for Outlying Aspect Mining Based on Genetic Algorithm

Zihao Chen, Lei Duan$^{(\boxtimes)}$, and Xinye Wang

School of Computer Science, Sichuan University, Chengdu, China
{chenzihao,wangxinye}@stu.scu.edu.cn, leiduan@scu.edu.cn

Abstract. Outlying aspect mining (OAM) aims to identify a feature subspace in which a given query object is dramatically distinctive from the rest data. The identified features can assist the formulation and optimization of decisions. Score-and-search methods are widely used in outlying aspect mining. However, limited by scoring instability and search inefficiency, studies using this strategy are unable to be comprehensive and accurate for mining outlying aspects. In this paper, it proposes a novel OAM method based on genetic algorithm, named OSIER, which can be applied in mining outlying aspects from multi-dimensional spaces. OSIER improves the search efficiency by analyzing the correlations between dimensions. By combining the genetic algorithm with the traditional beam search strategy, OSIER effectively improves the diversity of the searched aspects. As a result, the execution time for candidate outlying aspects search is controlled in an acceptable range. Experiments show that OSIER outperforms the benchmark methods in terms of effectiveness on the OAM task. Besides, OSIER is capable of providing valuable outlying aspect mining results for various types of datasets.

Keywords: Outlying aspect mining · Kernel density estimation · Genetic algorithm

1 Introduction

Outlying aspect mining aims to discover an aspect (i.e., a set of features or attributes) in which a query point has the most significant outlyingness. In many real-life application scenarios, it can provide interpretable information and decision support for downstream tasks [13]. For example, a recruitment team somehow highly interested in identifying what are the most outstanding merits or shortcomings of a particular candidate compared to others.

It is worth noting that the distribution of query points in different spaces varies significantly. As Fig. 1(a) shows, all the data samples are scattered in a 3-dimensional space. The outlyingness of the query point (red triangle) is not significant in the full space. After projecting the data into various 2-dimensional subspaces, the red triangle is more distinguishable from the other points (blue

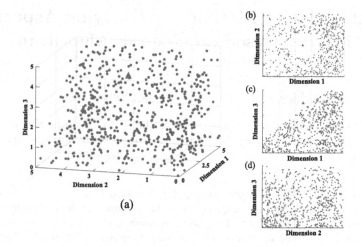

Fig. 1. An example of data distribution in different spaces. (a) Data distribution in full space; (b)(c)(d) Data distribution in three projection subspaces.

dots) in Fig. 1(b) and Fig. 1(c). The space that can exhibit significant differences between the query point and others is called the outlying aspect.

Technically, it is not feasible to enumerate all aspects due to the number of subspaces grows exponentially with the increasing data dimensionality. To find the optimal outlying aspect efficiently, we consider the following two challenges:

- **(C1) How to search aspects efficiently and comprehensively?** To reduce the enormous computational cost of enumerating aspects, it should select representative aspects in the search process. But the distributions of different spaces are not regular, making it challenging to find all representative aspects.
- **(C2) How to measure the outlyingness of different aspects impartially?** A scoring function needs to be designed to quantify the outlyingness of a query point in different aspects. However, calculating the outlyingness in different dimensional aspects may lead to biased results.

Current approaches have limitations in addressing the above challenges. Concerning the search strategy **(C1)**, the most advanced and general approach is the beam algorithm [10] using heuristic rules. This search strategy has an assumption that if a point scores high in an l-dimensional aspect, it generally generated by adding a dimension in a well-behaved $(l - 1)$-dimensional aspect. However, Fig. 1(b) and Fig. 1(c) show extreme cases that disprove this assumption. Higher-dimensional aspects with extreme distribution are not searchable. As for the outlyingness scoring function **(C2)**, Zhang et al. [17] used the distance from surrounding points to the query point as a metric, leading to results biased towards higher-dimensional aspects. The subsequent methods chose density as the evaluation criterion, but they still have shortcomings. The density ranking [5] loses the absolute degree of deviation, Z-score normalization [10] tends to aspects with high variance, and the method of generating hypersphere simulation densities by random sampling [12] produces a large instability.

In this paper, we propose a score-and-search method OSIER (short for outlying aspects mining based on genetic algorithm), which can effectively identify the candidate outlying aspects. OSIER makes improvements to overcome the shortcomings of existing methods from two perspectives. For **C1**, it retain the strategy of generating high-dimensional aspects from well-behaved aspects, while increasing the diversity of the searched aspect by replacing dimensions with a mutation operation. This search method has the probability of getting rid of the local optimum. For **C2**, it uses the strategy of stage comparison. It compares the absolute density of the query point in different aspects under the same dimension. Moreover, the optimal aspects under each dimension are compared with a normalized score. In this way, it retains the absolute density information of all data, avoiding biased results, and accelerates computational efficiency. The main contributions of this paper are summarized as follows:

- We propose a novel genetic algorithm-based method, named OSIER, to increase the diversity of searched aspects and generate more representative aspects efficiently.
- We use a simplified density estimation function in a multi-dimensional space and analyze an improved outlyingness measure guided by prior knowledge.
- We calculate the impact of each dimension on the outlying aspect by generating a maximum interval hyperplane, which can be used to guide the direction of evolution in genetic algorithms. Besides, we use a new comparison strategy preserving the absolute density of all data to avoid biased results.
- We demonstrate OSIER on multiple real-world and synthetic datasets. The experimental results show that OSIER has higher search efficiency and stability, which can be applied to some extreme cases.

The rest of the paper is organized as follows. The related work is reviewed in Sect. 2. Section 3 presents the details of OSIER. The experiments and results are provided in Sect. 4, and Sect. 5 shows the conclusion and future work.

2 Related Work

2.1 Outlying Aspect Mining

Assume that there exists a d-dimensional space $D = \{D_1, D_2, ..., D_d\}$ and a set of data points $X \in \mathbb{R}^d$. For a point $X_i \in X$, its feature under the full space is represented as $\{X_i.D_1, X_i.D_2, ..., X_i.D_d\}$.

We call the combination of multiple dimensions as aspects (a dimension can also be considered as an aspect), and the space composed of these dimensions is a subspace of the full space D. For an aspect $\mathcal{S} = \{D_{i_1}, D_{i_2}, ..., D_{i_{|S|}}\}$, we can define a measure of outlyingness scoring function $\rho_{\mathcal{S}}(q)$, which measures the outlyingness of query point q in the aspect $\mathcal{S}$. Based on the above definitions, we can formalize the *Outlying Aspect Mining* problem as follows:

Definition 1 (Outlying Aspect Mining). *Given a set of n instance X in d-dimensional space D, a query point q ∈ X, the outlying aspect mining is to identify the non-empty aspect S ⊆ D in which the query point q's outlying degree $\rho_S(q)$ is larger than any other aspect.*

The existing OAM methods are mainly divided into two categories, feature selection methods [4,9] and score-and-search methods [5,10,17]. Feature selection methods transform the OAM problem into a classical feature selection classification problem. More specifically, the two classes are defined as the query points (positive class) and the rest of the data (negative class). Therefore, there is a data imbalance problem when the model is trained. Besides, the interpretability of these methods is poor. The score-and-search methods are more widely and deeply researched than feature selection methods.

The frame of score-and-search methods is to search each candidate aspect, calculate outlyingness for query points, and select the aspect with the highest outlyingness as the optimal result. Zhang *et al.* [17] proposed a metric based on the idea of kNN, called outlying degree. Duan *et al.* [5] applied kernel density estimation to multidimensional space by using a product of univariate Gaussian kernels. Meanwhile, they used a boundary pruning-based search strategy.

Vinh *et al.* [10] considered that the use of density ranking would lose important information about the degree of absolute deviation. Thus they designed a standard scoring function and proposed the concept of dimensionality unbiasedness for outlying aspect mining measures.

Definition 2 (Dimensionality Unbiasedness). *If a density scoring function $\rho_S(\cdot)$ satisfies the formula:*

$$\frac{1}{n} \sum_{X_i \in X} \rho_S(X_i) = const.w.r.t.|S| \tag{1}$$

the function can be used to compare the outlyingness of query points in different dimensional spaces directly.

Dimensionality unbiasedness provides a desirable property for designing density scoring functions, thus avoiding bias due to different dimensions and making OAM more interpretable. Meanwhile, to prevent the problem of exploding the number of high-dimensional aspects, a beam search method was proposed to ensure that the number of search spaces is within a specific range by heuristic pruning. Wells *et al.* [15] analyzed the shortcomings of kernel density search and proposed SGrid density estimation instead, thus considerably speeding up the computational process. Samariya *et al.* [12] generated hyperspheres by random sampling to evaluate the outlyingness of query points to solve the complex problem of density computation in high-dimensional space.

It is worth noting that outlier detection and outlying aspect mining are different. Outlier detection aims to detect anomalous data that are exceptional with respect to the majority of objects in the databases. It can applied in various fields, such as disease detection [6], social media monitoring [19] and network intrusion supervision [1]. However, outlier detection is difficult to provide a reasonable and intuitive explanation for the identified objects. The OAM task was proposed for discovering aspects where the query instance exhibits the most outlying characteristics.

2.2 Genetic Algorithm

In genetic algorithms, each individual consists of a gene string that represents a feasible solution to that problem. Fitness is the metric used to evaluate the individual, and the fitness function is usually determined based on the objective function. The selection operation selects a parent based on fitness and inherits its genes to the next generation of individuals. The selected parent undergoes a crossover operation with a certain probability to produce the next individual. After many generations, the genetic algorithm jumps out of the loop with a defined threshold or number of iterations and obtains a better quality solution. Genetic algorithm has been used to solve a large variety of problems efficiently, including classification [3], credit risk assessment [8] and time-series analysis [11].

Zhu et al. [18] used genetic algorithms in the outlier detection problem. They used cell-based segmentation techniques, which resulted in a high outlyingness computation cost in a high-dimensional space. Zhang et al. [16] devised a method that does not depend on the upper and lower bound closure properties. Similar to [17], they chose distance as outlyingness, which is used to guide the evolution.

The genetic algorithm is an efficient optimization algorithm for intelligent global search, which is simple and robust. Thus, we can use these characteristics to discover outlying aspects efficiently.

3 Design of OSIER

In this section, we discuss the details of OSIER. It takes a query point q together with a dataset X of n points $\{X_1, ..., X_n\}$ as input, $X_i \in \mathbb{R}^d$, and outputs an aspect in which the given point has the highest outlyingness.

3.1 Outlying Scoring Function

In the choice of scoring function, we use a simplified version of the multidimensional density estimation function:

$$\rho_S(q) = \frac{1}{nh^{|S|}} \sum_{i=1}^{n} K(\frac{\|q - X_i\|_p}{h}) \tag{2}$$

where p denotes norm. In the absence of any prior knowledge, we choose the Euclidean norm ($p = 2$), adopt the Gaussian kernel as kernel function $K(\cdot)$, and calculate the bandwidth h follows Silverman's rule of thumb [14] is more general.

This default density estimation parameter can be improved by prior knowledge. One type of prior knowledge derived from the data description is the bound of each dimension. Suppose the dataset is restricted in a dimension to a range of values. It is not reasonable to have a density distribution for points outside the range. For example, age is a non-negative number. It is unreasonable to produce a probability distribution in the space where age is negative by the density estimation function. We use a reflection strategy to solve this problem. If the data has a minimum value boundary b in dimension D_i, for query point $q = \{q.D_1, ..., q.D_i, ..., q.D_{|S|}\}$, we set the symmetry point $q_{sym} = \{q.D_1, ..., 2b - q.D_i, ..., q.D_{|S|}\}$. The optimized scoring function is:

$$\rho'_S(q) = \rho_S(q) + \rho_S(q_{sym}) \tag{3}$$

For the case where there are boundaries on both sides, we only consider the first reflection point because the appropriate bandwidth ensures that the density distribution after multiple reflections is equal to 0 or infinitely close to 0.

In addition, if a dataset is composed of multiple datasets, resulting in far from normally distributed data, Silverman's rule of thumb [14] will result in a poor density estimate. We prefer to use an improved sheather algorithm [2] which can achieve better results when the dataset is distributed in multiple dense regions.

3.2 Dimensions Correlation Analysis

Before formally calculating the query points' outlyingness, each dimension of the dataset needs to be analyzed. Different dimensions provide different contributions to the generation of outlying aspects. Thus the analysis of a single dimension helps to guide subsequent search process.

To make the outlying aspects result more credible, we perform a deeper analysis of the density estimation method. For a query point, if the distribution of the projection on a dimension is remote and the probability density is minimal, the dimension can have a large impact on aspect generation. This kind of dimensions is called trivial outlying dimension, which is defined as follows:

Definition 3 (Trivial Outlying Dimension). *Given a query point q, an outlyingness scoring function $\rho(\cdot)$ and a threshold ϵ, a dimension D_i is called trivial outlying dimension if $\rho_{D_i}(q) \leq \epsilon$.*

An intuitive fact is that when a trivial outlying dimension is coupled with another dimension, the generated aspects may still have a good outlyingness score for the query point. Therefore, the trivial outlying dimensions should be pre-processed to reduce their impact on the search results. For a query point, if a trivial outlying dimension exists, aspects that may cause superior outlyingness will be replaced by different combinations of that trivial outlying dimension with other dimensions.

Example 1. Table 1 shows the shooting statistics for Los Angeles Lakers. We employ OAMiner [5] to figure out outlying aspects of *LeBron James*. The result shows that the top-5 outlying aspects are $\{3PM\}$, $\{3PM, 3PA\}$, $\{3PM, FGM\}$, $\{3PM, FTM\}$ and $\{3PM, 3PA, FGM\}$. When we ignore the effects of feature $3PM$, outlying aspect $\{FG\%, FT\%, 2P\%\}$ is revealed, which can offer more hidden information.

Table 1. Shooting statistics of six active players in Los Angeles Lakers

Name	FGM	FGA	FG%	3PM	3PA	3P%	FTM	FTA	FT%	2PM	2PA	2P%
LeBron James	11.1	21.3	52.3	2.9	8.0	36.2	4.6	6.0	76.6	8.3	13.4	61.8
Anthony Davis	9.2	17.2	53.7	0.3	1.8	18.2	4.4	6.1	70.9	8.9	15.4	57.8
Russell Westbrook	6.8	15.7	43.3	0.9	3.3	27.7	3.4	5.1	67.0	5.9	12.4	47.4
Stanley Johnson	2.3	4.8	47.0	0.6	2.0	31.9	0.9	1.2	70.7	1.6	2.8	57.9
Austin Reaves	2.3	4.8	47.6	0.8	2.6	31.8	1.3	1.5	83.8	1.4	2.1	67.3
Dwight Howard	1.9	3.1	60.9	0.1	0.2	66.7	1.3	2.0	62.9	1.8	3.0	60.6

Afterwards, we need to evaluate the correlation between dimensions and analyze the contribution of each dimension to the degree of query point outliers, represented as a set of weights $\mathbf{w} = \{w_1, w_2, ..., w_d\}$. We construct a hyperplane on the full space so that the query points are as separated as possible from other points to calculate w. Since the query point and the rest of the sample points are two classes of samples with extreme imbalance, we choose the one-class SVM method and set the query point as the origin. Solving this maximum geometric margin hyperplane is essentially a convex optimization problem:

$$\min \frac{1}{2}||\mathbf{w}||^2 - \tau + \frac{1}{\upsilon n}\sum_{i=1}^{n}\xi_i \tag{4}$$
$$s.t. \ \mathbf{w}^T X_i \geq \tau - \xi_i, \quad \xi_i \geq 0$$

where X_i denotes spatial vectors of the i-th point, τ denotes the hyperplane bias, ξ_i denotes the relaxation variable of the i-th point and υ denotes a trade-off parameter. We solve the quadratic programming problem by Lagrange Multiplier Method:

$$\mathcal{L}(\mathbf{w}, \tau, \xi, \alpha, \beta) = \frac{1}{2}||\mathbf{w}||^2 - \tau + \frac{1}{\upsilon n}\sum_{i=1}^{n}\xi_i - \sum_{i=1}^{n}\alpha_i(\mathbf{w}^T X_i - \tau + \xi_i) - \sum_{i=1}^{n}\beta_i\xi_i \tag{5}$$

In order to obtain a specific form for solving the dual problem, let the partial derivative of $\mathcal{L}(\mathbf{w}, \tau, \xi, \alpha, \beta)$ with respect to W, τ, and ξ equal to zero. We can obtain the following three conditions:

$$\mathbf{w} - \sum_{i=1}^{n}\alpha_i X_i = 0 \tag{6}$$

$$\sum_{i=1}^{n}\alpha_i - 1 = 0 \tag{7}$$

$$0 \leq \alpha_i \leq \frac{1}{\upsilon n} \tag{8}$$

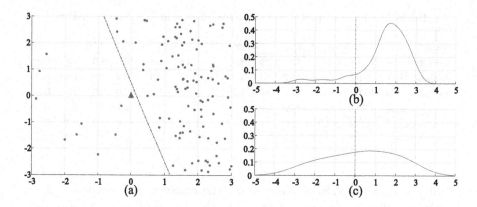

Fig. 2. Significance of the weights generated by hyperplane.

Equation 6 can be solved for $\mathbf{w}$, where $X = \{X_1, X_2, ..., X_d\}$ is known, and the optimal solution for α can be obtained by substituting Eq. (6)-(8) into Eq. 5:

$$\min \& \frac{1}{2} \sum_{i=1}^{n} \sum_{j=1}^{n} \alpha_i X_i^T X_j \alpha_j$$

$$s.t. \sum_{i=1}^{n} \alpha_i = 1, \quad 0 \le \alpha_i \le \frac{1}{vn} \tag{9}$$

In reality, it is reasonable to expect that if this hyperplane is more perpendicular to a dimension, the greater the contribution of this dimension in the classification. As Fig. 2(a) shows, red triangle indicates the query, and the red dashed line indicates the generated hyperplane. The hyperplane corresponds to a weight vector of $\mathbf{w} = [3, 1]$, indicating that the dimension corresponding to the x-axis has a greater influence on the query point becoming an outlier. Figure 2(b) and Fig. 2(c) denote the density distributions of the data after projection on the x-axis and y-axis, respectively, which also justify the analysis.

3.3   Outlying Aspect Generation

The strategy of searching candidate aspects is the core problem of computing the outlying aspects in a high-dimensional data set. When the dimensionality is large enough, the computation of traversing every aspect brings an unbearable computational cost. This cost is exponentially related to the number of dimensions. Taking OAMiner [5] as an example, it takes over 24 h on a dataset with 30 dimensions and 10,000 points, which is impracticable for many real-world high-dimensional datasets. OSIER uses a genetic algorithm, which includes recombination, mutation, and selection operations to search for representative aspects efficiently. The search strategy is given in Algorithm 1.

Procedure *Recombination*(·) generates new individuals by reorganizing parts of the structure of multiple parent individuals in Step 4. If an aspect performs

Algorithm 1. Pseudocode of OSIER
___
Input: a d-dimensional dataset X, a query point q, population P, mutation rate α.
Output: the outlying aspect of q.
1: Initialize the candidate dimension set $CSet = \{D_i | \rho_{D_i}(q) \leq \epsilon\}$ (optional, The full
   set can be used directly without considering trivial outlying dimension)
2: $C_1 \leftarrow CSet$
3: Best-scored aspect set $BS \leftarrow \{\underset{D_i \in CSet}{\arg\min}\, \rho_{D_i}(q)\}$
4: **for** $l \leftarrow 2$ to $|Cset|$ **do**
5:     $RC_l \leftarrow Recombination(C_{l-1}, P)$
6:     $MC_l \leftarrow Mutation(RC_l, P, \alpha)$
7:     **for** each candidate aspect S in $RC_l \cup MC_l$ **do**
8:         **if** S has not been considered **then**
9:             **if** $|C_l| < P$ **then**
10:                $C_l \leftarrow C_l \cup \{S\}$
11:            **else if** $\rho_S(q) < Max(\{\rho_{D_i}(q) | D_i \in C_l\})$ **then**
12:                replace the worst aspect in C_l by S
13:            **end if**
14:        **end if**
15:    **end for**
16:    $BS \leftarrow BS \cup \{\underset{A_i \in C_l}{\arg\min}\, \rho_{A_i}(q)\}$
17: **end for**
18: $BS \leftarrow Normalization(BS)$
19: **return** $bestAspect \leftarrow \underset{A_i \in BS}{\arg\min}\, \rho_{A_i}(q)$
___

better in the paternal generation, the dimensions that make up this aspect are more likely to participate in the generation of new individuals. Recombination can discover most of the high-scored aspects by the heuristic search strategy.

Procedure $Mutation(\cdot)$ in Step 5 can handle extreme cases (e.g. Figure 1(b) and Fig. 1(c)). OSIER use sampling to calculate the probability of each dimension participating in the mutation, which is calculated as:

$$pro(D_i) = \frac{W_{D_i}}{\sum\limits_{D_j \in H} W_{D_j}} \qquad (10)$$

where H is the set of dimensions not involved in the recombination, W_{D_i} indicates the weight of the dimension D_i on the outlyingness in full space, which is mentioned before. A dimension with a larger weight will have a larger opportunity to be selected for next generation to reproduce with modification. OSIER uses bit-wise mutation which randomly replacing one of the dimensions that make up an individual.

Moreover, in order to ensure a fair comparison between different dimensional aspects, the outlyingness score needs to be normalized after calculation. A well-known normalization method in the OAM task is Z-score [10]:

$$Z(\rho_S(q)) \triangleq \frac{\rho_S(q) - \mu_{\rho_S}}{\sigma_{\rho_S}} \qquad (11)$$

Table 2. Characteristics of the datasets

Data set	# objects(n)	# attributes(d)
Synthetic datasets	1000	10–100
Seed	210	7
Music emotion	400	50
Climate model	540	18
KSD	2856	71

where μ_{ρ_S} is the mean of the density of all points in the aspect $\mathcal{S}$, and σ_{ρ_S} is the standard deviation. The score obtained by this transformation satisfies the dimensional unbiasedness requirement of Eq. 1. However, the computational cost of normalizing each searched aspect is still large, so we consider a staged comparison. When generating the aspects in each dimensionality, the optimal aspect is obtained by comparing the original density evaluation score (Steps 6–18). After obtaining the optimal aspects under each dimension, the outliers are normalized (Step 19) and compared (Step 20) in these aspects. Compared to OAMiner [5] and Density Z-score [10], the overall search complexity is reduced from $O(n^2d \cdot Wd)$ to $O(nd \cdot Wd + n^2d \cdot d)$, where n and d are the size and dimensionality of data set, W is the average search width of each dimension.

4   Experiments and Result Discussion

4.1   Experimental Setting

Datasets. We use four real-world datasets from the UCI machine learning repository[1]. We also use the synthetic datasets provided by Keller *et al.* [7]. Table 2 shows the characteristics of these datasets. The attribute values in the dataset are all real numbers. For the sake of the subsequent description, we use incremental subscripts to record the attributes of the dataset.

Baselines. Four OAM methods are selected as baselines to demonstrate the efficiency and stability of OSIER, including kernel density rank (OAMiner) [5], Z-score normalized Kernel density (ZKDE) [10], sGrid [15] and SINNE [12]. We have made a brief summary in Table 3 for the outlyingness calculation used by each method, where ψ denotes the size of the random subsample, t denotes the number of ensemble models, and *Individual Complexity* represents the complexity of computing outlyingness in one aspect for a query point. There are three additional notes: (1) Although OAMiner introduces a boundary method to perform pruning operations, the amount of search space is uncertain and usually much larger than other methods. The exact number depends on the true distribution of the data. (2) SINNE is a sampling-based method. Given a query

---

[1] http://archive.ics.uci.edu/ml/index.php.

point and an aspect, the results will vary each time. In particular, the use of a heuristic search approach can also lead to a volatile search of the aspects. (3) The stability of OSIER depends on the mutation rate, which is usually small. It works by accepting the probability of generating an aspect that is worse than the current one, so it is possible to jump out of the local optimal solution. If there is no extreme case, the searched aspects are stable for the same dataset.

Table 3. Summary the characteristics of baselines

Methods	Individual complexity	Complexity	Interpretability	Stability
OAMiner	$O(n^2 d)$	–	High	High
ZKDE	$O(n^2 d)$	$O(n^2 d \cdot Wd)$	High	High
Sgrid	$O(n^2 d/\omega)$	$O(n^2 d/\omega \cdot Wd)$	Medium	High
SINNE	$O(\psi t^2 d)$	$O(\psi t^2 d \cdot Wd)$	Low	Low
OSIER	$O(nd)$	$O(nd \cdot Wd + nd^2)$	High	Medium

Experimental Setup. We use the parameters mentioned in Sect. 3.1. For the aforementioned methods, we apply the default parameters. For the scoring function, OAMiner and ZKDE use the Gaussian kernel. We set ψ to 8 and t to 100 in SINNE. For the search strategy, epsilon neighborhood range of OAMiner is set to 1. Search width W and maximum dimensionality of searched aspects d_{max} in beam search are set to 120 and 3, respectively.

Experiments were conducted on a PC with four Intel Xeon E5-2698 CPUs, four GeForce RTX 2080 Ti GPUs and 512 GB memory, running Ubuntu 20.04. The algorithms were implemented in Java and compiled by Java version 13.

4.2 Effectiveness

We use four real-world datasets from different domains to demonstrate the effectiveness of OSIER in the OAM problem. Since we do not have a standard to measure the quality of the found aspects, we display the visualization of outlier points in the aspects as shown in Fig. 3. We choose query points with aspects of three dimensions as examples for better visualisation. The red triangles indicate the query points which exhibit good outlyingness in each obtained subspaces. Visualization results show that OSIER can obtain well-behaved aspects.

We also compare our method with the baseline methods on 10, 20, 30, 40, 50, 75, and 100-dimensional synthetic datasets, respectively. The original datasets provide with 19–136 outliers and the ground-truth of their outlying aspects. For the fairness of the experimental criteria, we augmented the number of outliers. For example, there is an outlier X_1 and a normal point X_2 in a d-dimensional data set, and the outlying aspect of X_1 is $\{D_1, D_2\}$. By replacing the dimensions which are not in outlying aspect of X_1, we can generate a new outlier $X_3 = \{X_1.D_1, X_1.D_2, X_2.D_3, ...X_2.D_d\}$. Since OAMiner is committed to finding aspects ranked 1, it may find more than one aspect. We consider that the correct aspect has been found if the result contains the ground truth.

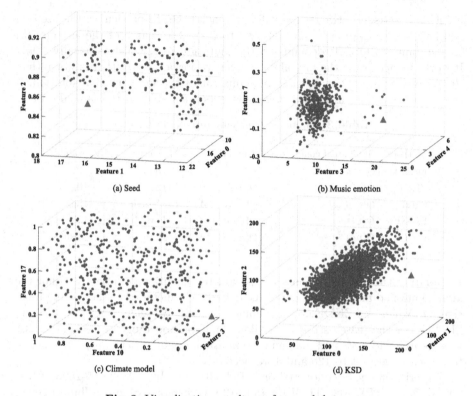

Fig. 3. Visualization results on four real data sets.

The results are shown in Table 4. Experiment shows that OSIER achieves the state-of-the-art performance on all datasets of different dimensions. There is little difference in the effectiveness of each method in the low-dimensional space. As the number of dimensions rises, the accuracy of OSIER improves more significantly. The trend shows that the search strategy plays a greater role in high-dimensional space. Besides, the heuristic rule pruning strategy (OSIER and ZKDE) can search for more representative aspects than the boundary pruning (OAMiner). ZKDE performs better than Sgrid and SINNE, which indicates that using partial data makes the searched aspects unstable.

Figure 4 shows the efficiency test on the synthetic datasets with varying number of dimensions d and data size n. The base OAMiner method [5] is chosen as the baseline method, which can reduce the impact caused by the different functions of calculating the outlyingness. We select several query points with a result subspace of no more than 3 dimensions to calculate the average running time. Experiment shows that the search efficiency of our method is much faster. Moreover, the execution time rises slower than the baseline as the dimensionality and size of the dataset expand.

Table 4. Overall Performances of Comparison with Baselines

Method	syn_10D	syn_20D	syn_30D	syn_40D	syn_50D	syn_75D	syn_100D
OAMiner	0.893	0.664	0.463	0.352	0.320	0.308	0.192
ZKDE	0.953	0.808	0.839	0.631	0.601	0.635	0.573
Sgrid	0.942	0.629	0.574	0.556	0.524	0.508	0.426
SINNE	0.879	0.709	0.558	0.640	0.609	0.648	0.595
OSIER	**0.967**	**0.856**	**0.843**	**0.770**	**0.694**	**0.671**	**0.654**

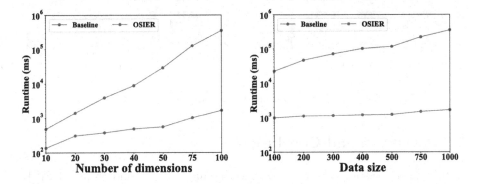

Fig. 4. Efficiency test w.r.t the number of dimensions d and data size n.

4.3   Parameter Analysis

For ease of understanding, we select the data under two dimensions of the synthetic datasets and visualize the results by contour lines.

The choice of norm comes in to play when $d \geq 2$. In the previous norm studies, the commonly used norms are $p = 1$ (Manhattan distance), $p = 2$ (Euclidean norm), and $p = \infty$ (Maximum norm). As shown in Fig. 5(a), the value of p has a tiny effect on the density distribution in a dense region. As the number of data points increases, the choice of p is less important. We recommend using the 2-norm for its stronger symmetry.

Figure 5(b) shows that the value of bandwidth cannot be static. A reasonable bandwidth should depend on the distribution of the data. A small bandwidth will result in relatively independent density estimates (e.g. $N = 100$, $h = 0.1$), and a large bandwidth will result in a dispersed density distribution, which makes it difficult to reflect the differences. Silverman's rule of thumb [14] is a variance-based bandwidth selection method. By this method, the value of h is taken closer to 0.25 when $N = 100$, and it can be observed that $h = 0.25$ is more representative compared with the others in visualization results. bandwidth essentially scales the kernel density estimation function in different dimensions to obtain better experimental results.

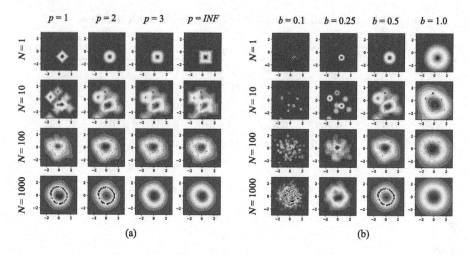

Fig. 5. Influence of different norm p and bandwidth b.

5   Discussion and Conclusion

In this paper, we study the outlying aspect mining problem and propose OSIER, which address the shortcomings of existing methods effectively and provide more interpretable and credible results. We analyze the application of kernel density estimation methods to outlying aspect mining and design an adaptive scoring function. In addition, we improve the commonly used aspects search strategy. We introduce the idea of a genetic algorithm to obtain the fitness of individuals in the process of genetic inheritance by analyzing the correlation among dimensions. Also, the mutation operation in the genetic algorithm can handle some extreme cases during the search process, thus avoiding getting trapped in a local optimum. Experimental results on several real and synthetic datasets demonstrate the effectiveness of the proposed method in outlying aspect mining.

Our future work will focus on applying OAM to hybrid or time-series data. We plan to design a rational outlying aspect structure for interpretable results.

Acknowledgments. This work was supported in part by the National Natural Science Foundation of China (61972268), the Sichuan Science and Technology Program (2020YFG0034), and the Med-X Center for Informatics funding project of SCU (YGJC001).

References

1. Beulah, J.R., Punithavathani, D.S.: An efficient mixed attribute outlier detection method for identifying network intrusions. Int. J. Inf. Secur. Priv. **14**(3), 115–133 (2020)
2. Botev, Z.I., Grotowski, J.F., Kroese, D.P.: Kernel density estimation via diffusion. Ann. Stat. **38**(5), 2916–2957 (2010)

3. Carvalho, E.D., Silva, R.R.V., Araújo, F.H.D., de A. L. Rabelo, R., de Carvalho Filho, A.O.: An approach to the classification of COVID-19 based on CT scans using convolutional features and genetic algorithms. Comput. Biol. Med. **136**, 104744 (2021)
4. Dang, X., Assent, I., Ng, R.T., Zimek, A., Schubert, E.: Discriminative features for identifying and interpreting outliers. In: ICDE, pp. 88–99 (2014)
5. Duan, L., Tang, G., Pei, J., Bailey, J., Campbell, A., Tang, C.: Mining outlying aspects on numeric data. Data Min. Knowl. Disc. **29**(5), 1116–1151 (2015). https://doi.org/10.1007/s10618-014-0398-2
6. Jenkinson, W.G., Li, Y.I., Basu, S., Cousin, M.A., Oliver, G.R., Klee, E.W.: Leaf-cuttermd: an algorithm for outlier splicing detection in rare diseases. Bioinform. **36**(17), 4609–4615 (2020)
7. Keller, F., Müller, E., Böhm, K.: HICS: high contrast subspaces for density-based outlier ranking. In: ICDE, pp. 1037–1048 (2012)
8. Lappas, P.Z., Yannacopoulos, A.N.: A machine learning approach combining expert knowledge with genetic algorithms in feature selection for credit risk assessment. Appl. Soft Comput. **107**, 107391 (2021)
9. Micenková, B., Ng, R.T., Dang, X., Assent, I.: Explaining outliers by subspace separability. In: ICDM, pp. 518–527 (2013)
10. Vinh, N.X., et al.: Discovering outlying aspects in large datasets. Data Min. Knowl. Disc. **30**(6), 1520–1555 (2016). https://doi.org/10.1007/s10618-016-0453-2
11. do Prado Ribeiro, K., Fontes, C.H., de Melo, G.J.A.: Genetic algorithm-based fuzzy clustering applied to multivariate time series. Evol. Intell. **14**(4), 1547–1563 (2021)
12. Samariya, D., Aryal, S., Ting, K.M., Ma, J.: A new effective and efficient measure for outlying aspect mining. In: WISE, pp. 463–474 (2020)
13. Samariya, D., Ma, J.: Mining outlying aspects on healthcare data. In: HIS, pp. 160–170 (2021)
14. Silverman, B.W.: Density estimation for statistics and data analysis (1986)
15. Wells, J.R., Ting, K.M.: A new simple and efficient density estimator that enables fast systematic search. Pattern Recognit. Lett. **122**, 92–98 (2019)
16. Zhang, J., Gao, Q., Wang, H.H.: A novel method for detecting outlying subspaces in high-dimensional databases using genetic algorithm. In: ICDM, pp. 731–740 (2006)
17. Zhang, J., Lou, M., Ling, T.W., Wang, H.H.: HOS-Miner: a system for detecting outlying subspaces of high-dimensional data. In: VLDB, pp. 1265–1268 (2004)
18. Zhu, C., Kitagawa, H., Faloutsos, C.: Example-based robust outlier detection in high dimensional datasets. In: ICDM, pp. 829–832 (2005)
19. Zrira, N., Mekouar, S., Bouyakhf, E.: A novel approach for graph-based global outlier detection in social networks. Int. J. Secur. Networks **13**(2), 108–128 (2018)

Effective Mining of Contrast Hybrid Patterns from Nominal-numerical Mixed Data

Min Fu[1], Lei Duan[1,2(✉)], and Zhenyang Yu[1]

[1] School of Computer Science, Sichuan University, Chengdu, China
leiduan@scu.edu.cn, {fu_min,yuzhenyang}@stu.scu.edu.cn
[2] Med-X Center for Informatics, Sichuan University, Chengdu, China

Abstract. Contrast pattern mining, which finds patterns describing differences between two classes of data, is an important task in various scenarios. As real-world data is usually a mixture of nominal and numerical attributes (e.g., electronic medical records), contrast pattern mining algorithms over nominal-numerical mixed data are in great demand. Existing algorithms on contrast pattern mining either can only handle a single type of attribute or transform numerical attributes into nominal attributes with prior knowledge. However, these algorithms may result in limited discrimination of contrast patterns due to the failure to exploit the original data information and inflexible pattern forms. In this paper, we propose a novel algorithm, CHPMiner, which mines a new kind of contrast pattern called contrast hybrid pattern (CHP) that contains nominal attributes and numerical relationships among numerical attributes based on extended gene expression programming (GEP). Specifically, CHPMiner develops two sub-expressions and a novel structure to combine nominal and numerical attributes. Moreover, CHPMiner leverages a specific fitness function to guide the evolution direction for mining CHPs that are highly discriminating. Experiments on four real-world datasets show that CHPMiner outperforms baselines. The case study further demonstrates the effectiveness of CHPMiner.

Keywords: Contrast pattern mining · Contrast hybrid pattern · Gene expression programming

1 Introduction

Contrast pattern mining (e.g., emerging pattern mining) finds patterns whose occurrences change significantly between two classes of data [3]. Such patterns can help domain experts understand data and solve specific tasks in various scenarios, such as biological data analysis [16], disease diagnosis [20], and network traffic events detection [1]. Since most real-world data is a mixture of nominal and numerical attributes (e.g., electronic medical records), contrast pattern mining algorithms over nominal-numerical mixed data are in great demand.

© The Author(s), under exclusive license to Springer Nature Switzerland AG 2022
W. Chen et al. (Eds.): ADMA 2022, LNAI 13725, pp. 352–367, 2022.
https://doi.org/10.1007/978-3-031-22064-7_26

Table 1. A nominal-numerical mixed dataset with two classes (C_1 and C_2)

Class	ID	A_1	A_2	A_3	A_4	A_5	A_6
C_1	1	Female	F	F	1	7	9
	2	Female	F	F	6	5	12
	3	Female	T	F	3	4	14
C_2	4	Female	F	T	2	5	8
	5	Male	T	T	3	6	17
	6	Female	F	F	4	6	9

*Nominal attributes: A_1, A_2, and A_3;
Numerical attributes: A_4, A_5, and A_6
* F: False; T: True

Existing contrast pattern mining algorithms either only deal with numerical attributes [4–6], or only handle nominal attributes [15], or discretize numerical attributes into nominal ones by using prior knowledge [2,9–11,18]. However, two limitations exist in the above algorithms: (1) *insufficient data utilization*: ignoring one type of attribute or discretizing numerical attributes may lose potentially helpful information of original data; (2) *limited discrimination of contrast patterns*: inflexible forms of contrast patterns (e.g., conjunctions of nominal attributes) could be limited in capturing data differences in complex scenarios.

To concretely illustrate these limitations, we take the dataset given in Table 1 as an example. Suppose we want to find contrast patterns that occur in all instances of C_1 but do not occur in any instance of C_2, which can fully capture the difference between C_1 and C_2. Unfortunately, it is difficult to find contrast patterns satisfying this condition by only considering numerical attributes (e.g., $A_4 + A_5 < A_6$), or by only handling nominal attributes (e.g., A_1 = Female AND A_3 = F), or by combining discrete numerical attributes with nominal ones (e.g., A_3 = F AND $A_6 < 15$). Nevertheless, we can find contrast patterns (e.g., A_3 = F AND $A_4 + A_5 < A_6$) that satisfy the condition when using both types of attributes without discretization.

We define the contrast pattern that satisfies the condition in the above example as *contrast hybrid pattern* (CHP), which contains nominal attributes and numerical relationships among numerical attributes. Moreover, CHPs are helpful in some real-world specific applications [17]. Therefore, it is significant to mine CHPs from nominal-numerical mixed data in some scenarios. To the best of our knowledge, there are few algorithms for mining CHPs, which motivates us to develop a novel algorithm to mine such contrast patterns from nominal-numerical mixed data without discretization.

Meanwhile, we need to address the following three challenges in mining CHPs:

- **(C1)** *How to leverage characteristics of nominal and numerical attributes?* Nominal attributes are represented by words semantically, while numerical attributes are expressed by real numbers. Due to their different characteristics, it is necessary to construct expressions under corresponding data types.

- **(C2)** *How to combine nominal and numerical attributes in a unified way?* It is difficult to directly combine nominal and numerical attributes, which vary greatly in expression forms. Thus, we need to design a unified approach for combining different expression forms of nominal and numerical attributes.
- **(C3)** *How to effectively generate CHPs with high discrimination?* The candidate solution space for CHPs is large because the generation of CHPs involves various combinations of attributes and operators. Therefore, it is challenging to effectively generate CHPs with high discrimination.

To address these three challenges above, we propose an extended GEP algorithm, called CHPMiner (short for <u>c</u>ontrast <u>h</u>ybrid <u>p</u>attern <u>miner</u>). GEP has the flexibility to generate expressions for various forms [7]. It performs a global search and uses evolution to generate the optimal solution efficiently. To tackle **C1**, CHPMiner constructs two kinds of sub-expressions, i.e., boolean expressions for nominal attributes and relational expressions for exploring numerical relationships among numerical attributes. To tackle **C2**, CHPMiner develops an extended chromosome structure in GEP named hybrid chromosome structure to combine sub-expressions for generating valid hybrid expressions. To tackle **C3**, CHPMiner leverages a specific fitness function to guide the evolution direction of sub-expressions for generating CHPs with high discrimination.

The main contributions of our work are as follows:

(1) We introduce the new data mining task, CHPs mining, which finds contrast patterns containing nominal attributes and numerical relationships among numerical attributes from nominal-numerical mixed datasets.
(2) We propose an extended GEP algorithm CHPMiner, which develops two types of sub-expressions and a novel hybrid chromosome structure to generate CHPs. Furthermore, CHPMiner leverages a specific fitness function for mining CHPs that are highly discriminating.
(3) We conduct experiments on four real-world datasets to evaluate CHPMiner, demonstrating that CHPMiner is more effective in mining distinguishing contrast patterns than baselines.

2 Related Work

2.1 Contrast Pattern Mining

Most existing studies on contrast pattern mining mainly focus on the item-based contrast patterns like emerging patterns [3]. Previous contrast pattern mining algorithms over nominal-numerical mixed data discretize numerical attributes into items and then combine these items with nominal attributes. Khade *et al.* used an adaptive binning strategy to find meaningful bin boundaries for numerical attributes for contrast set mining in mixed streaming data [9] and mixed data [10]. Alipourchavary *et al.* [2] used equal-frequency unsupervised discretization technique to discretize numerical attributes for finding most specific set contrast patterns. However, these algorithms only consider the conjunction of items, which lack flexibility in complex practical applications. Loekito *et al.* [18] leveraged Zero-Suppressed Binary Decision Diagrams to generalize emerging patterns by allowing disjunction and conjunction of items. Since discretization of numerical attributes could result in suboptimal results [8], some studies mine contrast patterns from numerical data directly. Duan *et al.* [5] used a GEP-based algorithm to mine contrast functions whose accuracies change significantly between two classes of data. Duan *et al.* [6] proposed contrast inequality to express the inequality relations among numerical attributes. DIFMiner and GEPDIF [4] were designed to leverage arithmetic-expression-based inter-attribute relationships to distinguish different classes of data. Nevertheless, these algorithms were designed to mine contrast patterns only from numerical attribute datasets, which cannot handle nominal-numerical mixed data.

2.2   Preliminary Concepts of GEP

GEP is an algorithm that simulates biological evolution to create and evolve computer programs, which is inspired by Genetic Algorithms (GAs) and Genetic Programming (GP) [7]. There are two main parts in GEP: chromosome and expression tree. A chromosome is a fixed-length string of symbols, while an expression tree contains genetic information. A chromosome consists of one or more genes, each with a head and a tail. The gene head contains symbols representing functions and terminals, while the gene tail only contains terminals. The length of gene head h is set by the user, whereas the length of gene tail t is a function of h and the number of arguments in this function with maximal arguments γ, i.e., $t = h(\gamma - 1) + 1$. Consider a gene for which the set of functions $F = \{+, -\}$. In this case, the maximum number of arguments of the element in F is $\gamma = 2$. Let $h = 3$, then $t = 3 \times (2 - 1) + 1 = 4$. In GEP, each gene can be parsed into an expression tree of different sizes and shapes. GEP constructs the genes' effective parts by parsing the expression tree from left to right and from top to bottom. Since the structural organization of GEP is flexible, any modification made in a chromosome will result in an efficient expression tree.

GEP starts with a random generation of the initial population. Each chromosome is evaluated by a fitness function. According to the fitness, some new chromosomes will be reproduced by modifying the selected chromosomes. Genetic operators, such as recombination and mutation, can guarantee the diversity of populations, which is essential to produce the optimal solution [7]. Chromosomes in each generation undergo the same evolution process as in the preceding generation. The evolution process repeats until some stop condition (e.g., the number of generations) is satisfied. Then the best-so-far solution is founded.

3   Problem Definition

We start with some preliminaries. Let D be a dataset with n attributes $\mathcal{A} = \{A_1, A_2, \ldots, A_n\}$. An attribute A_i can be nominal or numerical. For a nominal attribute, it can have m values, i.e., domain$(A_i) = \{v_{i1}, v_{i2}, \ldots, v_{im}\}$. For a numerical attribute, its value is a real number denoted by $a_i \in \mathbb{R}$.

Definition 1. Boolean Expression. *Given a nominal attribute set $\mathcal{A}$ and a logical operator set $F_b = \{\wedge, \vee, \neg, \oplus\}$, a boolean expression b is the combination of nominal attribute values and logical operators. Notably, a single nominal attribute value is also a boolean expression.*

Definition 2. Relational Expression. *Given a numerical attribute set $\mathcal{A}$, an arithmetic operator set $F_e = \{\times, \div, +, -\}$ and a relational operator set $L = \{<, >, \leqslant, \geqslant\}$, a relational expression e is the combination of numerical attribute values, arithmetic operators and a relational operator.*

Definition 3. Hybrid Expression. *Given a set of boolean expressions B, a set of relational expression E and a combination operator set $F_p = \{\wedge, \vee, \oplus\}$, a hybrid expression p is the combination of boolean expressions, relational expressions and combination operators.*

Example 1. According to Table 1, the set of nominal attribute values is $\{v_{11}, v_{12}, v_{21}, v_{22}, v_{31}, v_{32}\}$. For example, v_{31} represents A_3=F, and v_{32} represents A_3=T. And the set of numerical attribute values is $\{a_4, a_5, a_6\}$. Then, we can have $b = (v_{11} \wedge v_{21} \vee v_{32})$, $e = (a_4 \times a_6 > a_5)$, and $p = (v_{11} \wedge v_{21} \vee v_{32}) \wedge (a_4 \times a_6 > a_5)$.

Definition 4. Contrast Score. *Given a partition of D into two parts $D_1 \subset D$ and $D_2 = D \setminus D_1$. For any expression $s \in \{e, b, p\}$, the support of expression s in D_i is defined as:*

$$sup(s, D_i) = \frac{cnt(s, D_i)}{|D_i|} \tag{1}$$

where $cnt(s, D_i)$ denotes the number of samples in D_i holding $s \in \{e, b, p\}$. The contrast score of s targeting D_1 against D_2 is defined as:

$$cScore(s, D_1, D_2) = \frac{sup(s, D_1)}{sup(s, D_2) + \epsilon} \tag{2}$$

where $\epsilon > 0$ is a user-defined parameter to avoid zero-frequency problem. A larger value of $cScore(s, D_1, D_2)$ indicates expression s achieves greater contrast between D_1 and D_2.

Definition 5. Contrast Hybrid Pattern. *Given a partition of D into two parts $D_1 \subset D$ and $D_2 = D \setminus D_1$. Suppose we have a hybrid expression set P. A hybrid expression $p_i \in P$ is a contrast hybrid pattern (CHP) targeting D_1 against D_2 when $cScore(p_i, D_1, D_2) > \alpha$, where α is a user-defined parameter.*

Example 2. Consider Table 1, and suppose we have two hybrid expressions $p_1 = v_{31} \wedge (a_4 + a_5 < a_6)$ and $p_2 = (v_{21} \vee v_{31}) \wedge (a_4 + a_5 < a_6)$. Then, all instances of C_1 hold p_1 and p_2, and the instance $(ID = 4)$ of C_2 holds p_2. Given $\epsilon = 0.1$, $\alpha = 3$, then, $cScore(p_1, C_1, C_2) = 30 > 3$, and $cScore(p_2, C_1, C_2) = 2.7 < 3$. Thus, p_1 is a CHP but p_2 is not.

Task Statement. Given a partition of D into two parts $D_1 \subset D$ and $D_2 = D \setminus D_1$, CHPMiner aims to find a CHP with the highest contrast score among CHPs that satisfy user-defined conditions.

4   Design of CHPMiner

We now present our proposed CHPMiner, which mines the optimal CHP from nominal-numerical mixed data. There are three parts in CHPMiner: (1) sub-expression generation: creating the representation of boolean and relational expressions for fully leveraging nominal and numerical attributes (Sect. 4.1); (2) hybrid expression generation: leveraging an extended individual structure to generate hybrid expressions by using sub-expressions (Sect. 4.2); (3) fitness function design: designing a fitness function to guide the evolution process of sub-expressions for finding CHPs with high discrimination (Sect. 4.3).

4.1   Sub-Expression Generation

We note that most existing contrast patterns are conjunctions of items, i.e., the contrast pattern is in the form of a boolean expression where the nominal attribute values are connected by the function "AND". Therefore, we can capture contrasts for nominal attributes by generating boolean expressions. Nevertheless, this form is unsuitable for exploring the numerical relationship among numerical attributes because it involves arithmetic operators. In addition, it can be observed that the result of the contrast pattern expression is a boolean value, so we need to explore the relational expression of numerical attributes.

Through the above analysis, CHPMiner aims to generate boolean expressions and relational expressions for nominal and numerical attributes, respectively. For the convenience of description, we call these two kind of expressions sub-expressions. Since they contain different variables and operators, it is hard to determine specific forms of sub-expressions without prior knowledge. CHPMiner generates sub-expressions based on extended GEP to produce flexible forms of expressions with little prior knowledge.

For a boolean expression $b_i \in B$, we take nominal attribute values as GEP terminals. To make boolean expressions more expressive, CHPMiner considers not only the "AND" function, but also other logical functions, such as "OR", "NOT" and "XOR". A boolean expression has the following characteristics: (1) a boolean expression consists of one or more genes connected by a linking function from the logical operator set F_b; (2) in each gene, the head can contain logical operators from F_b and nominal attribute values, and the tail contains only nominal attribute values.

Taking Fig. 1 (a) as an example, the boolean expression b_i is generated by *gene1* and *gene2*. Each gene contains two logical operators in the head and three nominal attribute values in the tail. *gene1* and *gene2* are paired into two corresponding expression trees, and linked by the linking function $\vee$. Note that

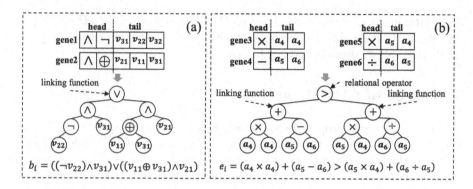

Fig. 1. Example of the generation of a boolean expression and a relational expression. (a) Boolean expression generation; (b) Relational expression generation.

$\neg$ in *gene1* is a unary operator, so v_{32} is redundant when *gene1* is parsed into its corresponding expression tree.

As for a relational expression $e_i \in E$, a simple way to generate it is to use a relational operator to connect two numerical attribute values. However, in this way, relational expressions only have two numerical attribute values, which lack the flexibility of expressions and may not be able to provide rich contrasts in complex practical applications. To avoid this limitation, inspired by [6], CHPMiner takes numerical attribute values as GEP terminals and arithmetic operators as GEP functions to generate arithmetic expressions, and a relational operator is used to connect two arithmetic expressions. A relational expression has the following characteristics: (1) a relational expression has two arithmetic expressions connected by a relational operator from the relational operator set L; (2) each arithmetic expression is generated by one or more genes, with the genes of each arithmetic expression connected by a linking function from the arithmetic operator set F_e; (3) the head of each gene can contain arithmetic operators from F_e and numerical attribute values, and the tail contains only numerical attribute values.

As shown in Fig. 1 (b), all genes contain one arithmetic operator in the head and two numerical attribute values in the tail. The relational expression contains two arithmetic expressions connected by a relational operator $>$. Each arithmetic expression consists of two genes, i.e., *gene3* and *gene4* are combined into one arithmetic expression, and *gene5* and *gene6* are combined into another arithmetic expression. Expression trees of genes of an arithmetic expression are connected by a linking function $+$.

4.2 Hybrid Expression Generation

To generate a hybrid expression, we need to combine boolean expressions and relational expressions in a unified way. Since the results of boolean expressions

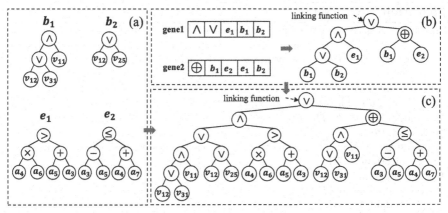

$$p_i = (((v_{12} \vee v_{31}) \wedge v_{11}) \vee (v_{12} \vee v_{25}) \wedge ((a_4 \times a_6) > (a_5 + a_3)) \vee (((v_{12} \vee v_{31}) \wedge v_{11}) \oplus ((a_3 - a_5) \leq (a_4 + a_7)))$$

Fig. 2. Example of a hybrid expression generation. (a) Expression trees of sub-expressions b_1, b_2, e_1, and e_2; (b) The hybrid expression and its corresponding expression tree; (c) Decoded expression tree of the hybrid expression based on the expression trees in (a) and (b).

and relational expressions are boolean values, we can naturally combine them by performing the logical operation to generate hybrid expressions.

We design a new chromosome structure to generate hybrid expressions, called *hybrid chromosome*, which takes sub-expressions as GEP terminals and logical operators as functions. A hybrid chromosome has the following characteristics: (1) a hybrid chromosome consists of one or more genes connected by a linking function from the combination operator set F_p; (2) in each gene, the head can contain combination operators from F_p and sub-expressions, and the tail contains only sub-expressions.

Thus, a hybrid chromosome contains several combination operators and sub-expressions. To ensure the diversity of individuals, we select sub-expressions randomly during the generation process. Moreover, the structure of the hybrid chromosome ensures that it can be decoded to a valid hybrid expression containing nominal and numerical attributes.

Here gives an example about the generation of a hybrid expression $p_i \in P$. Figure 2 (a) shows four expression trees of sub-expressions, which are generated in the same way as in Fig. 1. According to Fig. 2 (b), the hybrid expression is generated by *gene1* and *gene2*, which are paired into two corresponding expression trees and linked by a linking function $\vee$. The two genes have a head with the length of 2 and a tail with length of 3. In *gene1*, the head contains two combination operators $\wedge$ and $\vee$. In *gene2*, the head contains a combination operator $\oplus$ and a boolean expression b_1. The two genes have boolean expressions and relational expressions. Figure 2 (c) illustrates the decoded expression tree for the hybrid expression using the expression trees in Fig. 2 (a) and (b). Then, we can obtain the hybrid expression p_i according to the decoded expression tree.

Algorithm 1. CHPMiner(D_1, D_2, F_b, F_e, F_p, L, α, ε, η)

Input: D_1, D_2: two parts of data, F_b: logical operator set, F_e: arithmetic operator set, F_p: combination operator set, L: relational operator set, α: contrast score threshold, ε: parameter for avoiding zero-frequency problem, η: the maximal length of a expression

Output: *optimalCHP*: a CHP with the highest *cScore*

1: B, E ← SubExpressionGeneration(D_1, D_2, F_b, F_e, L);
2: ComputeFit(B, E, D_1, D_2, η);
3: *biggest_cScore* ← 0;
4: **while** the number of generations has not reached the predefined value **do**
5:     Selection(B, E);
6:     Mutation(B, E);
7:     Recombination(B, E);
8:     ComputeFit(B, E, D_1, D_2, η);
9:     P ← PGeneration($B \cup E$, F_p);
10:    **for** each $p \in P$ **do**
11:        Decode(p);
12:        $cScore(p, D_1, D_2)$ ← ComputeScore(p, D_1, D_2, ε);
13:        **if** $cScore(p, D_1, D_2) > \alpha$ **then**
14:            **if** $cScore(p, D_1, D_2) > biggest\_cScore$ **then**
15:                $biggest\_cScore$ ← $cScore(p, D_1, D_2)$;
16:                $optimalCHP$ ← p;
17:            **end if**
18:        **end if**
19:    **end for**
20: **end while**
21: **return** *optimalCHP*

4.3   Fitness Function Design

The fitness function in GEP determines the evolution direction of candidate solutions. As stated before, CHPMiner aims to find a CHP with the highest contrast score among CHPs, i.e., it appears more in D_1 and less in D_2. Inspired by the idea that the evolution of modular solutions can improve performance [12], sub-expressions are designed to be self-evolved to provide high-quality combinations of hybrid expressions. Hence, we design a specific fitness function to guide the evolution direction of sub-expressions. The fitness of a sub-expression s is calculated as follows.

$$fit(s) = cnt(s, D_1) * (|D_2| - cnt(s, D_2)) \tag{3}$$

where $|D_2|$ is the number of instances in D_2, $cnt(s, D_1)$ and $cnt(s, D_2)$ are the number of instances in D_1 and D_2 that hold s, respectively. Based on Eq. (3), a sub-expression that appears more in D_1 and less in D_2 will get high fitness value. Since sub-expressions with higher fitness values will get larger opportunities to survive and evolve, using Eq. (3) can obtain high-quality results.

Intuitively, shorter expressions are more concise and understandable. Therefore, for sub-expressions with the same fitness values, shorter ones are wanted. An modified fitness function based on Eq. (3) is designed.

$$fit^*(s) = \begin{cases} fit(s) + (1 - \dfrac{|s|}{\eta}), & |s| < \eta \\ fit(s), & |s| \geq \eta \end{cases} \tag{4}$$

where $|s|$ is the length (number of attribute values in s) of s, and $\eta > 0$ is the maximal length of an expression defined by the user.

During the evolution process, sub-expressions are first selected into the next generation according to their fitness values. Then, they are modified by mutation and recombination. Note that mutation for relational expressions is different from that for boolean expressions. Since the relational operator in a relational expression is not at the gene head, it is not changed during mutation. It means that the relational operator remains constant throughout evolution and reduces the diversity of relational expressions. Hence, CHPMiner sets the relational operator randomly change into another relational operator in L. And the change rate of the relational operator is the same as the mutation rate.

Finally, Algorithm 1 states the pseudo-code of CHPMiner. In Algorithm 1, Step 1 creates initial sub-expressions. Step 2 and 8 evaluate fitness of each sub-expression according to Eq. (4). From Step 5 to Step 7, selected sub-expressions are modified by genetic operators. Then, Step 9 generates hybrid expressions based on the hybrid chromosome structure by using the modified sub-expressions. Step 10 to Step 13 calculate the contrast score of each hybrid expression and find CHPs which meet the given condition. Step 14 to Step 16 select the optimal CHP so far from CHPs according to their contrast score. The source code of our implementation is available for download[1]

5 Experiment Evaluation

5.1 Experimental Settings

Datasets. We use four nominal-numerical mixed datasets from the UCI machine learning repository. Instances containing missing values are removed from the original datasets. Table 2 shows the statistical characteristics of the datasets.

- Credit Approval dataset[2]: The dataset is about credit card applications. D_1 contains instances with label "+", and D_2 contains instances with label "-".
- Cylinder Band dataset[3]: The dataset is about the cylinder banding in rotogravure printing. D_1 contains instances with label "Band", and D_2 contains instances with label "No band".

---

[1] Link to the source code: https://github.com/fumin-git/CHP-Miner.git.
[2] https://archive.ics.uci.edu/ml/datasets/Credit+Approval.
[3] https://archive.ics.uci.edu/ml/datasets/Cylinder+Bands.

- Census Income dataset[4]: The dataset is a collection of income from different states' instances. D_1 contains instances with label "$\leq$ 50K", and D_2 contains instances with label "> 50K".
- Allhypo dataset[5]: The dataset is a collection of thyroid disease data. D_1 contains instances with label "compensated hypothyroid", and D_2 contains instances with label "primary hypothyroid".

Table 2. The statistics of datasets in experiments

Dataset	#ins. in D_1	#ins. in D_2	#attributes
Credit approval	296	357	15
Cylinder band	99	178	35
Census income	22654	7508	14
Allhypo	106	50	21

Table 3. Parameter settings

Parameter	Values	Parameter	Values
α	1	Number of generations	500
ε	0.001	Selection operator	Tournament
η	15	Mutation rate	0.04
Population size	100	One-point recombination rate	0.4

Baselines. To the best of our knowledge, CHPMiner is the first work to mine CHPs. To verify the necessity of mining CHPs combining nominal and numerical attributes, we compare CHPMiner with the following algorithms:

- GEP-Nomi [7]: a classical GEP algorithm that can be used to find contrast patterns containing nominal attributes. Numerical attributes are leveraged by discretizing them into items using equal-length-bin technique [14].
- GEP-Nume [6]: a GEP-based algorithm for mining contrast patterns that can only deal with numerical attributes.
- CHPMiner-Nomi: a variant algorithm of CHPMiner that only deals with nominal attributes. Numerical attributes are leveraged by discretizing them into items using equal-length-bin technique [14].
- CHPMiner-Nume: a variant algorithm of CHPMiner that only deals with numerical attributes.

---

[4] https://archive.ics.uci.edu/ml/datasets/census+income.
[5] https://archive.ics.uci.edu/ml/datasets/thyroid+disease.

Experimental Setup. Table 3 shows the parameter settings of all algorithms. Besides, for CHPMiner and its two variants, the head length of the gene is set to be 1, the number of genes in boolean expressions, relational expressions, and hybrid expressions is set to be 2, 1, and 2, respectively. In order to have a fair comparison, for GEP-Nomi and GEP-Nume, the head length of the gene is set to be 2, and the number of genes is 4 for the former and 2 for the latter.

To compare the ability of CHPMiner and baselines to find the optimal pattern, we use the contrast score to measure the quality of contrast patterns, which is equivalent to the interest measure in EP mining [8]. For the convenience of description, the nominal attribute value and numerical attribute value in our experiments are denoted by v_i and a_i, respectively, and i starts from 0. We run all algorithms 100 times independently on each dataset.

Table 4. Comparison in maximum contrast score of contrast patterns found by all algorithms on four datasets

	Credit Approval	Cylinder Band	Census Income	Allhypo
GEP-Nomi	170	90	33	170
GEP-Nume	160	120	25	720
CHPMiner-Nomi	80	40	23	140
CHPMiner-Nume	35	50	6	450
CHPMiner	**186**	**150**	**57**	**740**

*The contrast score ranges from 0 to 1000

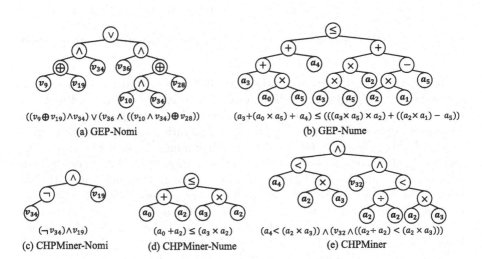

$((v_9 \oplus v_{19}) \wedge v_{34}) \vee (v_{36} \wedge ((v_{10} \wedge v_{34}) \oplus v_{28}))$

(a) GEP-Nomi

$(a_3 + (a_0 \times a_5) + a_4) \leq (((a_3 \times a_5) \times a_2) + ((a_2 \times a_1) - a_5))$

(b) GEP-Nume

$(\neg v_{34}) \wedge v_{19}$

(c) CHPMiner-Nomi

$(a_0 + a_2) \leq (a_3 \times a_2)$

(d) CHPMiner-Nume

$(a_4 < (a_2 \times a_3)) \wedge (v_{32} \wedge ((a_2 \div a_2) < (a_2 \times a_3)))$

(e) CHPMiner

Fig. 3. The optimal contrast patterns and corresponding expression trees of all algorithms on Credit Approval dataset

5.2   Performance Analysis

The proposed CHPMiner is compared with all baselines on the Credit Approval dataset, Cylinder Band dataset, Census Income dataset, and Allhypo dataset. We use these baselines to find the optimal contrast patterns to verify the ability of CHPMiner for mining CHPs with high discrimination. Table 4 records the results in contrast score (i.e., maximum contrast score of 100 experimental results) of the optimal contrast pattern mined by each algorithm on each dataset.

Table 4 shows that CHPMiner achieves the best performance. To further illustrate the superiority of CHPs in discrimination, Fig. 3 lists concrete expressions of the optimal contrast patterns found by all algorithms on the Credit Approval dataset (due to limited paper space, results on other datasets are not listed here, but can be found in the source code). The result shows that the optimal contrast patterns found by baselines only contain a single type of attribute. Comparatively, the optimal CHP found by CHPMiner contains nominal and numerical attributes simultaneously. The result indicates that CHPMiner can mine contrast patterns that are more distinguishing when exploiting the mathematical properties of nominal and numerical attributes in some scenarios.

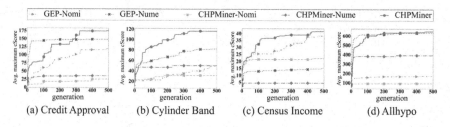

Fig. 4. Progression of average maximum contrast score of the optimal contrast pattern

In addition, to further demonstrate the ability of CHPMiner in finding the optimal CHP, Fig. 4 illustrates the change of contrast score of the optimal contrast pattern found by all algorithms in the evolution process. Note that the contrast score here is the maximum for each generation, and we use the average results of 100 experiments. Generally, CHPMiner mines the optimal contrast pattern with higher contrast score earlier than baselines. In particular, on datasets Cylinder Band and Census Income, we can see that from around the 50th generation, the contrast score of the optimal contrast pattern found by CHPMiner is already higher than that found by baselines. It can be observed that on the Allhypo dataset, the results of CHPMiner and GEP-Nume are relatively close. It may be that numerical attributes in this dataset are more likely to depict differences between different classes of data than nominal attributes.

Besides, Fig. 5 shows the runtime for CHPMiner to find the optimal CHP with respect to the number of instances. We can see that the runtime increases linearly with the number of instances in each synthetic dataset generated from the original datasets.

5.3   Case Study

To demonstrate the capability of CHPMiner in finding meaningful CHPs, we select a CHP randomly from the results of Allhypo dataset. Take $((v_{16} \oplus v_0) \vee (v_{19} \oplus v_9)) \wedge ((a_1 + a_1 + a_0) < (a_5 \div a_0 \times a_3))$ with the contrast score of 43.33 as an example, its *support* is 0.91 in D_1, and its *support* is 0.02 in D_2. According to the meanings of variables as shown in Table 5, the nominal part $(v_{16} \oplus v_0)$ can be translated as (Query Hypothyroid=t$\oplus$Sex=M). This part reflects the difference between compensated hypothyroid and primary hypothyroid, which is consistent with the results in existing studies [13,19] that women are more likely to suffer compensated hypothyroid than men. The numerical part computes the numerical relationship among TSH, FTI, TT4, and Age. These values have proven to be important for physicians when making decisions. The above analysis indicates the utility of CHPs in a real-world scenario, demonstrating the potential of CHPMiner in finding meaningful contrast between classes of data.

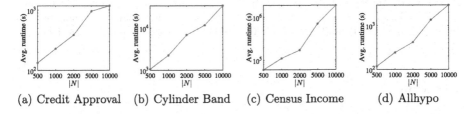

(a) Credit Approval   (b) Cylinder Band   (c) Census Income   (d) Allhypo

Fig. 5. Runtime of CHPMiner *w.r.t.* the number of instances $|N|$

Table 5. Variables and corresponding meanings of Allhypo dataset

Variable	Meaning	Variable	Meaning	Variable	Meaning
v_0	Sex = M	v_1	Sex = F	v_2	Thyroxine = t
v_3	On Thyroxine = f	v_4	Query On Thyroxine = t	v_5	Query On Thyroxine = f
v_6	On Antithyroid Medication = t	v_7	On Antithyroid Medication = f	v_8	Sick = t
v_9	Sick = f	v_{10}	Pregnant = t	v_{11}	Pregnant = f
v_{12}	Thyroid Surgery = t	v_{13}	Thyroid Surgery = f	v_{14}	131 Treatment = t
v_{15}	131 Treatment = f	v_{16}	Query Hypothyroid = t	v_{17}	Query Hypothyroid = f
v_{18}	Query Hyperthyroid = t	v_{19}	Query Hyperthyroid = f	v_{20}	Lithium = t
v_{21}	Lithium = f	v_{22}	Goitre = t	v_{23}	Ggoitre = f
v_{24}	Tumor = t	v_{25}	Tumor = f	v_{26}	Hypopituitary = t
v_{27}	Hypopituitary = f	v_{28}	Psych = t	v_{29}	Psych = f
a_0	Age	a_1	TSH	a_2	T3
a_3	TT4	a_4	T4U	a_5	FTI

6   Conclusion

In this paper, we introduce a novel contrast pattern mining task of CHP discovery from nominal-numerical mixed data. Different from previous contrast patterns, a CHP contains nominal attributes and numerical relationships among numerical attributes. We present CHPMiner, an extended GEP algorithm for mining CHPs by developing a novel hybrid chromosome structure to combine two kinds of self-evolved sub-expressions, i.e., boolean expressions of nominal attributes and relational expressions of numerical attributes. CHPMiner leverages a specific fitness function to guide the evolution direction for generating CHPs with high discrimination. Experiments on four real-world datasets show that CHPMiner outperforms baselines, demonstrating its effectiveness.

In the future, we plan to design a more comprehensive structure for CHPs. CHPMiner can also be applied to different fields such as financial data analysis to further evaluate its effectiveness.

Acknowledgments. This work was supported in part by the National Natural Science Foundation of China (61972268), the Sichuan Science and Technology Program (2020YFG0034), and the Med-X Center for Informatics funding project of SCU (YGJC001).

References

1. Chavary, E.A., Erfani, S.M., Leckie, C.: Mining rare recurring events in network traffic using second order contrast patterns. In: IJCNN, pp. 1–8 (2021)
2. Chavary, E.A., Erfani, S.M., Leckie, C.: Scalable contrast pattern mining over data streams. In: CIKM, pp. 2842–2846 (2021)
3. Dong, G., Li, J.: Efficient mining of emerging patterns: discovering trends and differences. In: KDD, pp. 43–52 (1999)
4. Duan, L., Dong, G., Wang, X., Tang, C.: Efficient mining of discriminating relationships among attributes involving arithmetic operations. Comput. Intell. **32**(1), 102–126 (2016)
5. Duan, L., Tang, C., Tang, L., Zhang, T., Zuo, J.: Mining class contrast functions by gene expression programming. In: ADMA, pp. 116–127 (2009)
6. Duan, L., Zuo, J., Zhang, T., Peng, J., Gong, J.: Mining contrast inequalities in numeric dataset. In: WAIM, pp. 194–205 (2010)
7. Ferreira, C.: Gene expression programming: a new adaptive algorithm for solving problems. Complex Syst. **13**(2) (2001)
8. Grosskreutz, H., Rüping, S.: On subgroup discovery in numerical domains. In: ECML PKDD, p. 30 (2009)
9. Khade, R., Lin, J., Patel, N.: Finding contrast patterns for mixed streaming data. In: EDBT, pp. 632–641 (2018)
10. Khade, R., Lin, J., Patel, N.: Finding meaningful contrast patterns for quantitative data. In: EDBT, pp. 444–455 (2019)
11. Komiyama, J., Ishihata, M., Arimura, H., Nishibayashi, T., Minato, S.I.: Statistical emerging pattern mining with multiple testing correction. In: SIGKDD, pp. 897–906 (2017)

12. Koza, J.R., Andre, D., Keane, M.A., Bennett III, F.H.: Genetic programming III: Darwinian invention and problem solving, vol. 3. Morgan Kaufmann (1999)
13. Li, J., et al.: Differential lipids in pregnant women with subclinical hypothyroidism and their correlation to the pregnancy outcomes. Sci. Rep. **11**(1), 1–9 (2021)
14. Li, J., Dong, G., Ramamohanarao, K.: Making use of the most expressive jumping emerging patterns for classification. In: PAKDD, pp. 220–232
15. Li, J., Liu, G., Wong, L.: Mining statistically important equivalence classes and delta-discriminative emerging patterns. In: SIGKDD, pp. 430–439 (2007)
16. Li, Q., Chen, X., Wu, R.: Mining contrast sequential patterns based on subsequence location distribution from biological sequences. In: DSIT, pp. 204–209 (2019)
17. Li, Y., Matzka, L., Flahive, J., Weber, D.: Potential use of leukocytosis and anion gap elevation in differentiating psychogenic nonepileptic seizures from epileptic seizures. Epilepsia Open **4**(1), 210–215 (2019)
18. Loekito, E., Bailey, J.: Fast mining of high dimensional expressive contrast patterns using zero-suppressed binary decision diagrams. In: KDD, pp. 307–316 (2006)
19. Redford, C., Vaidya, B.: Subclinical hypothyroidism: should we treat? Post Reprod. Health. **23**(2), 55–62 (2017)
20. Schmidt, J., et al.: Interpreting PET scans by structured patient data: a data mining case study in dementia research. Knowl. Inf. Syst. **24**(1), 149–170 (2010)

TPFL: Test Input Prioritization for Deep Neural Networks Based on Fault Localization

Yali Tao[1], Chuanqi Tao[1,2,3,4(✉)], Hongjing Guo[1], and Bohan Li[1,3]

[1] College of Computer Science and Technology, Nanjing University of Aeronautics
and Astronautics, Nanjing, China
{taoyali,taochuanqi,guohongjing,bhli}@nuaa.edu.cn
[2] Ministry Key Laboratory for Safety-Critical Software Development and
Verification, Nanjing University of Aeronautics and Astronautics, Nanjing, China
[3] Collaborative Innovation Center of Novel Software Technology and
Industrialization, Nanjing University, Nanjing, China
[4] State Key Laboratory for Novel Software Technology,
Nanjing University, Nanjing, China

Abstract. DNN testing is a critical way to guarantee the quality of
DNNs. To obtain test oracle information, DNN testing requires a huge
cost to label test inputs, which greatly affects the efficiency of DNN testing. To alleviate the labeling cost problem, the paper applies the idea
of the spectrum-based fault location technique to DNN testing and proposes a novel test input prioritization approach for DNNs based on fault
localization (called TPFL). TPFL first performs dynamic spectrum analysis on each neuron in the DNN. TPFL then proposes a suspiciousness
measure that uses the neuron spectrum to identify suspicious neurons
that cause the DNN to make wrong decisions. Finally, TPFL is based
on the following key insight: a test input makes the suspicious neurons
fully active, it indicates that this may be a bug-revealing input, so the
input should have a higher priority. To evaluate, we conduct an empirical
study on 3 widely used datasets and corresponding 8 DNN models. The
experimental results show that TPFL performs well in both classification
and regression models and overall outperforms most existing test input
prioritization techniques.

Keywords: Deep neural networks · Test input prioritization · Deep
learning testing

1 Introduction

In recent years, deep neural networks (DNNs) are increasingly being applied to
various domains of applications, e.g., image classification [12], medical diagnostics [13], and autonomous driving [6]. However, like traditional software, DNNs
also expose unexpected behaviors that can lead to serious consequences, such as
accidents caused by the self-driving cars of Google and Tesla.

© The Author(s), under exclusive license to Springer Nature Switzerland AG 2022
W. Chen et al. (Eds.): ADMA 2022, LNAI 13725, pp. 368–383, 2022.
https://doi.org/10.1007/978-3-031-22064-7_27

DNN testing is a critical way to guarantee the quality of DNNs. At present, the research on DNN testing mainly includes three parts: test metrics, test input generation, and test oracles. Recent research work [14,18,21] proposed a variety of metrics to measure the adequacy of the test, and some approaches were designed to generate adequate test inputs in [11,21,28]. However, to solve the test oracle problem in DNN testing, it is often necessary to manually label the test inputs, which is expensive and time-consuming. For example, the Imagenet dataset, the most widely used visual recognition dataset, was labeled by more than 49,000 workers from 167 countries over 9 years. The main reasons for the heavy cost of labeling work are as follows: First, the test set of DNN is often large-scale. Secondly, labeling requires certain domain knowledge, such as DNN for medical diagnosis, whose test inputs must be labeled by experts. Finally, human analysis is a subjective process, so multiple people are required to label a test input to ensure the correctness of the labeling result. Therefore, the expensive labeling cost remains a key challenge in DNN testing.

A feasible solution to alleviate this problem is to prioritize the test inputs so that under limited time and resources, as many test inputs that expose the defects of the DNN model can be labeled as possible. Therefore, [4,9,26,29, 32] proposed some test prioritization techniques for DNNs. However, most of these techniques [4,9,29,32] can only be applied to classification models and have limited application scenarios. Moreover, for example, DeepGini [9], the state-of-the-art prioritization approach for DNNs, prioritized test inputs based on the prediction probability of the classification models. When the model has low accuracy on a certain test set (such as an adversarial sample set), it will happen that the test input is misclassified with a high prediction probability. As in [10], the adversarial samples generated by adding invisible perturbations to the original images can allow the model to misclassify "Panda" as "Gibbon" with 99.3% confidence. In this case, the effectiveness of DeepGini is significantly reduced.

Therefore, to further improve the test prioritization technology for DNNs, this paper applies the idea of the spectrum-based fault location technique [1] to DNN testing and proposes a novel test input prioritization approach for DNNs based on fault localization (called TPFL). First, TPFL performs dynamic spectrum analysis on DNN neurons. Since different neurons in DNN often have different functional distributions (that is, the output range of neurons in the entire training set is different), TPFL adopts a dynamic activation threshold for each neuron, rather than a unified and predefined activation threshold. Then, TPFL proposes a suspiciousness measure based on neuron spectrum to identify suspicious neurons that may cause DNN to make incorrect decisions [8]. Finally, TPFL performs test input prioritization based on the following core insight: if a test input makes the suspicious neurons fully active (the neuron's output is large), it indicates that this may be a bug-revealing input, so the input should have a higher priority.

To evaluate the performance of TPFL, we conduct an empirical study on 3 widely used datasets and corresponding 8 DNN models. In particular, we consider DNN models for different tasks (classification and regression models). And different types of test inputs (natural test inputs and adversarial test inputs generated by the most widely used adversarial generation methods). The experimental results show that TPFL performs well in both classification and regression models, and the time consumption is completely acceptable. Especially on the regression model, TPFL outperforms all the compared approaches (with the average improvement of 6.15%–222.58% in terms of RAUC).

To summarize, this paper has the following major contributions:

- We propose a neuron spectral dynamic analysis approach and a novel suspiciousness measure approach to identify suspicious neurons in a DNN.
- We propose TPFL, which is an effective and novel test input prioritization approach based on suspicious neurons for classification and regression DNN models.
- Empirical studies on 3 widely used datasets and corresponding 8 DNN models demonstrate the effectiveness and efficiency of TPFL.

The remainder of this paper is organized as follows: Sect. 2 introduces the relevant basic concepts and the metric for evaluating the performance of test input prioritization. In Sect. 3, the approach proposed by this paper is elaborated. Section 4 presents the setting for the empirical study. Section 5 discusses the experimental results of the proposed research questions. Section 6 gives a summary of related work. Finally, Sect. 7 concludes this paper.

2   Background

DNN and DNN testing. DNNs consist of a large number of interconnected neurons located in different layers. Each neuron is a computational unit with weights. These weights are learned from the training data during the training process. The layers of DNN are mainly divided into input layer, hidden layer and output layer [17]. The input layer is used to obtain the input information. The hidden layer is mainly used to extract various features of the input. The output layer produces the prediction results of the model. Based on the type of prediction results, DNNs can be classified into two types: classification models and regression models.

DNN testing is one of the main methods to ensure the quality of DNNs [2,7]. DNN testing cannot be done without test cases. A test case is equal to a test input and a test oracle. The test input is the input that needs to be predicted by the DNN under test. The test oracle is check whether the test input is predicted correctly by the DNN under test.

Software Fault Localization. Traditional software fault location (FL) techniques can be classified into eight categories: spectrum-based, slice-based, program state-based, machine-learning-based, and other techniques [27]. Among them, the spectrum-based FL technology is the most widely studied. The spectrum-based FL technology will create a tuple (es, ef, ns, nf) for each program element p_i. es, ef (ns, nf) represent the p_i executed (not executed) frequency under the successful test cases and the failed test cases respectively. Then, the tuple of each p_i is used to calculate the suspiciousness by some measure to effectively find the code elements that may contain faults.

Test Input Prioritization and Evaluation Metrics. Test prioritization is a classic problem in traditional software [22]. It adjusts test inputs in a new order to form a prioritized test set to improve its effectiveness on some performance goals. The most widely used metric for evaluating test prioritization techniques in traditional software testing is the *Average Percent of faults detected* (APFD). However, in DNN testing, when a DNN under test has a low accuracy on a test set, the maximum value of APFD is much less than 1. This may affect comparisons between different test sets under the same model.

Therefore, [4,26] and this paper adopt the evaluation metric RAUC. Firstly, the prioritization result produced by a test prioritization technique is graphed in a figure. The x-axis of the figure is the number of the prioritized test cases and the y-axis is the number of the failed test cases in prioritized test cases. RAUC is the area under the actual prioritization result curve divided by the area under the ideal prioritization result curve. RAUC-n refers to the RAUC for the first n prioritized test cases. Therefore, a larger RAUC means closer to the optimal result of test prioritization.

3 Approach

In order to label as many test inputs that reveal the erroneous behavior of DNNs as possible in a limited time, this section will propose a test input prioritization approach for DNNs based on fault localization (TPFL), whose workflow is shown in Fig. 1. First, TPFL transposes the spectrum-based fault localization [1] approach to DNNs to identify suspicious neurons that may cause DNNs to make wrong decisions [8]. Then, TPFL is based on the following core insight: if a test input makes the suspicious neurons fully active (the output value of the neuron is large), indicating this may be an error-triggering input.

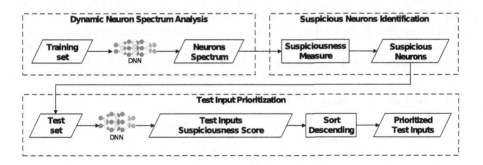

Fig. 1. The workflow of TPFL

3.1 Dynamic Neuron Spectrum Analysis

Like traditional software spectrum-based fault location methods, we need to establish a spectrum for neurons to record information about the internal activity of the DNNs. First, let D is a DNN model, the D with m layers and the i-th $(1 \leq i \leq m)$ layer has total l_i neurons, $n_{ij}(1 \leq i \leq m, 0 < j \leq l_i)$ is the j-th neuron in the i-th layer in D, x is an input of the D, $D(x)$ is the decision of the D and $O(n_{ij}, x)$ is the output of the neuron n_{ij} when given the input x. When $O(n_{ij}, x) \geq p$ (p is the activation threshold), it means that the neuron is active, otherwise, it is inactive.

Therefore, a neuron can have the following four execution properties. A_n^{as} and A_n^{us} represent the number of times neuron n was activated and inactivated when the D made correct decisions respectively. Similarly, A_n^{af} and A_n^{uf} represent the number of times neuron n was activated and inactivated when the D made incorrect decisions respectively. The four attributes of the neuron are combined into a hit spectrum (HS) to describe its behavior.

Definition 1: Given a test set T and a DNN model D, The HS of the neuron n in D is defined a tuple $HS_n^T = (A_n^{as}, A_n^{af}, A_n^{us}, A_n^{uf})$, and $A_n^{as} + A_n^{af} + A_n^{us} + A_n^{uf} = |T|$. Noticeably, the HS is not meaningful for the neurons in the input layer. Because the input layer is often a high-dimensional data to represent an input (e.g., each neuron in the input layer of an image classification model represents each pixel value of each color tunnel), its activation state is meaningless.

However, in recent research works [8,11,16,21,25], all neurons in a DNN use the predefined and same threshold to determine whether they are active or not. Differences in functional distribution between different neurons are not considered. Therefore, TPFL adopts a unique activation threshold for each neuron $(Thre_n)$ in the DNN, which is the average plus the standard deviation of the output values of each neuron on the test set T. The formula is as follows:

$$Thre_{n_{ij}} = Average(O(n_{ij}, T)) + StandardDeviation(O(n_{ij}, T)) \qquad (1)$$

In addition, TPFL can be used for classification and regression models. For classification models, when the label predicted by the DNN is not equal to the original label, it indicates that the DNN made a wrong decision. For regression models, we define that when $|Y - Y_{pre}| > a$, indicating the DNN making an incorrect decision. Among them, Y represents the correct output of an input, Y_{pre} represents the value predicted by the DNN, and a is the defined error range.

3.2 Suspicious Neurons Identification

The suspiciousness score (SPS) can measure the correlation between the behavior of a neuron and the decisions of the DNN. Therefore, TPFL proposes a spectrum-based suspiciousness measure to compute the suspiciousness score of neurons. The formula of SPS_n is defined as follows:

$$SPS_n = \frac{(A_n^{af})^x + A_n^{us}}{A_n^{as} + A_n^{uf}}(x > 0) \qquad (2)$$

The key insight underlying the suspiciousness measure is that the more frequently a neuron n is activated by test inputs where the DNN makes incorrect decisions, and the less frequently it is activated by test inputs where the DNN makes correct decisions, the SPS_n is higher. Since passed test cases are often much more than failed test cases in the training test set, to emphasize the effect of the A_n^{af}, TPFL performs the x-power operation on A_n^{af}. Referring to [8,20], in our experiments, x is set to 3. Although there is no guarantee that $x = 3$ is the optimal choice.

Since the output of a DNN is based on the aggregation effect of neurons, each neuron has its unique contribution to the overall computation [8]. The suspiciousness measure identifies a set of suspicious neurons rather than an individual suspicious neuron. Therefore, neurons are sorted in descending order of SPS. The top k neurons with SPS are selected as suspicious neurons SN_k. Because deeper neurons learn more abstract and concrete features [3,30], TPFL gives preference to neurons located at the deeper layer when there are different neurons with the same suspiciousness score. Noticeably, TPFL does not consider neurons in the input layer and layers without weights (e.g., flatten layers, activation layers).

3.3   Test Input Prioritization Based on Suspicious Neurons

First, TPFL sets corresponding weights for the k most suspicious neurons SN_k. We denote the i-th suspicious neuron in the SN_k as sn_i and the weight of the sn_i is W_i. When the SPS_n is higher, the greater the weight it has. The formula of W_i is as follows:

$$W_i = k - (i - 1)(1 \leq i \leq k) \tag{3}$$

Second, when given the test input x, $O(sn_i, x)$ is the output of the neuron sn_i. $P(x)$ represents the probability that the test input x can expose the DNN defects. The formula of $P(x)$ is as follows:

$$P(x) = \sum_{i=1}^{k} W_i * O(sn_i, x) \tag{4}$$

The test input x with a higher $P(x)$ indicates that it makes the suspicious neurons play a more important role in the DNN decision-making process. Therefore, TPFL will give it a higher priority.

4   Experiment Design

In this section, the experimental setup will be introduced. All experiments were performed on a Ubuntu 20.04.3 LTS server with one NVIDIA GTX 1080Ti GPU, one 8-core processor "Intel(R) Core(TM) i7-11700 @2.50GHz", and 128 GB physical memory. We implemented TPFL using Python 3.8.12, Keras 2.0.8, and Tensorflow 2.2.0.

4.1 Dataset and DNN Model

In our study, we used 36 pairs of data sets and DNN models, and the basic information is shown in Table 1. To comprehensively evaluate the effectiveness of the proposed approach, TPFL conducts experiments on DNN models for different tasks and different types of test inputs.

Different Tasks of DNN Models. Currently, many test input prioritization techniques for DNNs [9, 29, 32] can only be used for classification models. TPFL considers both classification and regression models. MNIST is an image dataset of handwritten digits ranging from 0 to 9, containing 60,000 training images and 10,000 test images. CIFAR-10 is a dataset including ten types of images. It has 50,000 training images and 10,000 testing images. MNIST_1, 2, 3, and CIFAR_1, 2, 3 are DNN models with different sizes trained by us. Driving is a dataset for self-driving cars consisting of 101,396 training camera images and 5,614 testing camera images from vehicles. The corresponding models we use are the pre-trained Dave_ori and Dave_dropout models from the literature [21]. Since the output of the Dave model is that the steering angle is a continuous value, 1-MSE (mean squared error) is used as the accuracy of the model.

Different Types of Test Input. TPFL considers not only the original test inputs but also the adversarial test inputs. We adopt the four most widely used adversarial input generation algorithms to construct mixed test sets for classification models, i.e., C&W [5], BIM [15], FGSM [10], and JSMA [19]. Specifically, for MNIST and CIFAR-10, we use each adversarial input generation algorithm to generate the same number of adversarial inputs as the original test inputs. Then, half of the original test set and half of each adversarial input set were randomly selected to form a mixed test set. However, for the Driving dataset, referring to the literature [25], we respectively adopt the contrast and brightness transformations to the original images to synthesize the same number of images. Then half of the synthetic image set is combined with half of the original test set to construct a mixed test set. The number of each test set is shown in Table 1.

Table 1. Details of the DNN models and datasets used and the corresponding test sets.

Task	Dataset	Model	Parameters	Layer	ACC or MSE	Test set
Classification	MNIST	MNIST_1	27,240	12	96.37%	Natural
		MNIST_2	22,975	14	96.05%	(10,000)
		MNIST_3	18,680	18	95.30%	C&W,FGSM
	CIFAR-10	CIFAR_1	1,250,858	17	77.60%	BIM,JSMA
		CIFAR_2	724,010	19	72.62%	Mixed
		CIFAR_3	411,434	25	63.12%	(4*10,000)
Regression	Driving	Dave_orig	2,116,983	13	0.0965	Natural (5,614)
						Contrast
		Dave_dropout	3,276,225	15	0.0819	Brightness
						mixed (2* 5,614)

4.2   Research Questions

RQ1. Layer Sensitivity: Does the selection of layers of suspicious neurons used for test input prioritization have any impact on effectiveness?

This section explores the test input prioritization based on (1) computing the suspiciousness scores of all neurons in a DNN, or (2) computing the suspiciousness scores of neurons in selected layers only, then getting a specified number of suspicious neurons. We use RAUC-n to answer RQ1. In our study, We consider n to be 100, 200, 300, 500, 1000, 2000, 3000, and 5000 respectively.

RQ2. Effectiveness: Can TPFL find a better permutation of tests than compared methods?

Similarly, RQ2 is answered by RAUC-n. This section considered DeepGini [9], LSA-based (Likelihood-based Surprise Adequacy [14]) test input prioritization, and random three comparison methods. Since DeepGini [9] is better than test input prioritization approaches based on different coverage criteria (e.g., NC, KMNC, NAC, etc.), excluding LSA, and DSA(Distance-based Surprise Adequacy [14]). However, DSA can only be used for classification models, so we consider comparing LSA-based methods.

RQ3. Efficiency: How efficient is TPFL?

The test input prioritization approach for DNNs is to label as many error-triggering inputs as possible in a limited time. Therefore, the efficiency of the approach itself needs to be considered, and if the approach takes a lot of time, it may not be practical. RQ3 is answered by the time cost of the approach.

5   Result and Discussion

5.1   RQ1. Layer Sensitivity

We select one model from each dataset as a representative to answer RQ1. Table 2 shows the RAUC results of test input prioritization based on the suspicious neurons located in different layers. These layers do not include layers without weights, and the number of suspicious neurons is the total number of neurons in a layer. The C&W mixed test set represents four mixed test sets of the classification model. The Contrast mixed test set represents two mixed test sets of the regression model. The bold numbers denote the best result under the same conditions.

As shown in Table 2, the RAUC of the deepest layer is much larger than the RAUC of other layers. For example, on CIFAR_1, RAUC-500 in the deepest layer is 0.615 and 0.861 for the natural test set and the C&W test set. However, RAUC-500 in the first layer is 0.279 and 0.610, an increase of 120.43% and 41.15%, respectively. Therefore, there is strong evidence that TPFL is more efficient when based on deeper suspicious neurons.

5.2   RQ2. Effectiveness

RQ1 demonstrates that TPFL is more efficient when based on deeper suspicious neurons. Therefore, the TPFL in this question is to select the layer before the output layer as the target layer, and then the number of suspicious neurons is the number of neurons in the target layer.

Overall Effectivenss. This study uses 36 pairs of data sets and the DNN models, so we first present the overall comparison results on different datasets in Table 3. For MNIST, TPFL performs best in 66.67% (80 cases in 120 cases). However, DeepGini performs best in 33.33%. Noticeably, when n < 5000, TPFL usually performs better than DeepGini. For CIFAR-10, DeepGini usually performs best. For Driving, DeepGini is not applicable. TPFL performs best in 100%, and the average RAUC-n is improved by 20.6%–134.7% compared with

Table 2. Comparison results when suspicious neurons located in different layers

Model	Test set	Layer Index	Suspicious Neurons num	RAUC-n							
				100	200	300	500	1000	2000	3000	5000
MNIST_1	Natural	All	160	0.058	0.098	0.099	0.099	0.130	0.197	0.266	0.382
		1	30	0.023	0.025	0.026	0.031	0.060	0.110	0.155	0.248
		2	30	0.083	0.108	0.107	0.102	0.131	0.185	0.237	0.345
		3	30	0.129	0.117	0.103	0.103	0.133	0.201	0.266	0.390
		4	30	0.110	0.092	0.077	0.072	0.111	0.204	0.286	0.412
		5	30	0.280	0.212	0.180	0.154	0.186	0.263	0.329	0.440
		6	10	**0.525**	**0.557**	**0.536**	**0.471**	**0.449**	**0.460**	**0.494**	**0.562**
	C&W	All	160	0.852	0.851	0.831	0.815	0.800	0.756	0.726	0.680
		1	30	0.661	0.632	0.627	0.612	0.609	0.590	0.577	0.558
		2	30	0.554	0.597	0.617	0.629	0.645	0.639	0.623	0.599
		3	30	0.752	0.746	0.722	0.698	0.671	0.648	0.633	0.611
		4	30	0.735	0.720	0.699	0.664	0.670	0.670	0.671	0.659
		5	30	0.914	0.913	0.894	0.880	0.843	0.781	0.742	0.689
		6	10	**1.000**	**0.988**	**0.985**	**0.984**	**0.983**	**0.979**	**0.979**	**0.947**
CIFAR_1	Natural	All	714	0.259	0.253	0.248	0.238	0.236	0.238	0.249	0.329
		1	(32,32,32)	0.348	0.299	0.286	0.279	0.278	0.261	0.257	0.321
		2	(30,30,32)	0.269	0.274	0.276	0.251	0.237	0.227	0.239	0.320
		3	(15,15,64)	0.303	0.306	0.294	0.272	0.247	0.224	0.231	0.304
		4	(13,13,64)	0.243	0.269	0.275	0.293	0.289	0.280	0.288	0.370
		5	512	0.154	0.161	0.178	0.192	0.210	0.218	0.236	0.320
		6	10	**0.602**	**0.629**	**0.636**	**0.615**	**0.578**	**0.532**	**0.515**	**0.593**
	C&W	All	714	0.621	0.615	0.614	0.615	0.616	0.616	0.616	0.615
		1	(32,32,32)	0.624	0.610	0.611	0.610	0.607	0.605	0.603	0.602
		2	(30,30,32)	0.629	0.625	0.617	0.604	0.606	0.612	0.615	0.614
		3	(15,15,64)	0.683	0.677	0.665	0.648	0.630	0.609	0.608	0.608
		4	(13,13,64)	0.610	0.614	0.618	0.628	0.630	0.628	0.631	0.628
		5	512	0.561	0.568	0.580	0.589	0.592	0.595	0.596	0.599
		6	10	**0.842**	**0.851**	**0.857**	**0.861**	**0.862**	**0.849**	**0.831**	**0.791**
Dave_orig	Natural	7	100	0.533	0.544	0.537	0.490	0.429	0.394	0.441	0.590
		8	50	0.662	0.590	0.555	0.512	0.438	0.392	0.436	0.586
		9	10	**0.945**	**0.881**	**0.825**	**0.734**	**0.608**	**0.513**	**0.550**	**0.675**
	Contrast	7	100	0.670	0.663	0.632	0.588	0.551	0.516	0.532	0.661
		8	50	0.770	0.681	0.637	0.582	0.539	0.513	0.531	0.660
		9	10	**1.000**	**0.993**	**0.982**	**0.947**	**0.843**	**0.701**	**0.675**	**0.760**

Table 3. Overall comparison results across all the subjects.

Dataset	Approach	Best case in RAUC-n								Average RAUC-n							
		100	200	300	500	1000	2000	3000	5000	100	200	300	500	1000	2000	3000	5000
MNIST	TPFL	**13**	**13**	**11**	**9**	**11**	**12**	**11**	0	**0.906**	**0.903**	**0.896**	**0.884**	0.878	0.872	0.886	0.871
	DeepGini	2	2	4	6	4	3	4	**15**	0.894	0.881	0.865	0.870	**0.885**	**0.919**	**0.938**	**0.959**
	LSA-based	0	0	0	0	0	0	0	0	0.856	0.856	0.841	0.774	0.831	0.814	0.808	0.712
	Random	0	0	0	0	0	0	0	0	0.390	0.408	0.375	0.421	0.425	0.437	0.447	0.469
CIFAR-10	TPFL	6	4	3	4	2	5	4	0	0.881	0.873	0.867	0.854	0.844	0.818	0.792	0.751
	DeepGini	**9**	**11**	**12**	**11**	**13**	**10**	**11**	**15**	**0.898**	**0.898**	**0.909**	**0.889**	**0.887**	**0.865**	**0.857**	**0.856**
	LSA-based	0	0	0	0	0	0	0	0	0.561	0.596	0.614	0.627	0.639	0.625	0.611	0.607
	Random	0	0	0	0	0	0	0	0	0.570	0.585	0.546	0.578	0.587	0.589	0.595	0.601
Driving	TPFL	**6**	**6**	**6**	**6**	**6**	**6**	**6**	**6**	**0.953**	**0.898**	**0.851**	**0.782**	**0.679**	**0.544**	**0.560**	**0.668**
	LSA-based	0	0	0	0	0	0	0	0	0.406	0.390	0.374	0.360	0.384	0.409	0.409	0.554
	Random	0	0	0	0	0	0	0	0	0.274	0.302	0.307	0.309	0.300	0.316	0.378	0.542

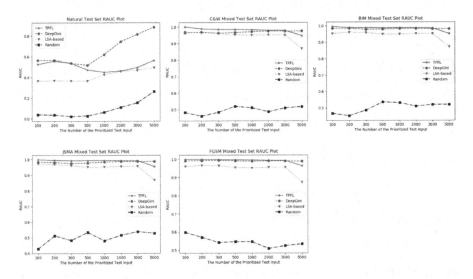

Fig. 2. The RAUC curves of different methods on MNIST_1 model under different test sets

LSA-based. Due to the limited space, Figs. 2, 3, and 4 only show the RAUC curves of different methods on the MNIST_1, CIFAR_1, and Dave_orig models. But the conclusions can hold for other models under the same dataset.

Effectiveness on Different Task of DNN Model. DeepGini can only be used for classification models, while TPFL can also be used for regression models. Moreover, TPFL performed well in regression models. As shown in Figs. 2 and 3, on the classification models MNIST_1 and CIFAR_1, TPFL and Deep-Gini perform better. On MNIST_1, the LSA-based is better than the random method. However, on CIFAR_1, LSA-based even performs worse than random. As shown in Fig. 4, on regression model Dave_orig, TPFL performs best. The RAUC results of TPFL range from 0.513 to 1.000 with the average improvements of 6.15%–222.58% compared with LSA-based.

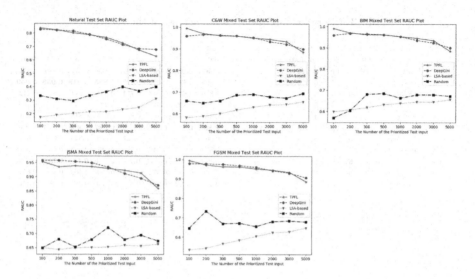

Fig. 3. The RAUC curves of different methods on CIFAR_1 model under different test sets

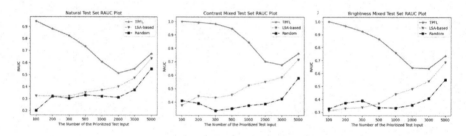

Fig. 4. The RAUC curves of different methods on Dave_orig model under different test sets

Effectiveness on Different Types of Test Inputs. TPFL considers not only the original test inputs but also the adversarial test inputs. Similarly, as shown in Fig. 2, for MNIST, on the Natural test set, TPFL performs slightly worse than DeepGini, but on the four mixed test sets, the overall performance of TPFL is better than DeepGini. For CIFAR-10, as shown in Fig. 3, on the Natural test set, TPFL performs slightly worse than DeepGini, but on the four mixed test sets, both perform well. This indicates that DeepGini is less effective when the model under test has low accuracy on a test set, because the model may misclassify the test input with high probability. However, TPFL does not have this challenge. For Driving, regardless of the type of test input, TPFL performs the best with clear improvements.

5.3   RQ3. Efficiency

Table 4 shows the average running time of different approaches on different datasets. Compared with DeepGini, the average time spent on TPFL and LSA-based is larger. Because TPFL needs to analyze the neuron spectrum and identify suspicious neurons. But this time is acceptable compared to the time and cost of manual labeling. Moreover, TPFL only needs to analyze the neuron spectrum and identify suspicious neurons once for each tested model. The results can be saved for next time, and it will take less than 1 min afterward.

Table 4. The average time of different approaches on different datasets (in seconds).

Dataset	Approach	Mean	Std	Min	Max
MNIST	TPFL	183.704	5.274	177.620	186.983
	DeepGini	0.611	0.119	0.512	0.743
	LSA-based	6.157	0.579	5.513	6.636
	Random	0.553	0.123	0.427	0.667
CIFAR-10	TPFL	666.607	88.848	582.135	755.650
	DeepGini	60.852	14.291	51.613	77.312
	LSA-based	248.224	250.325	94.712	537.084
	Random	57.872	13.411	44.559	71.379
Driving	TPFL	580.818	347.776	334.903	826.732
	LSA-based	209.2	12.935	200.053	218.346
	Random	13.238	0.013	13.229	13.247

5.4   Discussion

Extension of TPFL. Experiments have demonstrated that TPFL achieves great effectiveness for prioritizing test inputs. However, the input of DNNs also includes other domains, such as text, language, etc. Our work has only been verified in the image domain. However, TPFL has universality and there is no restriction on the input domain in the approach, so in the future, we can further apply TPFL to more domains.

Threats to Validity. The internal threat mainly lies in the implementation of TPFL and the other compared approaches. To reduce the threats, the implementation of the compared approaches use the code released by the original authors. At the same time, except for the difference in approaches, other parameters and variables are kept consistent to ensure the correctness of the comparison results.

The external threat mainly lies in the selection of the datasets and tested models. Therefore, our work adopts three well-known datasets and 8 pre-trained models of different sizes, structures, and complexity.

The construct threat mainly lies in the configurable parameters in TPFL and the evaluation metric in this paper. Firstly, RQ1 investigates the effect of the selection of target layer on the effectiveness of TPFL, and there is strong evidence that test input prioritization based on suspicious neurons in the deeper layers is more effective. Then, we set the number of suspicious neurons to the number of all neurons in the target layer, which may not be the best choice. Besides, the reason why we use the evaluation metric to be RAUC instead of APFD is also introduced in Sect. 2.

6   Related Work

DNN Testing. In recent years, many studies have focused on improving and ensuring DNN quality through testing [31]. Pei et al. [21] proposed neuron coverage to measure the adequacy of DNN testing. On this basis, Ma et al. [18] proposed multi-granularity test criteria. Other researchers [23,24], inspired by MC/DC testing, concolic testing, etc. in traditional software testing, have proposed various DNN coverage criteria. Kim et al. [14] proposed a novel test metric for a single input, called Surprise Adequacy including the LSA and the DSA. Zhang et al. [31], by constructing the decision structure of DNNs, proposed two different path coverage criteria to measure the adequacy of test cases in implementing decision logic. Meanwhile, it is necessary to use adequate test inputs to cover the corner cases of the DNN as much as possible. Therefore, [14,21] also proposed an approach to guide test input generation with coverage criteria. Guo et al. [11] adopted a differential algorithm to mutate the input to generate new test inputs with high coverage and expose the wrong behavior of DNNs. Lee et al. [16] adopted an adaptive strategy to dynamically adjust the neuron selection strategy to generate more diverse and higher coverage adversarial inputs. Xie et al. [28] proposed a diversity- and frequency-based seed selection strategy to select the seed inputs required in fuzzing and mutated the seed inputs using a metamorphic mutation strategy. Tian et al. [25] performed contrast, brightness changes, etc. on the original input to generate new test inputs to detect the wrong behavior of DNN-based autonomous cars.

Test Prioritization Techniques for DNNs. To alleviate the problem of labeling cost in DNN testing, Feng et al. [9] designed a test prioritization approach from a statistical perspective, DeepGini. They consider that a test input is predicted with similar class probabilities, indicating that the input is likely to be misclassified and should have a higher priority. Since traditional software test prioritization techniques follow an assumption that higher coverage inputs are more likely to detect faults. Therefore, DeepGini is compared with the test prioritization techniques based on various coverage criteria of DNNs. Experiment results prove that DeepGini is superior to coverage-based approaches. Wang et al. [26] prioritized the test inputs by model mutation and input mutation analysis. If an input kills the mutated model, or if a mutated input has a different prediction result than the original input, then it is more likely to be a bug-revealing input. Byun et al. [4] empirically studied the effectiveness of test prioritization approaches based on confidence, uncertainty, and surprise [14]. [29,32] utilized the difference between the neuron pattern from the given inputs and the neuron pattern extracted by the training set to prioritize test inputs.

7 Conclusion

The paper proposes a novel test input prioritization approach for DNNs called TPFL. TPFL applies the spectrum-based fault localization approach to DNNs to identify suspicious neurons that may cause DNNs to make wrong decisions. If a test input can make the suspicious neuron fully active, it indicates that this may be a bug-revealing input, TPFL will give it a high priority. Empirical studies on 3 well-known datasets demonstrated that TPFL outperforms other existing test prioritization approaches. In the future, we can further apply TPFL to more models in more domains. Besides, we will investigate whether the test input prioritized by TPFL can guide the model retraining and improve its quality in the future.

References

1. Abreu, R., Zoeteweij, P., Golsteijn, R.: A practical evaluation of spectrum-based fault localization. J. Syst. Softw. **82**(11), 1780–1792 (2009)
2. Aggarwal, A., Lohia, P., Nagar, S.: Black box fairness testing of machine learning models. In: Proceedings of the ACM Joint Meeting on European Software Engineering Conference and Symposium on the Foundations of Software Engineering, ESEC/SIGSOFT FSE 2019, Tallinn, Estonia, 26–30 August 2019, pp. 625–635. ACM (2019)
3. Bengio, Y., Mesnil, G., Dauphin, Y.N.: Better mixing via deep representations. In: Proceedings of the 30th International Conference on Machine Learning, ICML 2013, Atlanta, GA, USA, 16–21 June 2013. JMLR Workshop and Conference Proceedings, vol. 28, pp. 552–560. JMLR.org (2013)
4. Byun, T., Sharma, V., Vijayakumar, A.: Input prioritization for testing neural networks. In: IEEE International Conference On Artificial Intelligence Testing, AITest 2019, Newark, CA, USA, 4–9 April 2019, pp. 63–70. IEEE (2019)
5. Carlini, N., Wagner, D.A.: Towards evaluating the robustness of neural networks. In: 2017 IEEE Symposium on Security and Privacy, SP 2017, San Jose, CA, USA, 22–26 May 2017, pp. 39–57. IEEE Computer Society (2017)
6. Chen, Z., Huang, X.: End-to-end learning for lane keeping of self-driving cars. In: IEEE Intelligent Vehicles Symposium, IV 2017, Los Angeles, CA, USA, 11–14 June 2017, pp. 1856–1860. IEEE (2017)
7. Cheng, D., Cao, C., Xu, C.: Manifesting bugs in machine learning code: an explorative study with mutation testing. In: 2018 IEEE International Conference on Software Quality, Reliability and Security, QRS 2018, Lisbon, Portugal, 16–20 July 2018, pp. 313–324. IEEE (2018)
8. Eniser, H.F., Gerasimou, S., Sen, A.: DeepFault: fault localization for deep neural networks. In: Hähnle, R., van der Aalst, W. (eds.) FASE 2019. LNCS, vol. 11424, pp. 171–191. Springer, Cham (2019). https://doi.org/10.1007/978-3-030-16722-6_10

9. Feng, Y., Shi, Q., Gao, X.: DeepGini: prioritizing massive tests to enhance the robustness of deep neural networks. In: ISSTA 2020: 29th ACM SIGSOFT International Symposium on Software Testing and Analysis, Virtual Event, USA, 18–22 July 2020, pp. 177–188. ACM (2020)

10. Goodfellow, I.J., Shlens, J., Szegedy, C.: Explaining and harnessing adversarial examples. In: 3rd International Conference on Learning Representations, ICLR 2015, San Diego, CA, USA, 7–9 May 2015, Conference Track Proceedings (2015)

11. Guo, J., Jiang, Y., Zhao, Y.: DLFuzz: differential fuzzing testing of deep learning systems. In: Proceedings of the 2018 ACM Joint Meeting on European Software Engineering Conference and Symposium on the Foundations of Software Engineering, ESEC/SIGSOFT FSE 2018, Lake Buena Vista, FL, USA, 04–09 November 2018, pp. 739–743. ACM (2018)

12. He, K., Zhang, X., Ren, S.: Deep residual learning for image recognition. In: 2016 IEEE Conference on Computer Vision and Pattern Recognition, CVPR 2016, Las Vegas, NV, USA, 27–30 June 2016, pp. 770–778. IEEE Computer Society (2016)

13. Khope, S.R., Elias, S.: Critical correlation of predictors for an efficient risk prediction framework of ICU patient using correlation and transformation of MIMIC-III dataset. Data Sci. Eng. **7**(1), 71–86 (2022)

14. Kim, J., Feldt, R., Yoo, S.: Guiding deep learning system testing using surprise adequacy. In: Proceedings of the 41st International Conference on Software Engineering, ICSE 2019, Montreal, QC, Canada, 25–31 May 2019, pp. 1039–1049. IEEE/ACM (2019)

15. Kurakin, A., Goodfellow, I.J.: Adversarial examples in the physical world. In: 5th International Conference on Learning Representations, ICLR 2017, Toulon, France, 24–26 April 2017, Workshop Track Proceedings. OpenReview.net (2017)

16. Lee, S., Cha, S., Lee, D.: Effective white-box testing of deep neural networks with adaptive neuron-selection strategy. In: ISSTA 2020: 29th ACM SIGSOFT International Symposium on Software Testing and Analysis, Virtual Event, USA, 18–22 July 2020, pp. 165–176. ACM (2020)

17. Liu, W., Wang, Z., Liu, X.: A survey of deep neural network architectures and their applications. Neurocomputing **234**, 11–26 (2017)

18. Ma, L., Juefei-Xu, F., Zhang, F.: DeepQauge: multi-granularity testing criteria for deep learning systems. In: Proceedings of the 33rd ACM/IEEE International Conference on Automated Software Engineering, ASE 2018, Montpellier, France, 3–7 September 2018, pp. 120–131. ACM (2018)

19. Papernot, N., McDaniel, P.D., Jha, S.: The limitations of deep learning in adversarial settings. In: IEEE European Symposium on Security and Privacy, EuroS&P 2016, Saarbrücken, Germany, 21–24 March 2016, pp. 372–387. IEEE (2016)

20. Pearson, S., Campos, J., Just, R.: Evaluating and improving fault localization. In: Proceedings of the 39th International Conference on Software Engineering, ICSE 2017, Buenos Aires, Argentina, 20–28 May 2017, pp. 609–620. IEEE/ACM (2017)

21. Pei, K., Cao, Y., Yang, J.: DeepXplore: automated whitebox testing of deep learning systems. In: Proceedings of the 26th Symposium on Operating Systems Principles, Shanghai, China, 28–31 October 2017, pp. 1–18. ACM (2017)

22. Rothermel, G., Untch, R.H., Chu, C.: Prioritizing test cases for regression testing. IEEE Trans. Softw. Eng. **27**(10), 929–948 (2001)

23. Sun, Y., Huang, X., Kroening, D.: Testing deep neural networks. CoRR abs/1803.04792 (2018)

24. Sun, Y., Wu, M.: Concolic testing for deep neural networks. In: Proceedings of the 33rd ACM/IEEE International Conference on Automated Software Engineering, ASE 2018, Montpellier, France, 3–7 September 2018, pp. 109–119. ACM (2018)

25. Tian, Y., Pei, K., Jana, S.: DeepTest: automated testing of deep-neural-network-driven autonomous cars. In: Proceedings of the 40th International Conference on Software Engineering, ICSE 2018, Gothenburg, Sweden, May 27–June 03 2018, pp. 303–314. ACM (2018)

26. Wang, Z., You, H., Chen, J.: Prioritizing test inputs for deep neural networks via mutation analysis. In: 43rd IEEE/ACM International Conference on Software Engineering, ICSE 2021, Madrid, Spain, 22–30 May 2021, pp. 397–409. IEEE (2021)

27. Wong, W.E., Gao, R., Li, Y.: A survey on software fault localization. IEEE Trans. Softw. Eng. 42(8), 707–740 (2016)

28. Xie, X., Ma, L., Juefei-Xu, F.: DeepHunter: a coverage-guided fuzz testing framework for deep neural networks. In: Proceedings of the 28th ACM SIGSOFT International Symposium on Software Testing and Analysis, ISSTA 2019, Beijing, China, 15–19 July 2019, pp. 146–157. ACM (2019)

29. Yan, R., Chen, Y., Gao, H.: Test case prioritization with neuron valuation based pattern. Sci. Comput. Program. 215, 102761 (2022)

30. Zeiler, M.D., Fergus, R.: Visualizing and understanding convolutional networks. In: Fleet, D., Pajdla, T., Schiele, B., Tuytelaars, T. (eds.) ECCV 2014. LNCS, vol. 8689, pp. 818–833. Springer, Cham (2014). https://doi.org/10.1007/978-3-319-10590-1_53

31. Zhang, J.M., Harman, M., Ma, L.: Machine learning testing: survey, landscapes and horizons. IEEE Trans. Softw. Eng. 48(2), 1–36 (2022)

32. Zhang, K., Zhang, Y., Zhang, L.: Neuron activation frequency based test case prioritization. In: International Symposium on Theoretical Aspects of Software Engineering, TASE 2020, Hangzhou, China, 11–13 December 2020, pp. 81–88. IEEE (2020)

Mining Maximal Sub-prevalent Co-location Patterns Based on k-hop

Yingbi Chen[1], Lizhen Wang[1,2(✉)], and Lihua Zhou[1]

[1] School of Information Sciences and Engineering,
Yunnan University, Kunming, China
`lzhwang@ynu.edu.cn`
[2] Dianchi College of Yunnan University, Kunming 650213, China

Abstract. Mining spatial sub-prevalent co-location patterns (SPCPs) aim to discover the combinations of spatial features whose star feature instances co-occur frequently in the domain. This can in turn lay the foundation for deeper analysis of other important relationships between spatial objects (e.g., cause-effect relationships). However, no attempt was made to take into account the contribution of the k-hop neighbors. Therefore, we improve the partition-based algorithm in SPCPs mining, and propose a novel hierarchy-based algorithm based on k-hop neighbors (HPBA), which can discover more potentially valuable maximal sub-prevalent co-location patterns. HPBA is optimized by using multiple pruning strategies, for instance, k-core can filter features not present in k-size or higher size patterns. Lastly, we perform extensive experiments on real-world spatial datasets, and the results show the efficiency and effectiveness of our algorithm.

Keywords: Spatial data mining · Sub-prevalent co-location patterns (spcps) · k-hop · k-core

1 Introduction

Nowadays, all kinds of data are exploding, and spatial data is no exception. More and more spatial data mining techniques are being developed to discover hidden patterns of interest. Spatial co-location pattern [4] is one of the patterns of interest that its feature instances frequently co-occur in the domain. Many related research efforts for co-location patterns mining has been proposed, for instance, the join-less approach [9], etc. Based on the traditional co-location patterns, a novel pattern called sub-prevalent co-location pattern [5] is proposed to discover more potentially useful patterns. In the case of some special spatial distribution, such as patterns of instances distributed along the edges, they cannot be discovered because of not satisfying the proximity to each other. In Fig. 1, sub-prevalent co-location pattern{A,B,C,D} can be discovered when taking into account the 2-hop neighbors instances.

The main contributions of this paper are as follows:

(1). Considering the star neighborhood as 1-hop neighbors, we take into account higher hop. The k-hop neighbors relationship is proposed to measure the neighbor relationship between spatial instances.

(2). Based on the partition-based algorithm (PBA) in the traditional SPCPs mining, we propose a hierarchy-based algorithm (HPBA) which is a top-down method to process the candidate patterns grouped by the size of patterns.

(3). On the basis of the proposed mining algorithm (HPBA), two pruning strategies are proposed to reduce the search space of candidate patterns. First is based on k-core, and the second is based on the number of feature occurrences of the current-size candidate patterns.

The remainder of this paper is organized as follows. Section 2 reviews the related work. The basic concepts of SPCPs based on k-hop is noted in Sect. 3. The hierarchy-based SPCP mining algorithm (HPBA) and two pruning strategies are described in Sect. 4. The experimental results are demonstrated in Sect. 5. Lastly, the conclusions and future work are discussed in Sect. 6.

2   Related Work

Spatial data mining [3] is a popular research target for researchers, just like spatial keyword query [1], etc. Mining spatial co-location patterns [4] is to discover spatial patterns where spatial features frequently co-occurred together from spatial data.

For more efficient mining of co-location patterns, many approaches have been proposed, such as join-less [9], partial-join [10], and some Mapreduce-based algorithms [8]. For mining maximal co-location, some new structures were designed, like CPI-tree [6] and ordered ego-clique [7].

In order to find more potentially useful spatial patterns, the sub-prevalent co-location pattern [5] was first proposed, a prefix-tree based algorithm and a partition-based algorithm were proposed to discover maximal sub-prevalent co-location patterns.

The above works are all expanded based on the distance between the neighbor instances being less than the distance threshold, and they overlook the k-hop neighbors. We take into account the k-hop neighbors in the sub-prevalent co-location pattern mining, propose a novel mining framework based on PBA, and design two pruning strategies to reduce the search space.

3   Basic Concepts

In this section, the fundamentals of sub-prevalent co-location patterns (SPCPs) mining and k-hop are reviewed.

3.1 Maximal Sub-prevalent Co-location Patterns (MSPCPs)

Given a spatial data set $S = \{s_1, s_2, ..., s_n\}$, where each spatial **instance** belong to a specific feature in the set of spatial **features** $F = \{f_1, f_2, ..., f_m\}$. There is a **neighborhood relationship** $R(s_i, s_j)$ between two instances s_i and s_j if the distance between them is less than a given threshold d.

A k-size **sub-prevalent co-location pattern** $c = \{f_{l1}, ..., f_{lk}\}$ is a subset of the spatial feature set F, $c \in F$. The **star neighborhoods instances** (SNsI) of a center instance s is a collection of all instances that have a neighborhood relationship with s. The instances set of each feature f_l in c called **star participate instance set** $SPIns(c, f_l)$ when it meets the following conditions: (1) The features of $SNsI$ of star participate instance contain all features in c. (2) The star participation ratio of feature f_l in c is no less than the sub-prevalent threshold.

There are two metrics to measure the prevalence of SPCPs, star participation ratio and star participation index, which are defined by Wang [5]. Suppose f_l is a spatial feature in c, the **star participation ratio** of f_l in c denoted as SPR(c, f_l), is defined as follows:

$$SPR(c, f_l) = \frac{|SPIns(c, f_l)|}{|S_{f_l}|} \tag{1}$$

where $SPIns(c, f_l)$ is the star participation instances set of feature f_l in c, S_{f_l} is the whole instance set of f_l.

The **star participation index** of c is the minimum value of participation ratios for all features in c, which denoted as $SPI(c)$:

$$SPI(c) = min_{f_l \in c} SPR(c, f_l) \tag{2}$$

The pattern c is a **sub-prevalent co-location pattern** (SPCP) when the star participation index of c is not less than a given threshold min_sp, that is $SPI(c) \geq min\_sp$.

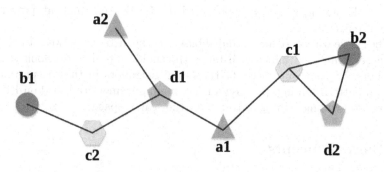

Fig. 1. A spatial dataset contains four features: A, B, C and D.

According to the monotonicity of SPR and SPI in SPCPs [5], let c and c' be two co-location patterns ($c' \subseteq c$), for each feature $f \in c'$, $SPR(c', f) \geq SPR(c, f)$, $SPI(c') \geq SPI(c)$ hold. Thus, the subsets of SPCPs are all sub-prevalent.

The pattern c is a **maximal sub-prevalent co-location pattern** (MSPCP) if c is sub-prevalent and any of c's super-patterns is not sub-prevalent. It's suitable to discover MSPCPs due to the large quantity of SPCPs.

3.2 k-hop

For two points (instances) s_i and s_j in the graph, k-**hop** represents the shortest path from s_i to another point s_j is k. Thus, the star neighborhoods instances (SNsI) of center instance s in SPCPs is equivalent to 1-hop neighbors instances $I_{1-hop}(s)$ because of the shortest path from the star neighborhoods instances to the center instance is 1.

Instead of 1-hop neighbors instances in SPCPs, we take into account k-hop neighbors instances. To mining MSPCPs based on k-hop, several definitions are given firstly as follows:

Definition 1. k-hop Neighborhood Relationship (R_{k-hop})

Given two points s_i and s_i, there exists a k-hop neighborhood relationship $R_{k-hop}(s_i, s_j)$ between them if the shortest path between them is less than k.

Definition 2. k-hop Star Neighborhood Instance (I_{k-hop})

$I_{k-hop}(s)$ is the k-hop neighbors instance set of s that each instance of $I_{k-hop}(s)$ has the k-hop neighbors relationship R_{k-hop} with the center instance s.

For instance, in Fig. 1, $I_{2-hop}(d_1) = \{b1, c2, a2, a1, c1\}$, due to the shortest path from each instance in $\{b1, c2, a2, a1, c1\}$ to d_1 is less than 2.

Definition 3. k-hop Star Participation Instance (KSPIns)

$KSPIns(c, f_l) = \{s_i \mid s_i$ is an instance of feature f_l and features of $I_{k-hop}(s_i)$ contain all features in $c\}$ is the star participation instances of feature f_l in c.

Definition 4. k-hop Star Participation Ratio (KSPR)

$KSPR(c, f_l) = |KSPIns(c, f_l)| / |S_{fl}|$ is the star participation ratio of feature f_l in a colocation pattern c, and S_{fl} is the instances set of feature f_l. Thus, $KSPR(c, f_l)$ is the ratio of k-hop star participation instances of feature f_l to the instances of feature f_l.

Definition 5. k-hop Star Participation Index (KSPI)

$KSPI(c) = \min_{f_l \in c} \{KSPR(c, f_l)\}$ is the k-hop star participation index of a colocation pattern c, which is the minimum value of the k-hop star participation ratio of all features in c.

Lemma 1. *(Monotonicity of KSPR and KSPI). Let c and c' be two co-location patterns ($c' \subseteq c$). For each feature $f \in c'$, $KSPR(c',f) \geq KSPR(c, f)$, $KSPI(c') \geq KSPI(c)$.*

Proof. KSPIns of feature f_l in c' include KSPIns of feature f_l in c due to $c' \subseteq c$, thus $|KSPIns(c', f_l)| \geq |KSPIns(c, f_l)|$, furthermore, $KSPR(c', f_l) \geq KSPR(c, f_l)$, $KSPI(c') \geq KSPI(c)$.

Based on Lemma 1, k-hop does not affect the monotonicity of SPCPs. The quantity of SPCPs based on k-hop is more than that based on 1-hop, because of the number of $KSPIns$ of feature f_l is more than the number of SPIns of feature f_l in c. To reduce the volume of the whole SPCPs, we are going to mining the MSPCPs based on k-hop. In the following sections, a novel algorithms to discover MSPCPs based on k-hop is proposed.

4 A Hierarchy-Based Algorithm (HPBA)

The basic algorithm framework of HPBA is based on PBA [5].

4.1 A Partition-Based Algorithm (PBA)

At first, all 2-size sub-prevalent co-location patterns $SPCP_2$ are obtained. Next as the processing of Partition 1 and Partition 2 are performed, $SPCP_2$ is partitioned into a set of strings (denoted as L_1) in lexicographical order, and the ordered strings in L_1 are divided by using Non_$SPCP_2$ until no Non_$SPCP_2$ in it, the ordered strings after Partition 2 are candidate patterns. Then as the processing of candidate patterns called Partition 3 is proceeded, a k-size candidate pattern c would be divided into two k-1 size patterns c^{k-1} by a 2-size pattern c^2 in c until k equals 3. The intersection operation is executed in the entire star participation instances set between these two patterns c^{k-1} and c^2 of c, whether c is sub-prevalent can be determined by the star participation instances set of c after the intersection operation.

4.2 Hierachy-Based Processing

In order to improve the processing of candidate patterns one by one in PBA, we added hierarchical operations on the candidate patterns after Partition 3. Firstly we obtained k-size candidate pattern set $c^k_{candidate}$ through dividing them by size of patterns, then process $c^k_{candidate}$ size by size from up to down, once there are SPCPs in $c^k_{candidate}$, we do not need to identify lower size candidate patterns, which can improve the efficiency of finding the maximum SPCPs. Furthermore, if a k-size candidate pattern c is not sub-prevalent by identify the two pattern c^{k-1} and c^2 of c, the rest of k-1 size patterns of c will be added to the k-1 size participation instances set instead of identify them instantly in PBA.

For instance, in Fig. 2, we have divided the candidate patterns after Partition 3, the maximal size of $c^k_{candidate}$ is 4, firstly we identify $c^4_{candidate} = \{\{A,B,C,D\}, \{A,B,C,E\}\}$, when $\{A,B,C,D\}$ is not SPCP and $\{A,B,C,E\}$ is SPCP, we do not need to identify $\{A,C,D\}$ and $\{B,C,D\}$ but it is necessary in PBA.

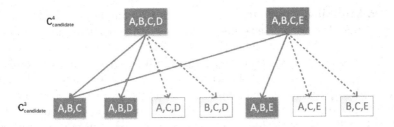

Fig. 2. Identify candidate patterns size by size.

4.3 Pruning Strategies

k-core. k-core is first defined by Dorogovtsev [2], which is a largest subgraph where vertices have at least k interconnections. In other words, a node must link to at least k nodes in the k-core, no matter how many the nodes linked to outside the k-core. For instance, in Fig. 3, node A is 1-core despite that A has four links.

We treat the features as the nodes in a graph. The prerequisite of $k+1$ size sub-prevalent co-location pattern c is each feature f_l in c is k-core. Feature f_l is k-core implies that feature f_l has the potential to form a $k+1$ size sub-prevalent pattern with other k-core features.

For instance, in Fig. 3, {A,B,C,D,E,F,G} represent the features of SPCPs. The line between two features means these two features can form a SPCP$_2$. Feature A is 1-kore so it cannot be the feature of 3-size sub-prevalent patterns. Thus, we can utilize k-core to reduce the search space of candidate patterns. Due to the iteration process of PBA is based on 3-size patterns and the enormous quantity of 2-size patterns, we can disregard the features with 1-core and lower.

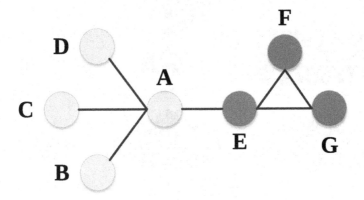

Fig. 3. Nodes with 1-core and 2-core.

Pattern Splitting Approach. The pattern splitting approach of a candidate pattern c may lead to worse result. Suppose a 4-size participation instances set $c_{candidate}^4 = \{\{A,B,C,D\}, \{B,C,D,E\}\}$, the dividing result is four candidate patterns $\{\{A,B,C\}, \{A,B,D\}, \{B,C,D\}, \{C,D,E\}\}$ if we choose the last two features to divide $c_{candidate}^4$. To reduce the search space, we sort and descend the current-size candidate patterns by the number of occurrences of features in current-size candidate pattern set. As for $c_{candidate}^4$, the sort order by occurrences of feature is $\{B,C,D,A,E\}$, thus the last two feature is relatively infrequent compare to other features. Then we divide $c_{candidate}^4$ by the last two features and obtain 3 candidate patterns $\{A,C,D\}$, $\{B,C,D\}$ and $\{C,D,E\}$. Compared with the previous dividing method, it can reduce the computation of a candidate pattern.

4.4 Mining Framework

A novel mining framework of maximal sub-prevalent co-location patterns is demonstrated in Fig. 4. The first step of mining framework is to preprocess spatial dataset to generate the k-hop neighbors set of all instances. We use a

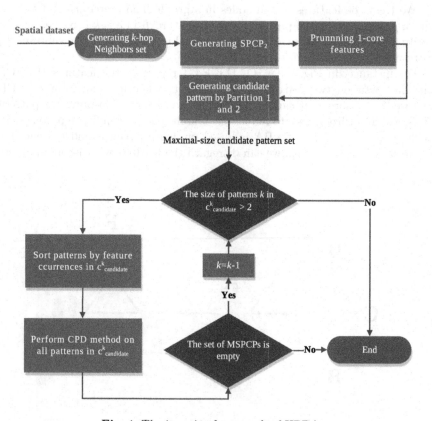

Fig. 4. The iterative framework of HPBA.

Algorithm 1. *gen_khop_neighborhood*

Input: $S = \{s_1, s_2, ..., s_n\}$: a set of spatial instance;
        d: a distance threshold;
1: $kdTree = $ buildKDTree(S)
2: $graph = $ gen1hopNeighbor(kdTree,d)
3: $khopNeighborMap = $ null
4: **for** each node $sourceNode$ in $graph$ **do**
5:    $khopNeighbor = $ dfs($graph, sourceNode, k$)
6:    $khopNeighborMap$.put($sourceNode, khopNeighbor$)
7: **end for**

Algorithm 2. Framework of HPBA.

Input: $F = \{f_1, f_2, ..., f_m\}$: a set of spatial feature;
    $khopNeighborMap$: a set of spatial instance and their khopNeighbor set;
    min_sp: a sub-prevalent threshold;
    $SPCP_2\_2core$: $SPCP_2$ without 1-core feature;
    $L2$: a set of ordered strings of $SPCP_2$ under Partition 1 and Partition 2;
    $candidatePatternBySize$: candidate patterns order by size;
    $c^k_{candidate}$: candidate patterns of k size;
    $c^k_{candidate}\_sort$: candidate patterns of k size order by occurNum;
    $NCoIns$: processed patterns and its star participation instance
Output: $MSPCPs$: a set of maximal-size sub-prevalent co-location patterns;
1: $NCoIns = gen\_2size\_colocation\_instance(F, khopNeighborMap)$;
2: $SPCP_2 = cut\_by\_kSPI(min\_sp, NCoIns.keySet)$;
3: $SPCP_2\_2core = get\_2\_core\_by\_SPCP_2$;
4: $L2 = PBA.partition\_1\_2(SPCP_2\_2core)$;
5: $candidatePatternBySize = hierarchy\_based\_processing(L2)$;
6: **for** $k = max\_size$ to 3 **do** //max_size is the maximal size of candidatePatterns
7:    $c^k_{candidate} = getCandidateBySize(k)$;
8:    $c^k_{candidate}\_sort = orderByOccurNum(c^k)$; //sort and descend by occurNum
9:    **while** ($c^k_{candidate}\_sort \neq \emptyset$) **do**
10:      get a pattern c from $c^k_{candidate}\_sort$;
11:      $PBA.CPD(c)$;
12:    **end while**
13:    **if** $MSPCPs.get(k).size! = 0$ **then**
        return $MSPCPs.get(k).keySet()$;
14:    **end if**
15: **end for**

---

depth-first search algorithm with depth limits to obtain the k-hop neighbors set, and it is shown in Algorithm 1:

At first, we index all points using the kd-tree, so that we can find neighborhood instances efficiently. In Step 3, the graph of neighborhood set is stored in an array-based adjacency table due to the graph is sparse. From Step 5 to Step 8, the depth-first search of depth k is performed for each node in graph as the source node. Then we can obtain the k-hop neighbors instance set. Next we mining maximal SPCPs by Algorithm 2 called HPBA.

Next, we obtain all 2-size sub-prevalent co-location patterns $(SPCP_2)$ by computing $KSPI(c)$, then pruning the 1-core features and obtain the $SPCP_2\_2core$ which is the subset of $SPCP_2$ in Step 3. After the process of Partition 1 and Partition 2 in PBA, we can finally get the set of candidate patterns without Non-$SPCP_2$ and 1-core features.

In Steps 6–15 of HPBA, we process candidate patterns size by size. NCoPatternInstance is a hash map storing processed patterns and their star participation instances, so that it can avoid duplicate operations for sub-patterns. Every time before the processing of k-size candidate pattern set L, to pruning the candidate patterns, the candidate pattern in L would be sorted by the order of occurrences of features in reverse order.

The procedure of CPD is recursively called by two k-1 size patterns of c until k equals to 3. 3-size patterns can calculate $KSPI(c)$ directly by three times intersection operations of $KSPIns(c)$. When pattern c is not sub-prevalent, PBA will calculate the rest of $(k-1)$-size sub-patterns of c instantly. On the contrary, we add them to k-1 size candidate pattern set, and process them after sorting by occurrences.

5   Experimental Result

In this section, we conducted a comparison experiment of the two algorithms, PBA [5] and HPBA. We reimplemented PBA and implemented HPBA by *java*. The experimental environment is as follows: cpu with Intel i5-8300H @ 2.30 GHz, memory with 16 GB.

We conducted experiments and evaluate the performance of PBA and HPBA on a real world dataset. Next, we examined other parameter effects of HPBA on a synthetic dataset. The real world dataset and synthetic dataset which used in the experiment is described in Table 1.

Table 1. The description of datasets.

Spatial dataset	Number of instances	Number of features
Beijing POI dataset	20000	200
Synthetic dataset	10000	100

5.1   Comparison of the Search Space Based on Candidate Patterns

We evaluated the effectiveness of two pruning strategies, k-core and pattern splitting approach in HPBA. k-core can reduce the number of features in $SPCP_2$, pattern splitting approach can pruning the number of processed candidate patterns. To eliminate the affect of k-hop, the k in k-hop is set to 1. As shown in Fig. 5, the number of features in $SPCP_2$ of HPBA is less than that in PBA. A little effect in reducing the search space of candidate patterns compare to the previous metrics, but it still works. The number of processed patterns in PBA and HPBA is 8267 and 8103 respectively when distance threshold is set to 800 m.

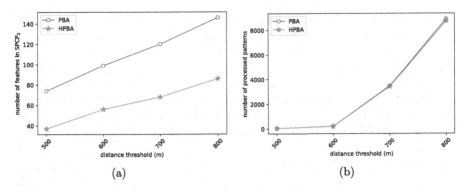

Fig. 5. Number of features in SPCP$_2$ and number of processed patterns by different distance threshold

5.2 Comparison of Running Time

The running time is an essential metrics for measuring the efficiency of an algorithm. Thus we compared the running time of HPBA and PBA on the BeiJing POI dataset when k is set to 1. In Fig. 6, by giving a lower distance threshold d, the running time of HPBA is about the same as that of PBA. As the increase of the distance thresholds, HPBA is much faster than PBA.

5.3 Comparison of the Size of MSCPs

By using k-hop neighbors instance set, we can find more potentially useful patterns. As illustrated in Fig. 7, higher size of MSCPs we obtained with the increase of k in k-hop.

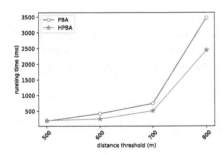

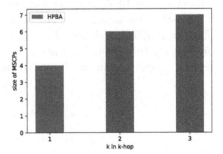

Fig. 6. Running time by different distance threshold.

Fig. 7. Size of MSCPs of HPBA by different k in k-hop.

5.4 Comparison of Parameter Effects

We performed comparison tests of parameter effects of HPBA which based on synthetic spatial datasets. The experiments are conducted on synthetic datasets whose x and y of instance' coordinates is generated in the range of 10000, and the distance threshold is set to 300, min_pre is set to 0.6. We assessed the impact of their following factors: the number of instances, the number of features, and the value of k in k-hop.

Firstly, we assessed the impact of the number of instances. The number of features is fixed to 100, and k is set to 1. The result is illustrated in Fig. 8, the execution time is continuously increased with the increasing of instances, and running time of HPBA is less than PBA except for the case of few instances.

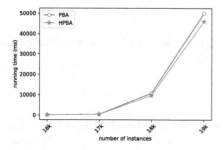 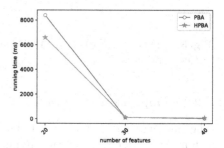

Fig. 8. Running time by number of instances.

Fig. 9. Running time by number of features.

Next, we assess the impact of the number of features. The number of instances is fixed to 10000, and k is given as 1. The result is illustrated in Fig. 9, the number of features is more, the number of instances of each feature is fewer, the execution time is continuously decreased with the increasing of features, and HPBA is more effective than PBA in larger dataset.

Lastly, we assess the impact of the value of k-hop. The number of instances is fixed to 10000, and the number of features is given as 100. The result is illustrated in Fig. 10, the execution time is continuously increased with the increasing of k due to the exponential growth in the neighbors instance set. Assume k is

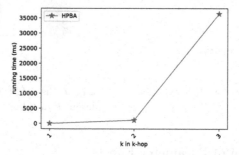

Fig. 10. Running time of HPBA by different k in k-hop.

large and equals the maximum value of k in graph, then each instance's k-hop neighbors include all other instances, thus large value of k is meaningless, there is no mining required based on a large value of k.

6   Conclusions and the Future Work

Previous works of sub-prevalent co-location pattern minning has only considered star neighborhood instances. So we relaxed the restrictions of neighbors, take into account the k-hop neighbors instances in SPCPs mining. We proposed the method of mining k-hop neighbors instances, which make it feasible to discover more potentially useful patterns.

To reduce the search space of candidate patterns, a hierarchy-based method HPBA and two pruning strategies are proposed to mining the all MSPCPs. Through the relative experiments, the comprehensive performance of the proposed HPBA is better. But the cost of time is expensive when the value of k is large, due to the exponential growth in the neighbors instance set. Thus in the future work, we will strive to solve the problem.

Acknowledgements. This work is supported by the National Natural Science Foundation of China (61966036, 62062066), the Project of Innovative Research Team of Yunnan Province (2018HC019), Yunnan Fundamental Research Projects (202201AS070015), and Yunnan University Postgraduate Technological Innovation Project (2021Y175).

References

1. Chen, L., Cong, G., Jensen, C.S., Wu, D.: Spatial keyword query processing: an experimental evaluation. Proc. LDB Endow. **6**(3), 217–228 (2013)
2. Dorogovtsev, S.N., Goltsev, A.V., Mendes, J.F.F.: K-core organization of complex networks. Phys. Rev. Lett. **96**(4), 040601 (2006)
3. Li, D., Wang, S., Li, D.: Spatial Data Mining. Springer, Heidelberg (2015). https://doi.org/10.1007/978-3-662-48538-5
4. Shekhar, S., Huang, Y.: Discovering spatial co-location patterns: a summary of results. In: Jensen, C.S., Schneider, M., Seeger, B., Tsotras, V.J. (eds.) SSTD 2001. LNCS, vol. 2121, pp. 236–256. Springer, Heidelberg (2001). https://doi.org/10.1007/3-540-47724-1_13
5. Wang, L., Bao, X., Zhou, L., Chen, H.: Mining maximal sub-prevalent co-location patterns. World Wide Web **22**(5), 1971–1997 (2019). https://doi.org/10.1007/s11280-018-0646-2
6. Wang, L., Bao, Y., Lu, J., Yip, J.: A new join-less approach for co-location pattern mining. In: 2008 8th IEEE International Conference on Computer and Information Technology, pp. 197–202. IEEE (2008)
7. Wu, P., Wang, L., Zou, M.: A maximal ordered ego-clique based approach for prevalent co-location pattern mining. Inf. Sci. **608**, 630–654 (2022)
8. Yang, P., Wang, L., Wang, X., Zhou, L., Chen, H.: Parallel co-location pattern mining based on neighbor-dependency partition and column calculation. In: Proceedings of the 29th International Conference on Advances in Geographic Information Systems, pp. 365–374 (2021)

9. Yoo, J.S., Shekhar, S.: A joinless approach for mining spatial colocation patterns. IEEE Trans. Knowl. Data Eng. **18**(10), 1323–1337 (2006)
10. Yoo, J.S., Shekhar, S., Smith, J., Kumquat, J.P.: A partial join approach for mining co-location patterns. In: Proceedings of the 12th Annual ACM International Workshop on Geographic Information Systems, pp. 241–249 (2004)

An Association Rule Mining-Based Framework for the Discovery of Anomalous Behavioral Patterns

Azadeh Sadat Mozafari Mehr$^{(\boxtimes)}$ [id], Renata M. de Carvalho[id],
and Boudewijn van Dongen[id]

Department of Mathematics and Computer Science, Eindhoven University of Technology,
Eindhoven, The Netherlands
{a.s.mozafari.mehr,r.carvalho,b.f.v.dongen}@tue.nl

Abstract. The identification of different risks and threats has become a top priority for organizations in recent years. Various techniques in both data and process mining fields have been developed to uncover unknown risks. However, applying them is challenging for risk analysts since it requires deep knowledge of mining algorithms. To help business and risk analysts to identify potential operational and data security risks, we developed an easy to apply automated framework which can discover anomalous behavioral patterns in business process executions. First, using a process mining technique, it obtains deviations in different aspects of a business process such as skipped tasks, spurious data accesses, and misusage of authorizations. Then, by applying a rule mining technique, it can extract anomalous behavioral patterns. Furthermore, in an automated procedure, our framework is able to automatically interpret anomalous patterns and categorize them into roles, users, and system deviating patterns. We conduct experiments on a real-life dataset from a financial organization and demonstrate that our framework enables accurate diagnostics and a better understanding of deviant behaviors.

Keywords: Rule mining · Anomalous behavioral pattern · Data privacy ·
Multi-perspective analysis

1 Introduction

Nowadays organizations face various risks and threats. The operational and data security threats are considered to be generated by different elements in an organization such as processes, users, and systems and are categorized into external and internal threats [17]. Among them, internal threats, such as the misbehavior of employees with legitimate identities, can often cause severe damages [26]. Thus, the organizations are currently shifting from prevention-only approaches to detection and response oriented strategies. Consequently, the detection of anomalies in process runtime behavior is crucial and a top priority for most organizations, as they can be indicative of various types of anomalies ranging from data security breaches and frauds to abnormalities resulting from special operating circumstances such as bottlenecks. To address this issue, a common approach would be to investigate the normal behaviors of employees and

© The Author(s), under exclusive license to Springer Nature Switzerland AG 2022
W. Chen et al. (Eds.): ADMA 2022, LNAI 13725, pp. 397–412, 2022.
https://doi.org/10.1007/978-3-031-22064-7_29

their interactions with the information systems through data/process mining on historical data and detect improper behaviors compared to the normal ones. Organizations on the one hand formulate their daily business processes as process models to detect deviations that cause operational risks [21]. On the other hand, by employing data access control mechanisms, they formulate information security policies [13] to discover spurious behaviors that could pose data security risks. However, these models and policies are based on identified risks and they cannot prevent unknown risks. Although there are various pattern mining techniques that might aid in the discovery of unknown risks, generally, these techniques return a large number of patterns. The analysis and investigation of these patterns can be difficult and very time consuming. Furthermore, the implementation of these techniques by the risk analysts is challenging since it requires proficiency in data mining and knowledge about the details of the mining algorithms. Therefore, implementing an easy to apply framework that can automatically extract anomalous behavioral patterns from historical data which considers different aspects of a business process to produce the results is still a challenge for organizations.

Previously, we proposed a method [19] that emphasizes on an investigation using a process mining technique to obtain deviations in different aspects of a business process. Such investigation deals with suspicious issues, i.e. skipped tasks, resource misrepresentation, spurious data accesses, and misusage of authorizations. In this paper, we propose a method for applying association rule mining technique to uncover anomalous behavioral patterns in a daily operations of an organization. It relies on a dataset which recorded all deviations presented by the process mining investigation. In an automated procedure our framework is able to automatically filter and interpret the useful rules obtained from applying association rule mining and provide comprehensive operational insights by categorizing the anomalous patterns into roles, users, and system deviating behaviors. Finally, the results can support business and risk analysts to identify a wide range of potential operational and data security risks and threats. The combination of the two methods results in a framework for discovery of anomalous behavioral patterns in business process executions.

The remainder of this paper is structured as follows. Section 2 explains background and the essential concepts of this paper. Section 3 introduces the methodology and explains different phases in our designed framework for extracting anomalous behavioral patterns. Section 4 presents the applicability of our approach through a real-life case study, discussing the experimental design and results. Section 5 reviews related work. At last, the conclusion of this paper is presented in Sect. 6.

2   Background

2.1   Real-Life Dataset: BPI Challenge 2017

Every year, the International Workshop on Business Process Intelligence (BPI) sets out a challenge for researchers and practitioners. Using the provided real-life datasets, many researchers and professionals demonstrate the usefulness of their novel tools, approaches, and algorithms in solving different challenges. In this paper, to evaluate the applicability of our approach to real-life settings, we used the event log recording the executions of a process for handling credit requests within a Dutch Financial Institute, which was made available for the 2017 BPI challenge [10]. The event log contains detailed information about applications submitted by clients, loan offers sent by the company, and work items processed by employees or by the system. It contains 26 types of events, divided into three categories. Application (A) state changes are the first type, Offer (O) state changes are the second, and Workflow (W) activities are the third.

We recognized that most of A-type and O-type events occurred within a few milliseconds of performing W type events. Thus, we considered these events as data operations that were executed while performing workflow activities. Consequently, we split the original dataset into process log which contains only workflow events and data log which shows offer and application state changes and accesses to customer data. After preprocessing, the resulting process log and data log contain 301,709 workflow events and 256,767 data operations, respectively. These logs were recorded from managing 26,053 loan applications. The activities and data operations were performed by 146 resources (employees or system).

By applying a process discovery technique [14] on the process log and based on the findings of other researchers [22, 24], we modeled the loan management process as shown in Fig. 1. In this model, there are four main milestones: receiving applications, negotiating offers, validating documents, and detecting potential fraud. In this process, the execution of activities may require performing certain mandatory or optional data operations. The data model of loan management process which presents the relationship between activities and data operations is shown in Table 1. The data model also indicates whether the user is allowed to repeat the execution of data operation.

As shown in the process model (Fig. 1), three roles are supposed to conduct the activities. Most of the activities are supposed to be done by the role *clerk*. Activities

Table 1. Data model of the loan management process. Type: Mandatory (M), Optional (O). Frequency: is allowed (True), is not allowed (False). A: Application, O: Offer, W: Workflow.

Activity	Data operation	Type	Frequency
A-Create Application	Create: (applicationID)	M	False
W-Shortened Completion start	Read: (applicationID, email)	M	False
A-Accepted	Create: (offerID)	M	False
	Read: (offerID)	M	False
	Read: (address, email)	M	False
	Read: (address)	O	False
A-Cancelled	Update: (OCancelledFlag)	M	True
W-Call after offer start	Update: (ACompletedFlag)	M	False
W-Call after offer complete	Update: (OAcceptedFlag)	M	False
W-Call after offer withdraw	Update: (OReturnedFlag)	M	False
W-Call after offer ate abort	Update: (OCancelledFlag)	M	False
W-Validate application start	Update: (AValidatedFlag)	M	False
A-Pending	Update: (OAcceptedFlag)	M	False
A-Denied	Update: (ORefusedFlag)	M	True
W-Validate application ate abort	Update: (OCancelledFlag)	M	True
W-Call incomplete files start	Update: (AInCompleteFlag)	M	False

related to fraud detection are supposed to be done by a *fraud analyst*. The activity "W Shortened completion" can only be executed by a *manager*. Managers also have the authority to perform all the activities related to a clerk. In the organizational model, the role of *User_11* and *User_43* is manager. The users *User_138*, *User_ 143* and *User_144* play the role of fraud analyst and all other users have the role of clerk.

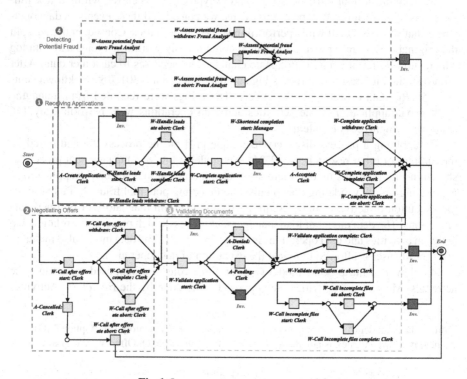

Fig. 1. Loan management process model

In Sect. 5, using the process, data, and organizational models along with historical data recorded by process log and data log, we present how our automated framework can extract and interpret the patterns of anomalous behaviors in the loan management process execution.

2.2 Process Mining

Process mining is an analytical discipline that sits between Business Intelligence and data mining on the one hand, and process modeling and analysis on the other [3]. It can be seen as a bridge between data mining and traditional model-driven BPM (Business Process Management) enabling process-centric analytics through automated process discovery, conformance checking, and model enhancement [3].

Many of process mining techniques have been implemented in the open source ProM platform[1]. In this paper, in order to find patterns of deviating behaviors in an organization, we first apply the multi-perspective conformance checking technique [19]. This technique was implemented as a package named "Multi-LayerAlignment" in ProM [20]. As shown in Fig. 2, it takes two sets of inputs. One is modeled behavior including a process model with role information, a data model and an organizational model. A process model illustrates the activities to be performed in a special order to reach a certain business goal. It also describes which role is allowed to perform which activities (tasks). The data model relates the process logic to the data layer by indicating which data operations on given data fields must be executed in order to complete a given activity. An organizational model represents the interaction between roles and users (actors). The other inputs are a process log and a data log. The process log records process executions including additional information related to each case such as the performed activities with timestamp and the resource (i.e., person or device) executing or initiating the activities. The data log contains data operations executed in each case showing which user accessed which data. These two inputs indicate observed behaviors. Afterwards, the technique compares the observed behavior with the modeled/expected behavior and can identify six types of deviating behaviors:

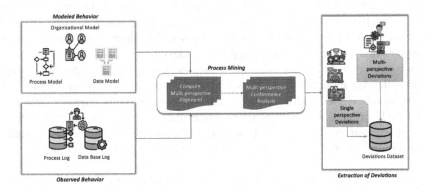

Fig. 2. Multi-perspective conformance analysis as an approach to generate a deviations dataset

- Deviation Type 1: Unexpected activities, by identifying in the process log, the completion of an activity which was not expected by the process model. As an example, according to the process model in Fig. 1, the activity *"W-Call after offers start"* should always be executed after the acceptance (A-Accepted activity) and completion of an application (W-Complete Application activity). If the activity *"W-Call after offers start"* happens before the completion of an application, this execution is suspicious and will be reported as unexpected activity deviation.
- Deviation Type 2: Not allowed data accesses, by identifying in the data log, the execution of a data operation which was not expected by the data model and executed

---

[1] https://www.promtools.org/.

with no clear context. According to the data model of the loan management process, Read:(phone number) is not expected, so in case an execution of this data operation is observed in the data log, it will be reported as not allowed data access.

- Deviation Type 3: Activities and data operations which were executed illegally by an authorized user. In this violation, although activity and its data operations are expected and performed, the user that performed the activity and had the authority to access the data does not have a legitimate role to do so (based on the role allocation in the process model). For instance, the activity *"W-Shortened Completion start"* along with its data opration *"Read:(applicationID, email)"* are supposed to be performed by a manager. Therefore, observing the execution of this activity in the process log and its related data operation in the data log by a clerk implies that the user accessed the data illegally by performing an unjustified activity.
- Deviation Type 4: Activities which were performed by an unexpected role. The same as Deviation Type 3, but no data was accessed. As an example, according to the loan management process model (Fig. 1), the activity *"W Assess potential fraud withdraw"* can only be done by fraud analyst and performing this activity is illegal for a manager.
- Deviation Type 5: Skipped tasks, pointing out the expected activity and data operation defined by the process and data model that were not performed. For instance, according to the process model in Fig. 1 and the data model in Table 1, performing the activity *"A-Accepted"* along with its data operations by the role *"Clerk"* is mandatory for completion of a process instance. The absence of these tasks in the process and data log shows the skipped tasks deviation.
- Deviation Type 6: Ignored data operations, indicating that an expected data operation defined by the data model was ignored in the context of the activity being performed. While handling a credit request, multiple offers might be sent to the client. If the application is denied, the active offers should be refused. Thus, observing the occurrence of activity *"A-Denied"* in the process log and the absence of data operation *"ORefusedFlag"* shows that the clerk ignored this data operation.

As the output, the multi-perspective process mining technique records each identified deviation as a set of attributes reflecting various dimensions (as shown in Table 2) to allow for further analysis.

Table 2. Different dimensions of deviations

Attribute	Description
Activity	The context in which the deviation happened
Data Operation	The data operation which was involved in the deviation
Expected Role	The role that was expected to perform the activity and execute the data operation
User (Actor)	The user that had deviating behavior
Execution Role	The role of the user who performed the deviation
Deviation	The type of violation that happened during the execution of the business process

2.3   Association Rule Mining

Association rule mining is a data mining approach designed to discover interesting and frequently occurring patterns, correlations, or associations between the data items of a dataset [11]. We now introduce some of the basic terminology of association rules derived from [5] and [6].

Given a set of data items $\mathcal{I} = \{ I_1, I_2, ..., I_n \}$, a transaction dataset is $\mathcal{D} = \{T_1, T_2, ..., T_m\}$, where $T \subseteq \mathcal{I}$. We say that a transaction T contains an itemset X if $X \subset T$.

An **association rule** is a rule of the form $X \Rightarrow Y$, where X and Y are itemsets $X \subset T, Y \subset T$, and $X \cap Y = \emptyset$. X is called the **antecedent** and Y is called the **consequent** of the rule. The intended meaning of this rule is that the presence of (all of the items of) X in a transaction implies the presence of (all of the items of) Y in the same transaction.

Each association rule $X \Rightarrow Y$ has two measures relative to a given set of transactions $\mathcal{D}$: its confidence and its support. The **confidence** of the rule is the fraction of transactions that contain both X and Y among transactions that contain X. The formula is defined as Eq. (1). The **support** of the rule is the fraction of transactions that contain both X and Y among all transactions in the input dataset $\mathcal{D}$, defined as Eq. (2).

$$conf(X \Rightarrow Y) = \frac{|X \cup Y|}{|X|} \tag{1}$$

$$sup(X \Rightarrow Y) = \frac{|X \cup Y|}{|\mathcal{D}|} \tag{2}$$

In other words, the confidence of a rule measures the degree of the correlation between itemsets, while the support of a rule measures the significance of the correlation between itemsets.

The basic concept of the association rule mining is to generate rules based on itemsets (transactions) with frequent occurrence in a dataset. This includes two main processes: the determination of frequent itemsets and the process of forming rules based on frequent itemsets. In both steps, a threshold is used to eliminate non-frequent itemsets/rules which is also known as the **minimum support**. Corresponding to this measurement, a valid rule is a rule such that its support is no less than the (user-defined) minimum support thresholds.

3   Methodology

In this section we describe a framework for the discovery of anomalous behavioral patterns in an organization using association rule mining. As shown in Fig. 3, our framework consists of four phases: data preparation, applying association rule mining technique, automatic interpretation of discovered rules and providing operational insights.

The first phase is **Data preparation**. By applying a multi-perspective process mining technique [19] on the event data stored by information systems, this technique can identify six types of deviating behaviors as mentioned in Sect. 2.1. The aforementioned technique records various dimensions of each kind of deviation as a set of related attributes

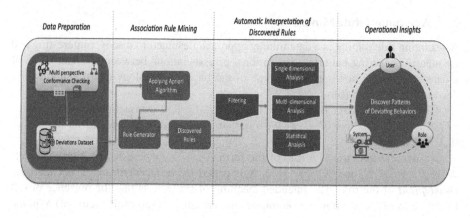

Fig. 3. Overview of the proposed approach

shown in Table 2. For instance, in case of data privacy violation (Deviation Type 3), along with the type of deviation, it records what data was accessed in the context of which activity and who accessed the data illegally (in terms of Role and User).

After data preparation, the next phase is applying the ***Association Rule Mining*** algorithm to discover hidden patterns of deviating behaviors. The applied Apriori algorithm [5] begins with finding frequent itemsets in a dataset of transactions by defining a threshold **S** for frequency (also known as the minimum support). Following this, association rules are generated for every frequent itemset l. Next, all non-empty subsets of l are generated. Then, for each non-empty subset k belonging to l, a rule of the form $k \Rightarrow (l - k)$ is generated only if it satisfies the minimum support threshold **S**.

Next phase in the framework is ***Automatic Interpretation of Discovered Rules***. An initial step in this phase is filtering and sifting through the rules to extract those that yielded useful information. For this goal, first, the framework allows the user to specify a support and/or confidence in order to further prune the returned rules of the previous step, stating the percentage of transactions in which the itemsets has to be present to be considered as a relevant association rule (as defined in Sect. 2.2). Second, as the goal is to discover patterns of deviating behaviors, the framework only looks for the association rules that are of the form $X \Rightarrow \{Deviation\}$ where $Deviation$ is one of the six types of deviations and X is a set of the other dimensions shown in the Table 2.

Once the list of discovered rules has been reduced by filtering and eliminating irrelevant ones, next, we interpret the rules. Based on the dimensions involved in the rule's antecedent, the type of deviation in the rule's consequent, and the rule's support and confidence measures, the framework facilitates conducting three types of analysis:

- **Single Dimensional Analysis:** The objective of this type of analysis is looking into business process from important aspects separately. Therefore, from the set of filtered rules, the focus of this analysis is the interpretation of the single dimensional rules. An association rule is single dimensional if the items (attributes) in its antecedent refer to only one dimension. As general examples, rules of the form:
  - $\{User\} \Rightarrow \{Deviation\}$, for showing the user's deviating behaviors.

- {*Expected Role*} ⇒ {*Deviation*} or {*Execution Role*} ⇒ { *Deviation*}, for representing the roles' deviating patterns.
- {*Activity*} ⇒ {*Deviation*}, for concluding which activities were involved in what kind of deviations.
- {*Data Operation*} ⇒ {*Deviation*}, for helping to analyse which data operations were ignored or illegally executed (based on the type of deviation involved in the consequent).

As a specific example, the rule {*Read(phonenumber)*} ⇒ {*DeviationType2*} can aid in identifying spurious accesses to personal information of the customers and points out which data operation led to illegal access of sensitive data fields.

- **Multi-Dimensional Analysis:** If a rule references more than one dimension in the antecedent, such as the dimensions activity, role, and user, then it is a multi-dimensional association rule. It helps to discover more complex hidden behavioral patterns that caused deviations and non-conformity. The following rule is an example of multi-dimensional rule in a loan management process: { *Read(address), Accept Application, user124* } ⇒ { *DeviationType3*}. It can be interpreted as "Employee with userID *User_124* accessed the address (of the customers) illegally while performing the activity Accept Application".

- **Statistical Analysis:** As the final step in this phase, the discovered associations will be further analyzed statistically, using the support and confidence measures. Statistical analysis is concerned with the organization and interpretation of data according to well-defined, systematic, and mathematical procedures and rules [1]. Descriptive statistics form the first level of statistical analysis and are used to reduce large sets of observations into more compact and interpretable forms [1]. Frequency analysis is a part of descriptive statistics. This type of analysis facilitates answering such questions as, "How much?", "How many?", "How long?", "How often?", and "How related?". Here, we use the support and confidence measures of the rules to uncover the frequency and severity of the discovered deviating patterns. For instance, the support of the above mentioned rule shows how often *User_124* misused his/her privilege to access the personal data of the customers.

Finally, the framework delivers ***Operational Insights*** by categorizing the findings into roles, users, and system deviating behaviors, and organizing the results in a hierarchical format. In the next section, we describe in details how the presented framework provides operational insights through a real-life case study.

4 Case Study

4.1 Experimental Setup

As mentioned in Sect. 2.1, to evaluate the applicability of our approach to real-life settings, we used the event log recording the loan management process of a Dutch Financial Institute provided by BPI 2017 challenge [10].

By applying multi-perspective process mining technique, 173,299 deviations were discovered and recorded during the whole execution of the loan management process[2].

---

[2] Available at https://github.com/AzadehMozafariMehr/Rule-mining.

Each deviation was recorded along with the related dimensions (the set of attributes mentioned in Table 2).

We conducted two sets of experiments. First, we set the minimum support to 1.0% and applied association rule mining on the entire process instances (26,053 cases which resulted in 173,299 deviations) to find most frequent deviating patterns. For more in-depth analysis of deviating patterns and extracting knowledge about users and roles deviating behaviors, we filtered the dataset to keep 5,000 cases including 34,660 deviations(see footnote 2). Then we applied association rule mining by setting the minimum support to 0.1%.

Note that, in addition to system errors the dataset also contains human errors and deviations. Generally, systems produce less errors but very frequent, whereas humans deviate in many different ways but with less frequency in comparison to systems. To extract frequent deviating patterns of both human and system at one run, we have selected such low values for minimum support in our experiments.

4.2   Results

Table 3 summarizes the results of our experiments. For the first experiment, using a support threshold of 1.0% for frequent itemsets, we obtained 48 frequent anomalous patterns which presented 16 frequent patterns of unexpected behaviors and 32 frequent patterns of skipped tasks.

Table 3. Results of experiments on the loan management deviations dataset

Experiment	Experiment setting (minimum support threshold)	Total number of discovered deviating patterns	Anomalous Behavioral Patterns			
			Unexpected		Skipped	
			Single dimension	Multi-dimension	Single dimension	Multi-dimention
Exp1- All cases (with 173,299 deviations)	1.0%	48	8	8	8	24
Exp2- 5,000 cases (with 34,660 deviations)	0.1%	208	23	122	17	46

Some of these frequent deviating patterns are shown in Table 4. Pattern 1 is extracted from single-dimensional rule {*Update(OfferCancelledFlag)*} $\Rightarrow$ {*DeviationType6*}. This pattern shows that the data operation *Update(OfferCancelledFlag)* was ignored *40,869* times during the execution of the loan management process. The other patterns in Table 4 are extracted from multi-dimensional rules. For instance, patterns 2, 4 and 6 are an automatic interpretation of the rules {*Update(OfferCancelledFlag)*, *Validate Application*} $\Rightarrow$ {*DeviationType6*}, {*Update(OfferCancelledFlag)*, *Call After Offers*} $\Rightarrow$ {*DeviationType6*}, and {*Update(OfferCancelledFlag)*, *Cancel Application*} $\Rightarrow$ { *DeviationType6*} showing in which contexts the data operation *Update(OfferCancelledFlag)* was ignored.

Rules with a {*DataOperation, Activity, ExpectedRole*} ⇒ {*Deviation*} structure provide more details about deviant behaviors. Patterns 3, 5, and 7 are examples of the interpretation of this structure. They reflect another important aspect of a business process which is the separation of duties indicating the roles whom expected to perform certain tasks during the execution of a process. Based on the type of deviation in the consequent of the rules, our framework is able to provide different interpretations of the rule automatically. For instance, three different types of deviating behavior can be extracted from the mentioned rule structure. Consider three rules with the same set of attributes in the antecedent, one with the consequent "DeviationType6" will be interpreted as if the role *ignored* the data operation while performing the activity, whereas the one with the consequent "DeviationType5" will be interpreted as if the role *skipped* the tasks (both activity and data operation), and the one with "DeviationType3" will be interpreted as if the role *illegally performed* the activity and the data operation.

As an example, in pattern 7 the rule {*Update(OfferCancelledFlag), Cancel Application, Clerk*} ⇒ {*DeviationType6*} with support 4.59% and confidence 100% is automatically interpreted as "The role Clerk while performing activity Cancel Application ignored mandatory data operation Update(OfferCancelledFlag) 7,950 times".

Table 4. Some of the anomalous patterns in system category obtained in experiment 1 showing *skipped* behaviors

Pattern	Deviating Behavior	Occurrence	Support %	Confidence %
1	• Mandatory data operation Update(OfferCancelledFlag) was ignored 40,869 times	40,869	23.58	100.0
2	+ 16,735 times in the context of activity Validate application	16,735	9.66	100.0
3	− 16,735 times by the role Clerk	16,735	9.66	100.0
4	+16,184 times in the context of activity Call after offers	16,184	9.34	100.0
5	−16,184 times by the role Clerk	16,184	9.34	100.0
6	+7,950 times in the context of activity Cancel Application	7,950	4.59	100.0
7	−7,950 times by the role Clerk	7,950	4.59	100.0

In order to gain detailed knowledge about the system, users and roles deviating behaviors, in a second experiment, we selected 5,000 instances (cases) of the loan management process and applied multi-perspective process mining which resulted in a deviations dataset including 34,660 deviations. Then, using a support threshold of 0.1% for frequent itemsets, we obtained 208 deviating patterns in total. The framework extracted 145 patterns for unexpected behaviors and 63 patterns for skipped behaviors. From the discovered anomalous patterns, 106 patterns show the users' unexpected behavior and 17 patterns indicate the roles unexpected behavior. Table 5 shows some of the patterns obtained in our second experiment. Pattern 1 is an interpretation of single dimensional rule {$Read(address, email)$} ⇒ {$DeviationType2$} (with support 5.67% and confidence 100%) indicating that unexpected access to personal information of the customers happened 1,966 times during the execution of the selected 5,000 process instances. Pattern 2 is an interpretation of the rule {$Read(address, email), Clerk$} ⇒ {$DeviationType2$} showing which role performed the deviation in the organisation.

Moreover, patterns 3 to 21 indicate the users whom misused their privilege to get such access. Using the support measure, the framework is able to show how often users/roles performed an specific deviating behavior. As shown in the hierarchical structure of the results in Table 5, only the role clerk had the mentioned pattern and among the users playing this role, user "$User\_28$" performed this deviating behavior more often. Note that the results show the patterns with high confidence (equal to 100%) which satisfied the minimum support threshold (0.1%) sorted by the number of occurrences.

Table 5. Some of the anomalous patterns in system category obtained in experiment 2 showing *unexpected* behaviors

Pattern	Deviating Behavior	Occurrence	Support %	Confidence %
1	• Unexpected data operation Read (address- email) was executed 1,966 times	1,966	5.67	100.0
2	+1,966 times by the role Clerk	1,966	5.67	100.0
3	−137 times by the user User_28	137	0.39	100.0
4	−133 times by the user User_85	133	0.38	100.0
5	−123 times by the user User_27	123	0.35	100.0
6	−99 times by the user User_73	99	0.28	100.0
7	−83 times by the user User_2	83	0.23	100.0
8	−72 times by the user User_5	72	0.21	100.0
9	−70 times by the user User_3	70	0.20	100.0
10	−61 times by the user User_4	61	0.17	100.0
11	−61 times by the user User_8	61	0.17	100.0
12	−60 times by the user User_14	60	0.17	100.0
13	−59 times by the user User_19	59	0.17	100.0
14	−57 times by the user User_10	57	0.16	100.0
15	−57 times by the user User_18	57	0.16	100.0
16	−53 times by the user User_38	53	0.15	100.0
17	−51 times by the user User_41	51	0.11	100.0
18	−47 times by the user User_25	47	0.14	100.0
19	−44 times by the user User_15	44	0.13	100.0
20	−43 times by the user User_35	43	0.12	100.0
21	−39 times by the user User_49	39	0.11	100.0

Another operational insight is the ability to identify the users that perform only one type of deviating behavior, and the users with multiple ones. The rules of the form $\{User\} \Rightarrow \{Deviation\}$ are employed to extract this knowledge. To provide such insight the confidence measure is used. Whenever such a rule has confidence 100%, the user (in the antecedent) performed only one type of deviation (in the consequent). Lower values of confidence indicate multiple types of deviations. The framework allows for setting the confidence measure in user/role category. The default confidence measure is set to 25% to find the users/roles with multiple types of deviating behaviors. Moreover, the framework reports all users with only one kind of deviating behavior since this information is also helpful for the analysis of users' anomalous behaviors. By employing all multi-dimensional rules which have user and/or role attributes in their antecedent along with their support and confidence measures, the framework is able to provide a comprehensive information about user/role patterns of deviating behaviors.

It facilitates the extraction of information such as what were the patterns of user/role behaviors in each type of deviation, how often they performed specific deviating patterns, which users skipped their tasks most, which users illegally accessed sensitive data, which users performed spurious activities and/or data operations, and etc.

Table 6 shows a part of the results in the user category in our second experiment. The results show the users *User_ 25, User_ 52, User_ 6* and *User_ 8* had only one type of deviation, while the users *User_ 2, User_ 114, User_ 30, User_ 28, User_ 112, User_ 113, User_ 35, User_ 5, User_ 116,* and *User_ 38* had more types of deviating behavior. Although user *User_ 25* had only one type of deviating behavior, the framework was also able to present different patterns for this user. Patterns 4 and 5 show spurious actions of *User_ 25* in creating offers for the customers. Pattern 6 shows this user misused his/her privilege 47 times and accessed personnel information of the customers illegally.

Table 6. Some of the anomalous patterns in user/role category

Pattern	Deviating behavior	Occurrence	Support %	Confidence %
1	User(s) {'User_2', 'User_114', 'User_30', 'User_28', 'User_112', 'User_113', 'User_35', 'User_5', 'User_14', 'User_116', 'User_38'} had the highest number of deviating behaviors and were involved in different types of deviations	–	–	–
2	User(s) {'User_25', 'User_52', 'User_6', 'User_8' } had only one kind of deviating behavior	–	–	–
3	• The user User_25 with role Clerk had only one kind of deviating behavior. This user performed unexpected data operations	185	0.53	100.0
4	−56 times performing data operation Create (OfferID)	56	0.16	100.0
5	−56 times performing data operation data operation Read (OfferID)	56	0.16	100.0
6	−47 times performing data operation data operation Read (address- email)	47	0.14	100.0

The results demonstrate that the patterns obtained from the proposed approach can highlight frequent and correlated anomalous behaviors. In particular, mined patterns enable more accurate diagnostics and a better understanding of deviant behaviors. The hierarchical presentation of the patterns provides analysts with a more comprehensive and compact representation of deviations.

5   Related Work

For business process anomaly detection, there are various techniques from areas of data mining and process mining. The process-oriented techniques for anomaly detection such as conformance checking in particular, are employed to discover process execution deviations from prescribed behaviors. In [4,7,15,18,19] different perspectives of a business process such as resources (users and role allocation), data dependencies and flow of activities are considered for discovery of anomalies.

In the data mining area, from technical view, there are several techniques that aid in mining behavioral rules. The classic association rule mining algorithms are Apriori [5], and FP-Growth [12]. The advantage of the Apriori algorithm is that it is easy to implement. However, an appropriate filtering and interpretation procedure is necessary for investigation of discovered rules. In this work, we applied the Apriori mining algorithm [5] on a dataset of violations in order to find patterns of anomalous behaviors. In recent years, association rule mining techniques have been improved in terms of data set processing capacity and speed. The techniques in [8,9] reflect these improvements.

From the application view, researchers have proposed the application of association rule mining in many different domains. For instance, in [2], the authors propose an anomaly detection approach based on association rule mining which helps the distinction into malign and benign anomalies. In addition to report if an execution is anomalous or not, considering the support and confidence measures of an anomaly, they provide additional information about the severity of reported anomalies. Similarly, we have the same approach for reporting the anomalous patterns. In [23], the authors present a hybrid association rule learning for fraud detection. Similar to our work, they also applied process mining technique in addition to rule mining to find anomaly patterns. However they can only detect and learn one type of anomalous behavioral patterns while our technique provides a wider range of anomalous patterns with more details. In contrast to our approach, their technique requires an expert to verify the detected anomalous patterns. In [16], the authors introduce an interactive approach for behavior rule mining and anomaly detection. Their focus is only on detecting user anomaly patterns in order to reduce internal risks while our focus is detecting users', roles' and system's anomaly patterns which help reducing both operational and data security risks (including internal risks). In [25], a technique was presented to determine the potential local behaviors which uses sequential rule mining. It focuses on the users' behavior and tries to mine daily relevant behavior and irregular behavior of a particular subject (worker). By contrast, our framework employs frequent itemset mining and its focus is not on normal behavior, but irregular and abnormal behaviors in the users, roles and system level.

6   Conclusion

In this work, we presented an automated association rule mining-based framework for the discovery of patterns of anomalous behaviors in an organization. One of the drawbacks of the association rule mining techniques is that they typically return a large set of valid rules. Investigating, filtering and interpreting of these rules to recognize relevant and useful ones are challenging and time consuming. In the proposed framework, we have automated these procedures. Additionally, our framework organizes related patterns in a hierarchical format and classifies them into users, roles, and system's anomalous behavioral patterns to provide operational insight. Using the proposed framework, an organization can uncover potential operational or data security risks.

We showed the applicability of our approach in a real-life case study. The experiments demonstrated the approach's capability to return useful patterns capturing deviant behaviors. As future step, we plan a qualitative analysis of how useful the rules are to the analysts to define a way to prioritize/order them by this usefulness measure.

Acknowledgement. The author has received funding within the BPR4GDPR project from the European Union's Horizon 2020 research and innovation programme under grant agreement No 787149.

Reproducibility. The source code and inputs required to reproduce the experiments can be found at https://github.com/AzadehMozafariMehr/Rule-mining.

References

1. DePoy, E., Gitlin, L.N. (eds.): Statistical analysis for experimental-type designs. In: DePoy, E., Gitlin, L.N. (eds.) Introduction to Research, 5th Edn., pp. 282–310. Mosby, St. Louis (2016). Chapter 20
2. Böhmer, K., Rinderle-Ma, S.: Mining association rules for anomaly detection in dynamic process runtime behavior and explaining the root cause to users. Inf. Syst. **90**, 101438 (2020). Advances in Information Systems Engineering Best Papers of CAiSE 2018
3. van der Aalst, W.: Process Mining: Discovery, Conformance and Enhancement of Business Processes. Springer, Heidelberg (2011). https://doi.org/10.1007/978-3-642-19345-3
4. Adriansyah, A., van Dongen, B.F., van der Aalst, W.M.P.: Towards robust conformance checking. In: zur Muehlen, M., Su, J. (eds.) BPM 2010. LNBIP, vol. 66, pp. 122–133. Springer, Heidelberg (2011). https://doi.org/10.1007/978-3-642-20511-8_11
5. Agrawal, R., Imieliński, T., Swami, A.: Mining association rules between sets of items in large databases. In: Proceedings of the 1993 ACM SIGMOD International Conference on Management of Data, pp. 207–216 (1993)
6. Agrawal, R., Srikant, R., et al.: Fast algorithms for mining association rules. In: Proceedings 20th International Conference Very Large Data Bases, VLDB, vol. 1215, pp. 487–499, Santiago, Chile (1994)
7. Alizadeh, M., Lu, X., Fahland, D., Zannone, N., van der Aalst, W.: Linking data and process perspectives for conformance analysis. Comput. Secur. **73**, 172–193 (2018)
8. Aqra, I., Abdul Ghani, N., Maple, C., Machado, J., Sohrabi Safa, N.: Incremental algorithm for association rule mining under dynamic threshold. Appl. Sci. **9**(24) (2019)
9. Chengyan, L., Feng, S., Sun, G.: DCE-miner: an association rule mining algorithm for multimedia based on the mapreduce framework. Multimedia Tools Appl. **79**(23), 16771–16793 (2020)
10. van Dongen, B.: BPI Challenge 2017 (2 2017). https://doi.org/10.4121/uuid:5f3067df-f10b-45da-b98b-86ae4c7a310b
11. Han, J., Pei, J., Kamber, M.: Data mining: concepts and techniques. Elsevier (2011)
12. Jiawei, H., Jian, P., Yiwen, Y., Runying, M.: Mining frequent patterns without candidate generation: a frequent-pattern tree approach. In: Data Mining and Knowledge Discovery, pp. 53–87 (2004)
13. Knapp, K.J., Morris, R.F., Jr., Marshall, T.E., Byrd, T.A.: Information security policy: an organizational-level process model. Comput. Secur. **28**(7), 493–508 (2009)
14. Leemans, S.J.J., Fahland, D., van der Aalst, W.M.P.: Discovering block-structured process models from event logs - a constructive approach. In: Colom, J.-M., Desel, J. (eds.) PETRI NETS 2013. LNCS, vol. 7927, pp. 311–329. Springer, Heidelberg (2013). https://doi.org/10.1007/978-3-642-38697-8_17
15. de Leoni, M., van der Aalst, W.M.P.: Aligning event logs and process models for multi-perspective conformance checking: an approach based on integer linear programming. In: Daniel, F., Wang, J., Weber, B. (eds.) BPM 2013. LNCS, vol. 8094, pp. 113–129. Springer, Heidelberg (2013). https://doi.org/10.1007/978-3-642-40176-3_10

16. Liu, K., Wu, Y., Wei, W., Wang, Z., Zhu, J., Wang, H.: An interactive approach of rule mining and anomaly detection for internal risks. In: Nigdeli, S.M., Kim, J.H., Bekdaş, G., Yadav, A. (eds.) ICHSA 2020. AISC, vol. 1275, pp. 365–376. Springer, Singapore (2021). https://doi.org/10.1007/978-981-15-8603-3_32

17. Loch, K.D., Carr, H.H., Warkentin, M.E.: Threats to information systems: today's reality, yesterday's understanding. MIS Q. **16**(2), 173–186 (1992)

18. Mannhardt, F., de Leoni, M., Reijers, H., van der Aalst, W.: Balanced multi-perspective checking of process conformance. BPMcenter.org (2014)

19. Mozafari Mehr, A.S., de Carvalho, R.M., van Dongen, B.: Detecting privacy, data and control-flow deviations in business processes. In: Nurcan, S., Korthaus, A. (eds.) CAiSE 2021. LNBIP, vol. 424, pp. 82–91. Springer, Cham (2021). https://doi.org/10.1007/978-3-030-79108-7_10

20. Mozafari Mehr, A.S., de Carvalho, R.M., van Dongen, B.: MLA: a tool for multi-perspective conformance checking of business processes. In: Jans, M., De Weerdt, J., Depaire, B., Dumas, M., Janssenswillen, G. (eds.) ICPM 2021 Doctoral Consortium and Demo Track 2021, pp. 35–36 (2021). http://ceur-ws.org/

21. Pika, A., van der Aalst, W.M., Wynn, M.T., Fidge, C.J., ter Hofstede, A.H.: Evaluating and predicting overall process risk using event logs. Inf. Sci. **352**, 98–120 (2016)

22. Rodrigues, A.M.B., et al.: Stairway to value: mining a loan application process (2017)

23. Sarno, R., Dewandono, R.D., Ahmad, T., Naufal, M.F., Sinaga, F.: Hybrid association rule learning and process mining for fraud detection. IAENG Int. J. Comput. Sci. **42**(2) (2015)

24. Scheithauer, G., Henne, R.L., Kerciku, A., Waldenmaier, R., Riedel, U.: Suggestions for improving a bank's loan application process based on a process mining analysis (2017)

25. Setiawan, F., Yahya, B.N.: Improved behavior model based on sequential rule mining. Appl. Soft Comput. **68**, 944–960 (2018)

26. Warkentin, M., Willison, R.: Behavioral and policy issues in information systems security: the insider threat. Eur. J. Inf. Syst. **18**(2), 101–105 (2009)

Mining ϵ-Closed High Utility Co-location Patterns from Spatial Data

Vanha Tran[1], Lizhen Wang[2,3]($\boxtimes$), Shiyu Zhang[2], Jinpeng Zhang[2,4], and SonTung Pham[1]

[1] FPT University, Hanoi 155514, Vietnam
hatv14@fe.edu.vn, tungpshe140670@fpt.edu.vn
[2] Yunnan University, Kunming 650091, China
lzhwang@ynu.edu.cn, zsy608118@mail.ynu.edu.cn
[3] Dianchi College of Yunnan University, Kunming 650213, China
[4] Yunnan University of Finance and Economics, Kunming 650221, China
zjp@ynufe.edu.cn

Abstract. Traditional spatial prevalent co-location pattern mining is discovering groups of spatial features whose instances frequently appear together in nearby areas. However, it is unsuitable for many real-world applications where the significance of these instances must be considered. High utility co-location pattern (HUCP) mining is developed to find highly beneficial patterns by considering the importance of spatial instances. However, the mining result typically contains many HUCPs, making it difficult for users to absorb, comprehend, and apply. This work proposes a compressed representation of HUCPs, ϵ-closed HUCPs, that allow for a user-specified small tolerance of the information between a pattern and its supersets. If the information difference is not larger than the small tolerance it only needs to keep the supersets. Moreover, an efficient algorithm is developed to discover ϵ-closed HUCPs. The proposed algorithm avoids examining many unnecessary candidates; therefore, the performance of mining ϵ-closed HUCPs is significantly improved. A set of different numbers of features, numbers of instances, and distribution of both synthetic and real data sets are employed to evaluate the performance of the proposed method completely. The experimental results show that ϵ-closed balances the compression rate and the UPI error rate and gives a large pattern compression rate within a relatively small range of error rates. Moreover, the proposed algorithm is high-performance on dense and large data sets.

Keywords: Prevalent co-location pattern · High utility co-location pattern · ϵ-closed pattern

1 Introduction

As an important direction of spatial data mining, discovering spatial prevalent co-location patterns (PCPs) from spatial data sets is finding groups of spatial features whose spatial instances frequently occur with each other in nearby

W. Chen et al. (Eds.): ADMA 2022, LNAI 13725, pp. 413–428, 2022.
https://doi.org/10.1007/978-3-031-22064-7_30

areas. For example, in a point of interest (POI) data set of a city, by using the PCP mining technique we find that banks, supermarkets, and bus stations often appear together within 200 m, i.e., {bank, supermarket, bus station} is a PCP. PCPs can expose relationship rules between features in spatial data sets and the rules have been applied in many fields such as public safety [10], disease control [11], agriculture [4,12], criminology [7,9], business [16], climate science [1], transportation and location-based services [20], and so on.

In PCP mining, spatial instances are treated equally, it only considers the prevalence of these instances that appear together in nearby of each other. However, in practice, each spatial instance has its importance, if the factor is not considered, PCPs may lose meaning in their practical applications. For example, the importance of a supermarket and a bank is not the same. The services covered by the supermarket will be more extensive than the bank in respect of the supply of goods and service objects. The impact on the surrounding populace of the supermarket will be greater than the bank. Therefore, the notion of high utility co-location patterns (HUCP) has been proposed [19]. In HUCPs, each spatial instance is assigned a utility value to reflect its significance or importance. In addition, the correlation of the spatial features in a pattern is also considered through the participating spatial instances of the pattern.

However, the mining result normally generates a large number of HUCPs and includes redundant information. It is difficult for users to understand, absorb and apply the mining result. Thus, it is very necessary to find a concise representation of the mining result. But it is not easy to find a good concise representation since it needs to meet three conditions at the same time: (1) the number of patterns should be as little as possible; (2) the redundancy information should be removed as much as possible; and (3) the error between the compressed and the original should be as small as possible.

This work focuses on devising a representation that satisfies the above three conditions. The main contributions of this paper are as follows:

- Propose an ϵ-closed high utility co-location pattern notion that is a concise representation of HUCPs. If there is a small amount of information difference between a HUCP and its supersets, the HUCP can be deleted, and it just only needs to remain the supersets. The number of HUCPs can be reduced significantly under losing a small amount of information.
- Users can control the number of output HUCPs through ϵ which defines the amount of information difference between a HUCP and its supersets.
- Design a neighboring instance-based mining algorithm. Different from the traditional generating-testing candidate mining algorithm, the proposed algorithm does not collect and validate every co-location instance of a pattern, it adopts an efficient query mechanism of a hash table.

The rest of this paper is organized as follows: Sect. 2 reviews related work. The formal definitions and properties of ϵ-closed HUCPs are described in Sect. 3. The neighboring instance-based mining algorithm is designed in detail in Sect. 4. Experiments on both synthetic and real data sets are shown in Sect. 5. Section 6 summarizes the paper and highlights the directions for future work.

2   Related Work

Compare with frequent itemset mining [13,14], PCP mining is more difficult, complex, and interesting since in PCPs, there are very complicated neighbor relationships between spatial instances in space. Thus, PCP mining has attracted many researchers. Join-based [8] has been known as the first algorithm for mining PCPs. It uses a join operation to collect co-location instances of each pattern. This operation becomes very expensive when dealing with long-size PCPs and/or dense spatial data sets. To avoid this disadvantage, many efficient PCPS mining algorithms have been proposed such as join-less [22], overlapping clique-based [17], clique-based [3], and candidate pattern tree [21].

How to efficiently discover PCPs from big data in limited execution time and computer resources has become a new challenge. Parallel PCP mining algorithms have been developed by executing on different platforms such as Hadoop and Map-reduce [15], the graphic processing unit (GPU) [2], and NoSQL [2].

The above algorithms focus on mining all complete and correct PCPs. However, the mining result normally contains too many redundant PCPs, this makes it difficult for users to understand, absorb and apply the mining result. Thus, mining concise representations of PCPs have been proposed such as maximal [16], closed [21], and non-redundant PCPs [18].

These mining algorithms mentioned above can be called traditional PCP mining algorithms. The traditional only considers the prevalence of spatial instances appearing together in nearby space, the importance or significance of spatial instances is ignored, this may cause the discovered PCPs to become meaningless. Therefore, HUCPs [19] that each spatial instance is assigned a utility value to reflect its importance are proposed. To discover HUCPs, a utility participation index (UPI) metric is designed to evaluate the interest of patterns. Unfortunately, UPI does not hold the downward closure property that is often used to reduce the candidate search space to improve mining efficiency.

Although HUCPs consider the importance of each instance, HUCPs have the same problem as the traditional PCPs, that is, the mining result contains too many patterns, making it difficult for users to absorb and apply. Therefore, in this work, we focus on finding a concise representation of HUCPs. First, an ϵ-closed HUCP notion, that relaxes the concise condition of the closure property by allowing a small difference ϵ in the UPI value between a HUCP and its supersets, is proposed. Second, a neighboring instance-based mining algorithm is designed. This algorithm is based on the neighboring spatial instances, i.e., it first finds all neighboring instances and stores the neighbor relationships into a compact hash table, then co-location instances of each pattern are obtained by querying on the hash table, it avoids verifying the neighbor relationship of each co-location instance in the generating-testing candidate mining mechanism.

3   Method

Given a set of spatial features $F = \{f_1, ..., f_m\}$ and a set of spatial instances $S = \{I_1, ..., I_m\}$ where $I_t = \{o_1, ..., o_q\}$ is the set of instances of feature $f_t \in F$.

Each spatial instance $o_i \in S$ is formed by a three-element vector, i.e., <feature type, instance identification, location>. We use $f(o_i) = f_t$ to represent the feature type of instance o_i is f_t. R is a user-defined neighbor relationship, if two spatial instances o_i and o_j have the neighbor relationship, they are denoted as $R(o_i, o_j)$. When R uses a distance metric (e.g., Euclidean, Manhattan, Minkowski distance metrics, and so on), if the distance between two instances o_i and o_j is not larger than a distance threshold d given by users, o_i and o_j satisfy the neighbor relationship R, i.e., $R(o_i, o_j) <=> distance(o_i, o_j) \le d$. $NR(o_i) = \{o_j | R(o_i, o_j) \land f(o_i) \neq f(o_j)\}$ is the set of neighboring instances of o_i under R.

Definition 1 *(Utility of a spatial instance). The utility of a spatial instance $o_i \in S$ is a value $v \le 0$ that is assigned to the instance to reflect its importance and expressed by the superscript on it, i.e., o_i^v. We denote the utility of a spatial instance o_i^v as $u(o_i^v) = v$.*

Definition 2 *(Utility of a spatial feature). The utility of a spatial feature $f_t \in F$ is the sum of the utilities of its instances and denote as*

$$u(f_t) = \sum_{i=1}^{q} u(o_i^v) \tag{1}$$

where q is the number of instances that belong to f_t.

For example, Fig. 1 shows an example of spatial data set with assigned instance utilities. $A1^{15}$ is an instance of feature A and its utility value is 15. The solid line connects two instances to represent that they have a neighbor relationship. Table 1 lists the instances with their utility values of each spatial feature. Moreover, the utility of each feature also is drawn in Table 1.

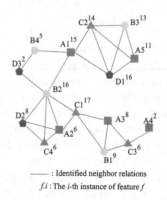

——— : Identified neighbor relations

$f.i$: The i-th instance of feature f

Fig. 1. An example of spatial data set with spatial instance utilities.

Table 1. The utilities of the instances and features of the data set in Fig. 1.

Feature	Instance	Feature	Instance	Feature	Instance	Feature	Instance
	$A1^{15}$		$B1^9$		$C1^{17}$		$D1^{16}$
	$A2^6$		$B2^{16}$		$C2^{14}$		$D2^8$
	$A3^8$		$B3^{13}$		$C3^6$		$D3^2$
	$A4^2$		$B4^5$		$C4^6$		
	$A5^{11}$						
$u(A)$	42	$u(B)$	43	$u(C)$	43	$u(D)$	26

Definition 3 *(Co-location pattern and co-location instance). A co-location pattern c is a subset of spatial feature set F, i.e., $c = \{f_1, ..., f_k\} \subseteq F$. The number of the features in c is k and it is called the size of c, i.e., c is a size k pattern. A co-location instance $CI(c)$ of a co-location pattern c is a set of spatial instances that includes the instances of all feature types in c and these instances have the neighbor relationship R, i.e., $CI(c) = \{o_1, ..., o_k\}$. The set of all co-location instances of c is the table instance of c and denoted as $TI(c) = \{CI_1(c), ..., CI_p(c)\}$.*

Definition 4 *(Utility of a feature in a pattern). The utility of a feature f_t in a co-location pattern $c = \{f_1, ..., f_k\}$, i.e., $f_t \in c$, is the sum of utilities of the instances belonging to f_t in the table instance of c and denoted as*

$$u(f_t, c) = \sum_{o_i^v \in TI(c), f(o_i^v) = f_t} u(o_i^v) \tag{2}$$

Definition 5 *(Intra-utility ratio). The intra-utility ratio of a feature f_t in a co-location pattern $c = \{f_1, ..., f_k\}$, i.e., $f_t \in c$, is the proportion of the utility of f_t in c to its utility in the data set and denoted as*

$$intraUR(f_t, c) = \frac{u(f_t, c)}{u(f_t)} \tag{3}$$

Definition 6 *(Inter-utility ratio). The inter-utility ratio of a feature f_t in a co-location pattern $c = \{f_1, ..., f_k\}$, i.e., $f_t \in c$, is defined as*

$$interUR(f_t, c) = \frac{\sum_{f_h \in c, f_h \neq f_t} u(f_h, c)}{\sum_{f_h \in c, f_h \neq f_t} u(f_h)} \tag{4}$$

Definition 7 *(Utility participation ratio). The utility participation ratio of a feature f_t in a pattern $c = \{f_1, ..., f_k\}$, i.e., $f_t \in c$, is the combination of the intra-utility ratio and the inter-utility ratio of the feature and is denoted as*

$$UPR(f_t, c) = \alpha \times intraUR(f_t, c) + \beta \times interUR(f_t, c) \tag{5}$$

where α and β represent the weighted values of the intra-utility ratio and the inter-utility ratio, respectively.

Definition 8 *(Utility participation index). The utility participation index of a co-location pattern $c = \{f_1, ..., f_k\}$ is the minimum utility participation ratio among all the features in c and denoted as*

$$UPI(c) = \min_{f_t \in c}\{UPR(f_t, c)\} \qquad (6)$$

Definition 9 *(High co-location pattern). A co-location pattern $c = \{f_1, ..., f_k\}$ is a high utility co-location pattern if and only if its utility participation index is not smaller than a minimum utility threshold μ given by users, i.e., $UPI(c) \geq \mu$.*

When a co-location pattern has not yet been determined to be a HUCP, the pattern is called candidate co-location pattern. For example, as shown in Fig. 1, examine candidate {A, B, D}, based on Fig. 1, its table instance is {A5^{11}, B3^{13}, D1^{16}} and {A2^6, B2^{16}, D2^8}, based on equations from (5)–(8), we can compute $UPI(\{A, B, D\}) = 0.49$. If a user sets $\mu = 0.2$, since $UPI(\{A, B, D\}) = 0.49 > \mu = 0.2$, thus {A, B, D} is a high utility co-location pattern.

Lemma 1. *UPI does not satisfy the downward closure property.*

Proof. In Fig. 1, when the minimum utility threshold is set to 0.2, i.e., $\mu = 0.2$, we can obtain {A, B, D} is a HUCP since $UPI(\{A, B, D\})=0.49$. Beside, its a superset {A, B, C, D} and its a subset {A, D} are also HUCPs since $UPI(\{A, B, C, D\}) = 0.5 > \mu$ and $UPI(\{A, D\})=0.74 > \mu$. It can be found that the UPI value of a pattern can be greater or less than its supersets.

Definition 10 *(ϵ-closed high utility co-location pattern). A co-location pattern $c = \{f_1, ..., f_k\}$ is an ϵ-closed HUCP if and only if c is a HUCP and there exists no HUCP c' such that $c' = \{f_1, ..., f_k\} \cup \{f_{k+1}\}$, i.e., $c \subset c'$, and $|UPI(c) - UPI(c')| \leq \epsilon$, where ϵ is a UPI tolerance value given by users.*

It is easy to find that, when $\epsilon = 0$, ϵ-closed high utility co-location patterns are closed high utility co-location patterns.

Definition 11 *(UPI error rate). Given HUCPS and ϵHUCPS are the set of all HUCPs and the set of ϵ-closed HUCPs, when using ϵ-closed HUCPs to represent HUCPs, the UPI error rate is computed as*

$$\sigma = \sum \frac{|UPI(c') - UPI_{c \in \chi}(c)|}{|HUCPS|} \qquad (7)$$

where χ is a set of HUCPs that are removed by ϵ-closed in Definition 10.

4 Neighboring Instance-Based Mining Algorithm

Based on Lemma 1, since the downward closure property is not held in the UPI metric, if we employ the level-wise search mining framework [19], unnecessary candidates cannot be effectively pruned, thus mining performance is extremely inefficient, especially when the data set is dense. Moreover, the key of discovering ϵ-closed HUCPs is collecting co-location instances of patterns and this step is the most expensive [19]. Thus, reducing candidate search space and quickly collecting co-location instances are the keys of improving mining ϵ-closed HUCPs. This section develops a neighboring instance-based mining ϵ-closed HUCP algorithm.

4.1   Maximal Clique Enumeration

Definition 12 *(Neighboring graph). Given a data set S and a neighbor relationship R, after materializing the neighbor relationship, we get a connected undirected graph $G(V, E)$ with V is the set of vertices of G that are the all instances in S and E is set of edges that are these lines connecting neighboring instances.*

Definition 13 *(Subgraph and remaining neighboring graph). Given an instance o_i and its neighboring instance set $NR(o_i)$, a subgraph of the neighboring graph $G(V, E)$ is $G_{o_i}(V_{o_i}, E_{o_i}) \in G(V, E)$, where $V_{o_i} = \{o_i\} \cup NR(o_i)$ and E_{o_i} is the set of lines that connect neighboring instances in V_{o_i}. The remaining neighboring graph is $G_{o_i^-}(V_{o_i^-}, E_{o_i^-})$ is the graph that contains all vertices and edges in $G(V, E)$ but not in $G_{o_i}(V_{o_i}, E_{o_i})$.*

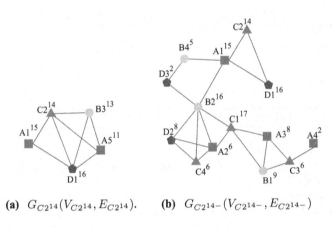

(a) $G_{C2^{14}}(V_{C2^{14}}, E_{C2^{14}})$.   **(b)** $G_{C2^{14-}}(V_{C2^{14-}}, E_{C2^{14-}})$

Fig. 2. The subgraph and remaining neighboring graph of $C2^{14}$.

For example, after materializing the spatial neighbor relationship, a graph is obtained as shown in Fig. 1. For $C2^{14}$, its subgraph and the remaining neighboring graph are plotted in Fig. 2.

After partitioning the neighboring graph by the subgraph and the remaining neighboring graph, a maximal algorithm is applied to the subgraph to enumerate maximal cliques (MCs) [5,6]. Here we do not focus on designing a new MC enumeration algorithm, this work is paid attention to how to employ MCs to improve the performance of mining ε-closed HUCPs. Algorithm 1 describes the pseudocode of a MC enumeration algorithm that is developed by Eppstein et al. [6]. This algorithm inputs a set of vertices and edges, then uses a recursive process to find maximal cliques. For more details, please reference [6].

Combining the partitioning strategy and MC enumeration algorithm, Algorithm 2 shows the pseudocode of listing all MCs from the input spatial data set under a neighbor relationship R. This algorithm first materializes the neighbor

Algorithm 1. Enumerating maximal cliques on a graph

Input: $G(V,E)$
Output: maximal cliques, MCs
$P \leftarrow V, R \leftarrow \emptyset, X \leftarrow \emptyset$
Proc: BronKerboschDegeneracy(V,E)

1: **for** each vertex v_i in a degeneracy ordering of $G(V,E)$ **do**
2:     $P \leftarrow NR(v_i) \cap \{v_{i+1}, ..., v_{n-1}\}$
3:     $X \leftarrow NR(v_i) \cap \{v_0, ..., v_{i-1}\}$
4:     BronKerboschPivot$(P, \{v_i\}, X)$
5: **end for**

Proc: BronKerboschPivot(P, R, X)

1: **if** $P \cup X = \emptyset$ **then**
2:     $MCs = MCs \cup R$ $\triangleright$ R is a maximal clique
3: **end if**
4: Choose a pivot $u \in P \cup X$ to maximize $|P \cup NR(u)|$
5: **for** $v \in P \setminus NR(u)$ **do**
6:     BronKerboschPivot$(P \cap NR(v), R \cup \{v\}, X \cap NR(v))$
7:     $P \leftarrow P \setminus v$
8:     $X \leftarrow X \cup v$
9: **end for**
10: **return** MCs

Algorithm 2. Enumerating MCs

Input: $G(V,E)$
Output: a set of maximal cliques, MCS

1: $NRS = \{..., NR(o_i), ...\} \leftarrow$ materializing_neighbor_relationship(S, R)
2: **while** $S \neq \emptyset$ **do**
3:     $o_i \leftarrow S.\text{pop}()$
4:     $G_{o_i}(V_{o_i}, E_{o_i}) \leftarrow$generate_subgraph$(o_i, NRS)$
5:     $G_{o_i^-}(V_{o_i^-}, E_{o_i^-}) \leftarrow$generate_ remaining_neighboring_graph$(G_{o_i}(V_{o_i}, E_{o_i}), S)$
6:     $S \leftarrow V_{o_i^-}$
7:     $MCs \leftarrow$ Algorithm 1
8:     $MCS \leftarrow MCS \cup MCs$
9: **end while**
10: **return** MCS

---

relationship between instances in S (Step 1). Then, for each instance $o_i \in S$, it constructs the subgraph $G_{o_i}(V_{o_i}, E_{o_i})$ (Step 4) and the remaining neighboring graph $G_{o_i^-}(V_{o_i^-}), E_{o_i^-})$ (Step 5). After that, Algorithm 1 is employed to find MCs in $G_{o_i}(V_{o_i}, E_{o_i})$ (Step 6). Finally, Algorithm 2 returns a set of MCs (Step 9).

Table 2 lists all MCs enumerated from the neighboring graph in the Fig. 1.

4.2   Co-location Instance Hash Table

Based on Definition 8, when calculating the utility participation ratio of a pattern $c = \{f_1, ...f_k\}$, it only needs to know the spatial instances of each feature

Table 2. The MCs listed from Fig. 1.

MC	MC	MC
{A1, C2, D1}	{B2, D3}	{A3, B1, C1}
{A5, B3, C2, D1}	{B4, D3}	{A3, B1, C3}
{A1, B4}	{A2, B2, C4, D2}	{A4, C3}
{A1, B2}	{A2, B2, C1}	

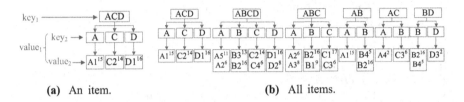

(a) An item.   (b) All items.

Fig. 3. The co-location instance hash table constructed based on MCs of the data set in Fig. 1.

participating in this pattern, it does not care which instance has a neighbor relationship with which instances. Therefore, given a pattern, if we can query the participation instances of each feature into the pattern, the utility participation ratio of the pattern can be easily calculated. We arrange these maximal cliques into a special hash table structure.

Definition 14 *(Co-location instance hash table). A co-location instance hash table is a two-level hash table, i.e., $(key_1, (key_2, value_2))$, where key_1 is the set of spatial feature types of these instances in a MC, key_2 is the spatial feature type of each instance in MC, and $value_2$ is the instance itself in MC. All MCs that instances in them belong to the same spatial feature type set are grouped into an item in the co-location hash table.*

Figure 3(a) shows an example of an item in the co-location instance hash table that is constructed from maximal clique $A1^{15}$, $C2^{14}$, $D1^{16}$. Based on Table 2, the co-location instance hash table of the input data set under the neighbor relationship is plotted in Fig. 3(b).

4.3   Mining ε-Closed HUCPs

After constructing the co-location instance hash table based on MCs of the input data set, the next step is how to quickly collect the instances that participate in the co-location instances of each pattern from the hash table structure.

Lemma 2. *Given a co-location pattern $c = \{f_1, ... f_k\}$, the instances of each feature type in c that participate in the table instance of c are collected from the values in the co-location instance hash table corresponding to the keys that are supersets of c (including if c is a key, the key is also a superset of c).*

Algorithm 3. Filtering ϵ-closed HUCPs

Input: the co-location instance hash table, $CIHash$; d, μ, ϵ
Output: the set of ϵ-closed HUPC, $\epsilon HUCPS$

1: $keyset \leftarrow CIHash.\text{getKeys}()$
2: **while** $keyset \neq \emptyset$ **do**
3:     $keyset \leftarrow \text{sort\_by\_size}(keyset)$
4:     $c \leftarrow keyset.\text{pop}()$
5:     **for** $item \in CIHash$ **do**
6:         **if** $c \subseteq item.\text{getKey}()$ **then**
7:             $TI(c) \leftarrow \text{get\_values}(item.\text{getValue})$
8:         **end if**
9:     **end for**
10:    $UPI(c) \leftarrow \text{calculate\_UPI}(TI(c))$
11:    **if** $UPI(c) \geq \mu$ **then**
12:        **for** $c' \in \epsilon HUCPS$ **do**
13:            **if** $(c \subseteq c')$ and $|UPI(c) - UPI(c')| > \epsilon$ **then**
14:                $\epsilon HUCPS.\text{add}(c)$
15:                **break**
16:            **end if**
17:        **end for**
18:    **end if**
19:    $subc \leftarrow \text{generate\_direct\_subsets}(c)$
20:    $keyset \leftarrow keyset \cup subc$
21: **end while**
22: **return** $\epsilon HUCPS$

Proof. From the definition of the co-location instance hash table, $value_2$ is a set of participating instances belonging to the feature type be key_2 in a co-location pattern named key_1. If a pattern $c \equiv key_1$, we can get the participating instances of c directly from $value_2$. However, the co-location instances of c may be the subsets of MCs that are built into an item in the co-location hash table, thus a part of co-location instances of c is queried from the values corresponding to the keys that are supersets of c.

For example, for co-location pattern $c = \{A, B\}$, the instances that participate in c of A and B are obtained from the values of keys AB, ABC, and ABCD, i.e., $\{A1^{15}, A2^6, A3^8, A5^{11}\}$ and $\{B2^{16}, B3^{13}, B4^5\}$, respectively.

Algorithm 3 describes the pseudocode of filtering ϵ-closed HUCPs from the co-location instance hash table. This algorithm first gets all keys in the co-location instance hash table and stores them in a set of keys, $keyset$ (Step 1). Next, Algorithm 3 executes a while loop. In the loop, $keyset$ is sorted in descending order size of the keys (Step 3) and it gets the first key as a co-location pattern c (Step 4). Then, it queries these keys that are a superset of c (Steps 5–6), and then takes the corresponding values and puts them into the table instance of c (Step 7). After that, the utility participation index of c is calculated (Step 8). If c is a HUCP (Step 9), it needs to verify whether the pattern is ϵ-closed (Steps 10–13). Next, the direct subsets of c are generated and put into *keyset* as new candidates that need to be examined (Steps 14–15). And finally, a set of ϵ-closed HUCPs is returned to users (Step 16).

5   Experiments

A set of experiments is constructed in this section to evaluate the effectiveness and efficiency of the proposed method. Our work is compared with the original work about mining HUCPs [19]. The two algorithms are implemented in C++ and run on a computer with CPU Intel(R) Core i7-3770 and 16 GB of RAM.

Table 3. A summary of the experimental data sets.

Name	Spatial frame size	# instances	# features	Property
Shenzhen [6]	28,800 × 89,900 (m^2)	35,220	20	Dense, clustering
Shanghai [6]	65,500 × 113,600 (m^2)	41,450	15	Dense, uniform
Vegetation [18]	71,930 × 66,440 (m^2)	503,340	16	Dense
Synthetic [22]	5,000 × 5,000	*	15	Dense

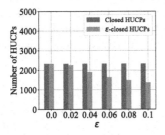
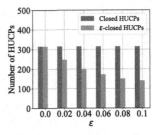

(a) Shenzhen (d=250m,$\mu = 0.1$).   **(b)** Shanghai (d=350m,$\mu = 0.1$).

Fig. 4. The number of HUCPs with the increase of the tolerance ϵ.

5.1   Spatial Data Sets

Both synthetic and real data sets are used in our experiments, in which the synthetic data sets are generated by a data generator [22] and the real data sets are listed in Table 3. As can be seen from Table 3, the experimental data sets are dense and have different numbers of instances, numbers of features, and distributions. This enables a more complete investigation of the performance of the proposed algorithm. The utility of the instances in these data sets is generated according to an exponential distribution under a hypothesis that the number of the low utility instances is much more than the high utility instances.

5.2   The Number HUCPs

The first experiment examines the number of HUCPs under different values of the tolerance ϵ. Since closed HUCPs are a lossless compression of all HUCPs, we only need to compare the proposed compression method with the closed. As shown in Fig. 4, many patterns can be reduced effectively by giving a relatively small tolerance value, ϵ, and as the value of the tolerance ϵ increases, more and more patterns are removed. Because the condition of the closed is too strict, and according to the UPI definition, this condition is difficult to achieve. It can be seen that reducing HUCPs can be effectively achieved by relaxing the closure condition under the tolerance ϵ controlled by users.

5.3   Compression Rate and UPI Error Rate

In this experiment, the compression rate and the UPI error rate of ϵ-closed HUCPs are surveyed to demonstrate the power of the ϵ-closed. The compression rate is calculated based on the closed HUCPs. We fix $\epsilon = 0.05$ (a sufficiently low tolerance in our opinion). Figure 5 plots the results on the two real data sets under different distance thresholds. As the distance threshold increases, the compression ratio increases, and the UPI error rate also increases. But the UPI error rate does not exceed the configured tolerance. The above results prove that the ϵ-closed can balance the compression rate and the UPI error rate, and it can give a large pattern compression rate within a relatively small range of the UPI error rate.

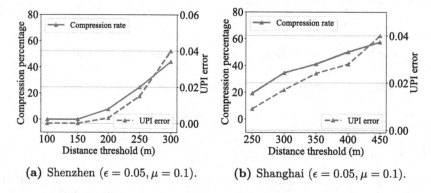

(a) Shenzhen ($\epsilon = 0.05, \mu = 0.1$).   (b) Shanghai ($\epsilon = 0.05, \mu = 0.1$).

Fig. 5. The compression rate and the error rate of the ϵ-closed method under different distance thresholds.

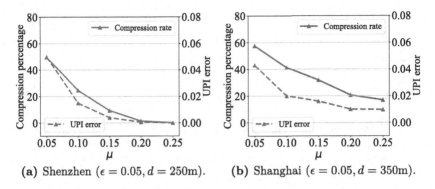

Fig. 6. The compression rate and the error rate of the ϵ-closed method under different minimum utility thresholds.

Moreover, the compression rate and the UPI error rate under different minimum utility thresholds are also examined and plotted in Fig. 6. As shown in Fig. 6, in the small value of minimum utility thresholds, the compression rate is large and the UPI error rate is also large. The larger the minimum utility threshold, the greater the difference in the UPI values of the patterns, so the number of HUCPs that can be reduced will be less.

From the above experimental results, we can conclude that the denser the HUCPs (large values of distance thresholds and small values of minimum utility thresholds), the stronger the ϵ-closed compression ability.

5.4 Scalability

In this section, the scalability of the proposed neighboring instance-based mining algorithm is evaluated with several workloads, i.e., different distance thresholds, minimum high utility thresholds, and numbers of spatial instances. For the convenience of description, the mining HUCP algorithm proposed by Wang et al. [19] (to the best of our knowledge, this is the only work on mining HUCP so far) is named level-wise-based since it adopts the traditional mining framework [22]. And our neighboring instance-based algorithm is named NI-based for short.

Figure 7 describes the execution time of the comparing algorithms on different distance thresholds. As can be seen that at a small distance threshold, the efficiency of the comparing algorithms is the same, the level-wise-based algorithm fails in the case of large distance thresholds, while the neighboring instance-based mining algorithm shows good scalability.

The execution of the proposed algorithm under different minimum high utility thresholds is also plotted in Fig. 8. As can be seen that our algorithm shows better performance at small thresholds.

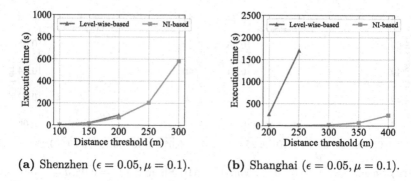

(a) Shenzhen ($\epsilon = 0.05, \mu = 0.1$).   **(b)** Shanghai ($\epsilon = 0.05, \mu = 0.1$).

Fig. 7. The execution time of the comparing algorithms on different distance thresholds.

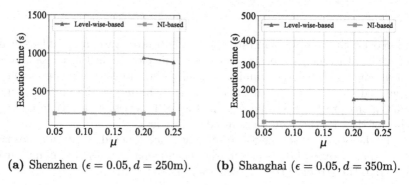

(a) Shenzhen ($\epsilon = 0.05, d = 250$m).   **(b)** Shanghai ($\epsilon = 0.05, d = 350$m).

Fig. 8. The execution time of the comparing algorithms on different minimum utility thresholds.

Moreover, Fig. 9 plots the execution of the two algorithms on different numbers of spatial instances. To generate different sizes of input data sets, for the synthetic data set, we fix the number of features and change the number of instances, whereas we perform random sampling on the real data set (Vegetation data set). As shown in Fig. 9, the execution time of the level-wise-based algorithm increases dramatically with the increase of the number of spatial instances. While the proposed mining algorithm gives scalability to large dense data sets.

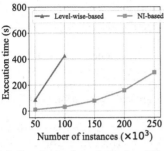
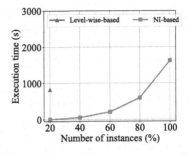

(a) Synthetic ($\epsilon = 0.05, d = 30$).   **(b)** Vegetation ($\epsilon = 0.05, d = 800\text{m}$).

Fig. 9. The execution time of the comparing algorithms on different numbers of spatial instances ($\mu = 0.1$ for all).

6   Conclusion

This work proposes a concise representation of HUCPs, ϵ-closed HUCPs. This representation allows a pattern and its superset to have a small tolerance ϵ that users can control according to their applications. Experimental results on the real data sets show that this representation can give a good compression rate under a small loss of information. Moreover, to avoid too many unnecessary candidates that need to be examined due to the UPI metric not satisfying the downward closure property, this work proposes a neighboring instance-based mining algorithm. The algorithm performs better than the existing algorithm on the experimental data sets.

The proposed mining algorithm can perform the parallel computation to improve the mining efficiency in giant spatial data sets.

Acknowledgements. This work is supported by the National Natural Science Foundation of China (61966036), the Project of Innovative Research Team of Yunnan Province (2018HC019), the Yunnan Fundamental Research Projects (202201AS070015).

References

1. Akbari, M., Samadzadegan, F., Weibel, R.: A generic regional spatio-temporal co-occurrence pattern mining model: a case study for air pollution. Journal of Geographical Systems **17**(3), 249–274 (2015)
2. Andrzejewski, W., Boinski, P.: Parallel approach to incremental co-location pattern mining. Inf. Sci. **496**, 485–505 (2019)
3. Bao, X., Wang, L.: A clique-based approach for co-location pattern mining. Inf. Sci. **490**, 244–264 (2019)
4. Cai, J., Liu, Q., Deng, M., Tang, J., He, Z.: Adaptive detection of statistically significant regional spatial co-location patterns. Comput. Environ. Urban Syst. **68**, 53–63 (2018)

5. Chang, L.: Efficient maximum clique computation over large sparse graphs. In: Proceedings of the 25th ACM SIGKDD International Conference on Knowledge Discovery & Data Mining, pp. 529–538 (2019)
6. Eppstein, D., Löffler, M., Strash, D.: Listing all maximal cliques in large sparse real-world graphs in near-optimal time. J. Exp. Algorithmics (JEA) **18**, 1–3 (2013)
7. He, Z., Deng, M., Xie, Z., Wu, L., Chen, Z., Pei, T.: Discovering the joint influence of urban facilities on crime occurrence using spatial co-location pattern mining. Cities **99**, 102612 (2020)
8. Huang, Y., Shekhar, S., Xiong, H.: Discovering colocation patterns from spatial data sets: a general approach. IEEE Trans. Knowl. Data Eng. **16**(12), 1472–1485 (2004)
9. Lee, I., Phillips, P.: Urban crime analysis through areal categorized multivariate associations mining. Appl. Artif. Intell. **22**(5), 483–499 (2008)
10. Leibovici, D.G., Claramunt, C., Le Guyader, D., Brosset, D.: Local and global spatio-temporal entropy indices based on distance-ratios and co-occurrences distributions. Int. J. Geogr. Inf. Sci. **28**(5), 1061–1084 (2014)
11. Li, J., Adilmagambetov, A., Mohomed Jabbar, M.S., Zaïane, O.R., Osornio-Vargas, A., Wine, O.: On discovering co-location patterns in datasets: a case study of pollutants and child cancers. GeoInformatica **20**(4), 651–692 (2016)
12. Liu, Q., Liu, W., Deng, M., Cai, J., Liu, Y.: An adaptive detection of multilevel co-location patterns based on natural neighborhoods. Int. J. Geogr. Inf. Sci. **35**(3), 556–581 (2021)
13. Luna, J.M., Fournier-Viger, P., Ventura, S.: Frequent itemset mining: a 25 years review. Wires Data. Min. Knowl. **9**(6), e1329 (2019)
14. Raj, S., Ramesh, D., Sreenu, M., Sethi, K.K.: Eafim: efficient apriori-based frequent itemset mining algorithm on spark for big transactional data. Knowl. Inf. Syst. **62**(9), 3565–3583 (2020)
15. Sheshikala, M., Rao, D.R., Prakash, R.V.: A map-reduce framework for finding clusters of colocation patterns-a summary of results. In: 2017 IEEE 7th International Advance Computing Conference (IACC), pp. 129–131. IEEE (2017)
16. Tran, V., Wang, L., Chen, H., Xiao, Q.: MCHT: a maximal clique and hash table-based maximal prevalent co-location pattern mining algorithm. Expert Syst. Appl. **175**, 114830 (2021)
17. Tran, V., Wang, L., Zhou, L.: A spatial co-location pattern mining framework insensitive to prevalence thresholds based on overlapping cliques. Distributed Parallel Databases 1–38 (2021)
18. Wang, L., Bao, X., Chen, H., Cao, L.: Effective lossless condensed representation and discovery of spatial co-location patterns. Inform. Sci. **436**, 197–213 (2018)
19. Wang, L., Jiang, W., Chen, H., Fang, Y.: Efficiently mining high utility co-location patterns from spatial data sets with instance-specific utilities. In: Candan, S., Chen, L., Pedersen, T.B., Chang, L., Hua, W. (eds.) DASFAA 2017. LNCS, vol. 10178, pp. 458–474. Springer, Cham (2017). https://doi.org/10.1007/978-3-319-55699-4_28
20. Yao, X., Jiang, X., Wang, D., Yang, L., Peng, L., Chi, T.: Efficiently mining maximal co-locations in a spatial continuous field under directed road networks. Inf. Sci. **542**, 357–379 (2021)
21. Yoo, J.S., Bow, M.: A framework for generating condensed co-location sets from spatial databases. Intell. Data Anal. **23**(2), 333–355 (2019)
22. Yoo, J.S., Shekhar, S.: A joinless approach for mining spatial colocation patterns. IEEE Trans. Knowl. Data Eng. **18**(10), 1323–1337 (2006)

Graph Mining

Implementation and Analysis of Centroid Displacement-Based k-Nearest Neighbors

Alex X. Wang$^{(\boxtimes)}$, Stefanka S. Chukova , and Binh P. Nguyen

School of Mathematics and Statistics, Victoria University of Wellington,
Wellington, New Zealand
{alex.wang,stefanka.chukova,binh.p.nguyen}@vuw.ac.nz

Abstract. k-NN is a widely used supervised machine learning method in different domains. Despite its simplicity, effectiveness, and robustness, k-NN is limited to the use of the Euclidean distance as the similarity metric, the arbitrarily selected neighborhood size k, the computational challenge from high dimensional data, and the use of the simple majority voting rule. Among different variants of k-NN in classification, we sought to address the last issue and proposed the Centroid Displacement-based k-NN (CDNN), where centroid displacement is used for class determination. In this study, we present an implementation of CDNN for `scikit-learn`, a well-known machine learning library for the Python programming language, and a comprehensive comparative performance analysis of CDNN with different variants of k-NN in `scikit-learn`. We open-source our algorithm to benefit the users, and to the best of our knowledge, no similar studies on performance analysis of k-NN and its variants in `scikit-learn` have been done. We also examine the effectiveness of different distance metrics on the performance of CDNN on different datasets. Extensive experiments on real-world and synthetic datasets verify the effectiveness of CDNN compared to the standard k-NN and other state-of-the-art k-NN-based algorithms. The results from the distance metrics comparison study also show that other distance metrics can further improve the classification performance of CDNN.

Keywords: Similarity · Nearest neighbors · Centroid displacement · k-NN · Scikit-learn

1 Introduction

The k-nearest neighbors (k-NN) algorithm is a simple yet efficient supervised machine learning (ML) algorithm. In classification, it is widely used in different fields, such as face recognition, text classification, graph analysis, disease prediction, anomaly detection, and time-series forecast, among others. The wide application of k-NN in different domains is primarily due to its advantages of conceptual simplicity, substantial theoretical foundation, strong generalization performance, and no assumptions on data distribution [1]. In addition to this,

© The Author(s), under exclusive license to Springer Nature Switzerland AG 2022
W. Chen et al. (Eds.): ADMA 2022, LNAI 13725, pp. 431–443, 2022.
https://doi.org/10.1007/978-3-031-22064-7_31

it is also adaptive and flexible for modification, where a number of variants in different aspects have been proposed [18].

Despite its simplicity, effectiveness, robustness, and applicability, k-NN is reported to have several limitations [14].

- As a distance-based algorithm, the core of k-NN depends on the distance or similarity metrics for quantifying the distances between the test data instance and instances in the training dataset [2]. Previous studies have been conducted to analyze the performance of k-NN using different distance metrics, given the default Euclidean distance is reported to provide non-optimal result. However, there is no consensus on which distance metric is the best one [22].
- In addition to a well-selected distance metric, a successful application of k-NN also requires a careful choice of the number of nearest neighbors k [7].
- As one from the lazy learning family, k-NN faces the computational challenges for big data [19]. Specifically, instead of learning a discriminative function from the training data, k-NN memorizes the whole training dataset. During prediction, for each time a new data point comes in, the k-NN algorithm will search for the nearest neighbors in the entire training set, which is computationally expensive for moderate datasets and prohibitive for large datasets.
- The majority vote system ignores the closeness between data, which is problematic especially when the distances between the test sample and nearest neighbors differ significantly.

Given these drawbacks, there are diverse k-NN variants have been developed to achieve competitive prediction performance [23]. Among them, we focused on the last issue and proposed the Centroid Displacement-based k-NN (CDNN) in our previous study [13], where centroid displacement is used for class determination. The CDNN algorithm is adaptive to noise and complex datasets. In many cases, it allows for the correct class label to be assigned to the test instance without extended tuning of the hyper-parameter k for the dataset. To accelerate future projects using our algorithm, we implement CDNN in the `scikit-learn` framework [24], a well-known ML library for the Python programming language, and open-source our algorithm at: https://github.com/ngphubinh/CDNN.

`Scikit-learn` is one of the most popular ML libraries, which enables an experimental pipeline where different ML algorithms can be run, evaluated, and compared rigorously and homogeneously [27]. To the best of our knowledge, no previous research attempts to make a comparative performance analysis of k-NN and its variants in `scikit-learn`. Therefore, to fill this gap and also evaluate the CDNN algorithm, in this study, we conduct a deep comparative performance analysis of the CDNN with these available in `scikit-learn`. In addition to this, we also examine the effect of different distance metrics on the performance of CDNN on different datasets.

2 k-NN Variants Considered in This Study

2.1 k-NN

The classic k-NN is one of the most fundamental and simple algorithm [8]. It is a non-parametric method with local approximation, which can be used for classification and regression. In classification, the problem can be defined as follows: we are given a set of points $P = \{(x_i, c_i) : x_i \in \mathbb{R}^d, c_i \in C\}$, where C is a set of class labels (for instance, a binary class labeling would be: $C = \{0, 1\}$) and x_i is a d-dimensional vector. The task at hand is, given a point $(q, c) \notin P$, and where c is unknown, attempt to determine c from C to the point based on the majority class of the k closest neighbors to q in P.

2.2   Weighted k-NN

Weighted k-NN (WKNN) is a simple and robust extension of k-NN, which leverages the distance information between neighbors [15]. As elaborated above, a majority voting is applied over the k nearest data points for classification whereas, in regression, the mean of k nearest data points is calculated as the output. This is problematic given it ignores the distance information between neighbors. To remedy this issue, [5] proposed WKNN so the vote of k nearest neighbors are weighted differently according to their distances to the test instance.

2.3   Radius NN

Radius NN (rNN) is a related algorithm constructed on the same idea as the classic k-NN. Instead of finding the k nearest neighbors, rNN works by finding all the neighbors within a given radius r [3]. So it makes predictions using all examples in the radius of a new example rather than the k-nearest neighbors. As such, the radius-based approach to select neighbors is more appropriate for sparse data, because it is able to prevent examples that are far away in the feature space from contributing to a prediction [6]. However, setting the radius r requires domain knowledge of the dataset. For example, if the data points are closely packed together in a dataset, it is better to use a smaller radius to avoid having nearly every point included for the prediction.

2.4   Nearest Centroid

The nearest neighbor selection is sensitive to the choice of k, mainly because it ignores the spatial pattern of each test instance and training instances on classification, where the same k is used for all the test instances [18]. Nearest centroid (NC) is proposed to remedy this issue, which considers the distribution of training instances on the neighborhood of the test instance. The NC algorithm works on a simple principle: given a data point (observation), the NC simply assigns it the class of the training samples whose local mean or centroid is closest to it [21]. During model training, the centroid for each target class is computed, where the training datasets are grouped into separated sets of the same class label. Out of all calculated centroids by each class, the class of the centroid to which the given test point's distance is minimum, is assigned to the given point.

2.5   CDNN

In [13], we addressed the simple majority voting rule of the classic k-NN and proposed the Centroid Displacement-based k-NN (CDNN) algorithm, where centroid displacement is used for class probability estimation. In addition to the classic k-NN algorithm, where the k nearest neighbors to q are obtained, the centroid for each target class is computed. We then calculate the displacement of the centroid if the test instance q is inserted into the set. The test instance is then assigned to the set in which the centroid displacement is minimal.

As illustrated in Fig. 1, class A (red dot) has more neighboring points to the test instance (the yellow diamond). Under the classic k-NN, the test instance will be labeled as class A for both small k ($k = 3$, where 2 out of 3 are red dots within the smaller dashed line circle) and large k ($k = 9$, where 5 out of 9 are red dots within the larger dashed line circle). However, under CDNN, the test instance will be assigned to class B (blue triangle) as the displacement of centroid of class B is the smallest. This conclusion is likely to be correct since the test instance is closer to the "triangle" neighbors.

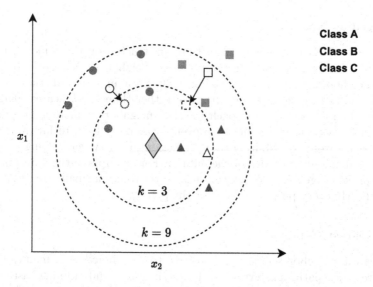

Fig. 1. Visual illustration of the CDNN algorithm

3   Distance Metrics

Like k-NN, various ML algorithms heavily rely on the distance metric for the input data patterns, where the distance metric is a function $d(x, y)$ that defines the distance between two given vectors x and y as a non-negative real number. Many studies have demonstrated, both empirically and theoretically, that the standard Euclidean distance may ignore some important properties available in the data set, and therefore the learning results might be non-optimal [26]. This is because when the instances have high similarity in some irrelevant features, the classification performance will be influenced by the noise from these irrelevant features [17]. To understand the patterns and sensitivity of using different distance metrics in CDNN, we conduct a comparative analysis with eight popular distance metrics. A high level information of all distance metrics considered in this study are summarized in Table 1, while details of all the distance metrics can be found in [4].

Table 1. A brief summary of distance metrics considered in this study

Distance metrics	Equation
Euclidean distance	$d_2(x, y) = \|x - y\|_2 = \sqrt{\sum_{i=1}^{n}(x_i - y_i)^2}$
Manhattan distance	$d_1(x, y) = \|x - y\|_1 = \sum_{i=1}^{n} \|x_i - y_i\|$
Chebyshev distance	$d_\infty(x, y) = \|x - y\|_\infty = \max_i \|x_i - y_i\|$
Canberra distance	$d_{cad}(x, y) = \sum_{i=1}^{n} \frac{\|x_i - y_i\|}{\|x_i\| + \|y_i\|}$
Cosine distance	$d_{cos}(x, y) = 1 - \frac{\sum_{i=1}^{n} x_i y_i}{\sqrt{\sum_{i=1}^{n} x_i^2}\sqrt{\sum_{i=1}^{n} y_i^2}}$
Jaccard distance	$d_{jad}(x, y) = 1 - \frac{\|x \cap y\|}{\|x \cup y\|}$
Jensen-Shannon divergence	$D_{JS}(P\|Q) = \frac{1}{2}D_{KL}(P\|\frac{P+Q}{2}) + \frac{1}{2}D_{KL}(Q\|\frac{P+Q}{2})$
Wasserstein distance	$W(P, Q) = \inf_{\gamma \sim \Pi(P,Q)} \mathbb{E}_{(x,y)\sim\gamma}[\|x - y\|]$

4   Experimental Evaluation

4.1   Data

We conduct extensive experiments on 12 datasets to compare CDNN with all k-NN variants available in the `scikit-learn` library [10]. The 6 real-world datasets are collected from the UCI Machine Learning Repository, where 3 have minor imbalance issues and 3 have severe imbalance issues [12]. The synthetic datasets are generated with the `scikit-learn` package in Python, using the make_classification function. All datasets are selected or generated carefully to cover varieties of data complexity.

The detailed information of datasets used in this study, is listed in Table 2. We also present the data complexity for each data set, by following [9]. Specifically, we report the average number of features per sample, the average number of principal components derived from the principal components analysis (PCA) per sample, the ratio of the original feature size to the PCA components size, and data sparsity, where a higher value suggests more complexity within the data set. For each data set, $\#C$ is the number of classes, $\#S$ is the number of samples, $\#F$ is the number of features, $\#F_{PCA}$ is the minimum number of principal components such that 95% of the variance is retained and sparsity means the percentage of data points that are zeros. We also report IR for all binary datasets.

Table 2. A brief list of datasets considered in this study

Name	Source	$\#C$	IR	$\#S$	$\#F$	$\#F/\#S$	$\#F_{PCA}/\#S$	$\#F/\#F_{PCA}$	Sparsity
Breastcancer	UCI	2	1.7:1	569	30	0.0527	0.0176	3.00	0.0047
Digits	UCI	10	–	1797	64	0.0730	0.0562	1.30	0.0000
Olivetti	UCI	40	–	400	4096	10.2400	0.3075	33.30	0.0000
Ecoli	UCI	2	8.6:1	336	7	0.0208	0.0179	1.17	0.0020
Optical_digits	UCI	2	9.1:1	5620	64	0.0114	0.0075	1.52	0.4961
Satimage	UCI	2	9.3:1	6435	36	0.0056	0.0009	6.00	0.0000
S1	Synthetic	2	1:1	100	2	0.0200	0.0200	1.00	0.0000
S2	Synthetic	2	1:1	1000	2	0.0020	0.002	1.00	0.0000
S3	Synthetic	3	–	1000	10	0.0100	0.0090	1.11	0.0000
S4	Synthetic	3	–	1000	50	0.0500	0.0420	1.19	0.0000
S5	Synthetic	10	–	5000	50	0.0100	0.0092	1.09	0.0000
S6	Synthetic	10	–	5000	100	0.0200	0.0182	1.10	0.0000

4.2   Experimental Settings and Evaluation

The primary performance metric used in this study is the F1 score, which is a harmonic mean of the precision and recall [11]. By following [20], we also report the ranking of each algorithm according to the average F1 score for each dataset. For multi-class classification, the F1 scores are computed first for each class, then the results are averaged to get the final performance measure.

The overall experimental framework is divided into two major parts:

- to get a comprehensive and concrete result, we follow an exhaustive method, where we use all k from 5 to 25, incremental by 2 with for all the k-NN queries [25]. For rNN and NC, we just use the default hyper-parameters for simplicity. All of our experiments are repeated ten times, with each trial using 5-fold cross validation, for a total of 50 results that are then averaged. The default Euclidean distance is used in all experiments in this step. To evaluate whether the performance improvements of the CDNN over k-NN are statistically significant or not, we further apply the 5×2cv paired t-test to compare the performances of the CDNN with these of k-NN over different k from 5 to 25, incremental by 2 [16].

– to find the best distance metrics for CDNN classifiers on different datasets. For simplicity, we use the default value $k = 5$ for CDNN with all distance metrics listed above. In this study, we define the best distance metric as the metric which performs with the highest average F1 score, given same experimental settings and procedures.

4.3   Comparative Study of k-NN Variants

To get a comprehensive and robust result for the k-NN variants comparison, we run our experiments on datasets with different complexity. The result is presented in Table 3. For CDNN, a clear improvement over k-NN and its variants has been observed in the results.

– The average F1 score is higher compared to k-NN and its variants available in the scikit-learn, especially for the high dimensional datasets with a high degree of noise.
– The classification performance of CDNN is more stable compared with competing algorithms, as we can see CDNN outputs the highest number of F1 scores throughout the datasets (8 times out of 12).
– For severe imbalance data with high dimensionality and noise, such as *optical_digits* and *satimage*, the classification performance improvement of CDNN is even more substantial.

Such results confirm that CDNN is more efficient and robust with datasets with high dimensionality and noise. The results of 5×2cv paired t-test are depicted in Fig. 2, where the darker the shade of green indicates lower the p value and higher level of statistical significance. From the significance test results given in Fig. 2, one can clearly see that at the 5% significance level, there is a significant classification performance improvement of the CDNN over the classic k-NN for those high dimensional datasets with large noise, especially when k is large.

It is worth noting that although NC is not the best classifier overall, it outperforms other algorithms significantly for the *olivetti* data set. Based on the data characteristics, *olivetti* is a high-dimensional image datasets with 40 classes and a small sample size. Therefore, there is a high ratio of features (pixel values) that are non-informative. This may suggest that centroid-based classifiers would work well on this type of data set.

438   A. X. Wang et al.

Table 3. The rank based on average F1 score of k-NN algorithms. Each row corresponding to a dataset and the last row contains the average F1 score and rankings over all the datasets. For each dataset, we highlight the algorithm with best performance.

Data set	KNN		WKNN		rNN		NC		CDNN	
	F1	Rank	F1	Rank	F1	Rank	F1	Rank	F1	Rank
Breastcancer	0.9571	3	0.9587	2	0.6193	5	0.9246	4	**0.9626**	1
Digits	0.9637	3	0.9672	2	0.1815	5	0.8889	4	**0.9704**	1
Olivetti	0.6914	4	0.7844	3	0.0527	5	**0.8775**	1	0.8222	2
Ecoli	0.7892	2	**0.7974**	1	0.7754	4	0.6926	5	0.7881	3
Optical_digits	0.9769	3	0.9773	2	0.4741	5	0.7652	4	**0.9791**	1
Satimage	0.8127	3	0.8149	2	0.6901	4	0.4727	5	**0.8222**	1
S1	0.8844	4	0.9023	2	**0.9095**	1	0.8243	5	0.9009	3
S2	**0.8992**	1	0.8945	2	0.8781	4	0.8408	5	0.8936	3
S3	0.979	3	0.9799	2	0.5319	5	0.8736	4	**0.9814**	1
S4	0.9642	3	0.9661	2	0.4655	5	0.7729	4	**0.9686**	1
S5	0.9026	3	0.9107	2	0.1802	5	0.5837	4	**0.9181**	1
S6	0.9075	3	0.9167	2	0.1802	5	0.5644	4	**0.9252**	1
Average	0.8947	3	0.9067	2	0.4949	5	0.7568	4	**0.9121**	1

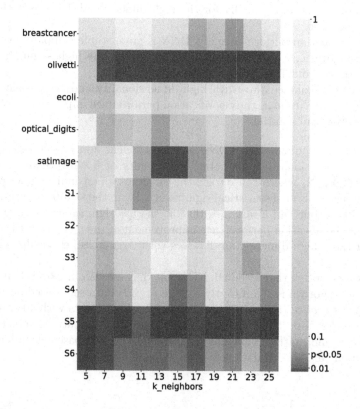

Fig. 2. Results of $5 \times 2cv$ paired t-test by datasets to show whether the performance improvement is statistical significantly between CDNN vs k-NN ($p < 0.05$).

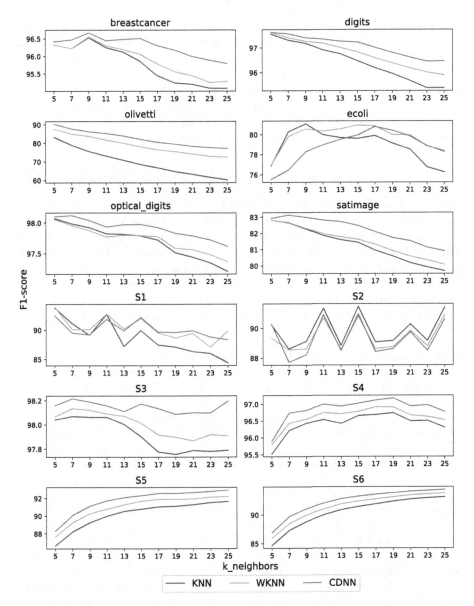

Fig. 3. F1 score of nearest neighbors based algorithms by neighborhood size k

To further explore the sensitivity of the neighborhood size k, we also conduct a comprehensive experiment to show the average F1 score of k-NN, WKNN, and CDNN in terms of k on all datasets. The result is presented in Fig. 3, where there are 12 subplots and each presents the experimental result for one data set. In each subplot, the average F1 score is displayed in y-axis and x-axis represents the number of nearest neighbors. A higher F1 score indicates better classification performance, while a flatter line means higher degree of stability of the algorithm over different neighborhood size k. It is shown that the classification performance varies with the neighborhood size and the degree of variations are significantly dependent on the data set. Overall, the CDNN achieves significant performance improvement and a high degree of stability over k-NN and WKNN for the majority of the datasets, except for `ecoil` and `S1`, where the data is simple. The robustness of CDNN is also demonstrated by its small variation over large k, where the performance of the other algorithms degraded significantly.

4.4   Comparative Study of CDNN with Different Distance Metrics

Eight predefined distance metrics, including the default Euclidean distance, are used in this set of experiments, to explore whether other distance metrics can help further improve the classification performance of CDNN on different datasets. For each data set, we repeat 5-fold cross validation 10 times to evaluate the classification performance of each distance measure, so we will have 50 results. The F1 score of CDNN with each distance on each data set is averaged and reported in Table 4, where cosine obtains the highest overall F1 score, followed by our baseline metric, Euclidean distance (0.9076 and 0.9013).

Table 4. Average F1 score of CDNN with different distance metrics on all datasets. For each data set, we highlight the distance metric with best performance and use * to flag significant improvement (5%) of top performer over the runner-up †.

Data set	Euclidean	Manhattan	Chebyshev	Canberra	Cosine	Jaccard	JSD	WD
Breastcancer	0.9634†	**0.9639**	0.9453	0.9603	0.9575	0.7890	0.6569	0.9071
Digits	0.9759†	**0.9766**	0.9456	0.9645	0.9674	0.9054	0.1348	0.6168
Olivetti	0.8586†	**0.8598**	0.4805	0.8444	0.8578	0.8548	0.0042	0.5857
Ecoli	0.7553†	0.7474	**0.7675**	0.7412	0.7524	0.6718	0.4572	0.7444
Optical_digits	**0.9810**	0.9768†	0.9671	0.9538	0.9727	0.8456	0.4752	0.7360
Satimage	0.8293	**0.8390**	0.8118	0.8254	0.8202	0.7918	0.4746	0.8334†
S1	0.8858†	0.8792	**0.8899**	0.8790	0.8676	0.5569	0.6492	0.8415
S2	0.8775†	0.8767	**0.8795**	0.8709	0.8526	0.7101	0.8067	0.8426
S3	0.9815	0.9820†	0.9804	0.9653	**0.9821**	0.2977	0.2977	0.6328
S4	0.9586†	0.938	0.8807	0.9033	**0.9753**	0.1975	0.1975	0.3998
S5	0.8816†	0.8455	0.7071	0.7656	**0.9360***	0.0453	0.0453	0.1319
S6	0.8676†	0.8244	0.5563	0.7483	**0.9494***	0.0362	0.0362	0.1221
Average	0.9013†	0.8924	0.8176	0.8685	**0.9076**	0.5585	0.3529	0.6162

Euclidean distance only outputs the best F1 score for the *optical_digits* data set, while for the rest of the datasets, there are always other distance metrics enable better predictions. Specifically, both Cosine distance and Manhattan distance achieve the best results on four datasets, while Chebyshev distance outperforms others on three datasets. However, the improvements over Euclidean distance are not substantial. Only Cosine distance outperforms Euclidean distance substantially (more than 5%), for datasets S5 and S6 (0.9360 vs 0.8816 and 0.9494 vs 0.8676). Therefore, we can conclude that other distance metrics can potentially increase classification performance. However, this conclusion is dependent on the data set.

5 Conclusions

In this study, we demonstrated the effectiveness of CDNN over k-NN variants available in `scikit-learn` for classification. We also analyzed the impacts of the choice k and distance metrics on CDNN. Overall, we have three outcomes:

- we develop and open-source CDNN as a Python library for performing multi-label classification tasks, based on the `scikit-learn` API. The open-source code provides native Python implementations of an enhanced k-NN algorithm alongside a flexible framework for adapting different distance metrics, which will accelerate future projects using our algorithm. We also aim to modify and extend CDNN for the study of regression in the near future.
- compared to all existing `scikit-learn` k-NN algorithms, our experimental results on various datasets have demonstrated the effectiveness of the CDNN algorithm, especially for the high dimensional sparse datasets. Therefore, we can conclude that the CDNN algorithm is generally applicable in a practical scenario where data are often sparse with high dimensionality.
- based on the distance metrics comparison study, we have shown the necessity of searching different distance metrics for different datasets, given the Euclidean distance normally provides sub-optimal results.

Our future work will focus on two directions. We will aim to analyze the interactive impacts of the choice k and distance metrics on the CDNN. Moreover, we will study the CDNN and distance metrics learning algorithm for the heterogeneous data. Specifically, how to treat and calculate the distance for both categorical features and missing values will be studied from the perspectives of distance metric learning.

Acknowledgments. The work of BPN was partly supported by a research programme funded by the New Zealand Ministry of Business Innovation and Employment (MBIE) under contract VUW RTVU1905 and the MBIE Strategic Science Investment Fund for Data Science under contract VUW RTVU1914.

References

1. Abu-Aisheh, Z., Raveaux, R., Ramel, J.Y.: Efficient k-nearest neighbors search in graph space. Pattern Recogn. Lett. **134**, 77–86 (2020)
2. Abu Alfeilat, H.A., et al.: Effects of distance measure choice on k-nearest neighbor classifier performance: a review. Big Data **7**(4), 221–248 (2019)

3. Bentley, J.L.: Survey of techniques for fixed radius near neighbor searching. Technical report, Stanford Linear Accelerator Center, Calif. (USA) (1975)
4. Cha, S.H.: Comprehensive survey on distance/similarity measures between probability density functions. City 1(2), 1 (2007)
5. Dudani, S.A.: The distance-weighted k-nearest-neighbor rule. IEEE Trans. Syst. Man Cybern. SMC 6(4), 325–327 (1976)
6. Elhamifar, E., Vidal, R.: Sparse manifold clustering and embedding. Adv. Neural Inf. Process. Syst. 24 (2011)
7. Ertuğrul, Ö.F., Tağluk, M.E.: A novel version of k nearest neighbor: dependent nearest neighbor. Appl. Soft Comput. 55, 480–490 (2017)
8. Fix, E., Hodges, J.: Discriminatory analysis, nonparametric discrimination: consistency properties. Technical report 4, USAF School of Aviation Medicine, Randolph Field 1951 (1951)
9. Ho, T.K., Basu, M.: Complexity measures of supervised classification problems. IEEE Trans. Pattern Anal. Mach. Intell. 24(3), 289–300 (2002)
10. Kramer, O.: Scikit-learn. In: Machine Learning for Evolution Strategies. SBD, vol. 20, pp. 45–53. Springer, Cham (2016). https://doi.org/10.1007/978-3-319-33383-0_5
11. Kumbure, M.M., Luukka, P., Collan, M.: A new fuzzy k-nearest neighbor classifier based on the Bonferroni mean. Pattern Recogn. Lett. 140, 172–178 (2020)
12. Lichman, M., et al.: UCI machine learning repository (2013)
13. Nguyen, B.P., Tay, W.L., Chui, C.K.: Robust biometric recognition from palm depth images for gloved hands. IEEE Trans. Hum.-Mach. Syst. 45(6), 799–804 (2015)
14. Pan, Z., Wang, Y., Pan, Y.: A new locally adaptive k-nearest neighbor algorithm based on discrimination class. Knowl.-Based Syst. 204, 106185 (2020)
15. Peterson, L.E.: K-nearest neighbor. Scholarpedia 4(2), 1883 (2009)
16. Raschka, S.: Model evaluation, model selection, and algorithm selection in machine learning. arXiv preprint arXiv:1811.12808 (2018)
17. Ruan, Y., Xiao, Y., Hao, Z., Liu, B.: A nearest-neighbor search model for distance metric learning. Inf. Sci. 552, 261–277 (2021)
18. Sengupta, S., Das, S.: Selective nearest neighbors clustering. Pattern Recogn. Lett. 155, 178–185 (2022)
19. Song, Y., Kong, X., Zhang, C.: A large-scale-nearest neighbor classification algorithm based on neighbor relationship preservation. Wireless Commun. Mob. Comput. 2022 (2022)
20. Sturm, B.L.: Classification accuracy is not enough. J. Intell. Inf. Syst. 41(3), 371–406 (2013)
21. Tibshirani, R., Hastie, T., Narasimhan, B., Chu, G.: Diagnosis of multiple cancer types by shrunken centroids of gene expression. Proc. Natl. Acad. Sci. 99(10), 6567–6572 (2002)
22. Todeschini, R., Ballabio, D., Consonni, V., Grisoni, F.: A new concept of higher-order similarity and the role of distance/similarity measures in local classification methods. Chemom. Intell. Lab. Syst. 157, 50–57 (2016)
23. Uddin, S., Haque, I., Lu, H., Moni, M.A., Gide, E.: Comparative performance analysis of k-nearest Neighbour (KNN) algorithm and its different variants for disease prediction. Sci. Rep. 12(1), 1–11 (2022)
24. Varoquaux, G., Buitinck, L., Louppe, G., Grisel, O., Pedregosa, F., Mueller, A.: Scikit-learn: machine learning without learning the machinery. GetMobile: Mob. Comput. Commun. 19(1), 29–33 (2015)

25. Xie, Z., Hsu, W., Liu, Z., Lee, M.L.: SNNB: a selective neighborhood based naïve Bayes for lazy learning. In: Chen, M.-S., Yu, P.S., Liu, B. (eds.) PAKDD 2002. LNCS (LNAI), vol. 2336, pp. 104–114. Springer, Heidelberg (2002). https://doi.org/10.1007/3-540-47887-6_10

26. Yang, L., Jin, R.: Distance metric learning: a comprehensive survey. Mich. State Universiy $2(2)$, 4 (2006)

27. Zhang, R.F., Urbanowicz, R.J.: A scikit-learn compatible learning classifier system. In: Proceedings of the 2020 Genetic and Evolutionary Computation Conference Companion, pp. 1816–1823 (2020)

EvAnGCN: Evolving Graph Deep Neural Network Based Anomaly Detection in Blockchain

Vatsal Patel[1], Sutharshan Rajasegarar[1], Lei Pan[1]($\boxtimes$), Jiajun Liu[2], and Liming Zhu[3]

[1] School of IT, Deakin University, Geelong, VIC, Australia
{vspatel,srajas,l.pan}@deakin.edu.au
[2] CSIRO Data61, Pullenvale, QLD, Australia
Ryan.Liu@data61.csiro.au
[3] CSIRO Data61, Eveleigh, NSW, Australia
Liming.Zhu@data61.csiro.au

Abstract. Detecting anomalous behaviors in the blockchain is important for maintaining its integrity. An imminent challenge is to capture the evolving model of transactions in the network. Representing the network with a dynamic graph helps model the system's time-evolving nature. However, as the graph evolves, real-world scenarios further stimulate the development of Graph Neural Networks (GNNs) to handle dynamic graph structures. In this paper, we propose a novel dynamic Graph Convolutional Network framework, namely EvAnGCN (Evolving Anomaly detection GCN), that helps detect anomalous behaviors in the blockchain. EvAnGCN exploits the time-based neighborhood feature aggregation of transactional features and the dynamic structure of the transaction network to detect anomalous nodes within the network. We conducted experiments on the Ethereum blockchain transaction dataset. Our experimental results demonstrate that EvAnGCH outperformed the baseline models.

Keywords: Anomaly detection · Blockchain transaction data · Evolving graph convolutional network · Dynamic graph convolutional network

1 Introduction

The blockchain network is prone to attacks through fraudulent transactions. It is essential to detect these anomalous behaviors on a real-time blockchain to protect the network. Representing the blockchain network transaction using evolving graphs will help capture the structural and temporal interactions. For static graphs, we have witnessed the success of graph neural networks (GNNs) for numerous applications such as node classification, edge prediction, and more. GNN models have proved successful by aggregating neighborhood nodes and learning the network structure. An extension was proposed in [10] to use an RNN (typically LSTM) model to inject dynamism into the GCN's parameters. Specifically, GNNs extract structural features of the network before RNNs conduct sequence learning from the extracted node embedding. The structure includes a GCN model for learning all graph parameters on all temporal axes. However, this setup suffers from poor flexibility because they require the knowledge of the nodes over the whole time span of the network for good performance; otherwise, it may perform poorly when new nodes are added in the future. To overcome this obstacle, recent research [11] proposed using an RNN to regulate the network parameters of GCN at each step. This approach effectively adopts the model itself and focuses on only node embeddings. Indeed, the model resembles good performance for newly added nodes. In this proposed method, the GCN parameters are imported from the RNN model without any training needed to reduce the number of parameters requiring training. To improve the performance over static graphs, we extend the work in [11] and apply it to a highly dynamic network like the Ethereum transaction network or Elliptic Bitcoin transaction network.

This paper proposes a dynamic GCN framework to detect anomalous nodes (transactions or accounts) in blockchain transaction networks by structuring the data as temporal graphs. We propose two novel evolving graph methods — EvAnGCN-H and EvAnGCN-O, and evaluate their performance against static models like Simplified GCN and One-Class Graph Convolutional Network (OCGCN). The Ethereum transaction blockchain dataset is used as a primary dataset for performance evaluation, while the Elliptic dataset is used for benchmarking. The F1 score is used as a primary metric, followed by MicroAvg F1, Precision, and Recall. Among all models, EvAnGCN-H claims higher performance from the benchmark model, followed by EvAnGCN-O and OCGCN. With the Ethereum dataset, we witnessed a considerable improvement of 10 to 12% in the F1 score, and the highest F1 score achieved is 87.02%.

Our Contributions. We propose a GCN-based novel learning framework for anomaly detection in dynamic transaction networks. In the framework, we devise a temporal feature aggregation mechanism at each time step to model the evolving nature of the node. For the blockchain transaction datasets studied, we calculate specific temporal and structural features that can add significant knowledge about the node's transaction pattern over different time steps. The graph features capture the transaction trends and node relationships with neighborhood nodes. We believe this addition improves the model's robustness and performance for a highly dynamic dataset. With a framework that includes dynamic GCN, feature engineering, and a temporal feature aggregations scheme, our model outperforms baseline methods with robustness against newly added nodes.

The rest of the paper is organized as follows. Section 2 covers related work in anomaly detection and temporal graph modeling. Section 3 formulates the problem, followed by a detailed architecture. Section 4 describes the experiment in detail, followed by discussion in Sect. 5. The paper concludes in Sect. 6.

2   Related Works

Dynamic and evolving graphs are often extensions to the static models, focusing more on temporal features and evolving parameters of the base model. The introduction of deep learning brought a series of unsupervised and supervised approaches for parameterizing the interest in Neural Networks. The set of approaches most relevant to this work combines GNN and recurrent architectures like LSTM [8,13]. The convolutional style of GNNs is extensively explored, and we call them Graph Convolutional Networks (GCNs). A novel unsupervised GNN framework was proposed in [14] that uses OCGCN for anomaly detection in attributed networks. The unsupervised learning task, based on hypersphere learning, has achieved good results in anomaly detection and is robust.

In recent years, we have witnessed an increased interest in anomaly detection in graphs [3,4,15,20], primarily because of their robust performance and ability to represent complex relationships and interactions between nodes. Several anomaly detection tasks have been performed on the Ethereum and Bitcoin network, which uses traditional anomaly detection algorithms which are distance-based [1,7], or through manual network exploitation [5]. There are implementations of anomaly detection tasks in graph data in different areas. Beladev et al. [2] proposed a novel temporal graph-level embedding method to globally embed both nodes of the graph and its representation at each time step which effectively performed temporal anomaly detection and trend analysis, which outperformed state-of-the-art anomaly detection models.

The Graph Convolutional Networks are used for detecting anomalies like anti-money launderers [16]. EvolveGCN dynamically updated the model parameters to capture the temporal aspects of the Bitcoin transaction dataset. Dou et al. [6] proposed CARE_GNN to enhance the GNN aggregation process with three unique modules against camouflages. The robustness of the model is important to improve the performance of the Graph convolutional network in anomaly detection. Zhang et al. [18] proposed a mechanism that learns to assign different importance to nodes and features for classification tasks with an interpretable explanation for prediction and increasing the model robustness.

Modeling a temporal graph is challenging yet essential to understand the dynamics and evolution in temporal graphs [10]. Combining LSTM and GCN helps learn long short-term dependencies and network structure with a significant improvement in F1 scores for vertex-based semi-supervised classification tasks. The STGCN model in [17] learns spatial and temporal patterns by adding all temporally connected nodes into temporal kernels and conducting a similar convolution as GCN on a skeleton graph with static topology.

3   Problem Formulation and the Proposed Model

3.1   Problem Formulation

Ethereum offers anonymity to its users. Identification of nodes (addresses) is crucial before a transaction is approved. Node identification becomes challenging when the network is highly dynamic and complex, where thousands of nodes are added and removed at every time. In this paper, we propose an Evolving Graph Convolutional Network for node classification by extracting temporal and structural graph features of the complex blockchain network at each time step individually. We make the node labels static because Ethereum regards that an anomalous node remains anomalous for n time steps till t_n. Our method can easily adopt cases where we have new nodes added or removed from the network through different time steps. In the training phase, we have a temporal graph G_{train} with ground truth labels Y_{train} obtained from https://etherscan.io/. The model is trained with the updated weights at different time steps, learns temporal and structural features through neighborhood aggregation at each discrete time t, and effectively performs a binary classification on nodes with unknown labels in future time steps t.

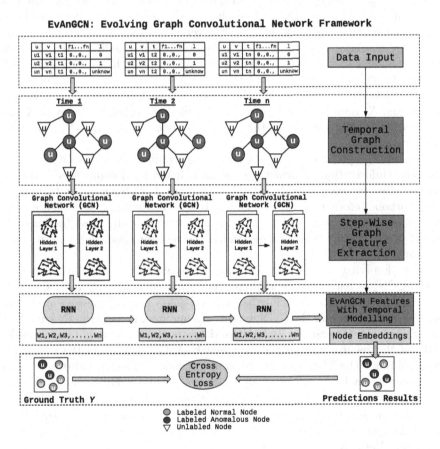

Fig. 1. The Framework: Using the Elliptic and Ethereum transactions, attributed graphs are created and split into discrete time steps. Features are aggregated for each time step to capture the temporal and structural information extracted using GCN. To update the GCN parameters, a recurrent unit is used over GCN. This architecture performs effective binary node classification.

We employ graph snapshot sequences by dividing our dataset into unique discrete time points, which represents temporal graphs. For t discrete time steps and N nodes, we divide the dataset into temporal graphs denoted as $G = (V, E, X)$, where V is a set of nodes appearing in a temporal graph and $V \in R^{t \times N \times N}$ is a tensor that encodes n-dimensional node attributes across t time points. E is a set of edges with $E \in R^{t \times N \times N}$ as a binary tensor with $E(t, u, v) = 1$, if an edge exists between node u (source node) and node v (target node) at time t. X is a feature tensor that incorporates calculated features taking into consideration the ecosystem of the Ethereum network. $X \in R^{t \times N \times N \times d_e}$ is a tensor that denotes d_e-dimensional edges attributed across t time points.

3.2   Our Proposed Method — EvAnGCN

The Evolving Anomaly Detection Graph Convolutional Network (EvAnGCN) learns the dynamics underlying a graph sequence using a recurrent model over GCN. In a dynamic network, like the Ethereum blockchain, the graphs created over time exhibit changes in the set of nodes and cardinality that require an evolving model to capture the dynamics and perform the classification task. Hence, at each time step t, the EvAnGCN framework models the data presented as a pair $(A_t \in R^{N \times N}, X_t \in R^{N \times D})$, where A_t represents the graph adjacency matrix and X_t represents the matrix of input node features. Each row of $\mathbf{X_t}$ is a d-dimensional feature vector of the respective node. The EvAnGCN framework is shown in Fig. 1.

In static graph-based methods, such as Simplified GCN and OCGCN, the structural information is learned considering neighborhood nodes close to the target node. In contrast, in the EvAnGCN framework, both temporal and structural neighborhood information are considered. It reveals which neighboring nodes are close to the target node and at what specific time they are close together. We aggregate the feature vectors across individual discrete time steps and use them for learning.

The Evolving Graph Convolutional Network for anomaly detection (EvAnGCN) is realized by combining the graph convolutional unit, GCONV, and a recurrent architecture. Further, two versions of this framework are formulated depending on how the GCN weights are updated, namely the EvAnGCN-H and EvAnGCN-O versions. EvAnGCN-H treats the GCN weights as a hidden state of the recurrent architecture, while EvAnGCN-O treats the weights as input/output. The EGCU performs graph convolutions along with the layers in both versions and evolves the weight matrices over time. By chaining the units bottom-up, a GCN with multiple layers is obtained. Then, by unrolling over time horizontally, the units form a lattice, on which the information $(H_t^{(l)}, W_t^{(l)})$ flows. We refer to the overall model as evolving anomaly detection Graph Convolutional Network (EvAnGCN).

Implementation of the EvAnGCN-H Version. We use a standard GRU to implement the $-H$ version with two extensions: (a) extending the inputs and hidden states from vectors to matrices (because the hidden state is now the GCN weight matrices); and (b) matching the column dimension of the input with that of the hidden state. In the matrix extension, the column vectors are placed side by side to form a matrix, in other words, using the same GRU to process each column of the GCN weight matrix. For completeness, we present the matrix version of GRU below.

$$1 \text{ :function } H_t = g(X_t, H_{t-1})$$
$$2: \quad Z_t = sigmoid(W_Z X_t + U_Z H_{t-1} + B_Z)$$
$$3: \quad R_t = sigmoid(W_R X_t + U_R H_{t-1} + B_R)$$
$$4: \quad \tilde{H}_t = tanh(W_H X_t + U_H(R_t \circ H_{t-1}) + B_H)$$
$$5: \quad H_t = (1 - Z_t) \circ H_{t-1} + Z_t \circ \tilde{H}_t$$
$$6 \text{ :end function}$$

The second requirement is to have the number of columns of the GRU input the same as that of the hidden state. The number of the hidden state is denoted by k. Our strategy summarizes all the node embedding vectors into k representative ones (each used as a column vector). The following pseudo-code shows a popular approach for this summarization. By convention, it takes a matrix X_t with multiple rows as input and produces a matrix Z_t with only k rows. The summarization requires a parameter vector $\mathbf{p}$ independent of the time index t, but it may vary for different graph convolution layers. This vector is used to compute weights for the rows, where the top k weights are selected and weighted for output.

$$1 \text{ :function } Z_t = summarize(X_t, k)$$
$$2: \quad y_t = X_t p/\|\mathbf{p}\|$$
$$3: \quad i_t = top - indices(y_t, k)$$
$$4: \quad Z_t = [X_t \circ tanh(y_t)]i_t$$
$$5 \text{ :end function}$$

With the above functions, g and $summarize$, we define the recurrent architecture as follows:

$$W_t^{(l)} = GRU(H_t^{(l)}, W_{t-1}^{(l)})$$
$$:= g(summarize(H_t^{(l)}, \#col(W_{t-1}^{(l)}))^T, W_{t-1}^{(l)}),$$

where $\#col$ denotes the number of columns of a matrix, and superscript T denotes matrix transpose. Effectively, it summarizes the node embedding matrix $H_t^{(l)}$ into one with appropriate dimensions and then evolves the weight matrix $W_{t-1}^{(l)}$ in the past time step to $W_t^{(l)}$ for the current time. Note that the recurrent hidden state can be realized by GRU or any other RNN architectures.

Implementation of the EvAnGCN-O Version. From the vector to matrix version, a straightforward extension of the standard LSTM is used to implement the $-O$ version. The pseudo-code is listed below.

```
1 :function  H_t = f(X_t)
2 :   The current input X_t is the same as the past output H_{t-1}
3 :   F_t = sigmoid(W_F X_t + U_F H_{t-1} + B_F)
4 :   I_t = sigmoid(W_I X_t + U_I H_{t-1} + B_I)
5 :   O_t = sigmoid(W_O X_t + U_O H_{t-1} + B_O)
6 :   C̃_t = tanh(W_C X_t + U_C H_{t-1} + B_C)
7 :   C_t = F_t ∘ C_{t-1} + I_t ∘ C̃_t
8 :   H_t = O_t ∘ tanh(C_t)
9 :end function
```

With the above function f, we specify the recurrent architecture as follows:

$$W_t^{(l)} = LSTM(W_{t-1}^{(l)}) := f(W_{t-1}^{(l)}).$$

Note that the recurrent input-output relationship may be realized using LSTM or other RNN architectures.

Cross Entropy Loss: This loss function is often used in classification problems for deep learning. We perform a binary classification task with a high class imbalance. It is a common issue with financial datasets where fraudulent/anomalous activities are in the minority class. The cross entropy loss function gives us advantages because the process does not affect the convergence speed while maintaining a high gradient state. The weighted cross entropy loss [9] is defined as $WCE = -\sum_{i=0}^{n} \alpha_i \times \rho \times \log \rho$, where ρ is a prediction vector. We train the GCN model using a weighted cross entropy loss to provide higher importance to the illicit class. A manual rescaling weight is assigned to each class, such that a tensor of size is equal to the number of classes with a reduction 'sum'. We optimized the class weights with multiple iterations and achieved a higher performance with [0.25, 0.75] for both datasets.

4   Experiments

We compare the traditional static GCN and OCGCN based on hypersphere learning with the EvAnGCN framework while considering temporal and structural features for effective classification. Hyperparameters are tuned using the validation set, and the test results are reported at the best validation epoch. We perform the temporal split of the training, validation, and test data. For the Elliptic dataset, we use 31 time steps for training, 5 for validation, and 13 for testing; and for the Ethereum dataset, we split the data in 5:4:4 for training, validation, and testing, respectively.

To compare the performance, we use the Elliptic Bitcoin transaction dataset[1], which is a publicly available benchmark dataset, and the Ethereum transaction dataset[2] that we have designed for the anomaly detection task.

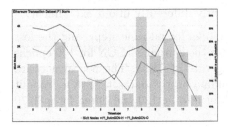

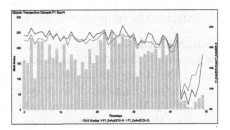

Fig. 2. F1 scores for the Ethereum transaction dataset are averaged over different seeds for each time step. The model shows different performances at each time step based on the number of illicit nodes.

Fig. 3. F1 scores for the Elliptic transaction dataset are averaged over different seeds for each time step. The model shows different performances at each time step based on the number of illicit nodes.

The Elliptic dataset is a benchmark dataset publicly available with hand-crafted features for the Bitcoin ecosystem. We use the Ethereum dataset without modifications to be consistent with related works. The Ethereum dataset is designed to capture the temporal aspect through the features created. For a node n, we aggregate the features at each discrete time step to capture the relationship and the node's behavior through interactions with other neighborhood nodes at different time steps. We capture 34 features, including average amount transferred, average transaction count, average transaction time, first and last instance of transactions, and many more. All the models leverage the graph structure of the data. We include various weighted graph features to understand the relationship between node interactions, including in-degree, out-degree, coefficient correlations, measures of strong connected components (SCC), and measures of weak connected components (WCC). Given a temporal weighted multidigraph denoted by $G = (V, E)$, V denotes the node set at time t and E denotes the edge set. In the Ethereum network, each edge is unique in this graph, denoted by a tuple (u, v, t, w), where u is the source node, v is the target node, t is the timestamp, and w is the weight associated with the transaction's number/value. This graph is subsequently used to calculate the features during the data preparation stage.

---

[1] https://www.kaggle.com/ellipticco/elliptic-data-set.

[2] http://xblock.pro/tx/.

Predicting the label of a node u at time t follows the same practice of a standard GCN. The activation function of the last graph convolution layer is a softmax layer such that h_t^u is a probability vector. We use the Elliptic dataset to investigate the dynamics of node classification. We leverage the information of nodes up to time t and classify the node based on various characteristics of the transactions finalized at different time steps. The model performs a binary classification to infer whether a specific node is normal or anomalous.

Machine learning models are set up slightly differently in our experiment.

Simplified GCN: We use a single GCN model for all time steps, and the loss is accumulated along the time axis. We train the simplified GCN for 100 epochs using the *Adam* optimizer with a learning rate of 0.001. The neural network architecture comprises 2 to 5 layers, depending on the dataset, with a dropout rate of 0.4. We set the size of node embeddings to be 120 and opted for a 30:70 ratio for the licit and illicit classes.

OCGCN [12]: Since OCGCN is an unsupervised learning algorithm, it is trained on only normal data points. Using the following hyper-parameters: Learning Rate=0.0001, nu=0.3, dropout=0.4, epoch=100, and num-layers=2, OCGCN is trained to perform the node classification task.

EvAnGCN-H and EvAnGCN-O: The models are trained using the hyperparameters as Learning Rate=0.001, num-layers=3, epoch=100, and dropout=0.4.

5   Results Discussion

Our primary measure for node classification tasks is the F1 score and MicroAvg F1 score. We first deployed the standard classification models for the binary node classification using two standard approaches: the Simplified GCN and OCGCN (with the Hypersphere learning). We evaluated these models using all the available/calculated features in the Ethereum and Elliptic datasets. The results are summarized in Tables 1 and 2 for both datasets. Table 1 shows the testing results in terms of F1, precision, recall, and the micro average for the illicit class. Among all the models, EvAnGCN-H outperforms the other models because the node features demonstrate specific knowledge of each node's transaction patterns and temporal patterns in their respective ecosystem for both Elliptic and Ethereum datasets. Results depict that the EvAnGCN-H version is more effective than the baseline models because it incorporates the informative feature matrix and node embeddings in a recurrent network framework that evolves with time. On the other hand, the graph structure plays a vital role for datasets where node features are not informative. The EvAnGCN-O may work better since it detects changes in the structural properties. The extracted features are listed in Table 3.

Simplified GCN and OCGCN learn the adjacent nodes' structural information with multiple iterations. However, it fails to exploit the temporal features of the proposed method. Ethereum and Bitcoin networks are highly dynamic and sensitive to time. The time-based features help the models learn the underlying transaction patterns. This setup explains why the F1 score for EvAnGCN-H is

Table 1. Results: ethereum dataset

Models	MicroAvg F1	F1	Precision	Recall
GCN	63.33	56.77	42.64	38.01
OCGCN	58.25	66.15	58.32	**64.09**
EvAnGCN -O	72.38	67.04	58.37	48.22
EvAnGCN -H	**77.88**	**87.02**	**72.84**	63.07

Table 2. Results: elliptic dataset

Models	MicroAvg F1	F1	Precision	Recall
GCN	78.36	71.03	74.94	61.04
OCGCN	**80.54**	**74.45**	72.82	52.09
EvAnGCN -O	68.38	68.64	58.37	**63.92**
EvAnGCN -H	72.88	73.12	**81.34**	63.31

high, and EvAnGCN-H works well with time-based informative features. Comparing the performance with baseline models, the EvAnGCN-H model delivers the highest F1 score of 0.8702, followed by EvAnGCN-O with 0.6704 considering it only focuses on changing network structure more. Between simplified GCN and OCGCN, OCGCN performance is higher in the Ethereum dataset with an F1 score of 0.6615 than 0.5677 in the other dataset because OCGCN considers both feature importance and interrelationship of nodes by learning the node embeddings.

Table 3. Features extracted

Features	Description
Count By Block	Total transaction count in each block
Amount By Block	Total amount in each block
Unique In Transaction Block Count	Total unique transaction from address in each block
Unique In Transaction Block Count	Total Unique transaction to address in each block
From In Transaction Count	Total in transaction count all blocks
From In Transaction Block Count	Total in transaction count by block
From Out Transaction Count	Total out transaction count all blocks
From Out Transaction Block Count	Total out transaction count by block
To In Transaction Count	Total incoming amount from users for all blocks
To In Transaction Block Count	Total incoming amount from users in each block
To Out Transaction Count	Total out transaction count all blocks
To Out Transaction Block Count	Total out transaction count by block
From In Transaction Value	Total in amount from users
From Out Transaction Value	Total out amount from users
To In Transaction Value	Total in amount to users
To Out Transaction Value	Total out amount to users
From Avg In Transaction Value	Total average in amount from user
From Avg Out Transaction Value	Total average out amount from user
To Avg In Transaction Value	Total average in amount to user
To Avg Out Transaction Value	Total average out amount to user
From In Degree	Total number of incoming transaction from users
From Out Degree	Total number of outgoing transaction from users
To In Degree	Total number of incoming transaction to users
To Out Degree	Total number of outgoing transaction to users
From Cluster Coefficient	Clustering coefficient from users
To Cluster Coefficient	Clustering coefficient to users

Comparing the performance between both datasets, all the models perform better on the Elliptic dataset for two primary reasons. The Elliptic dataset has 166 hand-crafted and illicit features, and it provides the opportunity to balance the dataset through temporal splits with higher time steps against the Ethereum dataset. The Ethereum dataset has low time steps with only 34 features and is a highly imbalanced dataset. The F1 score for the Elliptic dataset goes as high as 0.7312, but the micro average dips down due to the dark market shutdown event [16]. Figures 2 and 3 show the count of illicit nodes at each time step after performing the under-sampling technique, with the F1 score plotted across for both datasets. Interestingly, the F1 score dips when the node count is low, or the balance is not present between licit and illicit nodes. Specifically for the Elliptic Bitcoin transaction dataset, where the time step 43 shows a huge decrease in the performance because the transaction volume dived due to the dark market shutdown. However, both models demonstrate quick recovery once the volume increases. The decrease in the F1 score is witnessed specifically when there is a huge drop in the volume. Most importantly, the Simplified GCN has suffered the most with such a market event. Since OCGNN has proven its ability to outperform even with as low as 10% of normal training data, it shows higher performance even with decreasing node counts, which is reflected in Table 2, where the F1 score is high and has not significantly impacted by such uncertainty.

6   Conclusion and Future Work

The success of GNNs for anomaly detection has been witnessed through the recent research in this field. However, implementing the state of art methods like EvAnGCN, DynamicGCN, or others for the task of anomaly detection in real-world finance datasets like Ethereum is rare. We propose two novel frameworks EvAnGCN-H and EvAnGCN-O, for performing anomaly detection in a highly dynamic network. Experimental results confirm that the proposed approach outperforms related ones for the node classification task.

In the future, this work could be extended with an efficient/approximate optimization scheme that is fast enough to do real-time, iterative model training. As an extension to the current work, we focus on adding robustness to the model through implementing the state-of-the-art method named GraphSMOT [19] that can overcome the more significant issue of imbalanced data for node classification tasks, specifically in GNNs.

References

1. Agarwal, R., Barve, S., Shukla, S.K.: Detecting malicious accounts in permissionless blockchains using temporal graph properties. Appl. Network Sci. 6(1), 1–30 (2021)
2. Beladev, M., Rokach, L., Katz, G., Guy, I., Radinsky, K.: tdGraphEmbed: temporal dynamic graph-level embedding. In: Proceedings of the 29th ACM International Conference on Information & Knowledge Management, pp. 55–64. CIKM 2020, Association for Computing Machinery, New York, NY, USA (2020)

3. Cai, L., et al.: Structural temporal graph neural networks for anomaly detection in dynamic graphs. In: Proceedings of the 30th ACM International Conference on Information & Knowledge Management, pp. 3747–3756 (2021)
4. Chen, M., Wei, Z., Huang, Z., Ding, B., Li, Y.: Simple and deep graph convolutional networks. In: Proceedings of the 202 International Conference on Machine Learning, pp. 1725–1735. PMLR (2020)
5. Chen, T., et al.: Understanding ethereum via graph analysis. ACM Trans. Internet Technol. (TOIT) **20**(2), 1–32 (2020)
6. Dou, Y., Liu, Z., Sun, L., Deng, Y., Peng, H., Yu, P.S.: Enhancing graph neural network-based fraud detectors against camouflaged fraudsters. In: Proceedings of the 29th ACM International Conference on Information & Knowledge Management. ACM (2020)
7. Farrugia, S., Ellul, J., Azzopardi, G.: Detection of illicit accounts over the ethereum blockchain. Expert Syst. Appl. **150**, 113318 (2020)
8. Han, Y., Huang, G., Song, S., Yang, L., Wang, H., Wang, Y.: Dynamic neural networks: a survey. IEEE Trans. Pattern Anal. Mach. Intell. (2021)
9. Lu, S., Gao, F., Piao, C., Ma, Y.: Dynamic weighted cross entropy for semantic segmentation with extremely imbalanced data. In: Proceedings of the 2019 International Conference on Artificial Intelligence and Advanced Manufacturing (AIAM), pp. 230–233. IEEE (2019)
10. Manessi, F., Rozza, A., Manzo, M.: Dynamic graph convolutional networks. Pattern Recogn. **97**, 107000 (2020)
11. Pareja, A., et al.: Evolvegcn: evolving graph convolutional networks for dynamic graphs. In: Proceedings of the AAAI Conference on Artificial Intelligence, vol. 34, pp. 5363–5370 (2020)
12. Patel, V., Pan, L., Rajasegarar, S.: Graph deep learning based anomaly detection in ethereum blockchain network. In: Kutyłowski, M., Zhang, J., Chen, C. (eds.) NSS 2020. Lecture Notes in Computer Science(), vol. 12570, pp. 132–148. Springer, Cham (2020). https://doi.org/10.1007/978-3-030-65745-1_8
13. Skarding, J., Gabrys, B., Musial, K.: Foundations and modeling of dynamic networks using dynamic graph neural networks: a survey. IEEE Access **9**, 79143–79168 (2021)
14. Wang, X., Jin, B., Du, Y., Cui, P., Tan, Y., Yang, Y.: One-class graph neural networks for anomaly detection in attributed networks. Neural Comput. Appl. (2021)
15. Wang, Y., Wang, W., Liang, Y., Cai, Y., Hooi, B.: Progressive supervision for node classification. In: Hutter, F., Kersting, K., Lijffijt, J., Valera, I. (eds.) ECML PKDD 2020. LNCS (LNAI), vol. 12457, pp. 266–281. Springer, Cham (2021). https://doi.org/10.1007/978-3-030-67658-2_16
16. Weber, M., et al.: Anti-money laundering in bitcoin: experimenting with graph convolutional networks for financial forensics. arXiv preprint arXiv:1908.02591 (2019)
17. Yan, S., Xiong, Y., Lin, D.: Spatial temporal graph convolutional networks for skeleton-based action recognition. In: Proceedings of the 2018 AAAI Conference on Artificial Intelligence, vol. 32 (2018)
18. Zhang, L., Lu, H.: A feature-importance-aware and robust aggregator for GCN. In: Proceedings of the 29th ACM International Conference on Information & Knowledge Management, pp. 1813–1822. ACM (2020)
19. Zhao, T., Zhang, X., Wang, S.: Graphsmote: imbalanced node classification on graphs with graph neural networks. In: Proceedings of the 14th ACM International Conference on Web Search and Data Mining, pp. 833–841 (2021)

20. Zügner, D., Akbarnejad, A., Günnemann, S.: Adversarial attacks on neural networks for graph data. In: Proceedings of the 24th ACM SIGKDD International Conference on Knowledge Discovery & Data Mining, pp. 2847–2856 (2018)

Deterministic Graph-Walking Program Mining

Peter Belcak$^{(\boxtimes)}$ and Roger Wattenhofer

ETH Zürich, Rämistrasse 101, 8092 Zürich, Switzerland
{belcak,wattenhofer}@ethz.ch

Abstract. Owing to their versatility, graph structures admit representations of intricate relationships between the separate entities comprising the data. We formalise the notion of connection between two vertex sets in terms of edge and vertex features by introducing graph-walking programs. We give two algorithms for mining of deterministic graph-walking programs that yield programs in the order of increasing length. These programs characterise linear long-distance relationships between the given two vertex sets in the context of the whole graph.

Keywords: Graph walks · Complex networks · Program mining · Program induction

1 Introduction

While data has been stored in the form of tables since time immemorial, more complex data is often represented with graphs. This is because graph databases generalise conventional table-driven data storage methods, allowing for modelling of involved relationships among entities represented therein. As such, graph analysis and mining methods will be at the center of attention when it comes to contextual understanding of relationships between individual datapoints within a large database.

Here we investigate the identification of one type of such relationship between two groups of graph's vertices. For an illustrative example (Fig. 1), consider an individual who has just graduated from high school (starting qualifications S) and aims to reach a target career (target qualifications T) while being permitted only one study focus at the time – e.g. studying either social or biological sciences, but not both. What sequence of decisions with regards to their study foci should they take? Notice that the choice of focus made at every stage of the individual's education leads to a restriction on what qualifications they can obtain in the future. Thus, at each stage of their education, they will need to focus on qualifications that are pre-requisites for those that lead to T. Attempting to solve the problem on our own, we can search for a sequence of instructions that leads us from S to T either by naively tracing out paths from S and reading out instructions one by one, or by enumerating all possible instruction sequences and then verifying if they indeed do lead from S to T.

© The Author(s), under exclusive license to Springer Nature Switzerland AG 2022
W. Chen et al. (Eds.): ADMA 2022, LNAI 13725, pp. 457–471, 2022.
https://doi.org/10.1007/978-3-031-22064-7_33

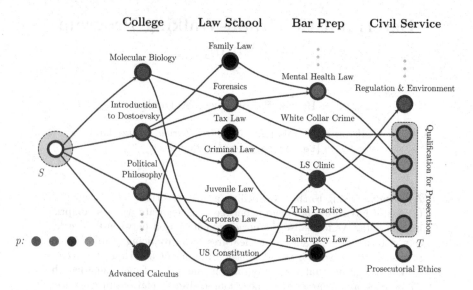

Fig. 1. A depiction of the qualification pathway p for an individual who has just graduated from high school (the starting qualification in S) and aims to become a prosecutor in the United States (target qualifications T). Vertices, colours, edges represent qualifications, qualification foci, and dependencies, respectively. Going into college, they will need to choose a focus that maximises their chances of being admitted to a law school (most likely social sciences). In law school, they will need to focus on criminal rather than tax or corporate law and prepare diligently for their barrister examinations. Finally, they will need to satisfy the necessary pre-requisites of civil service before becoming a prosecutor. (Color figure online)

Informally, let G be a graph where every vertex has some features, and let S, T be vertex sets. We ask the following question: "How is S connected to T in terms of the features of the vertices between them?" Or, alternatively: "What instructions should agents starting at the vertices of S follow in order to reach T"?

Revisiting the example above, if G is a map of qualifications, with G being qualifications and directed edge (v_1, v_2) denoting that v_1 is a pre-requisite qualification for obtaining v_2, one could iteratively ask "what type of qualifications from among the qualifications I am eligible for now should I achieve to eventually reach my target qualifications T"? A good answer would give a sequential list of characteristics of qualifications. Of course, getting all possible qualifications at every stage would likely lead to obtaining qualifications in T, but ideally one would not be doing more than absolutely necessary. We illustrate a variant of this example in which each qualification is only characterised by its type in Fig. 2.

We investigate this problem and aim to give answers in terms of lists of instructions for a single graph-walking agent that can be present at multiple vertices at the same time. We dub such lists of instructions *simple graph-walking programs*.

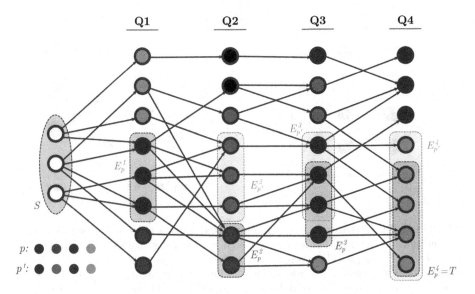

Fig. 2. An example of the qualification program problem with multiple starting qualifications. An individual possessing three different qualifications (vertices in S) of the same focus/type (colour) seeks to attain any of the qualifications in T such that they always do qualifications of the same type. Each of their qualifications, however, is a pre-requisite (directed edge) for a slightly different set of later qualifications, and it will take at least four steps to reach T from any qualification in S. p gives a program in which they first work towards green, then brown, then red, and finally yellow qualifications, and exactly the qualifications in T are achieved. p' gives a program *green-purple-red-yellow*, in which there is some overlap with T but an additional yellow qualification not in T is achieved. (Color figure online)

Let $G = (V, E)$ be a directed multi-graph, $\mathcal{F}_1, \ldots, \mathcal{F}_z$ spaces of the features appearing in G, and $\mathcal{F} := \prod_i \mathcal{F}_i \cup \{\emptyset_i\}$ their product, where $\emptyset_i$ denotes that a given graph element is not assigned feature i. Let $\phi_V : V \to \mathcal{F}, \phi_E : E \to \mathcal{F}$ be the vertex and edge feature mappings. Let $p : c_1 \cdots c_{2n}$ be a simple graph-walking program – a list of vertex movement selection instructions, i.e. functions on $c_t : \mathcal{F} \to \{0, 1\}$. Consider an agent, located at E_p^t for any time $t \geq 0$, that begins at the set of vertices $S = E_p^0$ and at each time $t \geq 1$ decides to proceed only to those vertices v in out-neighbourhood of E_p^{t-1} connected by edges e whose feature vectors $\phi_V(v), \phi_E(e)$ satisfy $c_{2t-1}(\phi_E(e)) = 1 = c_{2t}(\phi_V(v))$. The problem of *simple graph-walking program mining* is the problem of finding lists of instructions p such that an agent following it reaches T by the end of the program ($\emptyset \neq E_p^n \subseteq T$). See Fig. 3. Alternatively and more in line with the program synthesis literature, we can speak of *graph-walking program induction* or *synthesis*, with the triple (G, S, T) forming the inputs for the induction of the programs. We call programs that satisfy $\emptyset \neq E_p^n \subseteq T$ *feasible*, and programs that achieve the equality *exact*.

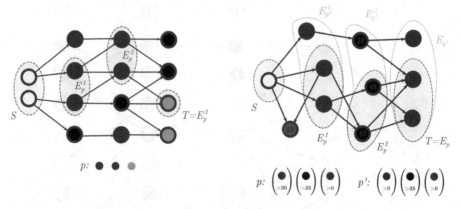

Fig. 3. Two illustrative examples with solutions. Recall the objective to find a sequence of conditions on features that leads imagined agent from the vertices of S to the vertices of T. *Left.* The simple colour program p can be used to instruct the agent starting at vertices of S to proceed towards T. On the first step E_p^1 is reached. On the second step, the agent proceeds to the red nodes marked by E_p^2. On the third step, the agent proceeds to T. *green-blue-yellow* would be another feasible program. *Right.* Two simple toset (totally-ordered set) programs p, p' are presented, conditioning on two feature dimensions of a general, unlayered graph: colour and integers. Agent starting in S reaches exactly T if following p (p is feasible and exact) but ends up at a strict superset $E_{p'}$ of T if she follows p' (hence p' is infeasible). (Color figure online)

The resulting programs are not frequent patterns, nor do they characterise the graphs locally; they characterize long-distance relationships between groups of vertices in G in terms of $\mathcal{F}$. Nevertheless, for a given pair of vertex sets S, T there are often many feasible programs, and our algorithms carry some characteristics of a priori graph pattern mining. We talk of "simple" programs because there are more elaborate program structures that could be studied in this setting (such as those that posses memory), and of "deterministic" programs since the instruction/criterion c_i always gives either 1 or 0 as firm directions to the agent walking the graph.

The difficulty of this problem dwells in it being a cunning composition of two necessary sub-tasks – *path search* and *classification* – well-understood and studied in graph theory and machine learning, respectively. This is because it is not enough to find a possible program walk from a vertex in S to a vertex in T – one has to choose from all possible walks for all possible choices of the pair $(s, t) \in S \times T$, and then find a subset of these walks for which a single graph-walking program can be used. In other words, it is necessary to both *discover* possible walks, and *discriminate* among them.

The mining algorithms we propose are *correct* (they return only valid programs) and *complete* (proceeding in stages, they always yield all valid programs up to some length ℓ before looking for longer programs).

This paper reviews related work (Sect. 2), describes the problem of simple colour program mining, gives an algorithm for the task (Sect. 3) – which, to the best of our knowledge, has never been addressed in the literature before – and extends simple colour program mining to simple totally-ordered-set programs (Sect. 4).

2   Related Work

Our effort lies at the intersection of two areas, namely graph program synthesis and analysis of complex networks.

Algorithmic program synthesis [4], traditionally considered a problem in deductive theorem proving, has recently been looked at as a search problem with constraints such as a logical specification of the program behaviour [11], syntactic template [1,9], and, most recently, previously discovered program fragments and utility functions [10,16]. Several new methods combine enumerative search with deduction, aiming to rule out infeasible sub-programs as soon as possible [12,13]. While relevant to us in their intent, the methods are domain-specific and do not extend to programs on graphs.

Under the paradigm of program search within a restricted graph context, Yaghmazadeh et al. [24] study the synthesis of transformations on tree-structured data and employ a combination of SMT solving and decision tree learning. Their synthesis system, HADES, outputs programs in a custom domain-specific language for tree-transformation. Their approach considers entire graphs at the same time and while it does provide insights into construction of programs for graphs, it does not extend to the graph walk scenarios.

Wang et al. [21] give a synthesis algorithm for queries q on a set of tables $T_1, T_2, \ldots, T_k$ that output records of a target table T_{out}. Their approach to query-walk search is syntactic (in contrast to treating the database schema as a graph) and relies on simple enumeration of possible table visits, something our algorithms avoid with further constraints on the search space.

Mendelzon and Wood [18] consider the problem of finding pairs of vertices in a graph connected by simple paths such that the trace of the labels of the vertices traversed satisfies a given regular expression. While being perhaps closest to our work, their goal is to find paths that satisfy a constraint, rather than finding constraints for which connecting paths exist, thus having one degree of freedom fewer.

A sub-branch of graph pattern mining considers the special case of mining frequent paths [14,15]. We note that while the knowledge of frequent paths in a graph might potentially accelerate the search for solutions for the graph-walking program mining problems, methods for frequent path mining are of little direct use since we seek programs that go between particular sets S, T.

Finally, the literature on analysis of complex networks frequently focuses on characterisation of elements of networks in terms of their interactions with their neighbourhoods. Among the examined characterisations are the notions of structural equivalence [5,7,17], regular equivalence [22], or other partitioning strategies [23]. Further, random walks of graphs are frequently employed to help with analysis of graphs as whole [8,19] or as sum of its communities [2,3], but to our knowledge, no work so far has investigated the identification of relationships between nodes related from beyond close neighbourhoods.

3   Simple Colour Programs

Without any loss of generality we restrict ourselves to program mining on simple directed graphs with featureless edges, to whom directed multi-graphs with edge features can be converted by replacing every edge with a vertex inheriting edge's features and retaining its endpoints as the only neighbours.

We focus on *simple colour programs* – programs $p : c_1 \cdots c_n$ such that c_i is the colour of the out-neighbours of the vertices reached by the prefix program $c_1 \cdots c_{i-1}$ whom the agent executing p should proceed to at step i. The program instructions (criteria) are thus colours.

3.1   Preliminaries

Let G be a simple directed k-coloured graph. The colouring does not have to be proper. Let $\emptyset \neq S, T \subseteq V$. Denote by $c(v)$ the colour of vertex v, by $c(A)$ the set of colours of the vertices in $A \subseteq V$, and by $C_c(A)$ the set of vertices of A with colour c. Call set of vertices A monochromatic if $c(A)$ is a singleton set. Denote the out- and in-neighbours of A by $N_o(A)$ and $N_i(A)$ respectively. Shorten n applications of N, i.e. $N_o(N...N(A)...))$, to $N_o^n(A)$, and similarly for N_i. For convenience, define $N_o^0(A) := A =: N_i^0(A)$.

Definition 1 (Simple Colour Program). $p : c_1 c_2 \cdots c_n$ *is a simple colour program (SCP) of length* n *iff* $1 \leq c_i \leq k$ *for all* $1 \leq i \leq n$.

Use ϵ for empty program – the unique program of length 0, p_i for c_i, $p_{\leq i}$ for the prefix $c_1 \cdots c_i$ of p for $1 \leq i \leq n$, and $p_{\geq i}$ for the suffix $c_i \cdots c_n$ of p.

Definition 2 (Program Endpoints). *For* $p : c_1 c_2 \cdots c_n$ *an SCP, define* $E_p^i(S)$ *for* $0 \leq i \leq n$ *recursively as follows:*

1. $E_p^0(S) = S$,
2. *For* $i > 0$, $E_p^i(S) = C_{c_i}(N_o(E_p^{i-1}(S)))$,

and denote $E_p^n(S)$ *by* $E_p(S)$.

Definition 3 (Feasible and Exact SCP). p *is feasible iff* $\emptyset \neq E_p(S) \subseteq T$, *and exact iff* $E_p(S) = T$.

Definition 4 (Partial Halting). p *partially halts (on* G*) if there exists an* $0 \leq i < n$ *and* v *such that* $v \in E_p^i(S)$ *but* $c_{i+1} \notin c(N_o(v))$.

In other words, p partially halts if it ever reaches a vertex from which it is impossible to proceed while still following p.

Lemma 1. *If p is a feasible program that does not partially halt, then for all $0 \leq i < n$ there exists a colour c_{to} such that $\emptyset \neq C_{c_{to}}(N_o(E_p^i(S))) \subseteq N_i^{n-i+1}(T)$.*

Proof. For each i, take $c_{to} = p_i$.

Definition 5 (Complete Halting). *p halts (completely) on G if there exists an $1 \leq i < n$ such that $c_{i+1} \notin c(N_o(E_p^i(S)))$. Equivalently, there is an $1 \leq i < n$ such that $E_p^{i+1}(S) = \emptyset$.*

Lemma 2. *Assume that a feasible SCP exists for S, T. Then*

1. *There exists a walk w from $s \in S$ to $t \in T$ such that the colours of the vertices from s to t give a feasible program for S, T.*
2. *If a program that does not partially halt exists, for every $s \in S$ there are $t \in T, w$ as in item 1.*
3. *If an exact program exists, for every $t \in T$ there are $s \in S, w$ as in item 1.*

Proof. Let $p : c_1 \cdots c_n$ be a feasible program.

1. Take any $w_n \in E_p(S)$. If $n = 1$ we are done. If not, prepend it by any $w_{n-1} \in N_i(w_n) \cap E_p^{n-1}(S)$ which is non-empty as p is feasible and therefore does not halt, and observe that $c(w_{n-1}) = p_{n-1}$. Repeat this process for a total of n times. Then $w_0 \in N_i(w_1) \cap E_p^0(S) \subseteq S$ and $w_1 \cdots w_n$ is a walk from a vertex in S to a vertex in T such that the colours of the vertices it visits give precisely the program p.
2. If p does not partially halt then for every $s \in S$, $E_p^i(s) \neq \emptyset$ and $E_p(s) \subseteq T$. So p is a feasible $\{s\}, T$-program, and hence item 1 applies.
3. If p is exact, $E_p(S) = T$ and the proof of item 1 also gives this stronger statement.

Definition 6 (Cover). *For $A, B \subseteq V$ we say that the vertices A cover B by c iff $C_c(N_o(A)) \supseteq B$.*

Definition 7 (Injection). *For $\emptyset \neq A, B \subseteq V$ we say that the vertices A inject B by c iff $\emptyset \neq C_c(N_o(A)) \subseteq B$. If that is the case, we call A a c-injection into B.*

Definition 8 (Spanning). *We say that A outspans B by c iff $C_c(N_o(A)) \backslash B \neq \emptyset$, and that A spans B by c iff A covers B by c but does not outspan B by c.*

Notice that A c-injects B iff A does not c-outspan B and A is not a c-halting point.

Lemma 3 (Cover-Inject Behaviour of Intermediate Endpoints). *For any program p decomposed as $\pi c d \sigma$, $E_\pi(S)$ spans $E_{\pi c}(S)$ by c but does not outspan $C_c(N_i(E_{\pi cd}(S)))$.*

Proof. Let $p : \pi c d \sigma$ be a feasible program.
Since $E_{\pi c}(S) = C_c(N_o(E_\pi(S)))$, E_π spans $E_{\pi c}$ by c.
Further, since $E_{\pi cd}(S) \subseteq N_o(E_{\pi c}(S))$, $C_c(N_o(E_\pi(S))) \subseteq C_c(N_i(E_{\pi cd}(S)))$, so E_π does not outspan $C_c(N_i(E_{\pi cd}(S)))$.

Proposition 1. *The problems of finding a feasible simple colour program and exact simple colour program are* NP.

Proof. Let G, S, T and a candidate simple colour program p be inputs.

To verify the the certificate p one can simulate the actions of a set graph-walking agent. Starting at $S = E_p^0(S)$, the agent visits vertices $N_o(E_p^0(S)) \subseteq V$ and compare their colour to c_1. Searching for edges originating at a vertex, searching for vertex colour, and comparing vertex colours to c_i is a polynomial-time operation. There are always at most $|V|$ vertices whose out-neighbours must be visited, and this operation is repeated for $1 \leq i \leq n$. Hence the verification of the certificate is a polynomial-time operation.

3.2   Viable Injection Basis Enumeration

We present a simple colour program mining algorithm constructing candidate programs from space of possibilities reduced by considering only those injections that cover "enough" vertices to have hope of reaching T. This is captured by the notion of *pseudo-basis*.

Definition 9 (Basis).   *We say that* $\mathcal{B}$ *is a c-basis for B iff* $\mathcal{B}$ *spans B by c and* $\mathcal{B}$ *is a minimal such set, i.e. for any* $v \in \mathcal{B}$, $\mathcal{B} - v$ *does not span B by c.*

Lemma 4. *A is a c-spanning set for B iff it contains a c-basis for B and does not c-outspan B.*

Proof. Let A be a c-spanning set for B. Then it has a basis (remove vertices until none can be removed without making it a non-spanning set) and by definition does not outspan B.

Conversely, let A be a set that contains a c-basis $\mathcal{B}$ but does not outspan B. Since $B = C_c(N_o(\mathcal{B})) \subseteq C_c(N_o(A))$, A covers B. Hence A c-spans B.

Definition 10 (c-Pseudo-Basis).   *We say that* $\mathcal{B}$ *is a c-pseudo-basis for* (B, M) *iff* $\mathcal{B}$ *c-covers B,* $\mathcal{B}$ *c-injects M, and* $\mathcal{B}$ *is a minimal such set, i.e. for any* $v \in \mathcal{B}$, $\mathcal{B} - v$ *does not c-cover B.*

Remark 1. Notice $\mathcal{B}$ is a c-basis for B if it is a c-pseudo basis for (B, B).

The utility of pseudo-bases comes from being a type of injection into M that covers all vertices of the out-neighbourhood designated as essential (B). This allows us to remove from our search space those injections that do not cover sufficiently many of its out-neighbours to fully reach T. More specifically, our strategy is to

1. consider all bases $\mathcal{B}_\bullet^1$ for T,
2. consider all sets $\mathcal{B}_\bullet^2$ covering each $\mathcal{B}_\bullet^1$ but not outspanning the corresponding monochromatic in-neighbourhoods $M_\bullet^1$ of T, i.e. the pseudo-bases for each $(\mathcal{B}_\bullet^1, M_\bullet^1)$,

3. do the same for pairings $(\mathcal{B}_\bullet^i, M_\bullet^i)$, $i \geq 2$. If the candidate pseudo-basis $\mathcal{B}_j^i$ lies in S and S is in turn fully contained in the in-neighbourhood of the appropriate c-span of $\mathcal{B}_j^i$, a valid program has been found.

In Algorithm 1, let Q and Q_{next} be queues of triples drawn from *programs* $\times$ *vertex-sets* $\times$ *monochromatic vertex-sets*, P be a set of *programs*. Each triple (p, B, M) in Q, Q_{next} represents a candidate program p, from where to begin B in order to reach T, and a monochromatic in-neighbourhood $M \subseteq N_i(E_{p_{\leq i}}(B))$.

Algorithm 1: (VIBE) Viable Injection Basis Enumeration

Initialisation: Q_{next} contains only (ϵ, T, T), Q and P are empty

1   **foreach** $\ell \geq 0$ *such that* $T \subseteq N_o^\ell(S)$ **do**
2     $Q \leftarrow Q_{\text{next}}$
3   empty Q_{next}
4   **while** Q *is not empty* **do**
5   pop (p, B, M) from Q
6       $n \leftarrow$ length of p
7
8   **if** $n = \ell$ **then**
9   **if** $B \subseteq S \subseteq N_i(E_{p_{\leq 1}}(B))$ **then**
10   add p to P
11   **end**
12   push (p, B, M) into Q_{next}
13   **continue**
14   **end**
15
16   **foreach** $c \in c(B)$ **do**
17         $N \leftarrow N_o^{\ell-n-1}(S) \cap N_i(B)$
18   **foreach** $d \in c(N)$ **do**
19   **if** $\ell \neq n + 1$ **then**
20             $N_d \leftarrow C_d(N)$
21   **else**
22             $N_d \leftarrow N$
23   **end**
24   **foreach** c-pseudo-basis $\mathcal{B} \subseteq N_d$ for (B, M) **do**
25   push $(cp, \mathcal{B}, N_d)$ into Q
26   **end**
27   **end**
28   **end**
29   **end**
30 **end**

Proposition 2. *The following hold.*

1. *Algorithm 1 is correct in the sense that all programs in P are exact programs for S, T.*

2. *Algorithm 1 is complete in the sense that whenever execution exists the loop closing at line 29, P contains all exact programs for S, T of length ℓ.*

Proof. Observe that for every (p, B, M) in Q, $B \subseteq M$, M is a monochromatic set, and the spans of B and M are the same.

1. First, we show inductively that every triple (p, B, M) in Q is such that for any set $B \subseteq A \subseteq M$, $E_p(A) = T$.

   The base case (stemming from the initialisation of Q_{next}) is straightforward as $E_\epsilon(T) = T$. Suppose $p \neq \epsilon$. Then there is a colour c and a shorter program q such that $p = cq$, as reaching line 25 is the only way for a non-empty program to enter Q.

   So there is a triple (q, B', M') and such that B is a pseudo-basis for (B', M') (cf. line 24). Thus $B' \subseteq E_c(B) \subseteq M'$ by Definition 10. But $E_p(B) = E_q(E_c(B))$, so by the inductive hypothesis $E_p(B) = T$.

   Now, every program p in P must have been added on line 10, so necessarily there is a triple (p, B, M) that was once in Q s.t. $B \subseteq S \subseteq N_i(E_{p_{\leq 1}}(B))$. Since $S \subseteq N_i(E_{p_{\leq 1}}(B))$ we have $E_{p_{\leq 1}}(S) = E_{p_{\leq 1}}(B)$, so

   $$E_p(S) = E_{p_{>1}}(E_{p_{\leq 1}}(S)) = E_{p_{>1}}(E_{p_{\leq 1}}(B)) = E_p(B) = T.$$

2. Let p be an exact program for S, T.

   Focusing on the combinatorial loop of line 4 and ignoring the caching by Q_{next}, we shall show inductively that for every prefix-suffix decomposition $\pi\sigma = p$ of p there is a triple (σ, B, M) s.t. $B \subseteq E_\pi(S) \subseteq M$ that appears on Q.

   For the base case with suffix $\sigma = \epsilon$, p is an exact program, so $T \subseteq E_p(S) \subseteq T$. This is the triple (ϵ, T, T) found in initialisation.

   Assume the inductive hypothesis holds for shorter suffixes and decompose p to $\pi'c\sigma$. Since by the inductive hypothesis there is a triple (σ, B, M) s.t. $B \subseteq E_{\pi'c}(S) \subseteq M$, $c \in c(B)$ and $E_{\pi'}(S) \subseteq N \neq \emptyset$ on line 17. Notice also that $E_{\pi'}$ is further monochromatic whenever $\pi' \neq \epsilon$, so by the branching on line 19 $E_{\pi'} \subseteq N_d$. Further, as a consequence of Lemma 3, $E_{\pi'}(S)$ covers $E_{\pi'c}$ but does not outspan M, so $E_{\pi'}(S)$ is a c-pseudo-basis for (B, M), proving the inductive hypothesis.

   Now, whenever $\ell = n$ execution will reach line 9 with various triples (p, B, M). Decompose $p = \epsilon p$. Then by the hereproven induction one of them will be such that $B \subseteq E_\epsilon(S) = S \subseteq M$, and by the aboveproven induction also $E_p(S) = T$. So p will be added to P on line 10. This completes the proof of completeness.

Algorithm 1 can be easily modified to also yield feasible programs. This can be done by altering line 24 to give c-injections into T if $n = 0$, and execute the present behaviour otherwise. Alternatively and equivalently, one can just precompute all viable c-injections, their monochromatic peers, and initialize Q_{next} to their set in arbitrary order.

The completeness of Algorithm 1 combined with Lemma 3 highlight the role of existence of appropriate pseudo-basis as a necessary and sufficient condition for local feasible program existence.

4 Simple Toset Programs

Extending on algorithms of Sect. 3 we now consider a more general setting where there are multiple features at each vertex, and the feature spaces admit a total order. See Fig. 3-*Right.* for an example.

Definition 11 (Criterion). *Let c be a triple (f, ω, ν) where f is a feature, ω is one of the operators $<, \leq, =, \geq, >$, and $\nu \in \mathcal{F}_f$. Then c is an atomic criterion. Inductively, c is a criterion if it is either an atomic criterion, a conjunction of criteria, or a disjunction of criteria.*

Definition 12 (Simple Toset Program). *We say that $p : c_1 c_2 \cdots c_n$ is a Simple Toset Program (STP) if each c_i is a criterion.*

Definition 13 (Criterion Satisfaction). *We say that $v \in V$ satisfies the atomic criterion $c = (f, \omega, \nu)$ iff $\phi_V(v) \omega \nu$. We then re-define $C_c(A)$ in the context of STPs to mean the set of all vertices in A that satisfy c. If c is a criterion, we say that $v \in V$ satisfies c iff*

- *c is an atomic criterion and v satisfies c in the sense for atomic criteria, or*
- *c is a disjunction of criteria $c_1 \vee \cdots \vee c_k$ and v satisfies at least one of $c_1, \ldots, c_k$, or*
- *c is a conjunction of criteria $c_1 \wedge \cdots \wedge c_k$ and v satisfies all of $c_1, \ldots, c_k$.*

All of the previous notions such as endpoints $E_p(\cdot)$, program feasibility, or program exactness, can be readily carried over from SCPs to STPs.

Proposition 3. *The problems of finding a feasible simple toset program and exact simple toset program are* NP.

Proof. See the proof of Proposition 1, with the difference that instead of comparing colours we verify whether a criterion (cf. Definition 11) is satisfied, which too is a polynomial-time operation.

Definition 14 (Out-Neighbour Consistency). *We say that $v \in V$ is a vertex with out-neighbours consistent with respect to A, B (where $A \cap B = \emptyset$ and $A, B \subseteq N_o(v)$) if there exists no pair of vertices $x \in A, y \in B$ such that $\phi_V(x) = \phi_V(y)$.*
We say that $S \subseteq V$ is a vertex set with out-neighbours consistent with respect to A, B (where $A \cap B = \emptyset$ and $A, B \subseteq N_o(S)$) if there exists no pair of vertices $x \in A, y \in B$ such that $\phi_V(x) = \phi_V(y)$.

Definition 15 (Building Criteria). *Let* COMPUTECRITERION*(B, M, E) be any algorithm that takes three vertex sets B, M, E as input and outputs a criterion such that every vertex in B is classified as "Yes", "Included" or 1, every vertex in E is classified as "No", "Excluded" or 0, and any vertex in M but not in B is classified as either.*

Such algorithms exist, with CART [6] and C4.5 [20] being two notable examples. In our case it is further important that when the tree pruning phase of these algorithms is initiated, pruning is done only if it does not break the guarantees of Definition 15 or omitted altogether.

Lemma 5 (Criterion existence for pseudo-bases with out-neighbours consistent). *Let B, M, E be vertex sets such that $B \subseteq M, E \cap M = \emptyset$. If there exists a pseudo-basis $\mathcal{B}$ with out-neighbours consistent for B, M then a criterion for B, M, E as per Definition 15 exists.*

Proof. Let $b \in B, e \in E$. Since $\mathcal{B}$ is a vertex set with out-neighbours consistent w.r.t B, E, b, e cannot have the same features. Hence there exists a feature f such that $\phi_V(b) \neq \phi_V(e)$. Thus, there exists a split $s_{b,e}$ on f that separates b, e. A conjunction of these splits for all b, e is a criterion for B, M, E as per Definition 15.

If a criterion exists, then both CART and C4.5, if left unterminated, will eventually build a decision tree that achieves the perfect separation. Thus, either can be used as the BUILDCRITERON routine.

Our strategy to tackle STP mining is to find pseudo-bases with out-neighbours consistent for each step of a potential program, and then to find criteria (out-neighbourhood classifiers) that correspond to those pseudo-bases. Lemma 5 shows that once an appropriate pseudo-basis has been found, the criterion can be found thanks to the consistency.

Multiple approaches can be taken to implement this strategy. If getting *a* solution is the priority, one can perform a depth-first search of pseudo-bases, and the moment the first valid sequence of pseudo-bases encapsulating S at the beginning and hitting exactly T at the end is found, find the step criteria and terminate. Since by Lemma 5 we know that for pseudo-bases with out-neighbours consistent a criterion always exists, there is no utility in computing criteria on the while pseudo-bases are still being determined, as this process presents no benefit to the determination of pseudo-bases. For the sake of consistency with earlier sections we chose to perform a breadth-first search of our pseudo-bases instead.

In Algorithm 2, let Q, Q_{next} be queues lists of triples drawn from *vertex-sets* $\times$ *vertex-sets* $\times \mathbb{Z}$, P be a set of *programs*. The subject of our study is Algorithm 2. Each list ℓ in Q is represents a chain of pseudo-bases that might trace out a program path from S to T. Every triple (B, M, n) in ℓ and Q_{next} represents from where to begin B to reach T, the in-neighbourhood with out-neighbours consistent $B \subseteq M \subseteq N_i(N_o(B))$, and the saved distance n from B to T.

Proposition 4. *The following hold.*

1. *Algorithm 2 is correct in the sense that all programs in P are exact programs for S, T.*
2. *Algorithm 2 is complete in the sense that whenever execution exists the loop closing at line 30, P contains all exact programs for S, T of length ℓ, up to criterion equivalence.*

Algorithm 2: (BPF) Basis-Path-Finding

Initialisation: Q_{next} contains only the singleton list $(T, T, 0)$,
Q, $\mathcal{L}$, P are empty

1   **for** $\ell \geq 1$ *such that* $T \subseteq N_o^\ell(S)$ **do**

2     $Q \leftarrow Q_{\text{next}}$ and empty Q_{next}

3   **while** Q *is not empty* **do**

4   pop the list ℓ from Q

5   let (B, M, m) be the head, n the length of ℓ

6

7   **if** $n - 1 = \ell$ **then**

8   **if** $B \subseteq S \subseteq M$ **then**

9   add ℓ to $\mathcal{L}$

10   **end**

11   push ℓ into Q_{next}

12   **continue**

13   **end**

14

15       $M' \leftarrow N_i(B) \cap N_o^{\ell-n}(S)$

16   remove from M' vertices with out-neighbours inconsistent w.r.t.
        B, $N_o(M') \backslash M$

17   **foreach** *pseudo-basis* $B' \subseteq M'$ *for* (B, M) **do**

18   push (B', M', n) into ℓ

19   **end**

20   **end**

21

22   **foreach** *list* ℓ *in* $\mathcal{L}$ **do**

23       $p \leftarrow \epsilon$

24   pop the head of ℓ and discard

25   **foreach** *element* (B, M, n) *in* $\mathcal{L}$ **do**

26         $c \leftarrow \textsc{ComputeCriterion}(B, M, N_o^{\ell-n}(S) \backslash M)$

27         $p \leftarrow cp$

28   **end**

29   add p to P

30   **end**

31 **end**

Proof. The proposition is analogous with the Proposition 2, and as such analogous proofs can be constructed.

Briefly, for correctness, if a candidate list

$$\ell = (B_0, M_0, \ell+1)(B_1, M_1, \ell) \cdots (B_\ell, M_\ell, 1)(T, T, 0)$$

has been added $\mathcal{L}$, then it must have been the case that S is a vertex set with out neighbours consistent spanning B_1 but not outspanning M_1. Inductively, it must have been the case that B_k spans B_{k+1} but does not outspan M_{k+1}. Now, the loop on line 22 ensures that at each step, the program always proceeds to a set of vertices S_k such that $B_k \subseteq S_k \subseteq M_k$. Hence any program in P is correct.

For completeness, just notice that following any $p \in P$ from the end there is always a pseudo-basis $B_k \subseteq E_{p_{\leq k}}(S)$ that spans B_{k+1} but (trivially) does not outspan $E_{p_{\leq k+1}}(S)$. So a valid chain of pseudo-bases exists, will be found and added to $\mathcal{L}$, and the corresponding chain of criteria logically equivalent to those of p (but not necessarily the same) will then be added to P.

As in the discussion of Algorithm 1, Algorithm 2 can easily be modified to search for feasible and not just exact programs.

The runtime of the algorithms we propose depends greatly on the network. In the worst case – on a fully connected graph, our algorithms perform a total enumeration of all possible programs. However, real-world labelled graph datasets tend to contain significant amounts of pattern structure, suggesting performance far better than that of the worst case.

5   Conclusion

We have pointed at the previously unaddressed problem of characterising relationships between groups of records in a database in terms of their long-distance connections within the database graph. We have identified the problem of graph-walking program mining as a simple case of this wider challenge, and investigated simple colour and totally-ordered set program mining. We addressed them by giving the Viable Injection Basis Enumeration and Basis-Path-Finding algorithms, and proved their correctness and completeness.

The main observation allowing us to sharply limit the search space is that the set of vertices through whom agent executing a simple program proceeds is bounded below by an appropriate basis and above by consistency with respect to out-vertices. The construct corresponding to these bounds in the process of mining is the criterion *pseudo-basis*, appearing as a necessary and sufficient condition on local existence of feasible programs. We have further shown that the problems of simple program mining are NP.

We have hinted that more complex program structures can be employed in graph-walking programs, and that the programs do not need to be deterministic. While adding to the structure of graph-walking programs would likely hinder their interpretation as relationships between sets of vertices, we believe that our current work can be extended by considering probabilistic graph-walking programs, further expanding the utility of graph-walking programs in the context of network analysis and beyond.

References

1. Alur, R., Singh, R., Fisman, D., Solar-Lezama, A.: Search-based program synthesis. Commun. ACM **61**(12), 84–93 (2018)
2. Andersen, R., Chung, F., Lang, K.: Local graph partitioning using pagerank vectors. In: 2006 47th Annual IEEE Symposium on Foundations of Computer Science (FOCS2006), pp. 475–486. IEEE (2006)

3. Avrachenkov, K., Gonçalves, P., Sokol, M.: On the choice of kernel and labelled data in semi-supervised learning methods. In: Bonato, A., Mitzenmacher, M., Prałat, P. (eds.) WAW 2013. LNCS, vol. 8305, pp. 56–67. Springer, Cham (2013). https://doi.org/10.1007/978-3-319-03536-9_5
4. Bodík, R., Jobstmann, B.: Algorithmic program synthesis: introduction (2013)
5. Breiger, R.L., Boorman, S.A., Arabie, P.: An algorithm for clustering relational data with applications to social network analysis and comparison with multidimensional scaling. J. Math. Psychol. 12(3), 328–383 (1975)
6. Breiman, L., Friedman, J.H., Olshen, R.A., Stone, C.J.: classification and regression trees. Statistics/Probability Series, The Wadsworth (1984)
7. Burt, R.S.: Positions in networks. Social forces 55(1), 93–122 (1976)
8. Cooper, C., Radzik, T., Siantos, Y.: Fast low-cost estimation of network properties using random walks. Internet Math. 12(4), 221–238 (2016)
9. Desai, A., et al.: Program synthesis using natural language. In: Proceedings of the 38th International Conference on Software Engineering, pp. 345–356 (2016)
10. Ellis, K., et al.: Dreamcoder: bootstrapping inductive program synthesis with wake-sleep library learning. In: Proceedings of the 42nd ACM SIGPLAN International Conference on Programming Language Design and Implementation, pp. 835–850 (2021)
11. Feng, Y., Martins, R., Bastani, O., Dillig, I.: Program synthesis using conflict-driven learning. ACM SIGPLAN Notices 53(4), 420–435 (2018)
12. Feng, Y., Martins, R., Van Geffen, J., Dillig, I., Chaudhuri, S.: Component-based synthesis of table consolidation and transformation tasks from examples. ACM SIGPLAN Notices 52(6), 422–436 (2017)
13. Feser, J.K., Chaudhuri, S., Dillig, I.: Synthesizing data structure transformations from input-output examples. ACM SIGPLAN Notices 50(6), 229–239 (2015)
14. Gudes, E., Pertsev, A.: Mining module for adaptive xml path indexing. In: 16th International Workshop on Database and Expert Systems Applications (DEXA2005), pp. 1015–1019. IEEE (2005)
15. Guha, S.: Efficiently mining frequent subpaths. In: Proceedings of the Eighth Australasian Data Mining Conference-vol. 101, pp. 11–15 (2009)
16. Huang, D., et al.: Neural program synthesis with query. In: International Conference on Learning Representations (2021)
17. Lorrain, F., White, H.C.: Structural equivalence of individuals in social networks. J. Math. Soc. 1(1), 49–80 (1971)
18. Mendelzon, A.O., Wood, P.T.: Finding regular simple paths in graph databases. SIAM J. Comput. 24(6), 1235–1258 (1995)
19. Page, L., Brin, S., Motwani, R., Winograd, T.: The pagerank citation ranking: bringing order to the web. Tech. rep, Stanford InfoLab (1999)
20. Quinlan, J.R.: C4. 5: programs for machine learning. Elsevier (2014)
21. Wang, C., Cheung, A., Bodik, R.: Synthesizing highly expressive SQL queries from input-output examples. In: Proceedings of the 38th ACM SIGPLAN Conference on Programming Language Design and Implementation, pp. 452–466 (2017)
22. White, D.R., Reitz, K.P.: Graph and semigroup homomorphisms on networks of relations. Social Netw. 5(2), 193–234 (1983)
23. Winship, C., Mandel, M.: Roles and positions: a critique and extension of the blockmodeling approach. Sociol. Methodol. 14, 314–344 (1983)
24. Yaghmazadeh, N., Klinger, C., Dillig, I., Chaudhuri, S.: Synthesizing transformations on hierarchically structured data. ACM SIGPLAN Notices 51(6), 508–521 (2016)

A Benchmarking Evaluation of Graph Neural Networks on Traffic Speed Prediction

Khang Nguyen Duc Quach[1], Chaoqun Yang[2], Viet Hung Vu[3],
Thanh Tam Nguyen[1(✉)], Quoc Viet Hung Nguyen[1], and Jun Jo[1]

[1] Griffith University, Gold Coast, Australia
thanhtamlhp@gmail.com
[2] Chongqing University of Posts and Telecommunications, Chongqing, China
[3] Hanoi University of Science and Technology, Hanoi, Vietnam

Abstract. Traffic speed prediction is the task of forecasting the average moving speed on certain roads and highways. Since transportation is an essential part of our daily life, traffic prediction helps reduce wasting time caused by traffic congestion and the negative environmental impacts of idling vehicles during traffic jams. Recently, graph neural network has achieved a certain level of success in forecasting traffic, thanks to the ability to capture both spatio-temporal features of traffic data. Although many publications propose and benchmark different traffic prediction methods, those models' performance and other profiles are hardly mentioned and evaluated, such as the running time and memories taken. This research aims to evaluate the performance of five mainstream GNN models in traffic prediction: DCRNN, Graph Wavenet, MTGNN, STGCN, and T-GCN. Training time, inference time, and other profiles of mentioned models are also further investigated and reported. During the experiments, we record various behaviors/factors from models, such as the memory allocation. Different models are believed to be better in specific scenarios in traffic forecasting based on their profiles. The finding shows that some prediction methods can only perform well if the training time is significant, such as T-GCN. The performances on single/multiple step forecasting vary with different models, where MTGNN outperforms other models. Additionally, the memory usage and inferencing time of GNN models are different, potentially impacting the selection of models for some limited memory machines and other scenarios.

Keywords: Traffic prediction · Graph neural network · Performance evaluation

1 Introduction

Thanks to the practical applicability in real-world traffic practice, traffic prediction has been a widespread study area over the last four decades. The need for

K. N. D. Quach and C. Yang—contributed equally to this research.

traffic forecasting is even more significant as the number of vehicles surges due to fast growth in economic output and other factors. As a result, idle vehicle emissions have led to a rise in greenhouse gases. Numerous articles suggesting various forecasting approaches have been submitted, and they have been very successful in both long-term and short-term traffic prediction. Those approaches share the similarity in model design and overcome the difficulties of the formers.

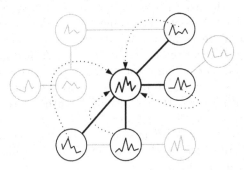

Fig. 1. The information on time series of surrounding neighbours can influence the time series of the linked central node.

Traffic prediction is the task of forecasting the state of current traffic based on past observations and collected events, such as traffic flow [1]. By using the traffic prediction technique, traffic managers can provide travelers with traffic forecasts to avoid heavy congestion [20]. Traffic prediction often involves mining and analyzing patterns from traffic situations, such as the number of vehicles and changes in speed on motorways. It is a time-series problem in which the traffic status trends are dynamic and influenced by many external variables. According to [21], the most challenging aspect of this task is to capture both traffic's spatial and temporal dependencies. Figure 1 illustrates the temporal and spatial features of a sensor in traffic data. Recently, the development of various deep learning architectures and machine learning models has improved the accuracy of traffic forecasting thanks to the current rise in computing capacity. This overcomes the constraint of statistical models, which can only reliably predict traffic over the long term. Additionally, additional traffic data sets from CCTV and roadway sensors have been made accessible to the general public. Furthermore, a massive quantity of data from taxis' locations is being produced thanks to the development of positioning technology [17,18]. This allows researchers to assess and conduct experiments on diverse data sets and models. Most of the papers are devoted to developing short-term traffic speed forecasting.

The graph neural network-based models have achieved particular success in traffic prediction. Many publications propose different architectures of GNNs to forecast traffic situations. This type of neural network model is helpful to traffic prediction because it can capture both spatial and temporal dependencies, in which classic statistical models and standard neural networks have failed. The traffic network can be considered a graph, where traffic sensors act as vertices and the connection of sensors on the same road is a weighted edge. Some popular GraphNN models are Graph Wavenet [12], STCNeT [8], and T-GCN [21]. Although they are based on graph theory, the performance of those models is different on multiple factors, such as training time and prediction accuracy. The differences are worth investigating to discover the underlying relationship between the dataset and the GraphNN models.

In 2002, [4] performed an evaluation of the short-term time-series traffic prediction model. In the experiment, various generated scenarios are based on the combination of model parameter settings and traffic conditions. In addition, a multivariate general linear model (GLM) was used to determine the factors that had the most significant influence on the model's performance. The study shows that travel time prediction errors increase dramatically under congested conditions. This is because the duration is not long enough for the prediction model to capture the steady pattern of change. Another performance evaluation article by [3] concluded that the MAE and RMSE are more significant during the day than at night, while the MRE is smaller during the daytime. Furthermore, there is still more performance evaluation comparing machine learning and deep learning models, such as [13] and [11].

There are five mainstream GNN frameworks in traffic speed prediction: Diffusion Convolution Recurrent Neural Network [7], Graph WaveNet [12], Multivariate Time Series With Graph Neural Network [15], Spatio-Temporal Graph Convolutional Network [19] and Temporal Graph Convolutional Network [21]. The major objective of this study is to provide a comprehensive comparison of all these GNN methods regarding the predictions' accuracy. Other models' profiles, such as training time, memory allocation, and inferencing time, are also evaluated and analyzed. This paper aims to deliver guidance on selecting appropriate models concerning different scenarios. The contributions of our work are highlighted as follows:

- We provide a concise definition and summary of the graph neural network models to give a deeper understanding of the models' architecture and their mathematical formulation. The technologies employed to capture spatial and temporal dependencies are clearly summarised.
- We select one of the most popular data sets for traffic speed prediction for the benchmark, namely PEMSBAY [7]. Furthermore, we propose a new data set based on PEMSBAY with half the size, namely PEMSBAY0.5. The purpose of the second data set is to investigate how the size of the traffic network affects the behaviors of GNN models.
- We propose an experiment where we fairly compare all the GNN models under the same circumstances, hyperparameters, and evaluation metrics. Other important profiles of models during the experiment are also recorded and analyzed.
- All deep insights and practical guidelines are reported to draw reliable conclusions. We believe the paper can be an applicable guideline for choosing suitable GNN models in specific scenarios.

The rest of this paper is organized as follows. Section 2 describes the formulation of traffic speed prediction. Section 3 gives an overview of five mainstream graph neural network models. In Sect. 4, we provide the detailed implementation for the benchmark, data sets, metrics, and settings. Section 5 discusses the results and findings, followed by Sect. 6 concluding this study.

2   Problem Formulation

The dataset in network-based traffic prediction contains both spatial and temporal features. The spatial features are the graph topology G, which includes road sensors as vertices. In Fig. 2, each slice represents the speed records of sensors in the road network G at time t. Let E be the collection of edges representing the connection between traffic sensors, which is an adjacency matrix of G. Let $V = \{1, 2, 3, ..., N\}$ be the set of traffic sensors in the dataset. Each traffic sensor contains historical observations of evenly distributed time steps of speed records on a highway during a certain period, such as six months. Let x_t^i be the record of traffic speed by the sensor i in the time step t. Then we have $x_t = \{x_t^1, x_t^2, x_t^3, ..., x_t^N\}$ is the observations of all traffic speed sensors at time step t, which can be seen as a slice of 2D traffic dataset ($x \in \mathbb{R}^{T \times N}$). The desired output from the model is the prediction target $\hat{x}_t$ at future time step t. The forecasting model learns the function $H(\cdot)$ to predict the traffic speed in the next single/multiple time steps. Given the graph G and the historical observations $X_t = \{x_1, x_2, ..., x_t\}$, the forecasting task is written as:

$$\hat{x}_t = H(G; X_t) \tag{1}$$

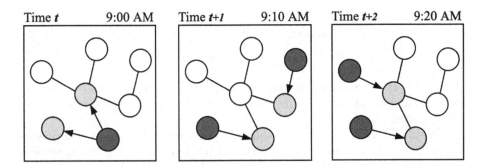

Fig. 2. Each time slice contains speed records of vertices in graph G at certain time step t. A node can only influence its neighbours at specific time steps. Thus the problem becomes more dynamic and more information on spatiotemporal features is given.

3   Benchmarked Models

3.1   Diffusion Convolutional Recurrent Neural Network (DCRNN)

Diffusion Convolutional Recurrent Neural Network was first introduced in a traffic prediction task by [7]. This model is a pioneer in utilizing graph convolution in traffic data forecasting. As [7] explained, this graph neural network comprises three components: diffusion convolution, sequence to sequence architecture (also known as RNN), and scheduled sampling. The development of this model is based on the ability to represent the traffic data as a process of diffusion on a directed graph, combined with the recurrent neural network as an encoder-decoder.

Graph diffusion is the process of continuously passing/diffusing the attention to the node's neighbors from each node in the graph. The diffusion process of a node can be expressed as $S = \sum_k \theta_k T^k$, where θ are the coefficients and T is the transition matrix. The diffusion convolutional layer mapping P-dimensional features to Q-dimensional outputs is thus represented as:

$$H_{:,q} = a(\sum_1^P X_{:,p} \star Gf_{\Theta_{q,p}})$$ (2)

where X is the input, H is the output, a is the activation function and $f_{\Theta_{q,p}}$ are the filters. After the diffusion process, the new sparse and weighted graph is obtained, which helps capture the spatial dependency from the traffic network.

The sequence-to-sequence architecture is based on leveraged recurrent neural network to model the temporal features of the traffic data. Gated Recurrent Units by [2] is implemented in the model by replacing matrix multiplication with the diffusion convolution.

$$r^{(t)} = \sigma(\Theta_r \star_G [X^t, H^{(t-1)}] + b_r)$$ (3)

$$u^{(t)} = \sigma(\Theta_u \star_G [X^t, H^{(t-1)}] + b_u)$$ (4)

$$C^{(t)} = tanh(\Theta_C \star_G [X^{(t)}, (r^{(t)} \odot H^{(t-1)}] + b_c)$$ (5)

$$H^{(t)} = u^{(t)} \odot H^{(t-1)} + (1 - u^{(t)}) \odot C^{(t)}$$ (6)

where $X^{(t)}$, $H^{(t)}$ are the input and output at time step t, $r^{(t)}$ and $u^{(t)}$ are the reset gate and update gate at time step t. $\Theta_r, \Theta_u, \Theta_C$ are the parameters for the corresponding filters.

Instead of only viewing relationship between direct neighbors as binary connection like standard graph convolution, DCRNN considers indirect neighbors of nodes through the diffusion process. As shown in Fig. 3, the neighbours' weights to the POI decrease as distance between them increases. However, this approach encounters a limitation, where graph diffusion convolution assumes homophily. As mentioned by [6], there is no straightforward solution to overcome the assumption of homophily. Overall, as stated by [7], the application of DCRNN can be helpful in various spatiotemporal prediction tasks thanks to the ability to capture spatiotemporal dependencies within time series data.

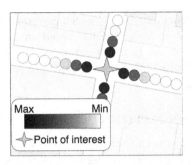
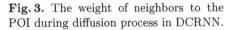

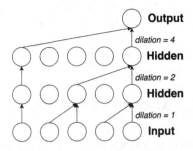

Fig. 3. The weight of neighbors to the POI during diffusion process in DCRNN.

Fig. 4. One dimensional dilated convolution in Graph WaveNet.

3.2 Graph WaveNet (GWNET)

Graph Wavenet is one of the first models to implement dilated causal convolution for capturing the temporal dependencies in traffic prediction tasks. It was first introduced in traffic prediction by [16], which is a graph version of the original WaveNet model proposed by [10]. Graph WaveNet uses stacked dilated causal convolution to capture the temporal dependencies and graph convolution to model the spatial features. Instead of transmitting the signals from limited time steps, stacked dilated causal convolution allows Graph WaveNet to cover a longer sequence of time steps by improving the model's receptive field. Therefore, the model can have more info on the data without needing to increase the number of neural network layers. Figure 4 shows the process of dilated convolution. Given a single dimension sequence $x \in \mathbb{R}^T$, a filter $f \in \mathbb{R}^K$ and d is the dilation controlling the skipping distance, the operation of dilated convolution on x with filter f is written as:

$$x \star f(t) = \sum_{s=0}^{K-1} f(s)x(t - d \times s) \tag{7}$$

This model also implements the Gated TCN mechanism to control the flow of information passed from temporal convolution layers to graph convolution layers. Given the input $X \in \mathbb{R}^{N \times D \times S}$, the output of Gated TCN h can be expressed as:

$$h = g(\Theta_1 \times X + b) \odot \sigma(\Theta_2 \star X + c) \tag{8}$$

where $g(\cdot)$ and $\sigma(\cdot)$ are the activation function $(tanh)$ and sigmoid function. Θ_1, Θ_2, b and c are the parameters of the model. Since the sigmoid function returns values between 0 and 1, the value of output h can be controlled by operation $\odot$ between activation function and sigmoid function.

This model overcomes two significant deficiencies of DCRNN. First, the DCRNN distance-based adjacency matrix assumes that geographically near sensors are always associated with one another. Compared to the static graph used in DCRNN, Graph WaveNet utilizes a self-adaptive network, which can preserve hidden spatial factors. Secondly, Graph WaveNet replaces GRU-based encoder-decoder systems of DCRNN with WaveNet blocks to avoid extensive training [12]. This framework is a CNN-based model, which also has the power of parallel computing and low memory consumption.

3.3   Multivariate Time Series with GNN (MTGNN)

Multivariate Time Series with Graph Neural Network (MTGNN) [15] is a GNN-Based and CNN-based model which performs graph learning on unstructured multivariate time series. The model consists of three main components: the graph learning layer, the graph convolution layer, and the temporal convolution layer. A graph learning layer computes a graph adjacency matrix, which is then utilized as input by all graph convolution modules to uncover hidden relationships between nodes. Given the adjacency matrix produced by the graph learning layer, a graph convolution module will handle the spatial relationship between variables. The graph convolution model utilizes the mix-hop propagation layer for information propagation and information selection. The formation for those two steps are respectively defined as

$$H^k = \beta H_{in} + (1 - \beta)\hat{A}H^{(k-1)} \tag{9}$$

$$H_{out} = \sum_{i=0}^{K} H^{(k)}W^{(k)} \tag{10}$$

where H_{in} and H_{out} represent input hidden states by the previous layer and output hidden states of the current layer, respectively. β is the hyper parameter controlling the retaining information from the root node's original states, and K is the propagation depth.

The temporal convolution module is implemented for capturing temporal patterns using modified 1D convolution. This process uses the dilated inception layer, which combines inception and dilation. Given a single dimensional sequence input z and the filter $f_{1 \times k}$ and d is the dilation factor, the dilated convolution is written as

$$z \star f_{1 \times k}(t) = \sum_{s=0}^{k-1} f_{1 \times k}(s)z(t - d \times s) \tag{11}$$

This framework automatically extracts the uni-directional relations among variables through a graph learning module, into which external knowledge may be easily integrated. MTGNN applies to both small and big graphs, short and long time series, with or without externally established graph topologies, which is a great advantage.

3.4   Spatio-Temporal Graph Convolutional Network (STGCN)

In contrast to the diffusion convolution in DCRNN, STGCN [19] uses spectral convolution. In a higher-level view, STGCN includes graph CNNs for extracting

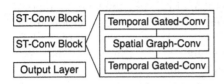

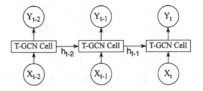

Fig. 5. STGCN's architecture.   **Fig. 6.** T-GCN's architecture.

spatial features and gated sequential convolutions for extracting temporal features. In [19], the authors introduce spatio-temporal convolutional block, which is the essential component in STGCN. Given the input frame $v^l \in \mathbb{R}^{M \times n \times C^l}$ of block l, M frames of the road graphs, C is the channel, the output frame v^{l+1} is expressed as

$$v^{l+1} = \Gamma_1^l *_\tau ReLU(\Theta^l *_G (\Gamma_1^l *_\tau v^l)) \tag{12}$$

where Γ_1 are the upper and lower kernel; $ReLU(\cdot)$ is the activation function; Θ^l is the spectral kernel of graph convolution.

The model consists of several spatio-temporal convolutional blocks as illustrated in Fig. 5, each of which has a "sandwich" structure consisting of two gated sequential convolution layers and one spatial graph convolution layer. The spatio-temporal convolutional block (ST-Convblock) handles graph-structured time series to merge features from the spatial and temporal domains. By downscaling and upscaling channels in the graph convolution layer, the "sandwich" topology also enables the network to utilize the bottleneck approach to achieve scale compression effectively and feature squeezing [19].

3.5   The Temporal Graph Convolutional Network (T-GCN)

The temporal graph convolutional network (T-GCN) model [21] is the combination of the graph convolutional network (GCN) and the gated recurrent unit (GRU). The graph convolutional network is employed to capture the spatial structure of the road network. The gated recurrent unit is ultilised to model the temporal dependency of traffic data on highways. The architecture of T-GCN is illustrated in Fig. 6. Given the graph convolution process $f(A, X_t)$, A is the adjacency matrix, and X_t is the feature matrix at time t, the formulation is inspired by standard GRU models, which is defined as

$$u_t = \sigma(W_u[f(A, X_t), h_{t-1}] + b_u) \tag{13}$$

$$r_t = \sigma(W_r[f(A, X_t), h_{t-1}] + b_r) \tag{14}$$

$$c_t = tanh(W_c[f(A, X_t), (r_t * h_{(t-1)})] + b_c) \tag{15}$$

$$h_t = u_t * h_{t-1} + (1 - u_t) * c_t \tag{16}$$

where W and b are the weights and biases of the model during training.

The result from [21] demonstrates a stable state over multiple prediction horizons, indicating that it may be suitable for both short-term and long-term traffic prediction tasks. However, as reported by [21], the limitation of T-GCN is poor prediction at local minima and maxima in the time series data set, which is caused by smooth filter in the Fourier domain. Table 1 summarises the technologies employed in five graph neural networks.

Table 1. Feature comparison of benchmarked models

	Spatial	Temporal	Extra
DCRNN	Diffusion-Conv	RNN	Schedule sampling^c
Graph WaveNet	Diffusion-Conv	Dilated-Conv	
MTGNN	Graph-Conv^a	Dilated-Conv	Graph learning layer^d
STGCN	Graph-Conv^b	1D-Conv+GLU	
T-GCN	Graph-Conv	GRU	

^{a} MTGNN's graph convolution method consists of two mix-hop propagation layers.
^{b} STGCN utilises Chebyshev Polynomials Approximation or 1st order Approximation to optimise the performance.
^{c} DCRNN implements schedule sampling to improve accuracy of long-term forecasting.
^{d} MTGNN uses graph learning module to capture hidden spatial dependencies among variables.

4   Benchmarking Procedure

4.1   Data Description

The traffic dataset used in this study is downloaded from the Caltrans Performance Measurement System (PeMS). The traffic data contains time series data from 325 road sensors recording the traffic speed every five minutes. Those sensors are located on the Bay Area roads, and the data are collected in four months. In addition, PEMSBAY [7] is used widely as a standard dataset for performance testing and comparison in many traffic prediction publications. This study also generates a variation of original PEMSBAY data by randomly selecting and eliminating traffic sensors by half. This results in a dataset having a reduced size to the original dataset with similar characteristics as the connection of sensors and traffic speed records.

4.2   Evaluation Criteria

The accuracy plays a vital role as a metric for model evaluation and comparison in any existing problem. Given y and f are the actual observation and the forecast from a model, respectively. The three most popular metrics for this purpose that have been widely used in other publications and this paper are:

- Mean Square Error (MSE) - the square of average of the difference between predictions and the observed values.

$$MSE = \frac{1}{N} \sum_{1}^{N} (y_i - f_i)^2 \tag{17}$$

- Root Mean Square Error (RMSE) - the square root of the mean of the square difference between observed and predicted values.

$$RMSE = \frac{1}{N} \sum_{1}^{N} \sqrt{(y_i - f_i)^2} \tag{18}$$

- Mean Absolute Error (MAE) - the mean of the absolute difference between predicted and observed values.

$$MAE = \frac{1}{N} \sum_{1}^{N} |y_i - f_i| \tag{19}$$

Also, training time and the other details during training will be recorded, such as validation loss and number of epochs.

4.3   Experiment Settings

The models are trained using the Graphical Processing Unit by the deep learning framework. The software used for the experiment is the modified version of LibCity proposed by [14]. The specification of the GPU is RTX 3070 with CUDA version of 11.4. The models are trained by passing through two primary datasets, namely PEMSBAY and PEMSBAY0.5. The number of max epochs for DCRNN, Graph WaveNet, MTGNN, and STGCN is set to 100, whereas the TGCN is set to 5000, as recommended by [5]. As Fig. 7 illustrated, the framework comprises of the data access layer, computing layer, and application layer.

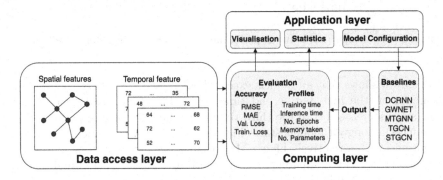

Fig. 7. Benchmark framework

During each epoch, the time taken and validation loss will be calculated and recorded for investigation purposes. It is noted that the total training time will not include the phase of reading data and constructing adjacency matrices from some models like STGCN. The baselines are also implemented with early stopping if it is converged. The batch size is set to 64, and the learning rate decay is applied to all training modes.

5   Results and Discussion

5.1   Effective Evaluation

Table 2 shows that the MTGNN nearly outperforms other approaches in the PEMSBAY0.5 dataset. The Recurrent Neural Network (RNN) is also included in the experiment as a baseline. The RNN model has the shortest training time but performs poorly compared to graph neural network models, and this shows how significant the spatial dependency is in the traffic dataset. Regarding the multistep prediction, MTGNN and Graph WaveNet are the top performers. This shows that diffusion convolution in Graph WaveNet can better memorize the relationships between nodes than other graph convolution methods. Dilated convolution seems to play an essential part in the high performance of MTGNN since the longer sequences of time series are considered thanks to the growth of the receptive field of the model.

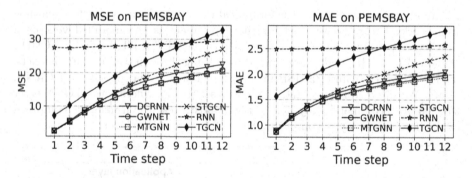

Fig. 8. The comparison between performance in multiple time steps predictions.

Figure 8 shows that all models except T-GCN and RNN have the same performance on the first and second time steps. The accuracy decreases as the time steps increase. Up to the third time step, the difference in prediction performances among all models starts increasing. The T-GCN model is the worst performer, with a relatively higher MSE than other models throughout 12 time steps. MTGNN is the apparent winner regarding both single and multiple time steps predictions in both data sets.

Table 2. Performance evaluation for multi-step prediction

Dataset	Model	3 steps			6 steps			12 steps		
		MAE	MSE	RMSE	MAE	MSE	RMSE	MAE	MSE	RMSE
PEMSBAY	DCRNN	1.38	8.75	2.96	1.73	15.9	3.99	2.05	22.52	4.75
	GWNET	**1.33**	8.14	2.85	1.67	**14.39**	3.79	1.95	20.92	4.57
	MTGNN	**1.33**	**8.08**	**2.84**	**1.65**	14.69	**3.78**	**1.94**	**20.48**	**4.55**
	STGCN	1.38	8.36	2.89	1.81	16.49	4.06	2.35	27.02	5.2
	RNN	2.50	27.43	5.24	2.52	27.98	5.29	2.58	29.51	5.43
	TGCN	1.95	13.38	3.66	2.33	21.26	4.61	2.86	32.7	5.72
PEMSBAY0.5	DCRNN	1.37	8.74	2.95	1.73	15.90	3.98	2.04	22.52	4.74
	GWNET	1.36	8.36	**2.89**	1.70	15.14	3.89	2.00	**21.27**	**4.61**
	MTGNN	**1.35**	**8.35**	**2.89**	**1.68**	**15.07**	**3.88**	**1.97**	21.29	**4.61**
	TGCN	1.62	11.17	3.10	1.91	20.14	3.91	2.26	31.32	4.92
	STGCN	1.43	9.11	3.01	1.85	17.27	4.15	2.35	27.23	5.21
	RNN	2.65	28.63	5.35	2.69	30.23	5.49	2.88	36.18	6.015

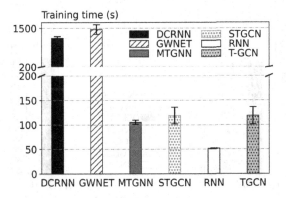

Fig. 9. The bar plot of training time on five baselines in logarithmic scale. The time amount of model training from DCRNN and GWNET takes significantly longer than the others. MTGNN, STGCN and T-GCN have similar average training time, range from 110 s to 125 s.

5.2 Efficiency Evaluation

Training Time. As illustrated in Fig. 9, the overall training time per epoch of Graph WaveNet takes the longest compared to other models in general. Although [15] claimed that this model is based on CNN architecture for better training optimization, it conflicted with the result we achieved in this experiment, in which the training time of Graph Wavenet is significantly higher than other popular graph neural networks in traffic prediction. It is also worth noting that T-GCN has a relatively low training time compared to other models, with an average of 112 s per epoch, but it also needs more than 5000 epochs to converge.

Overall, MTGNN requires the least amount of time for training among the six baselines.

Inference Time. Figure 11 shows that STGCN has the longest inferencing time among five baselines. DCRNN and Graph WaveNet have similar inferencing time, because they both ultilise diffusion convolution. MTGNN is the best performer regarding inferencing time.

5.3   Models and Profiling

Training Epochs. T-GCN requires more epochs to converge and reach optimal accuracy. The optimal model state of TGCN is at epoch 4972nd, where the allowed max training epoch is capped at 5000. This indicates that the T-GCN model can improve the prediction accuracy if there is more epoch. Figure 10 shows that DCRNN and MTGNN reach the optimized model's parameters at early epochs, which are the 20th and 5th epochs, respectively. Also. DCRNN tends to converge at $\frac{3}{4}$ of max allowed training epochs.

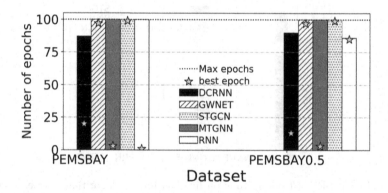

Fig. 10. Number of epochs reach convergence and the epochs that configures best model (marked as star).

Memory Allocation. Figure 14 illustrates that MTGNN and STGCN need more parameters in the more extensive data set, whereas DCRNN, GWNET, and TGCN use the same amount in PEMSBAY and PEMSBAY0.5. Figure 12 also shows that the memory allocation for training time and storing models are similar among six baselines. However, MTGNN and DCRNN need significantly higher memory for storing and loading the data during training. STGCN needs the least memory overall, this illustrates that Chebyshev Polynomials Approximation and/or 1st order Approximation can optimize the required memory for capturing spatial dependencies.

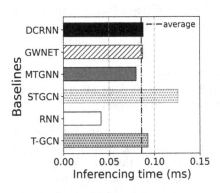

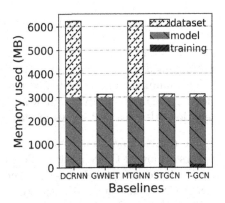

Fig. 11. Time for inference.

Fig. 12. Memory used.

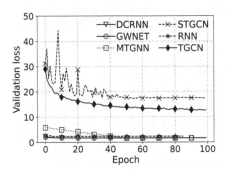

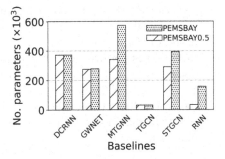

Fig. 13. Validation loss

Fig. 14. Number of parameters used

Validation Loss During Training. Regarding the validation loss during training phases, Fig. 13 shows that Graph WaveNet and DCRNN start the training with the lowest loss, whereas STGCN and TGCN have the higher loss and are approximately the same. Figure 13 also shows that the validation loss of STGCN frequently fluctuates during the first forty epochs, and the other models' validation loss slightly decreases over time. Interestingly, the validation loss of MTGNN follows a stair pattern, where the loss value decreases every five epochs.

6   Conclusion

This paper evaluates the performance of five mainstream graph neural networks in traffic prediction tasks, namely DCRNN, Graph WaveNet, MTGNN, TGCN, and STGCN. Although their architecture is based on graph theory, the way each approach captures the spatial information in traffic prediction is different. This paper also compares the standard recurrent neural network with the GNN models. An overview of five GNN models is given in this paper, discussing the characteristics and features of each model on the traffic data set. We try our best to ensure the benchmark and evaluation are fair for all models. Many important factors, including the models' performance and profiles, are also analyzed, such as the accuracy of prediction in multi-time steps, training time, memory used, and inference time, using the real-world data on traffic prediction.

The summary of our principle findings are shown below for the purpose of selecting the most suitable frameworks for traffic prediction depending on specific scenarios. Also, Table 3 shows models' performance on categories.

- The result shows that RNN models have a faster training time but perform worse than all GNN models. This shows that capturing spatial information in traffic data can help improve the accuracy of prediction significantly.
- STGCN needs the least memory allocation among five baselines, this shows Chebyshev Polynomials Approximation and/or 1st order Approximation better optimise the process of capturing spatial information than others.
- As shown in Table 3, MTGNN performs well on both single and multiple steps predictions with short inferencing time. As MTGNN utilizes a graph learning layer and mix-hop propagation, it can capture spatial dependencies and gain control of the information flow.
- All baselines perform well under short multi-step predictions, around three-time steps. The differences started from the fourth time step, where models like DCRNN, STGCN, and T-GCN begin having a higher prediction error.
- The result from the experiment also shows that Graph WaveNet has the longest average training time, despite the CNN-based structure for optimizing computation time.

Table 3. Summary of the performance on categories.

Category	Best	2nd best	Worst
Single step prediction	**MTGNN**	GWNET	TGCN
Multi-step prediction	**MTGNN**	GWNET	TGCN
Memory used	**STGCN**	GWNET	DCRNN
Training time	**MTGNN**	STGCN	GWNET
Inference time	**MTGNN**	GWNET	STGCN

Apart from traffic speed prediction, there are also different categories of traffic prediction, such as traffic flow prediction. Regarding future research, the benchmark of GNN models on traffic flow prediction can be conducted further to investigate the performance and behaviors of GNN frameworks [9]. As a concluding mark, we hope this research can deliver a meaningful and helpful guideline to the researchers in AI and data science community.

Acknowledgement. This work was supported by ARC Discovery Early Career Researcher Award (Grant No. DE200101465).

References

1. Bhanu, M., Priya, S., Dandapat, S.K., Chandra, J., Mendes-Moreira, J.: Forecasting traffic flow in big cities using modified tucker decomposition. In: Gan, G., Li, B., Li, X., Wang, S. (eds.) ADMA 2018. LNCS (LNAI), vol. 11323, pp. 119–128. Springer, Cham (2018). https://doi.org/10.1007/978-3-030-05090-0_10

2. Chung, J., Kastner, K., Dinh, L., Goel, K., Courville, A.C., Bengio, Y.: A recurrent latent variable model for sequential data. In: NIPS, vol. 28 (2015)

3. Duan, Y., Lv, Y., Wang, F.Y.: Performance evaluation of the deep learning approach for traffic flow prediction at different times. In: SOLI, pp. 223–227 (2016)

4. Ishak, S., Al-Deek, H.: Performance evaluation of short-term time-series traffic prediction model. J. Transp. Eng. **128**(6), 490–498 (2002)

5. Jiang, R., et al.: DL-TRAFF: survey and benchmark of deep learning models for urban traffic prediction. In: CIKM, pp. 4515–4525 (2021)

6. Klicpera, J., Weißenberger, S., Günnemann, S.: Diffusion improves graph learning. arXiv preprint arXiv:1911.05485 (2019)

7. Li, Y., Yu, R., Shahabi, C., Liu, Y.: Diffusion convolutional recurrent neural network: data-driven traffic forecasting. arXiv preprint arXiv:1707.01926 (2017)

8. Ma, M., Peng, B., Xiao, D., Ji, Y., Shi, C.: STCNet: spatial-temporal convolution network for traffic speed prediction. In: Yang, X., Wang, C.-D., Islam, M.S., Zhang, Z. (eds.) ADMA 2020. LNCS (LNAI), vol. 12447, pp. 315–323. Springer, Cham (2020). https://doi.org/10.1007/978-3-030-65390-3_24

9. Nguyen, T.T., et al.: Monitoring agriculture areas with satellite images and deep learning. Appl. Soft Comput. **95**, 106565 (2020)

10. Oord, A., et al.: WaveNet: a generative model for raw audio. arXiv preprint arXiv:1609.03499 (2016)

11. Ramchandra, N.R., Rajabhushanam, C.: Machine learning algorithms performance evaluation in traffic flow prediction. Mater. Today: Proc. **51**, 1046–1050 (2022)

12. Shleifer, S., McCreery, C., Chitters, V.: Incrementally improving graph wavenet performance on traffic prediction. arXiv preprint arXiv:1912.07390 (2019)

13. Vinayakumar, R., Soman, K., Poornachandran, P.: Applying deep learning approaches for network traffic prediction. In: ICACCI, pp. 2353–2358 (2017)

14. Wang, J., Jiang, J., Jiang, W., Li, C., Zhao, W.X.: Libcity: an open library for traffic prediction. In: SIGSPATIAL, pp. 145–148 (2021)

15. Wu, Z., Pan, S., Long, G., Jiang, J., Chang, X., Zhang, C.: Connecting the dots: multivariate time series forecasting with graph neural networks. In: KDD, pp. 753–763 (2020)

16. Wu, Z., Pan, S., Long, G., Jiang, J., Zhang, C.: Graph wavenet for deep spatial-temporal graph modeling. arXiv preprint arXiv:1906.00121 (2019)

17. Xu, X., Su, B., Zhao, X., Xu, Z., Sheng, Q.Z.: Effective traffic flow forecasting using taxi and weather data. In: Li, J., Li, X., Wang, S., Li, J., Sheng, Q.Z. (eds.) ADMA 2016. LNCS (LNAI), vol. 10086, pp. 507–519. Springer, Cham (2016). https://doi.org/10.1007/978-3-319-49586-6_35

18. Xu, X., Xu, Z., Zhao, X.: Traffic flow visualization using taxi GPS data. In: Li, J., Li, X., Wang, S., Li, J., Sheng, Q.Z. (eds.) ADMA 2016. LNCS (LNAI), vol. 10086, pp. 811–814. Springer, Cham (2016). https://doi.org/10.1007/978-3-319-49586-6_60

19. Yu, B., Yin, H., Zhu, Z.: Spatio-temporal graph convolutional networks: a deep learning framework for traffic forecasting. arXiv preprint arXiv:1709.04875 (2017)

20. Yu, S., Li, Y., Sheng, G., Lv, J.: Research on short-term traffic flow forecasting based on KNN and discrete event simulation. In: Li, J., Wang, S., Qin, S., Li, X., Wang, S. (eds.) ADMA 2019. LNCS (LNAI), vol. 11888, pp. 853–862. Springer, Cham (2019). https://doi.org/10.1007/978-3-030-35231-8_63
21. Zhao, L., et al.: T-GCN: a temporal graph convolutional network for traffic prediction. IEEE Trans. Intell. Transp. Syst. **21**(9), 3848–3858 (2019)

Multi-View Gated Graph Convolutional Network for Aspect-Level Sentiment Classification

Lijuan Wu, Guixian Zhang, Zhi Lei, Zhirong Huang, and Guangquan Lu[⊠]

Guangxi Key Lab of Multi-Source Information Mining and Security, Guangxi Normal
University, Guilin 541004, China
wulijuan@stu.gxnu.edu.cn, lugq@mailbox.gxnu.edu.cn,
{zgxcs,leiz,huangzr}@stu.gxnu.edu.cn

Abstract. Aspect-Level Sentiment Classification aims to identify the
sentiment polarity of each aspect in a sentence. Syntax-based graph
neural networks have been used to model dependencies between opinion words and aspects with good results. However, the analysis of these
works is highly dependent on the quality of the dependency graph and
may achieve suboptimal results for comments with ambiguous syntax.
We explore a novel Multi-View Gated Graph Convolutional Network
(MGGCN) to address the above problems. We utilize a Gated Graph
Convolutional Network (GateGCN) for a more reasonable interaction of
syntactic dependencies and semantic information, where we refine our
syntactic dependency graph by adding sentiment knowledge and aspect-
aware information to the dependency tree. We use the Inter-aspect Graph
Convolutional Network (InterGCN) to capture information about the
sentiment dependencies between multiple aspects that appear in a sentence. Finally, by adaptively learning multi-view sentiment information
through Simple Residual Multilayer Perceptron(SResMLP). Experimental results on four public datasets show that our proposed model outperforms state-of-the-art models.

Keywords: Aspect-level sentiment classification · Multi-view
learning · Gated graph convolution network

1 Introduction

Sentiment analysis is a fundamental task in natural language processing (NLP)
[10]. Its purpose is to mine sentiment information from user comment texts to
help companies or consumers provide valuable information and make reasonable
decisions. Aspect-Based Sentiment Analysis (ABSA) [25] is a fine-grained sentiment analysis task, in general, ABSA is divided into Aspect Extraction (AE)
[21] and Aspect-Level Sentiment Classification (ALSC). We focus only on the
ALSC task, which aims to identify the sentiment polarity (i.e., positive, negative, or neutral) of aspects that appear explicitly in a sentence. For example,

© The Author(s), under exclusive license to Springer Nature Switzerland AG 2022
W. Chen et al. (Eds.): ADMA 2022, LNAI 13725, pp. 489–504, 2022.
https://doi.org/10.1007/978-3-031-22064-7_35

"the ambiance was nice, but the food was dreadful !". In this sentence, the two aspects of the word "ambiance" and "food" have a positive and negative sentiment polarity respectively. Therefore, the main challenge of the ALSC task is to separate different opinion words for different aspects.

Previous work has proposed various attention mechanisms to extract aspect-related semantic information from the hidden states of contextual words with good results [7,16,30]. However, due to the complexity of the language, using the attention mechanism alone may not accurately capture the dependencies between aspects and corresponding opinion words. Take the sentence as an example, "So delicious was the vegetables but terrible noodles.", attention-based models usually take "terrible" as an opinion word for aspect "vegetables". Because "terrible" is closer to "vegetables", and "terrible vegetables" may also appear in other sentences, so that the attention mechanism will assign more weight to "terrible" in capturing the sentiment of "vegetables", which will limit the performance of the model. In recent years, graph neural networks (GNNs) in dependency trees have attracted extensive attention in ALSC. These works utilize syntactic dependency trees to shorten the distance between aspects and opinion words, alleviating the problem of long-term dependencies. As in the above example, the dependency tree can establish a path for the aspect "vegetables" and the opinion word "delicious" based on their syntactic dependencies, which solves the impact of long-distance dependency information on the ALSC task to a certain extent. However, GNNs-based models offer decent improvements over earlier neural network models, but they also have shortcomings. First, we observe that using only syntactic dependencies may have the opposite result for syntactically ambiguous sentence modeling. Second, when constructing dependency graphs, most only consider the syntactic dependencies of sentences. The edges of the dependency tree are edges with binary weights (i.e., 0 or 1), which cannot assign weights to words with different sentiments.

To address the above challenges, based on the idea that syntax is a complement to semantic information, the aim of both is to extract relationships between aspects and opinion words. We propose a GateGCN to project enhanced syntactic and semantic information into a subspace by a gated function without introducing additional parameters to achieve multi-view learning. We will construct a semantic graph using a multi-head self-attention [28] score matrix for the semantic. We argue that integrating external sentiment knowledge [3] into the dependency graph can provide the model with a more fine-grained sentiment signal and facilitate the model to extract relations between aspects and opinion words. In addition, to highlight the importance of aspects in the comments, we calculated the relative position of each word of the aspect to the context, which further enhances the relative dependencies of aspect words to the context. For multi-aspect sentences, the sentiment relations between aspects may provide context for aspect-specific sentiment polarity judgments [14,17]. For example, "I believe that the quality of a mac is worth the price. ", the positive sentiment polarity for "quality" is indirectly expressed through the "price" aspect, so we also try to connect the dependencies between different aspects as auxiliary information. Finally, we use SResMLP to adaptively perform an optimal fusion of different views so that its model learns the importance of different views.

The main contributions of our work can be summarized as follows:

- We propose that MGGCN networks incorporate multiple views of semantic, enhanced syntactic dependencies and inter-aspect dependencies information on the ALSC task.
- We redefine the syntactic dependency graph by adding sentiment knowledge and aspect-aware information to the dependencies.
- Our proposed method has extensively experimented on four publicly available datasets, and the experimental results validated the validity of the MGGCN model.

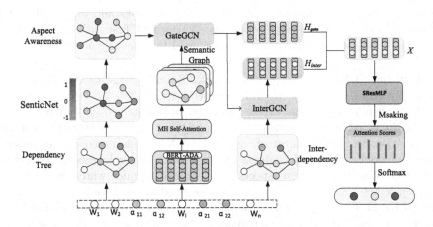

Fig. 1. A overall architecture for MGGCN, where domain-adaptive BERT-ADA is used as our sentence encoding, GateGCN is used to obtain enhanced syntactic dependencies and semantic information interactions. InterGCN is used to capture inter-aspect information, and multiple views of information are adaptively fused via SResMLP. The details of these components are described in the main text.

2   Related Work

In general, current ALSC methods can be divided into two main categories, namely semantic-based methods and syntax-based graph neural network methods.

Semantic-based Models. Earlier on aspect-based sentiment classification tasks, various attentional mechanisms [1] have been proposed to model the interaction between contextual words and aspects to predict the sentiment polarity of a particular aspect [5,7,16,23,30]. They can be considered as a way to exploit semantic information of sentences with good results. Among them, Wang et al. proposed attention based LSTM (ATAE-LSTM) [30], which concatenates aspects and sentences into the LSTM, generates aspect-aware sentence representations, and assigns weights to contexts through an attention mechanism. In [16], the aspect and context are separately input into the LSTM, and the context words are assigned values using the interactive attention mechanism. In [7], Fan et al. exploited the weighted summation of fine-grained attention mechanisms and coarse-grained attention mechanisms to capture word-level interactions between aspects and their contexts. Pre-training models such as BERT and RoBERTa are often used for sentiment analysis tasks [6,31]. Fine-tuned on BERT using a large general-purpose corpus, ease of use for most tasks and improved performance [31]. In [22], a two-stage process was performed, first with self-supervised in-domain fine-tuning and then with supervised task-specific fine-tuning.

Syntax-based Graph Neural Network. Recently, existing ALSC tasks are primarily based on Graph Neural Networks [2,9,26,33]. The graphs of these works are usually dependency trees generated for sentences using existing parsers and learning node representations from neighboring nodes through syntactic information. In [33],Zhang et al. building GCN models based on dependency trees to model aspects and contexts using syntactic dependencies can effectively reduce the distance between aspects and sentiment opinion words in a sentence. In [13], a heterogeneous sentence graph is constructed based on the interaction of the graph convolution of the focused aspects and the graph convolution between aspects. In [15], a heterogeneous graph neural network using syntactic relations, prior sentiment dictionary information and partial part-of-speech tagging information is proposed. In [26], Tang et al. builds syntactic and lexical graphs, designs two-layer interactive graph convolutions to fuse syntactic graphs and lexical graphs, and learns feature representations containing syntactic and semantic relations. In [29], Wang et al. introduces dependency type information, and prunes the dependency tree with the target aspect as the root node to construct a relational graph attention network. In [12], Liang et al. proposes to augment sentence dependency graphs with external sentiment knowledge according to specific aspects.

3   Proposed Approach

This section details the propose approach to model Multi-view Gated Graph Convolutional Network(MGGCN). The architecture is shown in Fig. 1. Our main idea is to use the information interaction from multiple views, such as syntactic features, semantic features, and inter-aspects features in a sentence, so as to better solve the problem of aspect level sentiment classification in the classification accuracy problem.

3.1 Embedding Module

Formally, we assume that there is a sentence $S = \{w_1, w_2, ..., w_i, w_{a_1}, w_{a_k}, ..., w_n\}$ consisting of n words, where w_i represents the i-th contextual word and w_{a_k} is the aspect term, where $w_{a_k} = \{a_{k1}, ..., a_{km}\}$ is a subsequence in S and a_{km} represents the m-th word of aspect k. We adopt the method mentioned in [22] to obtain domain-adapted BERT-ADA as our sentence encoder with "[CLS] sentence [SEP] aspect [SEP]" as input to get the aspect-aware hiddenness of sentences represents H.

3.2 Sentiment Knowledge Dependency Graph

For a given sentence S, we constructed a syntactic dependency tree for S using the spaCy toolkit[1]. Among them, each word in the sentence is regarded as a node, and the word-to-word dependency $r_{i,j}$ in the dependency tree is represented as an edge. Then, the adjacency matrix $D \in \mathbb{R}^{n \times n}$ of the sentence graph is derived as follows:

$$D_{i,j} = \begin{cases} 1 & \text{if} \quad i = j \text{ or } i \leftrightarrow j \\ 0 & \text{others} \end{cases} \tag{1}$$

where $i \leftrightarrow j$ indicates a dependency between w_i and w_j in the sentence.

We use the sentiment scores of words in SenteNet to enrich the adjacency matrix representation. Specifically, by querying the sentiment score of the word w_i in SenticNet is represented as SenticNet(w_i), where SenticNet$(w_i) \in [-1, 1]$. When the sentiment score of a query is close to -1, it indicates an extremely negative concept, the score close to 1 indicates an extremely positive concept, and the score of 0 indicates a neutral concept or the word does not exist in SenticeNet. Then, the corresponding dependency $r_{i,j}$ is extracted from the adjacency matrix. We will obtain the sentiment polarity value between the two dependent words by summing the sentiment scores of the parent node and the child node of the dependency tree:

$$S_{i,j} = \text{SenticNet}\,(w_i) + \text{SenticNet}\,(w_j). \tag{2}$$

In particular, aspect-specific words and local context information are also important in the ALSC task [13]. To highlight the importance of a specific aspect to the context in the dependency graph, we calculate the relative position weight of the specific aspect and get the other context words and the specific aspect-aware weight $R_{i,j}$:

$$R_{i,j} = \begin{cases} 1 & \text{if } w_i \in w_{a_i} \text{ and } w_j \in w_{a_i}, \\ 1/(|j - p^a| + 1) & \text{if } w_i \in w_{a_i}, \\ 1/(|i - p^a| + 1) & \text{if } w_j \in w_{a_i}, \\ 0 & \text{otherwise}, \end{cases} \tag{3}$$

---

[1] https://spacy.io/.

where, p^a is the starting position of a particular aspect, and $|\cdot|$ is the absolute value function. Then we construct a sentiment-enhanced syntactic graph, denoted as:

$$A_{sdep} = D_{i,j} \times (S_{i,j} + R_{i,j} + 1),\qquad(4)$$

we consider the dependencies between words to be mutual, adjusting each directed edge to an undirected edge.

3.3 Multi-Head Self-Attention Guided Semantic Graph

Some reviews sentences have ambiguous syntactic structures, which may have the opposite effect if the model only considers syntactically dependent information. Therefore, we will use the multi-head self-attention score to construct the semantic graph of the sentence, which can effectively capture the semantic correlation of each word in the sentence. It can be expressed as:

$$M_{se}^i = softmax\left(\frac{(HW^Q)\,(HW^K)^T}{\sqrt{d}}\right),\qquad(5)$$

$$M_{se} = \frac{\sum_{i=1}^{h}\left(M_{se}^1,\cdots,M_{se}^h\right)}{h},\qquad(6)$$

where $M_{se} \in \mathbb{R}^{n \times n}$, H is the word representation encoded by BERT-ADA, d is the dimensionality of the input node features. In addition, h is the number of heads, W^Q and W^k are trainable matrices.

3.4 Gated Graph Convolutional Network

Graph Convolutional Network use convolutional operations to encode local information for each node in the graph. We let the output of the l-th layer of node i be denoted by h_i^l, the l-th layer GCN is represented by H^l, that is, $H^l = \left[h_1^l, h_2^l, ..., h_n^l\right]$, and the operation of the graph convolution on the H^l representation are: $H^l = \sigma\left(AH^{l-1}W^l + b^l\right)$, σ is usually set to ReLU.

Gated Combination: We take the obtained A_{sdep} and M_{se} as input, and a gating mechanism to dynamically exchange the information after the two graph convolution operations, effectively combining the advantages of enhanced syntactic and semantic information. We consider both A_{sdep} and M_{se} through a gated combination I_{com} :

$$
\begin{aligned}
I_{\text{sdep}} &= A_{sdep}HW,\\
I_{\text{sem}} &= M_{se}HW,\\
\eta &= (1-\kappa)\,\text{sigmoid}\,(I_{\text{sem}}),\\
I_{com} &= (1-\eta)\odot I_{\text{sdep}} + \eta \odot I_{\text{sem}},\\
H_{gate}^l &= \text{GateGCN}\left(A_{sdep}, M_{sem}, H^{l-1}, W^l\right) = \sigma\left(I_{com}H^{l-1}W^l + b^l\right),
\end{aligned}
\qquad(7)
$$

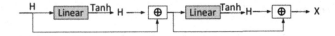

Fig. 2. Structure of SResMLP

where W and b are the convolution weight matrix and bias of the graph, respectively. η is a gating function, and κ [0,1] is a gating hyperparameter controlling the degree of mixing of the adjusted data through the gating coefficients. It is worth noting that we use the same graph convolution weights W for I_{sdep} and I_{sem} without introducing additional parameters, sharing latent variables to achieve multi-view learning. In fact, in the GateGCN module we use multiple GCN layers, for which the convolution parameters are different in the different GateGCN layers. Formally, the input of the l-th layer is H^{l-1}, the initialization input H^0 is a sentence representation H.

3.5 Inter-Aspect Dependent Graph Convolutional Network

For multiple aspects in a sentence, we will calculate the position weight of the specific aspect of the target and other aspects according to the relative distance between the aspects, $w = 1/\left(|p^a - p^o| + 1\right)$, where p^o is the starting position of another aspect, and short distance aspects will give more weight than far distance ones. On the basis of the adjacency matrix $D \in \mathbb{R}^{n \times n}$, we can obtain a representation of inter-aspect dependency information :

$$I = D + \frac{1}{k} \sum_{a \in a^o} (w * D_a),$$
(8)

where a^o represents the set of other aspects in a sentence, and k represents the number of aspects in this sentence. We use the inter-aspect dependency graph as auxiliary information and use the GCN to obtain the InterGCN representation H_{inter}. Among them, $H_{inter}^l = \sigma\left(IH^{l-1}W^l + b^l\right)$, the initial representation of H^0 is H_{gate}.

3.6 Feature Fusion Layer

We interact the output of the GateGCN and the output of the InterGCN, using the fusion coefficient ω to connect the two. However, to better reflect the importance of each part, we use the SResMLP layer to learn important weights while fusing multiple views adaptively:

$$X = \text{SResMLP}\left(H_{gate} + \omega H_{inter}\right).$$
(9)

SResMLP is inspired by [27], as shown in Fig. 2. We use two simple linear layers as projection and two residual operations to update the representation of multi-view information fusion.

3.7   Sentiment Classifier

We use the mask mechanism to mask the representation of the non-aspect vectors to highlight the important features of the aspect words to obtain X_{mask} = $\{0, ..., x_t, ..., x_{t+m-1}, ..., 0\}$, Where x_t is the aspect word representation with index t obtained by the feature fusion layer, and m is the length of the aspect word. Based on previous ASGCN [33] , we use the retrieval-based attention mechanism to capture important sentiment features from aspect-specific contextual representations:

$$\alpha_t = \sum_{i=t}^{t+m-1} \mathbf{h}_t \mathbf{x}_i \; ; \; \gamma_t = softmax\,(\alpha) \; ; \; f = \sum_{i=1}^{n} \gamma_i \mathbf{h}_i. \tag{10}$$

where $\mathbf{h}_t$ is the contextual representation of the t-th word obtained by the sentence encoder, α_t is the attention scores of the t-th context word with respect to the aspect, γ_t is normalized attention score. Finally, we linearly transform the aspect representation vector f, and use the softmax classifier to predict the probability of sentiment polarity $\mathbf{y} = softmax\,(W_o f + b_o)$, where W_o and b_o are the learned weights and biases.

3.8   Model Training

Our model uses the cross-entropy loss function as the objective function, and the training goal is to minimize the cross-entropy loss between true distribution $\hat{y}$ and the predicted distribution y:

$$\mathcal{L}\,(\Theta) = -\sum_{i=1}^{S}\sum_{j=1}^{C} \hat{y}_i^j \cdot \log\left(y_i^j\right) + \lambda\|\Theta\|_2, \tag{11}$$

where S is the number of training samples and C is the number of sentiment polarities. λ is the weight of the L_2 regularization term, and Θ denotes all trainable parameters.

Table 1. Dataset statistics

Domains	Positive		Neutral		Negative	
	Train	Test	Train	Test	Train	Test
LAP14	994	341	464	169	870	128
REST14	2164	728	637	196	807	196
REST15	1178	439	50	35	382	328
REST16	1620	597	88	38	709	190

4 Experiments

4.1 Dataset and Experiment Setting

We conduct experiments on four benchmark datasets, including the LAP14 and REST14 datasets from SemEval-2014 Task 4 [18], the REST15 dataset from SemEval2015 Task 12 [20], and the REST16 dataset from SemEval2016 Task 5 [19]. Each dataset includes reviews sentences, aspects, and their corresponding sentiment polarities. The statistics of each dataset are shown in Table 1. In our experiments, the number of GCN layers is set to 2 for the best performance, and the coefficient λ of the L_2 regularization term is set to 10^{-5}. The model is trained with Adam as the optimizer with a learning rate of 10^{-5} and the mini-batch size is 16.

Table 2. Main experimental results

Category	Method	REST14		LAP14		REST15		REST16	
		Acc.	F1.	Acc.	F1.	Acc.	F1.	Acc.	F1.
Sem.	IAN	79.26	70.09	72.05	67.38	78.54	52.65	84.74	55.21
	GCAE	75.74	62.45	71.98	68.71	77.56	56.03	83.7	62.29
	BERT	84.11	76.68	77.59	73.28	83.48	66.18	90.1	74.16
	BERT-PT	84.95	76.96	78.07	75.08	–	–	–	–
	BERT-ADA	87.14	80.09	80.23	75.77	–	–	–	–
Syn.	CDT	82.3	74.02	77.19	72.99	79.42	61.68	85.58	69.93
	ASGCN-DG	80.86	72.19	74.14	69.24	79.34	60.78	88.69	66.64
	RGAT- BERT	86.68	80.92	80.94	78.2	–	–	–	–
	DGEDT-BERT	86.3	80	79.8	75.6	84	71	91.9	79
	KumaGCN-BERT	86.43	80.3	81.98	78.81	86.35	70.76	92.53	79.24
	SenticGCN-BERT	86.92	81.03	82.12	79.05	85.32	71.28	91.97	79.56
	DualGCN-BERT	87.13	81.16	81.80	78.10	–	–	–	–
Ours	GateGCN	88.39	83.65	81.19	77.60	87.45	68.32	93.18	79.57
	MGGCN	**89.11**	**84.51**	**82.94**	**79.23**	**88.75**	**75.53**	**94.48**	**85.12**

4.2 Comparison Models

Our proposed model is compared with the following methods and the relevant baselines can in principle be divided into two categories.

Semantic-based Models:

- IAN [16] uses two LSTMs and an interactive attention mechanism to learn aspects and feature representations of sentences.

- GCAE [32] proposes a model based on CNN and gating mechanism capable of selectively outputting features based on a given aspect.
- BERT [6] is the ordinary BERT model that takes the sentence-aspect pair as input and uses a representation of $[CLS]$ for prediction.
- BERT-PT [31] is a bert joint post-training method to enhance domain knowledge explored aspect extraction and aspect sentiment classification.
- BERT-ADA [22] is a domain-adaptive pretrained model based on the Amazon laptop reviews dataset [8] and the Yelp Dataset Challenge review corpus[2].

Grammar-Based Model:

- CDT [24] utilizes Bi-LSTM to learn sentence representation, which is improved for learning aspect feature representation.
- ASGCN-DG [33] builds a graph convolutional network that exploits syntactic dependencies on sentence dependency trees.
- SenticGCN [12] proposes a graph convolutional neural network for enhancing syntactic dependency graphs using sentiment knowledge.
- RGAT-BERT [2] proposes a new relational graph attention network that makes full use of dependent label information.
- DGEDT-BERT [26] is based on a dual-transformer network, which fuses the planar representation learned by the transformer and the information learned from the dependency graph.
- KumaGCN-BERT [4] uses the HardKuma distribution to sample sentences and generate a specific latent graph structure. The latent graph and the dependency tree are combined with a gating mechanism
- DualGCN-BERT [11] uses two regularizations to aggregate semantic and syntactic information.

4.3   Main Results

Table 2 shows the results of our proposed MGGCN model compared to the baseline model.

Compared to semantic-based approaches such as IAN and GCAE, our results show a significant improvement. This suggests that our model, which combines complementary augmented syntactic and semantic information and inter-aspect dependency information, is more suitable for the ALSC task than using attention alone to model sentences. Compared to syntax-based approaches, such as CDT, BiGCN, our proposed GateGCN effectively exploits the semantic information of the sentence with the syntactic information of sentiment enhancement through a gating mechanism to achieve competitive results (no inter-aspect view information is used).

In addition, information between aspects can better assist the sentiment analysis of specific aspects for the unclear sentiment expression of aspects in the data. The results show that the MGGCN network has improved accuracy and

---

[2] https://www.yelp.com/dataset/challenge.

F1-scores compared to the GateGCN network on all datasets. It is worth noting that the domain-adapted BERT-ADA we used compared with other BERT-based models, all datasets have some improvement, which shows that using the data in the domain is very necessary for sentiment analysis tasks.

4.4 Ablation Study

The results are shown in Table 3. First, our model without SenticNet sentiment knowledge-enhanced dependencies (i.e., MGGCN w/o sentic) has unsatisfactory performance on all datasets, particularly on the LAP14 dataset, where the model accuracy dropped by 2.22%. This shows that dependency trees augmented with sentiment knowledge can facilitate models to exploit sentiment dependency information between opinion words and aspects to predict aspect-specific sentiment. After removing semantic information (i.e., MGGCN w/o H_{sem}), the model accuracy decreased by 2.32%, 2.06%, 1.11% and 1.3% on REST14, LAP14, REST15 and REST15 respectively, indicating that the multi-head self-attention mechanism can effectively capture semantic relevant information between specific-aspect and opinion words. Removing the dependencies between aspects (i.e., MGGCN w/o H_{inter}) also leads to a decrease in model performance, further suggesting that extracting the dependencies between aspects will improve model performance. Removing the fusion module also leads to a drop in experimental accuracy, which effectively shows that the SResMLP module can adaptively learn the emotional features that are most relevant to aspect words from multiple views.

Table 3. Ablation experiment results.

Model	REST14		LAP14		REST15		REST16	
	Acc.	F1.	Acc.	F1.	Acc.	F1.	Acc.	F1.
MGGCN w/o sentic	87.86	80.87	80.72	77.64	86.72	72.08	93.99	84.53
MGGCN w/o H_{sem}	86.79	81.54	80.88	76.84	87.64	66.58	93.18	77.76
MGGCN w/o H_{inter}	88.39	83.65	81.19	77.6	87.45	68.32	93.18	79.57
MGGCN w/o SRestMLP	87.23	81.34	81.25	77.55	88.56	74.75	93.34	81.26
MGGCN	**89.11**	**84.51**	**82.94**	**79.23**	**88.75**	**75.53**	**94.48**	**85.12**

4.5 Analysis and Discussion

Effect of Hyper-Parameter Head Number. To further analyze the influence of the number of multi-head self-attention heads on the performance of the model, we conducted experiments based on different values with REST15 and REST16 as examples, and the results are shown in Fig. 3(a). We can observe that the model performs best when the number of heads is 16. Intuitively, when the number of heads is too tiny, multi-head self-attention to extract semantic information will make the model random. On the other hand, when the number of heads increases by more than 16, the curve has a clear downward trend, which indicates that the model becomes overfitting.

Effect of Gating Coefficient. We further analyze the impact of the trade-off between sentiment-enhancing syntactic information and semantic information on model performance. We experiment by varying the value of κ in Eq. 7 from 0 to 1. When the value of κ is 0, GateGCN will learn syntactic information enhanced by sentiment knowledge. When the value of κ is 1, we can assume that GateGCN relies on the knowledge learned in the multi-head self-attention semantic graph. As shown in Fig. 3(b), in the REST16 dataset, the best performance was achieved when the value of κ was 0.3. In the REST15 data, a κ value of 0.4 works best. We also experimented on the REST14 and LAP14 datasets and found that REST14 performed best when κ was 0.8, and it performed best on the LAP14 dataset when κ was 0.3. This indicates that sentiment knowledge enhances syntactic information and the semantic information learned by multi-head self-attention is complementary.

Effect of Fusion Coefficient. We use the REST14 and LAP14 datasets as examples, and the results are shown in Fig. 3(c). In the REST14 dataset, as the value of ω increases from 0 to 0.3, the performance is steadily improved, but when the value of ω is greater than 0.3, the curve has a clear downward trend. In the LAP14 dataset, performance is best when ω is 0.4, implying that proper fusion of information between aspects can aid aspect-specific sentiment analysis. However, when ω is greater than 0.4, the curve has a significant downward trend. So it suggests that over-fusing inter-aspect information may hinder the representation of aspect-specific sentiment features.

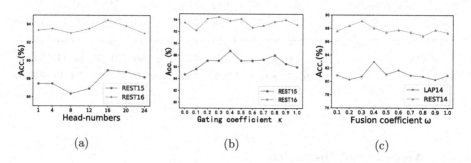

Fig. 3. The effect of hyperparameters on different datasets.

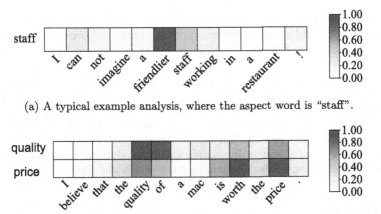

(a) A typical example analysis, where the aspect word is "staff".

(b) There are examples of multiple aspects in the analysis of the reviews, of which the aspect items are "quality" and "price".

Fig. 4. Attention score visualisation

4.6 Case Study

To better understand the performance of our proposed model, we show typical examples of attentional weighting visualizations. As shown in Fig. 4(a), in our proposed MGGCN, the model will consider the positive sentiment knowledgen of SenticNet(friendlier)=0.887 into the judgment of the model, and the emotional word "friendlier" has the highest attention weight. It can improve the attention scores of sentiment words while adaptively reducing the attention scores of other context-free words. Figure 4(b) shows the visualization of attentional weights for multiple aspects of the reviews. It can be observed that MGGCN uses the sentiment information of the related aspect "price" as an auxiliary knowledge connection when dealing with aspects "quality". In the case study, the effectiveness of MGGCN can be verified between multiple views from sentiment-enhanced syntactic information, semantic information, and inter-aspect dependency information.

5 Conclusion

This paper proposes a multi-view gated graph convolutional MGGCN approach. The new gating mechanism is dynamically complemented by enhanced sentiment syntactic and semantic information. We refine the syntactic dependency graph by adding sentiment knowledge and aspect-aware information to the dependency tree. In addition, for multiple-aspect reviews, the model uses the inter-aspect sentiment information as auxiliary knowledge for aspects whose sentiment expression is not clear. Finally, the sentiment features with the highest correlation with aspect words are adaptively learned from multiple views. Experimental results on four benchmark datasets show that the proposed method outperforms the baseline methods. In the future, we plan to study other approaches that fully exploit syntactic and semantic information and apply it to end-to-end sentiment analysis tasks, such as Aspect Sentiment Triplet Extraction (ASTE) task.

Acknowledgements. The work is supported partly by National Natural Science Foundation of China (No. 62166003), the Innovation Project of Guangxi Graduate Education (YCSW-2022124),the Project of Guangxi Science and Technology (GuiKeAD20159041), Intelligent Processing and the Research Fund of Guangxi Key Lab of Multi-source Information Mining and Security (No.20–A–01–01, MIMS20–M–01), the Guangxi Collaborative Innovation Center of Multi-Source Information Integration and the Guangxi "Bagui" Teams for Innovation and Research, China.

References

1. Bahdanau, D., Cho, K., Bengio, Y.: Neural machine translation by jointly learning to align and translate. arXiv preprint arXiv:1409.0473 (2014)
2. Bai, X., Liu, P., Zhang, Y.: Investigating typed syntactic dependencies for targeted sentiment classification using graph attention neural network. IEEE/ACM Trans. Audio Speech Lang. Process. **29**, 503–514 (2020)
3. Cambria, E., Poria, S., Hazarika, D., Kwok, K.: Senticnet 5: discovering conceptual primitives for sentiment analysis by means of context embeddings. In: Proceedings of the AAAI Conference on Artificial Intelligence, vol. 32 (2018)
4. Chen, C., Teng, Z., Zhang, Y.: Inducing target-specific latent structures for aspect sentiment classification. In: Proceedings of the 2020 Conference on Empirical Methods in Natural Language Processing (EMNLP), pp. 5596–5607 (2020)
5. Chen, P., Sun, Z., Bing, L., Yang, W.: Recurrent attention network on memory for aspect sentiment analysis. In: Proceedings of the 2017 Conference on Empirical Methods in Natural Language Processing, pp. 452–461 (2017)
6. Devlin, J., Chang, M.W., Lee, K., Toutanova, K.: BERT: pre-training of deep bidirectional transformers for language understanding. arXiv preprint arXiv:1810.04805 (2018)
7. Fan, F., Feng, Y., Zhao, D.: Multi-grained attention network for aspect-level sentiment classification. In: Proceedings of the 2018 Conference on Empirical Methods In Natural Language Processing, pp. 3433–3442 (2018)
8. He, R., McAuley, J.: Ups and downs: modeling the visual evolution of fashion trends with one-class collaborative filtering. In: proceedings of the 25th International Conference on World Wide Web, pp. 507–517 (2016)
9. Kipf, T.N., Welling, M.: Semi-supervised classification with graph convolutional networks. arXiv preprint arXiv:1609.02907 (2016)
10. Li, J., Hovy, E.: Reflections on sentiment/opinion analysis. In: Cambria, E., Das, D., Bandyopadhyay, S., Feraco, A. (eds) A Practical Guide to Sentiment Analysis. Socio-Affective Computing, vol 5. Springer, Cham (2017). https://doi.org/10.1007/978-3-319-55394-8_3
11. Li, R., Chen, H., Feng, F., Ma, Z., Wang, X., Hovy, E.: Dual graph convolutional networks for aspect-based sentiment analysis. In: Proceedings of the 59th Annual Meeting of the Association for Computational Linguistics and the 11th International Joint Conference on Natural Language Processing (vol. 1: Long Papers), pp. 6319–6329 (2021)

12. Liang, B., Su, H., Gui, L., Cambria, E., Xu, R.: Aspect-based sentiment analysis via affective knowledge enhanced graph convolutional networks. Knowl.-Based Syst. **235**, 107643 (2022)
13. Liang, B., Yin, R., Gui, L., Du, J., Xu, R.: Jointly learning aspect-focused and inter-aspect relations with graph convolutional networks for aspect sentiment analysis. In: Proceedings of the 28th International Conference on Computational Linguistics, pp. 150–161 (2020)
14. Lin, P., Yang, M., Lai, J.: Deep selective memory network with selective attention and inter-aspect modeling for aspect level sentiment classification. IEEE/ACM Trans. Audio Speech Lang. Process. **29**, 1093–1106 (2021)
15. Lu, G., Li, J., Wei, J.: Aspect sentiment analysis with heterogeneous graph neural networks. Inf. Process. Manage. **59**(4), 102953 (2022)
16. Ma, D., Li, S., Zhang, X., Wang, H.: Interactive attention networks for aspect-level sentiment classification. arXiv preprint arXiv:1709.00893 (2017)
17. Majumder, N., Poria, S., Gelbukh, A., Akhtar, M.S., Cambria, E., Ekbal, A.: IARM: inter-aspect relation modeling with memory networks in aspect-based sentiment analysis. In: Proceedings of the 2018 Conference on Empirical Methods In Natural Language Processing, pp. 3402–3411 (2018)
18. Pontiki, M., Galanis, D., Pavlopoulos, J., Papageorgiou, H., Manandhar, S.: Semeval-2014 task 4: aspect based sentiment analysis. In: Proceedings of International Workshop on Semantic Evaluation (2014)
19. Pontiki, M., et al.: Semeval-2016 task 5: aspect based sentiment analysis. In: International Workshop on Semantic Evaluation, pp. 19–30 (2016)
20. Pontiki, M., Galanis, D., Papageorgiou, H., Manandhar, S., Androutsopoulos, I.: Semeval-2015 task 12: aspect based sentiment analysis. In: Proceedings of the 9th International Workshop on Semantic Evaluation (SemEval 2015), pp. 486–495 (2015)
21. Poria, S., Cambria, E., Gelbukh, A.: Aspect extraction for opinion mining with a deep convolutional neural network. Knowl.-Based Syst. **108**, 42–49 (2016)
22. Rietzler, A., Stabinger, S., Opitz, P., Engl, S.: Adapt or get left behind: domain adaptation through bert language model finetuning for aspect-target sentiment classification. arXiv preprint arXiv:1908.11860 (2019)
23. Song, Y., Wang, J., Jiang, T., Liu, Z., Rao, Y.: Attentional encoder network for targeted sentiment classification. arXiv preprint arXiv:1902.09314 (2019)
24. Sun, K., Zhang, R., Mensah, S., Mao, Y., Liu, X.: Aspect-level sentiment analysis via convolution over dependency tree. In: Proceedings of the 2019 Conference on Empirical Methods in Natural Language Processing and the 9th International Joint Conference on Natural Language Processing (EMNLP-IJCNLP), pp. 5679–5688 (2019)
25. Tang, D., et al.: Learning sentiment-specific word embedding for twitter sentiment classification. In: ACL (1), pp. 1555–1565. CiteSeer (2014)
26. Tang, H., Ji, D., Li, C., Zhou, Q.: Dependency graph enhanced dual-transformer structure for aspect-based sentiment classification. In: Proceedings of the 58th Annual Meeting of the Association for Computational Linguistics, pp. 6578–6588 (2020)
27. Touvron, H., et al.: ResMLP: feedforward networks for image classification with data-efficient training. arXiv preprint arXiv:2105.03404 (2021)
28. Vaswani, A., et al.: Attention is all you need. In: Advances in Neural Information Processing Systems 30 (2017)
29. Wang, K., Shen, W., Yang, Y., Quan, X., Wang, R.: Relational graph attention network for aspect-based sentiment analysis. arXiv preprint arXiv:2004.12362 (2020)

30. Wang, Y., Huang, M., Zhu, X., Zhao, L.: Attention-based LSTM for aspect-level sentiment classification. In: Proceedings of the 2016 Conference on Empirical Methods in Natural Language Processing, pp. 606–615 (2016)
31. Xu, H., Liu, B., Shu, L., Yu, P.S.: Bert post-training for review reading comprehension and aspect-based sentiment analysis. arXiv preprint arXiv:1904.02232 (2019)
32. Xue, W., Li, T.: Aspect based sentiment analysis with gated convolutional networks. arXiv preprint arXiv:1805.07043 (2018)
33. Zhang, C., Li, Q., Song, D.: Aspect-based sentiment classification with aspect-specific graph convolutional networks. arXiv preprint arXiv:1909.03477 (2019)

Decentralized Graph Processing
for Reachability Queries

Joël Mathys[(✉)] [iD], Robin Fritsch, and Roger Wattenhofer [iD]

ETH Zürich, Zürich, Switzerland
{jmathys,rfritsch,wattenhofer}@ethz.ch

Abstract. Answering queries on large graphs is an essential part of data processing. In this paper, we focus on determining reachability between vertices. We propose a labeling scheme which is inherently distributed and can be processed in parallel. We study what properties make it difficult to find a good reachability labeling scheme for directed graphs. We focus on the genus of a graph. For graphs of bounded genus g, we design a labeling scheme of length $\mathcal{O}(g \log n + \log^2 n)$. We also prove that no labeling schemes with labels shorter than $\Omega(\sqrt{g})$ exist for this graph class.

Keywords: Reachability query · Labeling scheme · Data-mining-ready structures and pre-processing

1 Introduction

Databases and web services employ some variant of precomputation in order to answer queries reasonably quickly. Several applications, such as XML parsing, logical reasoning on semantic web RDF/OWL data, querying protein-protein interaction networks, lineage tracking for scientific workflows or routing work directly on graph data [10,18,28,30]. Determining the relationship between nodes in such graphs is a fundamental operation to efficiently answer queries. In this paper, we study the classical problem of determining reachability in graphs: Given a directed graph G, we want to answer whether a node u can reach a node v, for arbitrary u, v. Furthermore, it might be to inefficient to answer such reachability queries by running a naive linear time algorithm. If the graph is large, we can also not store all possible answers. Instead, we propose to construct a reachability oracle in the form of a labeling scheme: We store some additional information (a label) with each node. When we want to answer a reachability query for nodes u, v, we simply look up the labels of nodes u and v, and deduce the reachability just from these labels. Such an approach can also deal with graphs that are stored in a distributed fashion, since the labels of u and v do not need to be stored on the same machine. Furthermore, the labeling scheme is highly parallelizable as we only require read access to the labels of both nodes.

In the literature, such a distributed oracle is known as an informative labeling scheme [23]. A labeling scheme consists of an encoder l and a decoder d. The

© The Author(s), under exclusive license to Springer Nature Switzerland AG 2022
W. Chen et al. (Eds.): ADMA 2022, LNAI 13725, pp. 505–519, 2022.
https://doi.org/10.1007/978-3-031-22064-7_36

encoder l assigns a *label* (a bitstring) $l(u)$ to each vertex u of the graph. The decoder d takes the labels $l(u)$ and $l(v)$ as input and then directly decides if u can reach v in G, without using any other information.

The main objective when constructing a labeling scheme is to minimize the maximum label length assigned to any vertex. But how difficult is this task? For which graphs do we get a reasonable label length? Or more precisely, can we leverage the intrinsic properties of the given data to model graphs for which we can create short labeling schemes?

A straightforward idea is to use the sparsity and locality in the given data. The graph representation often only contains a small fraction of all possible connections. A first attempt is to consider that each element in the data is only directly related to a few other elements. Such an assumption translates naturally to the graph domain as they can be modeled with bounded degree graphs. However, such a local restriction is not strong enough. We show that even for sparse graphs that have a bounded outdegree Δ it is not possible to construct short schemes. We prove a lower bound of $\Omega(\sqrt{n\Delta})$, so already graphs with outdegree 2 require labels of length at least $\Omega(\sqrt{n})$.

In search for a more suitable characterization of the graphs, we consider a more global parameter, the genus. Graphs of bounded genus are a natural generalization of planar graphs, since planar graphs can be characterized as graphs of genus 0. For planar graphs, in a seminal paper, Thorup [27] constructed a labeling scheme of length $\mathcal{O}(\log^2 n)$. In the same paper, he posed as an open question whether this approach could be extended to graph classes excluding minors and bounded genus. Building upon Thorup's work, as well as on the work of Gilbert et al. [17] on separator sets in bounded genus graphs and the work of Kawarabayashi et al. [21] on labeling schemes for distances, we construct a new labeling scheme of length $\mathcal{O}(g \log n + \log^2 n)$. Furthermore, we provide a lower bound that applies to any labeling scheme designed for the class of bounded genus graphs of $\Omega(\sqrt{g})$, even if the genus is subquadratic in the size of the graph. This means that the genus of a graph is a possible indicator on the complexity of devising short labelings.

2   Related Work

Labeling schemes have been around for more than 50 years, making an appearance in the context of information theory by Breuer [9]. Ever since, labeling schemes have been studied for several different types of queries. These include, but are not limited to, labeling schemes for adjacency [2,6,8,20], lowest common ancestor [5,23] and distance [1,16,21,27]. All known to us online maps use distance labeling to compute shortest paths.

Labeling schemes relating to reachability in directed graphs, sometimes also referred to as ancestry queries, have been studied extensively on trees as they can be used for improving the performance of XML search engines [7,15]. The problem was first introduced by Kannan et al. [20], who presented a $\mathcal{O}(\log n)$ labeling scheme for reachability on trees. Schemes were improved [7,13,15] to match the lower bound of $\log n + \Omega(\log \log n)$ derived by Alstrup et al. [4]. There have also been publications on constructing labeling schemes for reachability for the class of planar graphs as they are very useful to answer routing queries in online maps. Working on distance oracles for approximate distances, Thorup [27] constructed a labeling scheme for reachability for directed planar graphs of length $\mathcal{O}(\log^2 n)$. His work also applied to labeling schemes of approximate distances for undirected planar graphs. Kawarabayashi et al. [21] then generalized the labeling scheme of Thorup for approximate distances in undirected graphs to bounded genus graphs. It uses a tree-cotree decomposition introduced by Eppstein [14] to reduce the problem to the planar case. More recently, there has been work on labeling schemes for other classes of directed graph, such as treewidth [19] or compressing the transitive closure [8]. But the relevance of reachability goes beyond trees and planar graphs. For instance, reachability labels are used for efficient lineage tracking [18].

There is an extensive list of publications, which focus on building implemented systems to answer reachability queries. Several approaches build on the ideas of chain covers [25], tree covers [3], hop labeling [11] or a combination of these techniques [10,28]. On the other hand, some of these systems allow for additional computation on the graph at querying time. These approaches are usually based on interval labeling [24,30], sampling linear extensions [12] or set containment [29]. Whenever the information given by the precomputation is not sufficient, they fall back to a linear search to answer the query. We refer to [29] for an in-depth coverage of existing work.

This paper analyses what parameters and restrictions are required for graph classes in order to construct short and efficient reachability schemes. These insights are important to construct new schemes and can further be used to estimate theoretical limitations of practical systems.

Table 1. An overview of the asymptotic length of reachability labeling schemes for different graph classes. The bounds printed in bold are shown in this paper. (The $\Omega(\log n)$ lower bounds are straightforward, since a labeling scheme for paths requires at least $\Omega(\log n)$ bits.)

Labeling schemes for reachability		
Graph class	Lower bound	Upper bound
General graphs	$\Omega(n)$ [8]	$\mathcal{O}(n)$ [8]
Trees	$\Omega(\log n)$	$\mathcal{O}(\log n)$ [13,15]
Outdegree $\Delta \geq 2$	$\boldsymbol{\Omega(\sqrt{n\Delta})}$	$\mathcal{O}(n)$ [8]
Planar	$\Omega(\log n)$	$\mathcal{O}(\log^2 n)$ [27]
Genus g	$\boldsymbol{\Omega(\sqrt{g})}$	$\boldsymbol{\mathcal{O}(g\log n + \log^2 n)}$

Table 1 shows known lower and upper bounds on the length of reachability labeling schemes for a number of graph classes. The bounds we prove in this paper are printed in bold.

In this paper, we use the ideas introduced by Kawarabayashi et al. [21] to construct a labeling scheme for reachability in directed graphs of bounded genus. However, instead of relying on the decomposition of Eppstein [14], we adapt the work of Gilbert et al. [17] on balanced vertex separators in bounded genus graphs. Gavoille et al. [16] proved that labeling schemes for exact distances in undirected graphs of bounded degree must be of length at least $\Omega(\sqrt{n})$. In our work, we show that this bound can be adapted to the case of reachability in directed graphs of bounded degree.

3   Graphs of Bounded Degree

When processing data represented as graphs, the data usually follows a given structure. Meaning that the number of edges is not quadratic, but roughly linear in the number of vertices. Therefore, we deal with very sparse graphs. However, is it enough to build a short labeling scheme by considering a sparsity constraint? We study a natural way to enforce sparsity by locally enforcing a maximum degree per vertex.

While restricting the degree of vertices of the graph seems like an appropriate choice to limit a graph's complexity, it turns out not to be the right parameter to explain the difficulty of constructing short labeling schemes for reachability. In the following, we prove a lower bound of $\Omega(\sqrt{n\Delta})$ on the length of a reachability labeling scheme for graphs with degree $\Delta \geq 2$. In order to achieve this, we introduce a transformation Ψ_Δ.

Note that if the outdegree is restricted to be 1, we can apply the labeling scheme for trees. Recall that for reachability, we always work with directed graphs. For a vertex v we denote its outdegree with $\deg^+(v)$ and its indegree with $\deg^-(v)$. Note that it does not matter whether we restrict either the in- or outdegree whenever we are interested in a labeling scheme for reachability. In fact, they are interchangeable. We can reverse the directions of all edges of a digraph G to swap in- and outdegree and obtain G'. We then have that $u \rightsquigarrow_G v$ (u reaches v in G) if and only if $v \rightsquigarrow_{G'} u$. Therefore, we usually only restrict the outdegree of a graph.

3.1   Degree Graph Transformation

The idea behind the transformation is that any digraph can be transformed into another digraph with smaller outdegree, while preserving the reachability relation. We replace each original vertex of the digraph with multiple vertices of smaller outdegree linked together by a chain. More formally, the transformation Ψ_Δ takes an arbitrary digraph G and transforms it into a digraph G' with maximum outdegree Δ.

Note that Gavoille et al. [16] used a similar technique to derive a $\Omega(\sqrt{n})$ lower bound for determining exact distances using labeling schemes in undirected bounded degree graphs. The lower bound technique and in particular the Ψ_Δ transformation can be applied to directed graphs and were found independently.

Recall that for general digraphs, there is a lower bound linear in the size of the graph.

Theorem 1 (Digraph Lower Bound [8]). *Let $\mathcal{G}$ be the class of digraphs on n vertices. Every labeling scheme for reachability for the class $\mathcal{G}$ has length at least $L = \frac{n}{4} = \Omega(n)$.*

Lemma 1. *Let G be an arbitrary digraph on n vertices with maximum outdegree ω. The transformation Ψ_Δ constructs a digraph $G' = \Psi_\Delta(G)$ with the following properties.*

1. *(Bounded Degree) G' has maximum outdegree $\Delta \geq 2$.*
2. *(Additional Vertices) G' has at most $n\lceil \frac{\omega}{\Delta-1}\rceil$ vertices.*
3. *(Reachability) $V(G) \subseteq V(G')$, furthermore for any $u, v \in V(G)$: u can reach v in G if and only if u can reach v in G'.*

Proof. We only outline the construction of the transformation here and refer to the full version of a paper for a more detailed construction and formal proof of the properties.

The transformation applies to each vertex independently. More specifically, in G' each vertex v of G is replaced by a chain of new vertices. Among these new vertices the outgoing edges are distributed in a way that ensures that all vertices in G' have outdegree at most Δ.

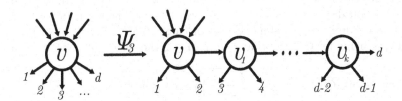

Fig. 1. The Ψ_Δ transformation applied to a vertex v of outdegree d with $\Delta = 3$. On the right side, $v = v_0$ and its virtual vertices $v_1, ..., v_k$ are displayed.

Note that the shape with which we replace each vertex with could be changed to resemble a tree instead of a chain. However, it is more convenient to have it be a simple path. Furthermore, we could also transform the indegree of the graph using the same technique. Finally, we need to relate the Ψ_Δ transformation to the labeling scheme (Fig. 1).

3.2   Bounded Degree Lower Bound

We can now use the transformation Ψ_Δ to get a lower bound for bounded degree graphs. The main idea behind the lower bound is that if there were a shorter labeling scheme, we could use it to construct a shorter scheme for general digraphs (Fig. 2).

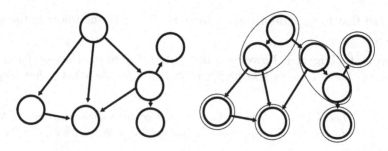

Fig. 2. On the left side a graph G that contains nodes with degree larger than Δ. On the right side the Ψ_Δ transformed graph G' with maximum outdegree $\Delta = 2$. The reachability information of G is preserved in G'.

Theorem 2 (Bounded Degree Lower Bound). *Let $\mathcal{G}_\Delta$ be the class of digraphs on n vertices with outdegree $\Delta \geq 2$. Every labeling scheme for reachability for the class $\mathcal{G}_\Delta$ has length at least $L = \Omega(\sqrt{n\Delta})$.*

Proof. Assume for the sake of contradiction that there exists a labeling scheme for reachability for the class $\mathcal{G}_\Delta$ and which has length $L = o(\sqrt{n\Delta})$. We now apply this scheme to $\mathcal{G}$, the class of digraphs on n vertices. Now take any digraph $G \in \mathcal{G}$ and Ψ_Δ transform G to the graph $G' \in \mathcal{G}'$. Note that due to the reachability property of the transformation any labeling for $G' = \Psi_\Delta(G)$ is also a valid labeling for G. Furthermore, G' now contains at most $n \cdot \lceil \frac{n}{\Delta-1} \rceil$ vertices and every vertex has outdegree at most Δ by the properties of the Ψ_Δ transformation.

We now apply the labeling scheme of length $o(\sqrt{n\Delta})$ to $\mathcal{G}'$, which also is a labeling scheme for $\mathcal{G}$. Note that due to $\Delta \geq 2$ the term $\lceil \frac{n}{\Delta-1} \rceil$ is at most $\frac{2n}{\Delta}$.

$$L = o\left(\sqrt{\Delta|V(G')|}\right) \leq o\left(\sqrt{\Delta n \cdot \left\lceil \frac{n}{\Delta-1} \right\rceil}\right) \leq o\left(\sqrt{\Delta n \cdot \frac{2n}{\Delta}}\right) = o\left(\sqrt{n^2}\right) = o(n) \quad (1)$$

However, we already know by Theorem 1 that any labeling scheme for $\mathcal{G}$ must use at least $\Omega(n)$ bits, a contradiction. Therefore, any labeling scheme for reachability for the class $\mathcal{G}_\Delta$ has length at least $L = \Omega(\sqrt{n\Delta})$.

4   Graphs of Bounded Genus

After studying the maximum degree of a graph, we now turn to another parameter: the genus of a graph. A graph of genus g can be embedded in a surface of genus g so that none of its edges are crossing. Graphs of bounded genus are a natural extension of planar graphs, since the latter are exactly the graphs of genus 0. We already know that low genus graphs, such as planar graphs, have very short labeling schemes. On the other hand, general directed graphs with many edges have a lot of crossings. In the following, we show that the genus of a graph is a useful indicator of how difficult it is to construct a short labeling.

Building upon the previous work of Thorup [27] on labeling schemes in planar graphs, the work of Gilbert et al. [17] on separator sets in bounded genus graphs and the work of Kawarabayashi et al. [21] on labeling schemes for distances, we construct a new labeling scheme of length $\mathcal{O}(g \log n + \log^2 n)$. Furthermore, we provide a lower bound on the length of any labeling scheme of $\Omega(\sqrt{g})$ for the class of bounded genus graphs. Moreover, the provided proof extends to graphs whose genus is linear in the size of the graph.

First, we recall the layering technique introduced by Thorup, as our construction heavily relies on it. We proceed with a slight variation of the separator proof of Gilbert et al. to reduce a given graph of bounded genus to a planar graph. Then we combine the tools to construct a new labeling scheme for bounded genus graphs. Before we begin, we briefly discuss a frequently used method (e.g. in [27]) that allows us to remove directed paths from the graphs.

4.1 Preliminaries

In the following, we describe two methods that we use as building blocks in our construction. The first method is used to handle a directed path p. It allows us to reduce the construction of the labeling to the graph without p by storing an additional $\mathcal{O}(\log n)$ bits per vertex.

Lemma 2 (Path Labeling [27]). *Let G be a digraph on n vertices, P a set of t directed paths in G and a labeling l' for the graph $G \setminus V(P)$. By storing $\mathcal{O}(t \log n)$ additional bits per vertex, we can extend the labeling l' to a labeling l for G.*

We leave the rigorous proof itself for the full version. The idea is that each vertex stores the first vertex it can reach on a path p and the last vertex on p that can reach it. Given two vertices we can then decide if there is a path between the two vertices that uses some vertex in p.

The second method, introduced by Thorup [27], reduces the problem of constructing a reachability scheme for the whole graph to a sequence of smaller graphs that are more local. By more local, we mean that it is possible that in the original graph, many parts of the graph will never interact with each other as there are no directed paths between them.

Definition 1 (t-layered Spanning Tree [27]). *A t-layered spanning tree T in a digraph G is a disoriented rooted spanning tree such that any path in T starting from the root is the concatenation of at most t directed paths in G.*

Lemma 3 (Digraph Layering [27]). *Given an arbitrary digraph G, there exists a series of digraphs $G_0, ..., G_{k-1}$ with $k \leq n$ so that*

1. *(Reachability) Each vertex v has an index $\tau(v)$, so that a vertex w is reachable from v in G if and only if w is reachable from v in $G_{\tau(v)}$ or $G_{\tau(v)-1}$.*
2. *(Spanning Tree) Each $G_i = (V_i, E_i)$ is a digraph with a 2-layered spanning tree T_i rooted at r_i.*

3. *(Minor closed) G_i is a minor of G. If G has genus at most g, then G_i also has genus at most g.*
4. *(Linear Size) The total number of edges and vertices over all G_i is linear in the number of edges and vertices in G. Furthermore, every vertex v appears as a non-root vertex only in $G_{\tau(v)}$ and $G_{\tau(v)-1}$.*

We now relate the digraph layering construction to labeling schemes. Recall that each vertex v only appears in two digraphs $G_{\tau(v)-1}, G_{\tau(v)}$. Furthermore, every vertex w which is reachable from v is also reachable from v in one of the two digraphs. The idea is to construct a labeling for each digraph separately. Then to obtain the labeling for the original graph, we concatenate the sublabelings of the two graphs $G_{\tau(v)-1}, G_{\tau(v)}$. This way, the length of the labeling is at most twice the length of the sublabelings. We refer to the full version of the paper for the proof.

Lemma 4 (Digraph Layering Labeling [27]). *Given an arbitrary digraph G, its sequence of digraphs $G_0, ..., G_{k-1}$ obtained by digraph layering and a labeling l_i for each digraph G_i. We can construct a labeling l for G. Furthermore, l is of length at most $2 \log n + 2 \max |l_i|$.*

4.2   Planarizing Technique

In this section, we are given a bounded genus graph, and we want to reduce its genus to make it planar. The idea is to relate the problem of finding a reachability scheme for bounded genus graphs to the labeling scheme for planar graphs.

Given a digraph and a spanning tree, we construct a set of undirected paths so that the removal of these paths makes the graph planar. Moreover, if the tree is a 2-layered spanning tree, we can guarantee that the set of paths consists of not too many directed paths.

Given an arbitrary digraph G of genus g with a spanning tree T, we want to reduce the genus of G by removing certain paths of T. More formally, we are given T and build a set P consisting of paths of T starting at the root. The goal is that $G \setminus V(P)$ is a planar graph, not necessarily connected. Furthermore, if the given spanning tree T is a k-layered spanning tree, the set P consists of at most $\mathcal{O}(kg)$ directed paths.

Lemma 5 (Planarizing Tree). *Given a digraph G of genus at most g and a spanning tree T with root r, there exists a set of paths P so that*

1. *(Planarizing) Each connected component of $G \setminus V(P)$ is planar.*
2. *(Pathset) P consists of at most $\mathcal{O}(g)$ undirected paths in T starting at the root vertex r.*
3. *(Layered Tree) If T is a k-layered spanning tree, then P consists of at most $\mathcal{O}(kg)$ directed paths of T.*

We only outline the construction of the set P and refer to [17] for a more in-depth coverage. The main difference between the original proof of Gilbert et al.

and our proof is the choice of the spanning tree T. In the original proof the spanning tree had to be of small depth to bound the lengths (and number of vertices) of the undirected paths of P. This leads to a small vertex separator set. However, we only require that the set consists of a small number of paths which might consist of many vertices. We can therefore take an arbitrary spanning tree T as the origin of the construction instead of a low depth tree.

Instead of directly constructing the set P in the graph G, we remove parts of the graph that do not belong to P. The part that remains after removing certain edges and vertices is the desired set P. Everything that we remove in this process, namely $G \setminus V(P)$, is planar.

First, the graph is embedded in a surface of genus g with a spanning tree T. Then, non-tree edges are repeatedly removed to reduce the number of faces while preserving the embedding. Afterwards, all vertices of degree one are repeatedly removed until none are left. What remains is a graph G' spanned by $T' \subseteq T$. Due to Euler's Theorem $(v - e + f = 2 - 2g)$ the number of non-tree edges in G' and as a consequence also the number of leaves of T can be upper bounded by $\mathcal{O}(g)$. Thus, P can be covered by $\mathcal{O}(g)$ undirected paths. In particular, if T is a k-layered spanning tree P consists of $\mathcal{O}(kg)$ directed paths.

4.3   Bounded Genus Labeling Scheme

Now that we have introduced the necessary technical tools, we put them together to construct the labeling scheme for bounded genus graphs.

We remark that the labeling scheme of Kawarabayashi et al. [21] for distances in bounded genus graphs builds on the same idea. However, for reachability we work with directed graphs. Therefore, the technique is not directly applicable. Furthermore, instead of using the more complex tree-cotree decomposition of Eppstein [14], we rely on the separator construction of Gilbert et al. [17] to planarize the graph.

The idea of the construction is as follows. First, we construct a sequence of digraphs according to the digraph layering. Due to the construction, the genus does not increase for these layered digraphs. We then use the planarizing technique from the previous section to reduce them to planar graphs. Finally, we can apply the scheme devised by Thorup [27] for the planar remainder of the graph.

Theorem 3 (Bounded Genus Labeling). *Let $\mathcal{G}$ be the class of digraphs on n vertices with genus at most g. There exists a labeling scheme for reachability for the class $\mathcal{G}$ of length $L = \mathcal{O}(g \log n + \log^2 n)$.*

Proof. Let $G \in \mathcal{G}$ be an arbitrary digraph on n vertices of genus at most g. First we construct a sequence of layered digraphs $G_0, ..., G_{k-1}$ according to Lemma 3. Due to Lemma 4 it is sufficient for us to construct a labeling scheme of length $\mathcal{O}(g \log n + \log^2 n)$ for each G_i separately.

Recall that each G_i has a 2-layered spanning tree T_i rooted at r_i. Furthermore, G_i is a minor of G, therefore the genus of G_i is at most g.

Let H be an arbitrary G_i of the layered digraph construction. We can apply Lemma 5 on H using the 2-layered spanning tree. This gives us a set P of at most $2 \cdot 4g = \mathcal{O}(g)$ directed paths whose removal yields a planar graph H'.

Using Lemma 2, we can store $\mathcal{O}(g \log n)$ bits for the paths in P and reduce the problem to constructing a labeling scheme for $H \setminus V(P) = H'$. As the remaining graph H' is planar, we can apply the labeling scheme of Thorup [27] to construct a labeling of length $\mathcal{O}(\log^2 n)$.

Therefore, we have constructed a labeling scheme for the graph G using a total of $\mathcal{O}(g \log n + \log^2 n)$ bits.

4.4 Bounded Genus Lower Bound

In this section, we want to prove a lower bound for the length of any labeling scheme for reachability in bounded genus graphs. We do this by relating the class of bounded degree graphs to the class of bounded genus graphs. We first need two technical lemmas to relate the genus to the number of edges in a graph and to get a bound on the genus of the complete graph.

Lemma 6 (Genus Upper Bound). *Let G be an arbitrary graph on v vertices, $e \geq 1$ edges and genus g. Then the genus is at most the number of edges: $g \leq e$.*

Proof. Let G be a graph of genus g with v vertices and $e \geq 1$ edges. Let f be the number of faces of an embedding of G in a surface of genus g. Then, according to Euler's formula

$$v - e + f = 2 - 2g. \tag{2}$$

Rearranging the terms gives the desired bound on the genus. Note that v, e and f are all nonnegative integers.

$$g = \frac{2 - v + e - f}{2} \leq 1 + \frac{e}{2} \leq e \tag{3}$$

Lemma 7 (Genus of complete graph [22]). *The complete graph K_g has genus $\Theta(g^2)$ and as a consequence every graph with a K_g minor must have genus at least $\lceil \frac{(g-4)(g-5)}{12} \rceil = \Theta(g^2)$.*

Now we can to relate the lower bound of the bounded degree graphs to the bounded genus graphs.

Theorem 4 (Bounded Genus Labeling Lower Bound). *Let $\mathcal{G}$ be the class of digraphs on n vertices of genus at most g. Every labeling scheme for reachability for the class $\mathcal{G}$ has length at least $L = \Omega(\sqrt{g})$.*

Proof. Assume for the sake of contradiction that there exists a labeling scheme for reachability for the class $\mathcal{G}$ and has length $L = o(\sqrt{g})$. The idea is to take this scheme and apply it to the class of bounded degree graphs $\mathcal{G}_\Delta$. More formally, $\mathcal{G}_\Delta$ is the class of digraphs on n vertices with outdegree at most Δ. These graphs have at most $n \cdot \Delta$ edges. By Lemma 6, their genus is bounded by $n \cdot \Delta$. Now we use the labeling scheme for bounded genus graphs of length $o(\sqrt{g})$ for $\mathcal{G}_\Delta$.

$$L = o(\sqrt{g}) \le o(\sqrt{n\Delta}) \tag{4}$$

However, we already know by Theorem 2 that any labeling scheme for $\mathcal{G}_\Delta$ must use at least $\Omega(\sqrt{n\Delta})$ bits, a contradiction. Therefore, any labeling scheme for reachability for the class $\mathcal{G}$ has length at least $L = \Omega(\sqrt{g})$.

Note that the $\Omega(\sqrt{g})$ bound can be derived by arguing that the genus is always at most the number of vertices squared. However, the proof presented here is stronger in the following sense. The simple proof only shows the result for the class of all graphs, which includes the graphs of genus of order up to $\Theta(n^2)$. However, the proof can't be applied if the class of graphs is more restricted, i.e. graphs of genus $\Theta(n^2)$ are not included. Whereas with our proof, the result also follows for more general classes of graphs including the graphs of genus linear in the size of the graph.

Note that we could have also derived the square root bound by arguing that the genus is always at most the number of vertices squared. However, we show a specific class of graphs with genus linear in the size of the graph to which the square root bound applies, instead of only graphs with genus of order $\Theta(n^2)$.

Finally, we study that the lower bound for bounded genus graphs can be transferred to the class of minor excluded graphs. We say that a graph G contains a minor H if H can be obtained through a series of edge contractions, vertex deletions, or edge deletions. The class of minor excluded graphs consists of the graphs which exclude all graphs H of a family $\mathcal{H}$ as a minor. This is of interest as many classes of graphs can be characterized through excluding minors. In particular, the class of planar graphs can be defined as the graphs which exclude $K_{3,3}$ and K_5, known as Kuratowski's Theorem [26]. Note that for reachability, we work with directed graphs. In the context of minor excluded graphs, we work with the undirected graph underlying G whenever necessary.

Theorem 5 (Minor Excluded Lower Bound). *Let $\mathcal{G}_H$ be the class of digraphs on n vertices, which excludes the graph H on h vertices as a minor. Every labeling scheme for reachability for the class $\mathcal{G}_H$ has length at least $L = \Omega(h)$.*

Proof. First, let us consider the class $\mathcal{G}_{K_h}$ which excludes the complete graph on h vertices as a minor. Note that $\mathcal{G}_{K_h}$ is a subclass of $\mathcal{G}_H$ as any graph that excludes H as a minor can not have K_h as a minor. Therefore, any lower bound we derive for $\mathcal{G}_{K_h}$ must also apply to $\mathcal{G}_H$.

Assume for the sake of contradiction that there exists a labeling scheme for the class $\mathcal{G}_{K_h}$ of length $L = o(h)$. We will now take this scheme and apply it to the class of bounded genus graphs $\mathcal{G}$ of genus at most g. Recall that due to Lemma 6, the genus does not increase for any minor. Furthermore, the complete graph $K_{g'}$ where $g' = \lceil 12(\sqrt{g+1} + 5) \rceil$ has genus at least $g+1$ due to Lemma 7.

As a consequence, any graph with a $K_{g'}$ minor has genus at least $g+1$. Therefore, $\mathcal{G}$ excludes $K_{g'}$ and we can apply the labeling scheme for excluded graph minors of length $o(h)$ on $\mathcal{G}$.

$$L = o(h) = o(g') \le o\left(\lceil 12(\sqrt{g+1}+5)\rceil\right) = o\left(\sqrt{g}\right) \tag{5}$$

However, we already know by Theorem 4 that any labeling scheme for $\mathcal{G}$ must use at least $\Omega(\sqrt{g})$ bits, a contradiction. Therefore, any labeling scheme for reachability for the class $\mathcal{G}_{K_h}$ and $\mathcal{G}_H$ has length at least $L = \Omega(h)$.

In this section we have strengthened existing results and showed how minor excluded, bounded genus and bounded degree graphs are intertwined. We view this as a first step to obtain tighter lower bounds, which might be harder to obtain and require new insights.

Finally, a short remark on the tightness of our results. For undirected graphs which exclude H as a minor there is a property called path separability studied by Abraham et al. [1]. How path separability behaves for the directed case is not yet as well understood. If such graphs were $\mathcal{O}(h)$ path separable, this would further imply the existence of a labeling scheme of length $\mathcal{O}(h\log^2(n))$ which could also be applied to graphs of bounded genus and graphs of bounded degree. In this case, the derived bounds would be tight up to $\log(n)$ factors.

5   Conclusion

Processing graph data is essential for many applications. Furthermore, determining the relationship between vertices quickly and efficiently is central to get good performance. We study how to answer reachability queries using a labeling scheme. A labeling scheme can be stored in a distributed fashion by design as each node only needs to store his own label. Therefore, they are well suited to be applied to large graphs. Furthermore, they allow for parallel processing as only read operations on the labels are needed. However, a key challenge is how to design short labeling schemes and determine whether it is even possible.

In order to determine an appropriate parameter to characterize the difficulty of designing short labeling schemes for reachability we study the degree and genus of a graph as a measure. We first study sparse graphs, where we enforce local sparseness through limiting the maximum degree. It turns out, that even for constant degree graphs there are no short schemes as we prove a $\Omega(\sqrt{n\Delta})$ lower bound for graphs of outdegree Δ. On the other hand, we present a novel labeling scheme for graphs of bounded genus of length $\mathcal{O}(g\log n + \log^2 n)$.

Furthermore, any labeling scheme for graphs of bounded genus must use at least $\Omega(\sqrt{g})$ bits, even if the genus is subquadratic in the size of the graph. This means that the genus of a graph is a possible indicator on the complexity of devising short labelings. Moreover, the result can be generalized to minor excluded graphs. However, it remains an open question whether there exists a polylogarithmic labeling scheme. Similar results for undirected graphs [1] suggest that this might be possible. However, adapting existing results to the directed case has not been achieved so far and might need novel insights.

References

1. Abraham, I., Gavoille, C.: Object location using path separators. In: Proceedings of the Twenty-Fifth Annual ACM Symposium on Principles of Distributed Computing, PODC 2006, pp. 188–197. Association for Computing Machinery, New York (2006). https://doi.org/10.1145/1146381.1146411

2. Adjiashvili, D., Rotbart, N.: Labeling schemes for bounded degree graphs. In: Esparza, J., Fraigniaud, P., Husfeldt, T., Koutsoupias, E. (eds.) ICALP 2014. LNCS, vol. 8573, pp. 375–386. Springer, Heidelberg (2014). https://doi.org/10.1007/978-3-662-43951-7_32

3. Agrawal, R., Borgida, A., Jagadish, H.V.: Efficient management of transitive relationships in large data and knowledge bases. In: Proceedings of the 1989 ACM SIGMOD International Conference on Management of Data, SIGMOD 1989, pp. 253–262. Association for Computing Machinery, New York (1989). https://doi.org/10.1145/67544.66950

4. Alstrup, S., Bille, P., Rauhe, T.: Labeling schemes for small distances in trees. SIAM J. Discrete Math. **19**(2), 448–462 (2005). https://doi.org/10.1137/S0895480103433409

5. Alstrup, S., Gavoille, C., Kaplan, H., Rauhe, T.: Nearest common ancestors: a survey and a new algorithm for a distributed environment. Theor Comput. Syst. **37**(3), 441–456 (2004). https://doi.org/10.1007/s00224-004-1155-5

6. Alstrup, S., Kaplan, H., Thorup, M., Zwick, U.: Adjacency labeling schemes and induced-universal graphs. In: Proceedings of the Forty-Seventh Annual ACM Symposium on Theory of Computing, STOC 2015, pp. 625–634. Association for Computing Machinery, New York (2015). https://doi.org/10.1145/2746539.2746545

7. Alstrup, S., Rauhe, T.: Improved labeling scheme for ancestor queries. In: Proceedings of the Thirteenth Annual ACM-SIAM Symposium on Discrete Algorithms, SODA 2002, pp. 947–953. Society for Industrial and Applied Mathematics, USA (2002)

8. Bonamy, M., Esperet, L., Groenland, C., Scott, A.: Optimal labelling schemes for adjacency, comparability, and reachability, p. 1109–1117. Association for Computing Machinery, New York (2021). https://doi.org/10.1145/3406325.3451102

9. Breuer, M.: Coding the vertexes of a graph. IEEE Trans. Inf. Theory **12**, 148–153 (1966)

10. Cheng, J., Huang, S., Wu, H., Fu, A.W.C.: TF-label: a topological-folding labeling scheme for reachability querying in a large graph. In: Proceedings of the 2013 ACM SIGMOD International Conference on Management of Data, SIGMOD 2013, pp. 193–204. Association for Computing Machinery, New York (2013). https://doi.org/10.1145/2463676.2465286

11. Cohen, E., Halperin, E., Kaplan, H., Zwick, U.: Reachability and distance queries via 2-hop labels. SIAM J. Comput. **32**(5), 1338–1355 (2003). https://doi.org/10.1137/S0097539702403098

12. da Silva, R.F., Urrutia, S., Hvattum, L.M.: Extended high dimensional indexing approach for reachability queries on very large graphs. Exp. Syst. Appl. **181**, 114962 (2021). https://doi.org/10.1016/j.eswa.2021.114962. https://www.sciencedirect.com/science/article/pii/S0957417421004036

13. Dahlgaard, S., Knudsen, M.B.T., Rotbart, N.: A simple and optimal ancestry labeling scheme for trees. In: Halldórsson, M.M., Iwama, K., Kobayashi, N., Speckmann, B. (eds.) ICALP 2015. LNCS, vol. 9135, pp. 564–574. Springer, Heidelberg (2015). https://doi.org/10.1007/978-3-662-47666-6_45

14. Eppstein, D.: Dynamic generators of topologically embedded graphs. CoRR cs.DS/0207082 (2002). https://arxiv.org/abs/cs/0207082

15. Fraigniaud, P., Korman, A.: An optimal ancestry labeling scheme with applications to xml trees and universal posets. J. ACM **63**(1) (2016). https://doi.org/10.1145/2794076

16. Gavoille, C., Peleg, D., Pérennes, S., Raz, R.: Distance labeling in graphs. In: Proceedings of the Twelfth Annual ACM-SIAM Symposium on Discrete Algorithms, SODA 2001, pp. 210–219. Society for Industrial and Applied Mathematics, USA (2001)

17. Gilbert, J.R., Hutchinson, J.P., Tarjan, R.E.: A separator theorem for graphs of bounded genus. J. Algorithms **5**(3), 391–407 (1984). https://doi.org/10.1016/0196-6774(84)90019-1. https://www.sciencedirect.com/science/article/pii/0196677484900191

18. Heinis, T., Alonso, G.: Efficient lineage tracking for scientific workflows. In: Proceedings of the 2008 ACM SIGMOD International Conference on Management of Data, SIGMOD 2008, pp. 1007–1018. Association for Computing Machinery, New York (2008). https://doi.org/10.1145/1376616.1376716

19. Jain, R., Tewari, R.: Reachability in high treewidth graphs. In: Lu, P., Zhang, G. (eds.) 30th International Symposium on Algorithms and Computation (ISAAC 2019). Leibniz International Proceedings in Informatics (LIPIcs), vol. 149, pp. 12:1–12:14. Schloss Dagstuhl-Leibniz-Zentrum fuer Informatik, Dagstuhl (2019). https://doi.org/10.4230/LIPIcs.ISAAC.2019.12. https://drops.dagstuhl.de/opus/volltexte/2019/11508

20. Kannan, S., Naor, M., Rudich, S.: Implicit representation of graphs. In: Proceedings of the Twentieth Annual ACM Symposium on Theory of Computing, STOC 1988, pp. 334–343. Association for Computing Machinery, New York (1988). https://doi.org/10.1145/62212.62244

21. Kawarabayashi, K., Klein, P.N., Sommer, C.: Linear-space approximate distance oracles for planar, bounded-genus and minor-free graphs. In: Aceto, L., Henzinger, M., Sgall, J. (eds.) ICALP 2011. LNCS, vol. 6755, pp. 135–146. Springer, Heidelberg (2011). https://doi.org/10.1007/978-3-642-22006-7_12

22. Liu, Y.: Topological theory of graphs: de Gruyter (2017). https://doi.org/10.1515/9783110479492

23. Peleg, D.: Informative labeling schemes for graphs. In: Nielsen, M., Rovan, B. (eds.) MFCS 2000. LNCS, vol. 1893, pp. 579–588. Springer, Heidelberg (2000). https://doi.org/10.1007/3-540-44612-5_53

24. Seufert, S., Anand, A., Bedathur, S., Weikum, G.: FERRARI: flexible and efficient reachability range assignment for graph indexing. In: 2013 IEEE 29th International Conference on Data Engineering (ICDE), pp. 1009–1020 (2013). https://doi.org/10.1109/ICDE.2013.6544893

25. Simon, K.: An improved algorithm for transitive closure on acyclic digraphs. In: Kott, L. (ed.) ICALP 1986. LNCS, vol. 226, pp. 376–386. Springer, Heidelberg (1986). https://doi.org/10.1007/3-540-16761-7_87

26. Thomassen, C.: Kuratowski's theorem. J. Graph Theor. **5**(3), 225–241 (1981). https://doi.org/10.1002/jgt.3190050304. https://onlinelibrary.wiley.com/doi/abs/10.1002/jgt.3190050304

27. Thorup, M.: Compact oracles for reachability and approximate distances in planar digraphs. J. ACM **51**(6), 993–1024 (2004). https://doi.org/10.1145/1039488.1039493

28. Wang, H., He, H., Yang, J., Yu, P., Yu, J.: Dual labeling: answering graph reachability queries in constant time. In: 22nd International Conference on Data Engineering (ICDE 2006), pp. 75–75 (2006). https://doi.org/10.1109/ICDE.2006.53
29. Wei, H., Yu, J.X., Lu, C., Jin, R.: Reachability querying: an independent permutation labeling approach. VLDB J. **27**(1), 1–26 (2017). https://doi.org/10.1007/s00778-017-0468-3
30. Yildirim, H., Chaoji, V., Zaki, M.J.: GRAIL: scalable reachability index for large graphs. Proc. VLDB Endow. **3**(1–2), 276–284 (2010). https://doi.org/10.14778/1920841.1920879

Being Automated or Not? Risk Identification of Occupations with Graph Neural Networks

Dawei Xu, Haoran Yang, Marian-Andrei Rizoiu, and Guandong Xu[✉]

University of Technology Sydney, Sydney, Australia
{dawei.xu,haoran.yang-2}@student.uts.edu.au,
{marian-andrei.rizoiu,guandong.xu}@uts.edu.au

Abstract. The rapid advances in automation technologies, such as artificial intelligence (AI) and robotics, pose an increasing risk of automation for occupations, with a likely significant impact on the labour market. Recent social-economic studies suggest that nearly 50% of occupations are at high risk of being automated in the next decade. However, the lack of granular data and empirically informed models have limited the accuracy of these studies and made it challenging to predict which jobs will be automated. In this paper, we study the automation risk of occupations by performing a classification task between automated and non-automated occupations. The available information is 910 occupations' task statements, skills and interactions categorised by Standard Occupational Classification (SOC). To fully utilize this information, we propose a graph-based semi-supervised classification method named **A**utomated **O**ccupation **C**lassification based on **G**raph **C**onvolutional Networks (**AOC-GCN**) to identify the automated risk for occupations. This model integrates a heterogeneous graph to capture occupations' local and global contexts. The results show that our proposed method outperforms the baseline models by considering the information of both internal features of occupations and their external interactions. This study could help policymakers identify potential automated occupations and support individuals' decision-making before entering the job market.

Keywords: Automated occupation identification · Graph convolutional network · Semi-supervised classification

1 Introduction

In light of rapid development in the fields of artificial intelligence and robotics, the number of jobs for certain occupations is decreasing alarmingly. Also, massive applications of automation technologies, especially in the workplace, are raising fears of large-scale technological unemployment and a renewed appeal for policymakers to handle the consequences of technological change. From a historical perspective, every disappearance of occupations comes with increasing unemployment [2,16] which brings individual existence crisis, social instability,

© The Author(s), under exclusive license to Springer Nature Switzerland AG 2022
W. Chen et al. (Eds.): ADMA 2022, LNAI 13725, pp. 520–534, 2022.
https://doi.org/10.1007/978-3-031-22064-7_37

and techno-phobia that may hinder the future development of science and technologies. Therefore, occupations' automated risk identification is critical to both policymakers and individuals.

Generally, an occupation at risk of being automated has attributes containing routine and rule-based information, which are prone to be replaced by automation technologies [5,6]. Predicting the automated risk of occupations can be performed by capturing the routine and rule-based features in their attributes, more specifically, their tasks and skills information [4]. Therefore, this prediction task can be conducted as a binary classification between automated and non-automated occupations by mining their internal attributes and external interactions (Fig.1).

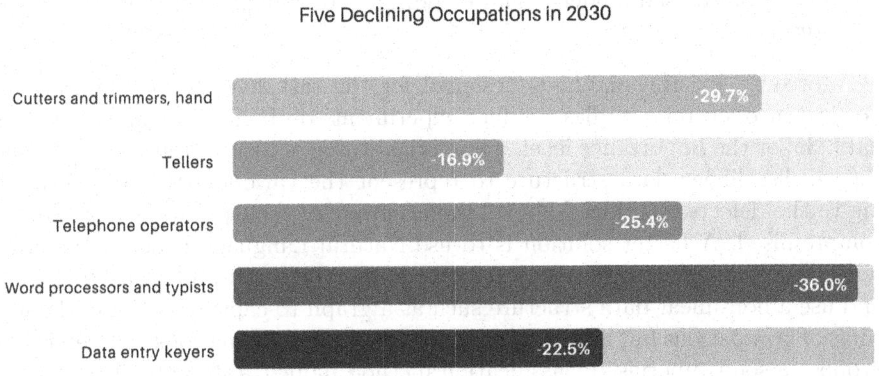

Five Declining Occupations in 2030

Fig. 1. BLS projected declining occupations in 2030

However, this task becomes more challenging due to the following three characteristics of the data:

- **Limited Data Source.** Predicting automated risk of occupations requires tasks and skills data that are at their most granular level to reflect micro differences among occupations. Standard labor data from U.S. Bureau of Labor Statistics (BLS) or Australian Bureau of Statistics (ABS) focus on demographic, economic or other aggregate statistics that lack resolutions into the defining features to distinguish different occupations. Extracting granular tasks and skills data from job recruitment platforms is also not feasible since they are dynamic, computing expensive and not all jobs are posted online. The numbers of job posts are also various across occupations which may lead to imbalanced data [8]. Furthermore, the lack of ground truth data is also a barrier hindering the prediction of automated risk.
- **Non-numerical Data Type.** Most of the tasks and skills data are in text form, which are single words, grouped words or sentences. These text data cannot be directly processed by statistical models. Current studies [2,6] use importance or level value attached to each task or skill as its numerical representations; however, this kind of representation is aggregated and coarse

which may obfuscate the differential impact of various technologies and skill requirements then further affects models' performance in distinguishing occupations. Therefore, choosing a method that can not only encode the text data into a numerical format but maintain the features hidden in the context are necessary.

- **Complex Data Structure.** Since occupations are likely to interact by attributes, e.g., they may share similar skillsets, the prediction of automated risk cannot be limited to occupations' internal attributes but also the connections with other occupations [4]. Moreover, various attributes may lead to different data dimension sizes, and the relationships among occupations can be multiple paths [17]. Therefore, linear data structures are not capable of capturing all the information, and a more complicated non-linear structure and computational model should be designed to recognise the patterns in the information.

Moreover, existing methods designed for the task have several limitations. For instance, current studies conduct experiments on BLS wages and education data [35] or the importance level, a numerical value of task statements [2,3], and they utilise linear data structure to represent the various attributes. According to the defects mentioned above, the accuracy of their results is unavoidable compromised. A better solution is to use Natural Language Processing (NLP) technologies to generate numerical embeddings for tasks and skills text data and use a non-linear data structure such as a graph to capture not only the features of occupations but also the interactions between occupations. Therefore, we propose a semi-supervised classification method named Automated Occupation Classification based on Graph Convolutional Networks (AOC-GCN) to identify the automated risk for occupations. We will be using tasks and skills datasets extracted from the O*NET database, which is a publicly available occupational information network containing hundreds of job definitions, to understand the labour market.

In summary, the contributions of our work are listed below:

- We use Natural Language Processing (NLP) technologies – Word2Vec and Doc2Vec to generate embeddings for tasks and skills data, which is, to the best of our knowledge, the first work in this domain using NLP to generate numerical representations for text data.
- We propose a GCN-based heterogeneous graph classification method that aggregates both node features and network-wide behaviours, which is also the first work in this domain.
- Our proposed method significantly outperforms baseline models, and the results are verified through comparison with government statistical data.

The rest of the paper is organized as follows. Section 2 lists the related work. We elaborate on the proposed AOC-GCN method in Sect. 3. Section 4 compares and analyzes the experimental results of our method. In Sect. 5, we discuss the results and compare them with government statistical data. In Sect. 6, we discuss limitations and future work, and the paper is concluded in Sect. 7.

2   Related Work

2.1   Occupations Automated Risk Identification

Most existing studies about automated occupation identification are based on expert-based assessments, or basic statistical models focusing on probability space. Frey et al. [2] assessed the automated probability of occupations using a Gaussian process classification and concluded that 47% of current US employment is at high risk of being automated. Similar studies have been conducted at the impact of automation on occupations in other countries and reached cautionary conclusions: automated occupations rate will be 35% in Finland [10], 59% in Germany [9], and 45 to 60 % across Europe [11]. On the contrary, Arntz et al. [3] followed the experiments in [2] but used job-level data and a different coefficient estimates method, and concluded that a less alarming 9% of employment is at risk. Vermeulen et al. [6] and Arntz et al. [12] also conducted job-level automatability scorings and get similar findings that about 5% of jobs have automated risks. However, the data utilized in Arntz et al. are job survey data conducted on different workers. Arntz et al. considers this data as job level data compared to occupation level data without considering the duplicates of same tasks and skills performed by different workers' job. Therefore, the results may be underestimated. That is to say, job-level studies consider each task to be independent and unique which masked the similarity between tasks. Furthermore, current studies failed to discover the semantic information hidden in text data and ignored the interactions between occupations.

2.2   Graph Neural Networks

Recent years have witnessed a growing application in using graph neural network-based algorithms to solve graph structure problems [13, 14, 18, 23–25, 36–38]. These methods include both supervised [7] and unsupervised methods [15, 20]. Especially, Graph Convolutional Networks (GCN) [7] have achieved significant performance compared to previous methods, which is an efficient variant of Convolutional Neural Networks (CNN) [22] operating directly on graphs. In GCN, both the graph structure information and node features are also aggregated from neighbours during convolution. The graph convolution operation is defined as feature aggregations of neighbours, and through iterative convolutions, the whole graph's information can be propagated and aggregated to each part of the graph. After GCN, GraphSAGE [21] is proposed, which is a simpler but efficient inductive learning model that breaks the limitation of applying GCN in transductive learning. Furthermore, Graph Attention Networks (GAT) [26] introduces an attention mechanism into GCN and allows nodes to focus on the most relevant neighbours during training.

In this paper, a GCN-based method is first applied to automated occupation classification task through building an occupation-skill graph. Our proposed method can encode not only the features of occupation but also the interaction between occupations.

3   Methodology

3.1   Preliminaries

Generally, GCN-based models [7,21,26] consist of multiple propagation layers and all the nodes are updated simultaneously in each propagation layer. A propagation layer can be divided into two sub-layers: *aggregation* and *combination*. Let $G = (V, E)$ be a graph with node $v \in V$, edge $(v, v') \in E$, and node feature $x_v = h_v^0 \in \mathbf{R}^{d_0}$ for $v \in V$ where d_0 denotes the feature dimension of the node and $h_v^l \in \mathbf{R}^{d_l}$ denotes the hidden state of node v learned by the l-th layer of the model. For a GCN with L layers, aggregation and combination sub-layers at l-th layer($l = 1, 2, ... L$) can be written as:

$$h_{N(v)}^l = \sigma(W^l \cdot \text{AGG}(\{h_{v'}^{l-1}, \forall v' \in N(v)\})) \tag{1}$$

$$h_v^l = \text{COMBINE}(h_v^{l-1}, h_{N(v)}^l) \tag{2}$$

where $N(v)$ is a set of nodes adjacent to v, AGG is a function used to aggregate embeddings from neighbor nodes of v. W^l is a trainable matrix shared among all nodes at layer l. σ is a non-linear activation function e.g., $RELU$. $h_{N(v)}^l$ denotes the aggregated feature of all neighbors of node v at l-th layer. $COMBINE$ function is used to combine the embedding of node itself and the aggregated embeddings of neighbors.

For heterogeneous graph with different types of nodes and edges, it always comes with different sizes of embeddings. For each edge connecting two different nodes, we generally concatenate the edge embedding from the last propagation layer together with embeddings of the two nodes this edge links to. Similarly, for each node, besides the information from neighbor nodes, the features of edges connected to them are also collected. Therefore the aggregation sub-layer for an edge e can be defined as:

$$h_e^l = \sigma(W_E^l \cdot \text{AGG}_E^l(h_e^{l-1}, h_{U(e)}^{l-1}, h_{I(e)}^{l-1})) \tag{3}$$

where $h_{U(e)}^{l-1}$ and $h_{I(e)}^{l-1}$ denote the hidden states of two types of nodes U and I from the previous propagation layer and

$$\text{AGG}_E^l(h_e^{l-1}, h_{U(e)}^{l-1}, h_{I(e)}^{l-1}) = \text{concat}(h_e^{l-1}, h_{U(e)}^{l-1}, h_{I(e)}^{l-1}) \tag{4}$$

For the two nodes $u \in U$ and $i \in I$, the aggregated neighbor embedding $h_{N(u)}^l$ and $h_{N(i)}^l$ can be calculated as:

$$
\begin{aligned}
h_{N(u)}^l &= \sigma(W_U^l \cdot \text{AGG}_U^l(\text{concat}(h_i^{l-1}, h_e^{l-1}), \forall e = (u, i) \in E(u))) \\
h_{N(i)}^l &= \sigma(W_I^l \cdot \text{AGG}_I^l(\text{concat}(h_u^{l-1}, h_e^{l-1}), \forall e = (u, i) \in E(i)))
\end{aligned}
\tag{5}
$$

Here the two types of nodes maintain different parameters (W_U^l, W_I^l) and different aggregation functions (AGG_U^l, AGG_I^l).

After aggregating the neighbors' information, we follow the concatenation in [6] for the two types of nodes as

$$h_u^l = \text{concat}(V_U^l \cdot h_u^{l-1}, h_{N(u)}^l)$$
$$h_i^l = \text{concat}(V_I^l \cdot h_i^{l-1}, h_{N(i)}^l) \tag{6}$$

where V_U^l and V_U^l denote trainable weight matrix for node U and node I, and the h_u^l and h_i^l are the hidden states of l-th layer for the two types of nodes.

3.2 Problem Setup

Our purpose is to identify the automated risk for occupations listed in Standard Classification Code (SOC). The attributes used in this problem are task statements and skills extracted from O*NET database. According to the properties of data, each task statement is unique for each occupation while each skill data are a list of skills which are overlapped and shared among other occupations. So we can build a bipartite graph $G(O, S, E)$ where O is the set of occupation nodes, S is the set of skill nodes, and E is the set of edges between occupations and skills. An edge $e \in E$ from an occupation $o \in O$ to a skill $s \in S$ exists if o contains s. It is worth noting that the task statements data will be used as occupation nodes' feature representations which will be introduced in Sect. 3.4. Therefore, the task can be formulated to a node classification problem on an undirected bipartite graph with two types of attributed nodes.

3.3 Graph Convolutional Networks on Occupation-skill Graph

The occupation-skill graph is a bipartite graph with two types of nodes and one type of edge. Moreover, there is no attributes on edges in our problem, so applying a heterogeneous GCN to solve this problem can be excessive. In fact, we take a shortcut way by applying GCN to treat this bipartite graph as a homogeneous graph by setting identical dimension sizes for both nodes features. Therefore, the propagation rule at layer l is this problem can be defined as:

$$h_v^l = \text{COMBINE}(h_v^{l-1}, \sigma(W^l \cdot \text{AGG}(\{h_{v'}^{l-1}, \forall v' \in N(v)\}))) \tag{7}$$

where $v \in O \cup S$, $N(v)$ is a set of nodes adjacent to v, h_v^l denote the hidden states at l-th layer for each nodes.

3.4 Nodes Feature Representation

Text data of task statements and skills should be converted into an embedding before being encoded as nodes' features. Here we apply two word embeddings technologies – Word2vec [27] and Doc2vec [28] to generate feature representations for occupation nodes and skill nodes. Firstly, we use Glove pretrained word embeddings to initialize the Word2vec model and retrain the model on task statements data to aggregate the context and semantic similarity information in this domain. The output from the Word2vec model is a word embeddings table containing vectors for each word. Since the skill data are single word or grouped words, we represent each skill's feature by simply applying the word vector or making additions of multiple word vectors from word embeddings. Then, we initialize our Doc2vec model with word embeddings from last step and train on task statements data to generate document-level embeddings as occupation nodes feature representations. In detail,

$$h_s^0 = \text{fusion}(\boldsymbol{w}_1, \boldsymbol{w}_2, ..., \boldsymbol{w}_n) \tag{8}$$

$$h_o^0 = \text{doc2vec}(t_0) \tag{9}$$

where h_s^0, h_o^0 denote the initial embedding of a skill node and an occupation node. $\boldsymbol{w}_i$ represents the i-th word vector in a skill generated from Word2vec. t_0 is the task statement document of an occupation. Therefore, the parameters of Word2vec and Doc2vec are trained together with others in the model described in Sect. 3.3.

3.5   Semi-supervised Automated Occupation Classification Model

The final embeddings of AOC-GCN are the concatenation of nodes embeddings learned from Occupation-skill Graph, so we can get

$$y = \text{softmax}(f(\text{concat}(z_o, z_s))) \tag{10}$$

where z_o and z_s denote the embeddings of O and S learned by the proposed GCN method. $f(\cdot)$ is the mapping function to map the embeddings to lower dimension space before putting into $softmax$ which is the classifier used to classify automated and non-automated occupations. This method is semi-supervised because the whole graph is updated during training process which is on a partially labeled dataset. The whole pipeline of the study is shown as in Fig. 2.

4   Experiment

4.1   Dataset and Metrics

Experiments are conducted on occupation lists, task statements and skills in Standard Classification Code (SOC) from O*NET database. The label information utilized in the experiments is generated from expert-based assessments on O*NET occupations conducted by [2,6] with automated occupations being labeled as 1 and non-automated occupations being labeled as 0. However, the original label information has defects that the number of two classes are unbalanced, and some occupations are too close e.g., web developer and software programmer. We preprocess the data and finally select 112 labeled occupations with 56 automated occupations and 56 non-automated occupations which also have distance between each other. Then, all the data will form an undirected graph structure with 910 occupation nodes, 135 skill nodes and 13222 edges. Training, validation and test set are randomly split with a ratio of 8:1:1. Given the prediction and ground truth, we evaluate the model's performance using accuracy, precision, recall and F1 score.

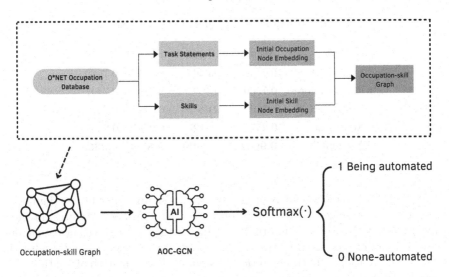

Fig. 2. The whole system pipeline

4.2 Baselines

We compare our model with traditional classification models. Specifically, we design features of each occupation by concatenate the embeddings generated from Sect. 3.4 for both task statements and skills. Then we use zero padding to make sure all the occupations have identical feature dimension sizes before putting into baseline models. The chosen baseline models are as following:

- **Decision Tree**. DT splits the dataset as a tree based on a set of rules and conditions which is a supervised learning algorithm that can be used for both classification and regression [32].
- **Random Forest**. Random Forest is an ensemble method which combines the output of multiple decision trees to reach a single result [30].
- **Adaptive Boost**. AdaBoost (Adaptive Boosting) is a boosting technique that aims at combining multiple weak classifiers to build one strong classifier and is widely used in both classification and regression problem [29].
- **Light Gradient Boosting Machine**. LightGBM is a gradient boosting framework based on decision trees to increase the efficiency of the model and reduce memory usage [31].

4.3 Results and Analysis

Evaluation. We evaluate the results of AOC-GCN model on O*NET dataset with above metrics and compare them with baseline models. The results of the comparison are shown in Table 1 below.

Table 1. Preprocessed results of O*NET dataset

Method	Accuracy	Precision	Recall	F1
Decision Tree	0.7143	0.6667	0.8571	0.7500
Random Forest	0.7500	0.7692	0.7143	0.7407
LightGBM	0.7500	0.6842	0.9286	0.7879
AdaBoost	0.8571	0.8125	0.9286	0.8667
AOC-GCN	**0.9091**	**1.0000**	**0.8750**	**0.9333**

We find that our AOC-GCN method performs significantly better than other baseline models, since our model not only uses external node features, but also capture the graph structural information which is the interactions between nodes. Besides, baseline models using raw features only already achieve good results, the GCN's combination of both semantic and structural information of relations successfully attains the best performance. Furthermore, our proposed model ensures high model precision to avoid disturbing non-automated occupations.

Parameter Experiments. In this section, we explore the effects on the model performance of different parameters. The key parameters used here are the dimensions of initial node embedding sizes and the dimension sizes of GCN layer. The dimension sizes of node embeddings range from 50 to 300 and the dimension sizes of GCN layer range from 16 to 512.

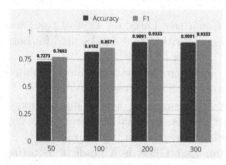 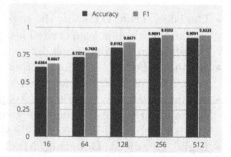

(a) Model performance on different initial node embeddings size

(b) Model performance on different GCN layer size

Fig. 3. Parameters effect on model performance

We can see from the Fig. 3 that with the increase of sizes of initial node embeddings from 50 to 200, both the accuracy and F1 scores rise and reach the maximum value. When the size is greater than 200, both of the values keep stable. The same pattern can be found in GCN layer dimension sizes, when the size increases from 16 to 256, both of the accuracy and F1 increase. When the size is greater than 256, the two scores keep unchanged. Therefore, we can conclude that the model gets the best performance when the initial node embeddings size is greater than 200 and the GCN layer dimension size is greater than 256.

Graph Embeddings Visualization. In this section, we visualize the embeddings generated from the GCN layer using t-Distributed Stochastic Neighbor Embedding (t-SNE) [33] which is a feature reduction technique mitigating the effects of the "Curse of Dimensionality". We also conduct an unsupervised K-means clustering on the dataset and visualize the result using a dimensionality reduction algorithm Principal component analysis (PCA) [34]. Through comparison, we can find out the difference between applying unsupervised learning and semi-supervised learning. The results are showing in Fig. 4 and 5.

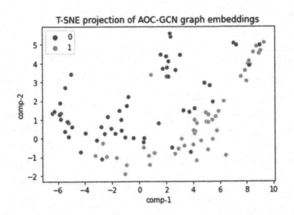

Fig. 4. t-SNE projection of GCN embeddings

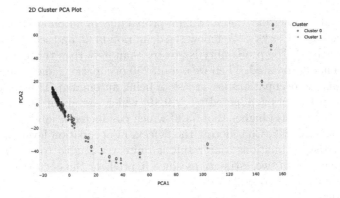

Fig. 5. PCA projection of K-means clustering

We discover that there is a distinct boundary between label 1 and label 0 which are colored in blue and red in t-SNE projection of GCN embeddings in Fig. 4. However, there is no distinguishable separation of two K-means clusterings colored in red and blue from PCA projection Fig. 5, in which labels are annotated in the graph as 1 and 0. It means that GCN semi-supervised method achieved much better performance compared to the unsupervised training by using only a small portion of labelled training data.

5   Results and Discussion

In this section, we apply the model with the best performance to conduct predictions on the rest of 798 unlabelled occupation nodes to get the probabilities of their automated risk. Part of the results is shown in Table 2.

Table 2. Top 10 occupations in high risk of being automated

Occupation	Risk
File Clerks	0.7002
Real Estate Brokers	0.6968
Word Processors and Typists	0.6943
Payroll and Timekeeping Clerks	0.6941
Data Entry Keyers	0.6940
Transportation Engineers	0.6938
Credit Analysts	0.6938
Insurance Appraisers, Auto Damage	0.6938
Insurance Underwriters	0.6938
Tax Examiners and Collectors, and Revenue Agents	0.6938

Our results show that the maximum probability of automated risk is 0.7002 which is the "File Clerks". We follow the definition in [2] and set 69% as the cutoff point so that if the probability is greater than 69% then this occupation has a risk of being automated. Then we obtain 230 occupations and a percentage of 25.3% among all occupations are at risk of being automated. This result is lower than the experiments conducted by [2,9,10] which are about 50% occupations are at risk and is also higher than [3,6] which conducted a job-level estimation of 9% automated risk. Considering the defects in occupation-level and job-level automated risk estimations, our result is quite reasonable to reach a median value between these two estimations since it predicts the occupation-level risk using granular task and skills data.

To further validate our results, we compare them with "U.S. Bureau of Labor Statistics (BLS) Occupations with the largest job declines, 2020 and projected 2030". It projected out 29 occupations that have declining employments in 2030. The outcome is visualized in Fig. 6.

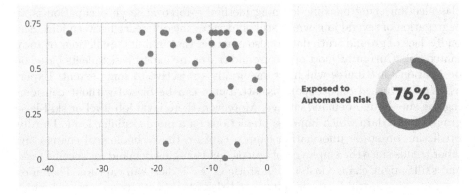

Fig. 6. BLS projected declining occupations in 2030

We assume that automated risk greater than 50% would cause employment declining and we can find that 76% of the BLS projected declining occupations have automated risks higher than 50% except for "customer service representatives", "computer programmers", "inspectors, testers, sorters, samplers and weighters" which have an automated risk probability of 0.1%, 8.15% and 7.5% that are obviously automated safe but are projected declining in 2030. This discrepancy can be caused by rules used in expert assessment of the ground truth data. According to [2], experts assess the risks based on three attributes of an occupation which are "perception and manipulation", "creative intelligence tasks" and "social intelligence tasks". i.e., "customer service representatives" have tasks including persuading and dispute resolution which requires higher social intelligence and perception, therefore, they are not easy to be automated according to the analysis in [2]. BLS projection made predictions based on employment change in time series. And employment decline may be caused by various reasons e.g., supply and demand change [6], other than automation technologies which mean the declining employment of "customer service representatives" can be caused by the demand change, i.e. some the customers' basic demands can be satisfied by AI dialogue agents. That is to say, expert assessment based on occupations' attributes only can be biased since some occupations may still decline even if they have tasks and skills that automation technologies cannot perform because of the supply and demand change. Overall, the performance and evaluation demonstrate the effectiveness of our AOC-GCN method in occupation's automated risk prediction.

6   Limitations and Future Work

We acknowledge several limitations in our method and the results presented in this paper. Firstly, data limitations hindered the performance of our model. O*NET is the only globally admitted database in labour market, however, it only classified 1016 occupations, and this data size limited the performance of classification using machine-learning method. Moreover, each occupation is an aggregation of several jobs which still obfuscates the difference between jobs. Secondly, lack of ground truth data is also a barrier that inhibits predictions of automated risk. Currently, most of the ground truths are expert assessments based on occupation's attributes which are inevitably subjective to some extent. Expert assessments based on occupations' attributes can be biased without considering the supply and demand change. Moreover, there is no job-level or skill-level ground truth data which impedes predictions on a more granular level. Thirdly, predicting on static information cannot capture the technological change and labor trends since the employment trends and changing demand for specific tasks and skills might change faster than static O*NET data can capture. Therefore, our future work will focus on automated risk identifications on job-level which requires more granular ground truth data such as which skills or tasks decide a specific occupation being automated [19]. Moreover, the study will be conducted on real-world datasets that capture technological change, employment trends and supply and demand change.

7   Conclusion

In this paper, we applied the AOC-GCN model to the Occupation-skill graph to identify the automated risk of a specific occupation. The initial node embeddings are generated from Word2vec and Doc2vec. GCN learns the graph structure and interactions between occupations. Extensive experiments and comparisons with real-world projections demonstrate the effectiveness of our method. Our study paves the way for understanding where to concentrate upskilling efforts in the following decades [1]. In the future, we will further extend our model to solve more granular-level tasks in labour market.

Acknowledgement. This work is supported by the Australian Research Council (ARC) under Grant No. DP220103717, and LE220100078.

References

1. Ahadi, A., Kitto, K., Rizoiu, M.A., Musial, K.: Skills taught vs skills sought: using skills analytics to identify the gaps between curriculum and job markets. In: International Conference on Educational Data Mining, pp. 538–542. Durham, UK (2022). https://doi.org/10.5281/zenodo.6853121
2. Frey, C., Osborne, M.: The future of employment: how susceptible are jobs to computerisation? Oxford Martin. **114** (2013). https://doi.org/10.1016/j.techfore.2016.08.019
3. Arntz, M., Gregory, T., Zierahn, U.: Revisiting the risk of automation. Econ. Lett. **159**, 157–160 (2017). https://doi.org/10.1016/j.econlet.2017.07.001
4. Frank, M., et al.: Toward understanding the impact of artificial intelligence on labor. Proc. Natl. Acad. Sci. U.S.A. **116**, 6531–6539 (2019). https://doi.org/10.1073/pnas.1900949116

5. Cortes, M., Jaimovich, N., Siu, H.: Disappearing routine jobs: who, how, and why? J. Monetary Econ. **91** (2017). https://doi.org/10.1016/j.jmoneco.2017.09.006

6. Vermeulen, B., Pyka, A., Saviotti, P.-P., Kesselhut, J.: The impact of automation on employment: just the usual structural change? Sustainability **10** (2018). https://doi.org/10.3390/su10051661

7. Kipf, T., Welling, M.: Semi-supervised classification with graph convolutional networks (2016). https://doi.org/10.48550/arXiv.1609.02907

8. Nakamura, A.O., Shaw, K.L., Freeman, R.B., Nakamura, E., Pyman, A.: Jobs online. In: Studies of Labor Market Intermediation, pp. 27–65. University of Chicago Press (2009). https://doi.org/10.7208/chicago/9780226032900.003.0002

9. Brzeski, C., Burk, C.: Die Roboter kommen: Folgen der Automatisierung für den deutschen Arbeitsmarkt [The Robots are coming: Consequences of Automation for the German Labor Market]. INGDiBa Econ Res **30**, 1–7 (2015). German

10. Pajarinen, M., Rouvinen, P., Ekeland, A.: Computerization threatens one-third of Finnish and Norwegian employment. ETLA Brief **34** (2015). https://www.etla.fi/wp-content/uploads/ETLA-Muistio-Brief-34.pdf

11. Bowles, J.: The computerisation of European jobs. Bruegel Blog, 24 July 2014. https://www.bruegel.org/2014/07/the-computerisation-of-european-jobs

12. Arntz, M., Gregory, T., Zierahn, U.: The risk of automation for jobs in OECD countries: a comparative analysis; OECD social, employment and migration working papers. OECD Publishing, Paris (2016). https://doi.org/10.1787/1815199X

13. Yang, H., Chen, H., Pan, S., Li, L., Yu, P. S., Xu, G.: Dual space graph contrastive learning. In: Proceedings of the ACM Web Conference 2022 (2022). https://doi.org/10.1145/3485447.3512211

14. Li, A., Qin, Z., Liu, R., Yang, Y., Li, D.: Spam review detection with graph convolutional networks, pp. 2703–2711 (2019). https://doi.org/10.1145/3357384.3357820

15. Grover, A., Leskovec, J.: node2vec: scalable feature learning for networks. In: Proceedings of the 22nd ACM International Conference on Knowledge Discovery and Data Mining, SIGKDD, pp. 855–864. ACM (2014). 1607.00653

16. Dawson, N., Molitorisz, S., Rizoiu, M.A., Fray, P.: Layoffs, inequity and COVID-19: a longitudinal study of the journalism jobs crisis in Australia from 2012 to 2020. J., 146488492199628 (2021). https://doi.org/10.1177/1464884921996286

17. Dawson, N., Rizoiu, M.A., Johnston, B., Williams, M.A.: Adaptively selecting occupations to detect skill shortages from online job ads. In: 2019 IEEE International Conference on Big Data (Big Data), pp. 1637–1643. IEEE, Los Angeles, December 2019. https://doi.org/10.1109/BigData47090.2019.9005967

18. Wang, X., Li, Q., Yu, D., Wang, Z., Chen, H., Xu, G.: MGPolicy. In: Proceedings of the 45th International ACM SIGIR Conference on Research and Development in Information Retrieval (2022). https://doi.org/10.1145/3477495.3532021

19. Dawson, N., Williams, M.A., Rizoiu, M.A.: Skill-driven recommendations for job transition pathways. PLOS ONE **16**(8), e0254722 (2021). https://doi.org/10.1371/journal.pone.0254722

20. Perozzi, B., Al-Rfou, R., Skiena, S.: DeepWalk: online learning of social representations. In: Proceedings of the 20th ACM International Conference on Knowledge Discovery and Data Mining, SIGKDD, pp. 701–710. ACM (2014). 1403.6652

21. Hamilton, W., Ying, Z., Leskovec, J.: Inductive representation learning on large graphs. In: Advances in Neural Information Processing Systems, NIPS, pp. 1024–1034 (2017)

22. LeCun, Y., Haffner, P., Bottou, L., Bengio, Y.: Object recognition with gradient-based learning. In: Shape, Contour and Grouping in Computer Vision. LNCS, vol. 1681, pp. 319–345. Springer, Heidelberg (1999). https://doi.org/10.1007/3-540-46805-6_19. https://yann.lecun.com/exdb/publis/pdf/lecun-99.pdf

23. Mihaita, A.S., Li, H., He, Z., Rizoiu, M.A.: Motorway traffic flow prediction using advanced deep learning. In: 2019 IEEE Intelligent Transportation Systems Conference (ITSC), pp. 1683–1690. IEEE, Auckland, October 2019. https://doi.org/10.1109/ITSC.2019.8916852

24. Mihaita, A.S., Liu, Z., Cai, C., Rizoiu, M.A.: Arterial incident duration prediction using a bi-level framework of extreme gradient-tree boosting. In: Proceedings of the 26th ITS World Congress, pp. 1–12, Singapore, May 2019

25. Mihaita, A.S., Papachatgis, Z., Rizoiu, M.A.: Graph modelling approaches for motorway traffic flow prediction. In: 2020 IEEE 23rd International Conference on Intelligent Transportation Systems (ITSC), pp. 1–8. IEEE, Rhodes, September 2020. https://doi.org/10.1109/ITSC45102.2020.9294744

26. Veličković, P., Cucurull, G., Casanova, A., Romero, A.,Lió, P., Bengio, Y.: Graph attention networks. In: International Conference on Learning Representations, ICLR (2018). 1710.10903

27. Mikolov, T., et al.: Efficient estimation of word representations in vector space (2013). 1301.3781

28. Le, Q., Mikolov, T.: Distributed representations of sentences and documents. In: 31st International Conference on Machine Learning, ICML 2014 (2014). 1405.4053

29. Freund, Y., Schapire, R.E.: Experiments with a new boosting algorithm. In: ICML (1996). https://dl.acm.org/doi/10.5555/3091696.3091715

30. Ho, T.K.: Random decision forests (PDF). In: Proceedings of the 3rd International Conference on Document Analysis and Recognition, Montreal, QC, 14–16 August 1995, pp. 278–282. Archived from the original (PDF) on 17 April 2016 (1995). https://doi.org/10.1109/ICDAR.1995.598994. Accessed 5 June 2016

31. Meng, Q.: LightGBM: a highly efficient gradient boosting decision tree (2018). https://proceedings.neurips.cc/paper/2017/file/6449f44a102fde848669bdd9eb6b76fa-Paper.pdf

32. Quinlan, J.R.: "Induction of decision trees" (PDF). Mach. Learn. 1, 81–106 (1986). https://doi.org/10.1007/BF00116251.S2CID189902138

33. Roweis, S.: Hinton, Geoffrey (January 2002). Stochastic neighbor embedding (PDF). Neural Information Processing Systems (2018). https://www.cs.toronto.edu/~hinton/absps/sne.pdf

34. Hotelling, H.: Analysis of a complex of statistical variables into principal components. J. Educ. Psychol. 24(417–441), 498–520 (1933). https://doi.org/10.1037/h0071325

35. Beaudry, P., Green, D.A., Sand, B.M.: The great reversal in the demand for skill and cognitive tasks. J. Labor Econ. 34, S199–S247 (2016)

36. Chen, H., Yin, H., Wang, W., Wang, H., Nguyen, Q.V.H., Li, X.: PME. In: Proceedings of the 24th ACM SIGKDD International Conference on Knowledge Discovery & Data Mining (2018). https://doi.org/10.1145/3219819.3219986

37. Chen, H., YIN, H., Sun, X., Chen, T., Gabrys, B., Musial, K.: Multi-level graph convolutional networks for cross-platform anchor link prediction. In: Proceedings of the 26th ACM SIGKDD International Conference on Knowledge Discovery & Data Mining (2020). https://doi.org/10.1145/3394486.3403201

38. Chen, H., Yin, H., Chen, T., Nguyen, Q.V.H., Peng, W.-C., Li, X.: Exploiting centrality information with graph convolutions for network representation learning. IEEE Xplore (2019). https://doi.org/10.1109/ICDE.2019.00059

Author Index

Printed in the United States
by Baker & Taylor Publisher Services